Sheffield Hallam University
Learning and IT Services
Adsetts Centre City Campus
Sheffield S1 1WS

101 937 541 8

D1338787

Design

With an Introduction to the Verilog HDL

ONE WEEK LOAN

SHEFFIELD HALLAM UNIVERSITY
LEARNING CENTRE
WITHDRAWN FROM STOCK

Digital Design

With an Introduction to the Verilog HDL

FIFTH EDITION

M. Morris Mano
Emeritus Professor of Computer Engineering
California State University, Los Angeles

Michael D. Ciletti
Emeritus Professor of Electrical and Computer Engineering
University of Colorado at Colorado Springs

International Edition contributions by
B. R. Chandavarkar
Assistant Professor, Department of Computer Science and Engineering
National Institute of Technology Karnataka, Surathkal

PEARSON

Upper Saddle River Boston Columbus San Francisco New York
Indianapolis London Toronto Sydney Singapore Tokyo Montreal
Dubai Madrid Hong Kong Mexico City Munich Paris Amsterdam Cape Town

Vice President and Editorial Director, ECS:
Marcia J. Horton
Executive Editor: Andrew Gilfillan
Vice-President, Production: Vince O'Brien
Executive Marketing Manager: Tim Galligan
Marketing Assistant: Jon Bryant
Permissions Project Manager: Karen Sanatar
Senior Managing Editor: Scott Disanno
Production Project Manager/Editorial Production Manager:
Greg Dulles

Publisher, International Edition: Angshuman Chakraborty
Acquisitions Editor, International Edition: Somnath Basu
Publishing Assistant, International Edition: Shokhi Shah
Print and Media Editor, International Edition: Ashwitha Jayakumar
Project Editor, International Edition: Jayashree Arunachalam
Composition: Jouve India Private Limited
Full-Service Project Management: Jouve India Private Limited

Pearson Education Limited
Edinburgh Gate
Harlow
Essex CM20 2JE
England

and Associated Companies throughout the world

Visit us on the World Wide Web at:
www.pearsoninternationaleditions.com

© Pearson Education Limited 2013

The rights of M. Morris Mano and Michael D. Ciletti are to be identified as authors of this work has been asserted by them in accordance with the Copyright, Designs and Patents Act 1988.

Authorized adaptation from the United States edition, entitled Digital Design, With an Introduction to the Verilog HDL, 5th edition, *ISBN 978-0-13-277420-8 by M. Morris Mano and Michael D. Ciletti published by Pearson Education © 2013.*

All rights reserved. No part of this publication may be reproduced, stored in a retrieval system, or transmitted in any form or by any means, electronic, mechanical, photocopying, recording or otherwise, without either the prior written permission of the publisher or a licence permitting restricted copying in theUnited Kingdom issued by the Copyright Licensing Agency Ltd, Saffron House, 6–10 Kirby Street, London EC1N 8TS.

All trademarks used herein are the property of their respective owners. The use of any trademark in this text does not vest in the author or publisher any trademark ownership rights in such trademarks, nor does the use of such trademarks imply any affiliation with or endorsement of this book by such owners.

Microsoft® and Windows® are registered trademarks of the Microsoft Corporation in the U.S.A. and other countries. Screen shots and icons reprinted with permission from the Microsoft Corporation. This book is not sponsored or endorsed by or affiliated with the Microsoft Corporation.

British Library Cataloguing-in-Publication Data
A catalogue record for this book is available from the British Library

10 9 8 7 6 5 4 3 2
14 13 12

Typeset in Times Ten 10/12 by Jouve India Private Limited
Printed and bound by Courier Westford in The United States of America

The publisher's policy is to use paper manufactured from sustainable forests.

PEARSON

ISBN-13: 978-0-273-76452-6
ISBN-10: 0-273-76452-7

Contents

Preface

Since the fourth edition of *Digital Design*, the commercial availability of devices using digital technology to receive, manipulate, and transmit information seems to have exploded. Cell phones and handheld devices of various kinds offer new, competing features almost daily. Underneath the attractive graphical user interface of all of these devices sits a digital system that processes data in a binary format. The theoretical foundations of these systems have not changed much; indeed, one could argue that the stability of the core theory, coupled with modern design tools, has promoted the widespread response of manufacturers to the opportunities of the marketplace. Consequently, our refinement of our text has been guided by the need to equip our graduates with a solid understanding of digital machines and to introduce them to the methodology of modern design.

This edition of *Digital Design* builds on the previous four editions, and the feedback of the team of reviewers who helped set a direction for our presentation. The focus of the text has been sharpened to more closely reflect the content of a foundation course in digital design and the mainstream technology of today's digital systems: CMOS circuits. The intended audience is broad, embracing students of computer science, computer engineering, and electrical engineering. The key elements that the book focuses include (1) Boolean logic, (2) logic gates used by designers, (3) synchronous finite state machines, and (4) datapath controller design—all from a perspective of designing digital systems. This focus led to elimination of material more suited for a course in electronics. So the reader will not find here content for asynchronous machines or descriptions of bipolar transistors. Additionally, the widespread availability of web-based ancillary material prompted us to limit our discussion of field programmable gate arrays (FPGAs) to an introduction of devices offered by only one manufacturer, rather than two. Today's designers rely heavily on hardware description languages

(HDLs), and this edition of the book gives greater attention to their use and presents what we think is a clear development of a design methodology using the Verilog HDL.

MULTI-MODAL LEARNING

Digital Design supports a multimodal approach to learning. The so-called VARK characterization of learning modalities identifies four major modes by which humans learn: (V) visual, (A) aural, (R) reading, and (K) kinesthetic. In hindsight, we note that the relatively high level of illustrations and graphical content of our text addresses the visual (V) component of VARK; discussions and numerous examples address the reading (R) component. Students who exploit the availability of free simulators to work assignments are led through a kinesthetic (K) learning experience, including the positive feedback and delight of designing a logic system that works. The remaining element of VARK, the aural/auditory (A) experience, is left to the instructor. We have provided an abundance of material and examples to support classroom lectures. Thus, a course in digital design, using *Digital Design*, can provide a rich, balanced learning experience and address all the modes identified by VARK.

For those who might still question the presentation and use of HDLs in a first course in digital design, we note that industry has largely abandoned schematic-based design entry, a style which emerged in the 1980s, during the nascent development of CAD tools for integrated circuit (IC) design. Schematic entry creates a representation of functionality that is implicit in the layout of the schematic. Unfortunately, it is difficult for anyone in a reasonable amount of time to determine the functionality represented by the schematic of a logic circuit without having been instrumental in its construction, or without having additional documentation expressing the design intent. Consequently, industry has migrated to HDLs (e.g., Verilog) to describe the functionality of a design and to serve as the basis for documenting, simulating, testing, and synthesizing the hardware implementation of the design in a standard cell-based ASIC or an FPGA. The utility of a schematic depends on the careful, detailed documentation of a carefully constructed hierarchy of design modules. In the old paradigm, designers relied upon their years of experience to create a schematic of a circuit to implement functionality. In today's design flow, designers using HDLs can express functionality directly and explicitly, without years of accumulated experience, and use synthesis tools to generate the schematic as a by-product, automatically. Industry practices arrived here because schematic entry dooms us to inefficiency, if not failure, in understanding and designing large, complex ICs.

We note, again in this edition, that introducing HDLs in a first course in designing digital circuits is not intended to replace fundamental understanding of the building blocks of such circuits or to eliminate a discussion of manual methods of design. It is still essential for a student to understand *how hardware works*. Thus, we retain a thorough treatment of combinational and sequential logic devices. Manual design practices are presented, and their results are compared with those obtained with a HDL-based paradigm. What we are presenting, however, is an emphasis on *how hardware is designed*, to better prepare a student for a career in today's industry, where HDL-based design practices are dominant.

FLEXIBILITY

The sequence of topics in the text can accommodate courses that adhere to traditional, manual-based, treatments of digital design, courses that treat design using an HDL, and courses that are in transition between or blend the two approaches. Because modern synthesis tools automatically perform logic minimization, Karnaugh maps and related topics in optimization can be presented at the beginning of a treatment of digital design, or they can be presented after circuits and their applications are examined and simulated with an HDL. The text includes both manual and HDL-based design examples. Our end-of-chapter problems further facilitate this flexibility by cross referencing problems that address a traditional manual design task with a companion problem that uses an HDL to accomplish the task. Additionally, we link the manual and HDL-based approaches by presenting annotated results of simulations in the text, in answers to selected problems at the end of the text, and in the solutions manual.

NEW TO THIS EDITION

This edition of *Digital Design* uses the latest features of IEEE Standard 1364, but only insofar as they support our pedagogical objectives. The revisions and updates to the text include:

- Elimination of specialized circuit-level content not typically covered in a first course in logic circuits and digital design (e.g., RTL, DTL, and emitter-coupled logic circuits)
- Addition of "Web Search Topics" at the end of each chapter to point students to additional subject matter available on the web
- Revision of approximately one-third of the problems at the end of the chapters
- A printed solution manual for entire text, including all new problems
- Streamlining of the discussion of Karnaugh maps
- Integration of treatment of basic CMOS technology with treatment of logic gates
- Inclusion of an appendix introducing semiconductor technology

DESIGN METHODLOGY

This text presents a systematic methodology for designing a state machine to control the datapath of a digital system. Moreover, the framework in which this material is presented treats the realistic situation in which status signals from the datapath are used by the controller, i.e., the system has feedback. Thus, our treatment provides a foundation for designing complex and interactive digital systems. Although it is presented with an emphasis on HDL-based design, the methodology is also applicable to manual-based approaches to design.

JUST ENOUGH HDL

We present only those elements of the Verilog language that are matched to the level and scope of this text. Also, correct syntax does not guarantee that a model meets a functional specification or that it can be synthesized into physical hardware. So, we introduce students to a disciplined use of industry-based practices for writing models to ensure that a behavioral description can be synthesized into physical hardware, and that the behavior of the synthesized circuit will match that of the behavioral description. Failure to follow this discipline can lead to software race conditions in the HDL models of such machines, race conditions in the test bench used to verify them, and a mismatch between the results of simulating a behavioral model and its synthesized physical counterpart. Similarly, failure to abide by industry practices may lead to designs that simulate correctly, but which have hardware latches that are introduced into the design accidentally as a consequence of the modeling style used by the designer. The industry-based methodology we present leads to race-free and latch-free designs. It is important that students learn and follow industry practices in using HDL models, independent of whether a student's curriculum has access to synthesis tools.

VERIFICATION

In industry, significant effort is expended to verify that the functionality of a circuit is correct. Yet not much attention is given to verification in introductory texts on digital design, where the focus is on design itself, and testing is perhaps viewed as a secondary undertaking. Our experience is that this view can lead to premature "high-fives" and declarations that "the circuit works beautifully." Likewise, industry gains repeated returns on its investment in an HDL model by ensuring that it is readable, portable, and reusable. We demonstrate naming practices and the use of parameters to facilitate reusability and portability. We also provide test benches for all of the solutions and exercises to (1) verify the functionality of the circuit, (2) underscore the importance of thorough testing, and (3) introduce students to important concepts, such as self-checking test benches. Advocating and illustrating the development of a *test plan* to guide the development of a test bench, we introduce test plans, albeit simply, in the text and expand them in the solutions manual and in the answers to selected problems at the end of the text.

HDL CONTENT

We have ensured that all examples in the text and all answers in the solution manual conform to accepted industry practices for modeling digital hardware. As in the previous edition, HDL material is inserted in separate sections so that it can be covered or skipped as desired, does not diminish treatment of manual-based design, and does not dictate the sequence of presentation. The treatment is at a level suitable for beginning students who are learning digital circuits and a HDL at the same time. The text prepares students to work on signficant independent design projects and to succeed in a later course in computer architecture and advanced digital design.

Instructor Resources

Instructors can download the following classroom-ready resources from the publisher's website for the text (www.pearsoninternationaleditions.com/mano):

- Source code and test benches for all Verilog HDL examples in the test
- All figures and tables in the text
- Source code for all HDL models in the solutions manual
- A downloadable solutions manual with graphics suitable for classroom presentation

HDL Simulators

The Companion Website identifies web URLs to two simulators provided by SynaptiCAD. The first simulator is *VeriLogger Pro*, a traditional Verilog simulator that can be used to simulate the HDL examples in the book and to verify the solutions of HDL problems. This simulator accepts the syntax of the IEEE-1995 standard and will be useful to those who have legacy models. As an interactive simulator, *Verilogger Extreme* accepts the syntax of IEEE-2001 as well as IEEE-1995, allowing the designer to simulate and analyze design ideas before a complete simulation model or schematic is available. This technology is particularly useful for students because they can quickly enter Boolean and D flip-flop or latch input equations to check equivalency or to experiment with flip-flops and latch designs. Students can access the Companion Website at www.pearsoninternationaleditions.com/mano.

Chapter Summary

The following is a brief summary of the topics that are covered in each chapter.

Chapter 1 presents the various binary systems suitable for representing information in digital systems. The binary number system is explained and binary codes are illustrated. Examples are given for addition and subtraction of signed binary numbers and decimal numbers in binary-coded decimal (BCD) format.

Chapter 2 introduces the basic postulates of Boolean algebra and shows the correlation between Boolean expressions and their corresponding logic diagrams. All possible logic operations for two variables are investigated, and the most useful logic gates used in the design of digital systems are identified. This chapter also introduces basic CMOS logic gates.

Chapter 3 covers the map method for simplifying Boolean expressions. The map method is also used to simplify digital circuits constructed with AND-OR, NAND, or NOR gates. All other possible two-level gate circuits are considered, and their method of implementation is explained. Verilog HDL is introduced together with simple examples of gate-level models.

Chapter 4 outlines the formal procedures for the analysis and design of combinational circuits. Some basic components used in the design of digital systems, such as adders and code converters, are introduced as design examples. Frequently used digital logic functions such as parallel adders and subtractors, decoders, encoders, and multiplexers are explained, and their use in the design of combinational circuits is illustrated. HDL examples are given in gate-level, dataflow, and behavioral models to show the alternative ways available for describing combinational circuits in Verilog HDL. The procedure for writing a simple test bench to provide stimulus to an HDL design is presented.

Chapter 5 outlines the formal procedures for analyzing and designing clocked (synchronous) sequential circuits. The gate structure of several types of flip-flops is presented together with a discussion on the difference between level and edge triggering. Specific examples are used to show the derivation of the state table and state diagram when analyzing a sequential circuit. A number of design examples are presented with emphasis on sequential circuits that use D-type flip-flops. Behavioral modeling in Verilog HDL for sequential circuits is explained. HDL Examples are given to illustrate Mealy and Moore models of sequential circuits.

Chapter 6 deals with various sequential circuit components such as registers, shift registers, and counters. These digital components are the basic building blocks from which more complex digital systems are constructed. HDL descriptions of shift registers and counter are presented.

Chapter 7 deals with random access memory (RAM) and programmable logic devices. Memory decoding and error correction schemes are discussed. Combinational and sequential programmable devices such as ROMs, PLAs, PALs, CPLDs, and FPGAs are presented.

Chapter 8 deals with the register transfer level (RTL) representation of digital systems. The algorithmic state machine (ASM) chart is introduced. A number of examples demonstrate the use of the ASM chart, ASMD chart, RTL representation, and HDL description in the design of digital systems. The design of a finite state machine to control a datapath is presented in detail, including the realistic situation in which status signals from the datapath are used by the state machine that controls it. This chapter is the most important chapter in the book as it provides the student with a systematic approach to more advanced design projects.

Chapter 9 outlines experiments that can be performed in the laboratory with hardware that is readily available commercially. The operation of the ICs used in the experiments is explained by referring to diagrams of similar components introduced in previous chapters. Each experiment is presented informally and the student is expected to design the circuit and formulate a procedure for checking its operation in the laboratory. The lab experiments can be used in a stand-alone manner too and can be accomplished by a traditional approach, with a breadboard and TTL circuits, or with an HDL/synthesis approach using FPGAs. Today, software for synthesizing an HDL model and implementing a circuit with an FPGA is available at no cost from vendors of FPGAs, allowing students to conduct a significant amount of work in their personal environment before using prototyping boards and other resources in a lab.

Circuit boards for rapid prototyping circuits with FPGAs are available at a nominal cost, and typically include push buttons, switches, seven-segment displays, LCDs, keypads, and other I/O devices. With these resources, students can work prescribed lab exercises or their own projects and get results immediately.

Chapter 10 presents the standard graphic symbols for logic functions recommended by an ANSI/IEEE standard. These graphic symbols have been developed for small-scale integration (SSI) and medium-scale integration (MSI) components so that the user can recognize each function from the unique graphic symbol assigned. The chapter shows the standard graphic symbols of the ICs used in the laboratory experiments.

ACKNOWLEDGMENTS

We are grateful to the reviewers of *Digital Design*, 5e. Their expertise, careful reviews, and suggestions helped shape this edition.

Dmitri Donetski, Stony Brook University
Ali Amini, California State University, Northridge
Mihaela Radu, Rose Hulman Institute of Technology
Stephen J Kuyath, University of North Carolina, Charlotte
Peter Pachowicz, George Mason University
David Jeff Jackson, University of Alabama
A. John Boye, University of Nebraska, Lincoln
William H. Robinson, Vanderbilt University
Dinesh Bhatia, University of Texas, Dallas

We also wish to express our gratitude to the editorial and publication team at Prentice Hall/Pearson Education for supporting this edition of our text. We are grateful, too, for the ongoing support and encouragement of our wives, Sandra and Jerilynn.

M. Morris Mano
Emeritus Professor of Computer Engineering
California State University, Los Angeles

Michael D. Ciletti
Emeritus Professor of Electrical and Computer Engineering
University of Colorado at Colorado Springs

The publishers would like to thank Sourav Sen Gupta of the Indian Statistical Institute, Kolkata for reviewing the content of the International Edition.

Chapter 1

Digital Systems and Binary Numbers

1.1 DIGITAL SYSTEMS

Digital systems have such a prominent role in everyday life that we refer to the present technological period as the *digital age*. Digital systems are used in communication, business transactions, traffic control, spacecraft guidance, medical treatment, weather monitoring, the Internet, and many other commercial, industrial, and scientific enterprises. We have digital telephones, digital televisions, digital versatile discs, digital cameras, handheld devices, and, of course, digital computers. We enjoy music downloaded to our portable media player (e.g., iPod Touch™) and other handheld devices having high-resolution displays. These devices have graphical user interfaces (GUIs), which enable them to execute commands that appear to the user to be simple, but which, in fact, involve precise execution of a sequence of complex internal instructions. Most, if not all, of these devices have a special-purpose digital computer embedded within them. The most striking property of the digital computer is its generality. It can follow a sequence of instructions, called a program, that operates on given data. The user can specify and change the program or the data according to the specific need. Because of this flexibility, general-purpose digital computers can perform a variety of information-processing tasks that range over a wide spectrum of applications.

One characteristic of digital systems is their ability to represent and manipulate discrete elements of information. Any set that is restricted to a finite number of elements contains discrete information. Examples of discrete sets are the 10 decimal digits, the 26 letters of the alphabet, the 52 playing cards, and the 64 squares of a chessboard. Early digital computers were used for numeric computations. In this case, the discrete elements were the digits. From this application, the term *digital* computer emerged. Discrete elements of information are represented in a digital system by physical quantities called signals. Electrical signals such as voltages and currents are the most common. Electronic devices called transistors predominate in the circuitry that implements these signals. The signals in most present-day electronic digital systems use just two discrete values and are therefore said to be *binary*. A binary digit, called a *bit*, has two values: 0 and 1. Discrete elements of information are represented with groups of bits called *binary codes*. For example, the decimal digits 0 through 9 are represented in a digital system with a code of four bits (e.g., the number 7 is represented by 0111). How a pattern of bits is interpreted as a number depends on the code system in which it resides. To make this distinction, we could write $(0111)_2$ to indicate that the pattern 0111 is to be interpreted in a binary system, and $(0111)_{10}$ to indicate that the reference system is decimal. Then $0111_2 = 7_{10}$, which is not the same as 0111_{10}, or one hundred eleven. The subscript indicating the base for interpreting a pattern of bits will be used only when clarification is needed. Through various techniques, groups of bits can be made to represent discrete symbols, not necessarily numbers, which are then used to develop the system in a digital format. Thus, a digital system is a system that manipulates discrete elements of information represented internally in binary form. In today's technology, binary systems are most practical because, as we will see, they can be implemented with electronic components.

Discrete quantities of information either emerge from the nature of the data being processed or may be quantized from a continuous process. On the one hand, a payroll schedule is an inherently discrete process that contains employee names, social security numbers, weekly salaries, income taxes, and so on. An employee's paycheck is processed by means of discrete data values such as letters of the alphabet (names), digits (salary), and special symbols (such as $). On the other hand, a research scientist may observe a continuous process, but record only specific quantities in tabular form. The scientist is thus quantizing continuous data, making each number in his or her table a discrete quantity. In many cases, the quantization of a process can be performed automatically by an analog-to-digital converter, a device that forms a digital (discrete) representation of a analog (continuous) quantity.

The general-purpose digital computer is the best-known example of a digital system. The major parts of a computer are a memory unit, a central processing unit, and input–output units. The memory unit stores programs as well as input, output, and intermediate data. The central processing unit performs arithmetic and other data-processing operations as specified by the program. The program and data prepared by a user are transferred into memory by means of an input device such as a keyboard. An output device, such as a printer, receives the results of the computations, and the printed results are presented to the user. A digital computer can accommodate many input and output devices. One very useful device is a communication unit that provides interaction with other users through the Internet. A digital computer is a powerful instrument that can perform not only arithmetic computations, but also logical operations. In addition, it can be programmed to make decisions based on internal and external conditions.

There are fundamental reasons that commercial products are made with digital circuits. Like a digital computer, most digital devices are programmable. By changing the program in a programmable device, the same underlying hardware can be used for many different applications, thereby allowing its cost of development to be spread across a wider customer base. Dramatic cost reductions in digital devices have come about because of advances in digital integrated circuit technology. As the number of transistors that can be put on a piece of silicon increases to produce complex functions, the cost per unit decreases and digital devices can be bought at an increasingly reduced price. Equipment built with digital integrated circuits can perform at a speed of hundreds of millions of operations per second. Digital systems can be made to operate with extreme reliability by using error-correcting codes. An example of this strategy is the digital versatile disk (DVD), in which digital information representing video, audio, and other data is recorded without the loss of a single item. Digital information on a DVD is recorded in such a way that, by examining the code in each digital sample before it is played back, any error can be automatically identified and corrected.

A digital system is an interconnection of digital modules. **To understand the operation of each digital module, it is necessary to have a basic knowledge of digital circuits and their logical function.** The first seven chapters of this book present the basic tools of digital design, such as logic gate structures, combinational and sequential circuits, and programmable logic devices. Chapter 8 introduces digital design at the register transfer level (RTL) using a modern hardware description language (HDL). Chapter 9 concludes the text with laboratory exercises using digital circuits.

A major trend in digital design methodology is the use of a HDL to describe and simulate the functionality of a digital circuit. An HDL resembles a programming language and is suitable for describing digital circuits in textual form. It is used to simulate a digital system to verify its operation before hardware is built. It is also used in conjunction with logic synthesis tools to automate the design process. Because **it is important that students become familiar with an HDL-based design methodology,** HDL descriptions of digital circuits are presented throughout the book. While these examples help illustrate the features of an HDL, they also demonstrate the best practices used by industry to exploit HDLs. Ignorance of these practices will lead to cute, but worthless, HDL models that may simulate a phenomenon, but that cannot be synthesized by design tools, or to models that waste silicon area or synthesize to hardware that cannot operate correctly.

As previously stated, digital systems manipulate discrete quantities of information that are represented in binary form. Operands used for calculations may be expressed in the binary number system. Other discrete elements, including the decimal digits and characters of the alphabet, are represented in binary codes. Digital circuits, also referred to as logic circuits, process data by means of binary logic elements (logic gates) using binary signals. Quantities are stored in binary (two-valued) storage elements (flip-flops). The purpose of this chapter is to introduce the various binary concepts as a frame of reference for further study in the succeeding chapters.

1.2 BINARY NUMBERS

A decimal number such as 7,392 represents a quantity equal to 7 thousands, plus 3 hundreds, plus 9 tens, plus 2 units. The thousands, hundreds, etc., are powers of 10 implied by the position of the coefficients (symbols) in the number. To be more exact, 7,392 is a shorthand notation for what should be written as

$$7 \times 10^3 + 3 \times 10^2 + 9 \times 10^1 + 2 \times 10^0$$

However, the convention is to write only the numeric coefficients and, from their position, deduce the necessary powers of 10 with powers increasing from right to left. In general, a number with a decimal point is represented by a series of coefficients:

$$a_5a_4a_3a_2a_1a_0.\ a_{-1}a_{-2}a_{-3}$$

The coefficients a_j are any of the 10 digits (0, 1, 2, . . . , 9), and the subscript value j gives the place value and, hence, the power of 10 by which the coefficient must be multiplied. Thus, the preceding decimal number can be expressed as

$$10^5a_5 + 10^4a_4 + 10^3a_3 + 10^2a_2 + 10^1a_1 + 10^0a_0 + 10^{-1}a_{-1} + 10^{-2}a_{-2} + 10^{-3}a_{-3}$$

with $a_3 = 7$, $a_2 = 3$, $a_1 = 9$, and $a_0 = 2$.

The decimal number system is said to be of *base*, or *radix*, 10 because it uses 10 digits and the coefficients are multiplied by powers of 10. The *binary* system is a different number system. The coefficients of the binary number system have only two possible values: 0 and 1. Each coefficient a_j is multiplied by a power of the radix, e.g., 2^j, and the results are added to obtain the decimal equivalent of the number. The radix point (e.g., the decimal point when 10 is the radix) distinguishes positive powers of 10 from negative powers of 10. For example, the decimal equivalent of the binary number 11010.11 is 26.75, as shown from the multiplication of the coefficients by powers of 2:

$$1 \times 2^4 + 1 \times 2^3 + 0 \times 2^2 + 1 \times 2^1 + 0 \times 2^0 + 1 \times 2^{-1} + 1 \times 2^{-2} = 26.75$$

There are many different number systems. In general, a number expressed in a base-r system has coefficients multiplied by powers of r:

$$a_n \cdot r^n + a_{n-1} \cdot r^{n-1} + \cdots + a_2 \cdot r^2 + a_1 \cdot r + a_0 + a_{-1} \cdot r^{-1} + a_{-2} \cdot r^{-2} + \cdots + a_{-m} \cdot r^{-m}$$

The coefficients a_j range in value from 0 to $r - 1$. To distinguish between numbers of different bases, we enclose the coefficients in parentheses and write a subscript equal to the base used (except sometimes for decimal numbers, where the content makes it obvious that the base is decimal). An example of a base-5 number is

$$(4021.2)_5 = 4 \times 5^3 + 0 \times 5^2 + 2 \times 5^1 + 1 \times 5^0 + 2 \times 5^{-1} = (511.4)_{10}$$

The coefficient values for base 5 can be only 0, 1, 2, 3, and 4. The octal number system is a base-8 system that has eight digits: 0, 1, 2, 3, 4, 5, 6, 7. An example of an octal number is 127.4. To determine its equivalent decimal value, we expand the number in a power series with a base of 8:

$$(127.4)_8 = 1 \times 8^2 + 2 \times 8^1 + 7 \times 8^0 + 4 \times 8^{-1} = (87.5)_{10}$$

Note that the digits 8 and 9 cannot appear in an octal number.

It is customary to borrow the needed r digits for the coefficients from the decimal system when the base of the number is less than 10. **The letters of the alphabet are used to supplement the 10 decimal digits when the base of the number is greater than 10.** For example, in the *hexadecimal* (base-16) number system, the first 10 digits are borrowed from the decimal system. The letters A, B, C, D, E, and F are used for the digits 10, 11, 12, 13, 14, and 15, respectively. An example of a hexadecimal number is

$$(B65F)_{16} = 11 \times 16^3 + 6 \times 16^2 + 5 \times 16^1 + 15 \times 16^0 = (46{,}687)_{10}$$

The hexadecimal system is used commonly by designers to represent long strings of bits in the addresses, instructions, and data in digital systems. For example, B65F is used to represent 1011011001010000.

As noted before, the digits in a binary number are called *bits*. When a bit is equal to 0, it does not contribute to the sum during the conversion. Therefore, the conversion from binary to decimal can be obtained by adding only the numbers with powers of two corresponding to the bits that are equal to 1. For example,

$$(110101)_2 = 32 + 16 + 4 + 1 = (53)_{10}$$

There are four 1's in the binary number. The corresponding decimal number is the sum of the four powers of two. Zero and the first 24 numbers obtained from 2 to the power of n are listed in Table 1.1. In computer work, 2^{10} is referred to as K (kilo), 2^{20} as M (mega), 2^{30} as G (giga), and 2^{40} as T (tera). Thus, $4K = 2^{12} = 4{,}096$ and $16M = 2^{24} = 16{,}777{,}216$. Computer capacity is usually given in bytes. A *byte* is equal to eight bits and can accommodate (i.e., represent the code of) one keyboard character. A computer hard disk with four gigabytes of storage has a capacity of $4G = 2^{32}$ bytes (approximately 4 billion bytes). A terabyte is 1024 gigabytes, approximately 1 trillion bytes.

Arithmetic operations with numbers in base r follow the same rules as for decimal numbers. When a base other than the familiar base 10 is used, one must be careful to use only the r-allowable digits. Examples of addition, subtraction, and multiplication of two binary numbers are as follows:

augend:	101101	minuend:	101101	multiplicand:	1011
addend:	+100111	subtrahend:	−100111	multiplier:	× 101
sum:	1010100	difference:	000110		1011
					0000
				partial product:	1011
				product:	110111

Table 1.1
Powers of Two

n	2^n	*n*	2^n	*n*	2^n
0	1	8	256	16	65,536
1	2	9	512	17	131,072
2	4	10	1,024 (1K)	18	262,144
3	8	11	2,048	19	524,288
4	16	12	4,096 (4K)	20	1,048,576 (1M)
5	32	13	8,192	21	2,097,152
6	64	14	16,384	22	4,194,304
7	128	15	32,768	23	8,388,608

The sum of two binary numbers is calculated by the same rules as in decimal, except that the digits of the sum in any significant position can be only 0 or 1. Any carry obtained in a given significant position is used by the pair of digits one significant position higher. Subtraction is slightly more complicated. The rules are still the same as in decimal, except that the borrow in a given significant position adds 2 to a minuend digit. (A borrow in the decimal system adds 10 to a minuend digit.) Multiplication is simple: The multiplier digits are always 1 or 0; therefore, the partial products are equal either to a shifted (left) copy of the multiplicand or to 0.

1.3 NUMBER-BASE CONVERSIONS

Representations of a number in a different radix are said to be equivalent if they have the same decimal representation. For example, $(0011)_8$ and $(1001)_2$ are equivalent—both have decimal value 9. The conversion of a number in base r to decimal is done by expanding the number in a power series and adding all the terms as shown previously. We now present a general procedure for the reverse operation of converting a decimal number to a number in base r. If the number includes a radix point, it is necessary to separate the number into an integer part and a fraction part, since each part must be converted differently. The conversion of a decimal integer to a number in base r is done by dividing the number and all successive quotients by r and accumulating the remainders. This procedure is best illustrated by example.

EXAMPLE 1.1

Convert decimal 41 to binary. First, 41 is divided by 2 to give an integer quotient of 20 and a remainder of $\frac{1}{2}$. Then the quotient is again divided by 2 to give a new quotient and remainder. The process is continued until the integer quotient becomes 0. The *coefficients* of the desired binary number are obtained from the *remainders* as follows:

	Integer Quotient		**Remainder**	**Coefficient**
$41/2 =$	20	+	$\frac{1}{2}$	$a_0 = 1$
$20/2 =$	10	+	0	$a_1 = 0$
$10/2 =$	5	+	0	$a_2 = 0$
$5/2 =$	2	+	$\frac{1}{2}$	$a_3 = 1$
$2/2 =$	1	+	0	$a_4 = 0$
$1/2 =$	0	+	$\frac{1}{2}$	$a_5 = 1$

Therefore, the answer is $(41)_{10} = (a_5a_4a_3a_2a_1a_0)_2 = (101001)_2$.

The arithmetic process can be manipulated more conveniently as follows:

Integer	Remainder
41	
20	1
10	0
5	0
2	1
1	0
0	1 101001 = answer

Conversion from decimal integers to any base-r system is similar to this example, except that division is done by r instead of 2.

■

EXAMPLE 1.2

Convert decimal 153 to octal. The required base r is 8. First, 153 is divided by 8 to give an integer quotient of 19 and a remainder of 1. Then 19 is divided by 8 to give an integer quotient of 2 and a remainder of 3. Finally, 2 is divided by 8 to give a quotient of 0 and a remainder of 2. This process can be conveniently manipulated as follows:

153	
19	1
2	3
0	2 = $(231)_8$

The conversion of a decimal *fraction* to binary is accomplished by a method similar to that used for integers. However, multiplication is used instead of division, and integers instead of remainders are accumulated. Again, the method is best explained by example.

■

EXAMPLE 1.3

Convert $(0.6875)_{10}$ to binary. First, 0.6875 is multiplied by 2 to give an integer and a fraction. Then the new fraction is multiplied by 2 to give a new integer and a new fraction. The process is continued until the fraction becomes 0 or until the number of digits has sufficient accuracy. The coefficients of the binary number are obtained from the integers as follows:

	Integer		Fraction	Coefficient
$0.6875 \times 2 =$	1	+	0.3750	$a_{-1} = 1$
$0.3750 \times 2 =$	0	+	0.7500	$a_{-2} = 0$
$0.7500 \times 2 =$	1	+	0.5000	$a_{-3} = 1$
$0.5000 \times 2 =$	1	+	0.0000	$a_{-4} = 1$

Therefore, the answer is $(0.6875)_{10} = (0.\, a_{-1}\, a_{-2}\, a_{-3}\, a_{-4})_2 = (0.1011)_2$.

To convert a decimal fraction to a number expressed in base *r*, a similar procedure is used. However, multiplication is by *r* instead of 2, and the coefficients found from the integers may range in value from 0 to $r - 1$ instead of 0 and 1.

■

EXAMPLE 1.4

Convert $(0.513)_{10}$ to octal.

$$0.513 \times 8 = 4.104$$
$$0.104 \times 8 = 0.832$$
$$0.832 \times 8 = 6.656$$
$$0.656 \times 8 = 5.248$$
$$0.248 \times 8 = 1.984$$
$$0.984 \times 8 = 7.872$$

The answer, to seven significant figures, is obtained from the integer part of the products:

$$(0.513)_{10} = (0.406517\ldots)_8$$

The conversion of decimal numbers with both integer and fraction parts is done by converting the integer and the fraction separately and then combining the two answers. Using the results of Examples 1.1 and 1.3, we obtain

$$(41.6875)_{10} = (101001.1011)_2$$

From Examples 1.2 and 1.4, we have

$$(153.513)_{10} = (231.406517)_8$$

■

1.4 OCTAL AND HEXADECIMAL NUMBERS

The conversion from and to binary, octal, and hexadecimal plays an important role in digital computers, because shorter patterns of hex characters are easier to recognize than long patterns of 1's and 0's. Since $2^3 = 8$ and $2^4 = 16$, each octal digit corresponds to three binary digits and each hexadecimal digit corresponds to four binary digits. The first 16 numbers in the decimal, binary, octal, and hexadecimal number systems are listed in Table 1.2.

The conversion from binary to octal is easily accomplished by partitioning the binary number into groups of three digits each, starting from the binary point and proceeding to the left and to the right. The corresponding octal digit is then assigned to each group. The following example illustrates the procedure:

$$\begin{array}{ccccccccccc} (10 & 110 & 001 & 101 & 011 & \cdot & 111 & 100 & 000 & 110)_2 & = (26153.7406)_8 \\ 2 & 6 & 1 & 5 & 3 & & 7 & 4 & 0 & 6 & \end{array}$$

Table 1.2
Numbers with Different Bases

Decimal (base 10)	Binary (base 2)	Octal (base 8)	Hexadecimal (base 16)
00	0000	00	0
01	0001	01	1
02	0010	02	2
03	0011	03	3
04	0100	04	4
05	0101	05	5
06	0110	06	6
07	0111	07	7
08	1000	10	8
09	1001	11	9
10	1010	12	A
11	1011	13	B
12	1100	14	C
13	1101	15	D
14	1110	16	E
15	1111	17	F

Conversion from binary to hexadecimal is similar, except that the binary number is divided into groups of *four* digits:

$$\underset{2}{(10}\quad\underset{C}{1100}\quad\underset{6}{0110}\quad\underset{B}{1011}\quad\cdot\quad\underset{F}{1111}\quad\underset{2}{0010)_2} = (2C6B.F2)_{16}$$

The corresponding hexadecimal (or octal) digit for each group of binary digits is easily remembered from the values listed in Table 1.2.

Conversion from octal or hexadecimal to binary is done by reversing the preceding procedure. Each octal digit is converted to its three-digit binary equivalent. Similarly, each hexadecimal digit is converted to its four-digit binary equivalent. The procedure is illustrated in the following examples:

$$(673.124)_8 = (\underset{6}{110}\quad\underset{7}{111}\quad\underset{3}{011}\quad\cdot\quad\underset{1}{001}\quad\underset{2}{010}\quad\underset{4}{100})_2$$

and

$$(306.D)_{16} = (\underset{3}{0011}\quad\underset{0}{0000}\quad\underset{6}{0110}\quad\cdot\quad\underset{D}{1101})_2$$

Binary numbers are difficult to work with because they require three or four times as many digits as their decimal equivalents. For example, the binary number 111111111111 is equivalent to decimal 4095. However, digital computers use binary numbers, and it is sometimes necessary for the human operator or user to communicate directly with the machine by means of such numbers. One scheme that retains the binary system in the computer, but reduces the number of digits the human must consider, utilizes the relationship between the binary number system and the octal or hexadecimal system. By this method, the human thinks in terms of octal or hexadecimal numbers and performs the required conversion by inspection when direct communication with the machine is necessary. Thus, the binary number 111111111111 has 12 digits and is expressed in octal as 7777 (4 digits) or in hexadecimal as FFF (3 digits). During communication between people (about binary numbers in the computer), the octal or hexadecimal representation is more desirable because it can be expressed more compactly with a third or a quarter of the number of digits required for the equivalent binary number. Thus, **most computer manuals use either octal or hexadecimal numbers to specify binary quantities.** The choice between them is arbitrary, although hexadecimal tends to win out, since it can represent a byte with two digits.

1.5 COMPLEMENTS OF NUMBERS

Complements are used in digital computers to **simplify the subtraction operation** and for logical manipulation. Simplifying operations leads to simpler, less expensive circuits to implement the operations. There are two types of complements for each base-r system: the radix complement and the diminished radix complement. The first is referred to as the r's complement and the second as the $(r - 1)$'s complement. When the value of the base r is substituted in the name, the two types are referred to as the 2's complement and 1's complement for binary numbers and the 10's complement and 9's complement for decimal numbers.

Diminished Radix Complement

Given a number N in base r having n digits, the $(r - 1)$'s complement of N, i.e., its diminished radix complement, is defined as $(r^n - 1) - N$. For decimal numbers, $r = 10$ and $r - 1 = 9$, so the 9's complement of N is $(10^n - 1) - N$. In this case, 10^n represents a number that consists of a single 1 followed by n 0's. $10^n - 1$ is a number represented by n 9's. For example, if $n = 4$, we have $10^4 = 10{,}000$ and $10^4 - 1 = 9999$. It follows that the 9's complement of a decimal number is obtained by subtracting each digit from 9. Here are some numerical examples:

The 9's complement of 546700 is $999999 - 546700 = 453299$.

The 9's complement of 012398 is $999999 - 012398 = 987601$.

For binary numbers, $r = 2$ and $r - 1 = 1$, so the 1's complement of N is $(2^n - 1) - N$. Again, 2^n is represented by a binary number that consists of a 1 followed by n 0's. $2^n - 1$ is a binary number represented by n 1's. For example, if $n = 4$, we have $2^4 = (10000)_2$ and $2^4 - 1 = (1111)_2$. Thus, the 1's complement of a binary number is obtained by subtracting each digit from 1. However, when subtracting binary digits from 1, we can have either $1 - 0 = 1$ or $1 - 1 = 0$, which causes the bit to change from 0 to 1 or from 1 to 0, respectively. Therefore, **the 1's complement of a binary number is formed by changing 1's to 0's and 0's to 1's.** The following are some numerical examples:

The 1's complement of 1011000 is 0100111.

The 1's complement of 0101101 is 1010010.

The $(r - 1)$'s complement of octal or hexadecimal numbers is obtained by subtracting each digit from 7 or F (decimal 15), respectively.

Radix Complement

The r's complement of an n-digit number N in base r is defined as $r^n - N$ for $N \neq 0$ and as 0 for $N = 0$. Comparing with the $(r - 1)$'s complement, we note that the r's complement is obtained by adding 1 to the $(r - 1)$'s complement, since $r^n - N = [(r^n - 1) - N] + 1$. Thus, the 10's complement of decimal 2389 is $7610 + 1 = 7611$ and is obtained by adding 1 to the 9's complement value. The 2's complement of binary 101100 is $010011 + 1 = 010100$ and is obtained by adding 1 to the 1's-complement value.

Since 10 is a number represented by a 1 followed by n 0's, $10^n - N$, which is the 10's complement of N, can be formed also by leaving all least significant 0's unchanged, subtracting the first nonzero least significant digit from 10, and subtracting all higher significant digits from 9. Thus,

the 10's complement of 012398 is 987602

and

the 10's complement of 246700 is 753300

The 10's complement of the first number is obtained by subtracting 8 from 10 in the least significant position and subtracting all other digits from 9. The 10's complement of the second number is obtained by leaving the two least significant 0's unchanged, subtracting 7 from 10, and subtracting the other three digits from 9.

Similarly, the 2's complement can be formed by leaving all least significant 0's and the first 1 unchanged and replacing 1's with 0's and 0's with 1's in all other higher significant digits. For example,

the 2's complement of 1101100 is 0010100

and

the 2's complement of 0110111 is 1001001

The 2's complement of the first number is obtained by leaving the two least significant 0's and the first 1 unchanged and then replacing 1's with 0's and 0's with 1's in the other four most significant digits. The 2's complement of the second number is obtained by leaving the least significant 1 unchanged and complementing all other digits.

In the previous definitions, it was assumed that the numbers did not have a radix point. If the original number N contains a radix point, the point should be removed temporarily in order to form the r's or $(r - 1)$'s complement. The radix point is then restored to the complemented number in the same relative position. It is also worth mentioning that **the complement of the complement restores the number to its original value.** To see this relationship, note that the r's complement of N is $r^n - N$, so that the complement of the complement is $r^n - (r^n - N) = N$ and is equal to the original number.

Subtraction with Complements

The direct method of subtraction taught in elementary schools uses the borrow concept. In this method, we borrow a 1 from a higher significant position when the minuend digit is smaller than the subtrahend digit. The method works well when people perform subtraction with paper and pencil. However, when subtraction is implemented with digital hardware, the method is less efficient than the method that uses complements.

The subtraction of two n-digit unsigned numbers $M - N$ in base r can be done as follows:

1. Add the minuend M to the r's complement of the subtrahend N. Mathematically, $M + (r^n - N) = M - N + r^n$.
2. If $M \geq N$, the sum will produce an end carry r^n, which can be discarded; what is left is the result $M - N$.
3. If $M < N$, the sum does not produce an end carry and is equal to $r^n - (N - M)$, which is the r's complement of $(N - M)$. To obtain the answer in a familiar form, take the r's complement of the sum and place a negative sign in front.

The following examples illustrate the procedure:

EXAMPLE 1.5

Using 10's complement, subtract 72532 − 3250.

$$\begin{aligned} M &= \quad 72532 \\ \text{10's complement of } N &= +\ \underline{96750} \\ \text{Sum} &= \quad 169282 \\ \text{Discard end carry } 10^5 &= -\ \underline{100000} \\ \textit{Answer} &= \quad 69282 \end{aligned}$$

Note that M has five digits and N has only four digits. Both numbers must have the same number of digits, so we write N as 03250. Taking the 10's complement of N produces a 9 in the most significant position. The occurrence of the end carry signifies that $M \geq N$ and that the result is therefore positive.

■

EXAMPLE 1.6

Using 10's complement, subtract 3250 − 72532.

$$\begin{aligned} M &= 03250 \\ \text{10's complement of } N &= +\ \underline{27468} \\ \text{Sum} &= 30718 \end{aligned}$$

There is no end carry. Therefore, the answer is −(10's complement of 30718) = −69282.

Note that since 3250 < 72532, the result is negative. Because we are dealing with unsigned numbers, there is really no way to get an unsigned result for this case. When subtracting with complements, we recognize the negative answer from the absence of the end carry and the complemented result. When working with paper and pencil, we can change the answer to a signed negative number in order to put it in a familiar form.

Subtraction with complements is done with binary numbers in a similar manner, using the procedure outlined previously.

■

EXAMPLE 1.7

Given the two binary numbers $X = 1010100$ and $Y = 1000011$, perform the subtraction **(a)** $X - Y$ and **(b)** $Y - X$ by using 2's complements.

(a)

$$\begin{aligned} X &= 1010100 \\ \text{2's complement of } Y &= +\ \underline{0111101} \\ \text{Sum} &= 10010001 \\ \text{Discard end carry } 2^7 &= -\ \underline{10000000} \\ \textit{Answer: } X - Y &= 0010001 \end{aligned}$$

(b)

$$\begin{aligned} Y &= 1000011 \\ \text{2's complement of } X &= +\ \underline{0101100} \\ \text{Sum} &= 1101111 \end{aligned}$$

There is no end carry. Therefore, the answer is $Y - X = -$(2's complement of 1101111) = −0010001.

■

Subtraction of unsigned numbers can also be done by means of the $(r - 1)$'s complement. Remember that the $(r - 1)$'s complement is one less than the r's complement. Because of this, the result of adding the minuend to the complement of the subtrahend produces a sum that is one less than the correct difference when an end carry occurs. Removing the end carry and adding 1 to the sum is referred to as an *end-around carry*.

EXAMPLE 1.8

Repeat Example 1.7, but this time using 1's complement.

(a) $X - Y = 1010100 - 1000011$

$$\begin{aligned} X &= 1010100 \\ \text{1's complement of } Y &= +\ \underline{0111100} \\ \text{Sum} &= 10010000 \\ \text{End-around carry} &= +\ \underline{1} \\ \textit{Answer: } X - Y &= 0010001 \end{aligned}$$

(b) $Y - X = 1000011 - 1010100$

$$\begin{aligned} Y &= 1000011 \\ \text{1's complement of } X &= +\ \underline{0101011} \\ \text{Sum} &= 1101110 \end{aligned}$$

There is no end carry. Therefore, the answer is $Y - X = -$(1's complement of 1101110) $= -0010001$.

■

Note that the negative result is obtained by taking the 1's complement of the sum, since this is the type of complement used. The procedure with end-around carry is also applicable to subtracting unsigned decimal numbers with 9's complement.

1.6 SIGNED BINARY NUMBERS

Positive integers (including zero) can be represented as unsigned numbers. However, to represent negative integers, we need a notation for negative values. In ordinary arithmetic, a negative number is indicated by a minus sign and a positive number by a plus sign. Because of hardware limitations, computers must represent everything with binary digits. It is customary to represent the sign with a bit placed in the leftmost position of the number. The convention is to make the sign bit 0 for positive and 1 for negative.

It is important to realize that both signed and unsigned binary numbers consist of a string of bits when represented in a computer. The user determines whether the number is signed or unsigned. If the binary number is signed, then the leftmost bit represents the sign and the rest of the bits represent the number. If the binary number is assumed to be unsigned, then the leftmost bit is the most significant bit of the number. For example, the string of bits 01001 can be considered as 9 (unsigned binary) or as +9 (signed binary) because the leftmost bit is 0. The string of bits 11001 represents the binary equivalent of 25 when considered as an unsigned number and the binary equivalent of −9 when considered as a signed number. This is because the 1 that is in the leftmost position designates a negative and the other four bits represent binary 9. Usually, there is no confusion in interpreting the bits if the type of representation for the number is known in advance.

The representation of the signed numbers in the last example is referred to as the *signed-magnitude* convention. In this notation, the number consists of a magnitude and a symbol (+ or −) or a bit (0 or 1) indicating the sign. This is the representation of signed numbers used in ordinary arithmetic. When arithmetic operations are implemented in a computer, it is more convenient to use a different system, referred to as the *signed-complement* system, for representing negative numbers. In this system, a negative number is indicated by its complement. Whereas the signed-magnitude system negates a number by changing its sign, the signed-complement system negates a number by taking its complement. Since positive numbers always start with 0 (plus) in the leftmost position, the complement will always start with a 1, indicating a negative number. The signed-complement system can use either the 1's or the 2's complement, but the 2's complement is the most common.

As an example, consider the number 9, represented in binary with eight bits. +9 is represented with a sign bit of 0 in the leftmost position, followed by the binary equivalent of 9, which gives 00001001. Note that all eight bits must have a value; therefore, 0's are inserted following the sign bit up to the first 1. Although there is only one way to represent +9, there are three different ways to represent −9 with eight bits:

signed-magnitude representation:	10001001
signed-1's-complement representation:	11110110
signed-2's-complement representation:	11110111

In signed-magnitude, −9 is obtained from +9 by changing only the sign bit in the leftmost position from 0 to 1. In signed-1's-complement, −9 is obtained by complementing all the bits of +9, including the sign bit. The signed-2's-complement representation of −9 is obtained by taking the 2's complement of the positive number, including the sign bit.

Table 1.3 lists all possible four-bit signed binary numbers in the three representations. The equivalent decimal number is also shown for reference. Note that the positive numbers in all three representations are identical and have 0 in the leftmost position. The signed-2's-complement system has only one representation for 0, which is always positive. The other two systems have either a positive 0 or a negative 0, something not encountered in ordinary arithmetic. Note that all negative numbers have a 1 in the leftmost bit position; that is the way we distinguish them from the positive numbers. With four bits, we can represent 16 binary numbers. In the signed-magnitude and the 1's-complement representations, there are eight positive numbers and eight negative numbers, including two zeros. In the 2's-complement representation, there are eight positive numbers, including one zero, and eight negative numbers.

The signed-magnitude system is used in ordinary arithmetic, but is awkward when employed in computer arithmetic because of the separate handling of the sign and the magnitude. Therefore, the signed-complement system is normally used. The 1's complement imposes some difficulties and is seldom used for arithmetic operations. It is useful as a logical operation, since the change of 1 to 0 or 0 to 1 is equivalent to a logical complement operation, as will be shown in the next chapter. The discussion of signed binary arithmetic that follows deals exclusively with the signed-2's-complement

Table 1.3
Signed Binary Numbers

Decimal	Signed-2's Complement	Signed-1's Complement	Signed Magnitude
+7	0111	0111	0111
+6	0110	0110	0110
+5	0101	0101	0101
+4	0100	0100	0100
+3	0011	0011	0011
+2	0010	0010	0010
+1	0001	0001	0001
+0	0000	0000	0000
−0	—	1111	1000
−1	1111	1110	1001
−2	1110	1101	1010
−3	1101	1100	1011
−4	1100	1011	1100
−5	1011	1010	1101
−6	1010	1001	1110
−7	1001	1000	1111
−8	1000	—	—

representation of negative numbers. The same procedures can be applied to the signed-1's-complement system by including the end-around carry as is done with unsigned numbers.

Arithmetic Addition

The addition of two numbers in the signed-magnitude system follows the rules of ordinary arithmetic. If the signs are the same, we add the two magnitudes and give the sum the common sign. If the signs are different, we subtract the smaller magnitude from the larger and give the difference the sign of the larger magnitude. For example, $(+25) + (-37) = -(37 - 25) = -12$ is done by subtracting the smaller magnitude, 25, from the larger magnitude, 37, and appending the sign of 37 to the result. This is a process that requires a comparison of the signs and magnitudes and then performing either addition or subtraction. The same procedure applies to binary numbers in signed-magnitude representation. In contrast, the rule for adding numbers in the signed-complement system does not require a comparison or subtraction, but only addition. The procedure is very simple and can be stated as follows for binary numbers:

The addition of two signed binary numbers with negative numbers represented in signed-2's-complement form is obtained from the addition of the two numbers, including their sign bits. A carry out of the sign-bit position is discarded.

Numerical examples for addition follow:

+ 6	00000110	− 6	11111010
+13	00001101	+13	00001101
+19	00010011	+ 7	00000111
+ 6	00000110	− 6	11111010
−13	11110011	−13	11110011
− 7	11111001	−19	11101101

Note that negative numbers must be initially in 2's-complement form and that if the sum obtained after the addition is negative, it is in 2's-complement form. For example, −7 is represented as 11111001, which is the 2s complement of +7.

In each of the four cases, the operation performed is addition with the sign bit included. Any carry out of the sign-bit position is discarded, and negative results are automatically in 2's-complement form.

In order to obtain a correct answer, we must ensure that the result has a sufficient number of bits to accommodate the sum. If we start with two n-bit numbers and the sum occupies $n + 1$ bits, we say that an overflow occurs. When one performs the addition with paper and pencil, an overflow is not a problem, because we are not limited by the width of the page. We just add another 0 to a positive number or another 1 to a negative number in the most significant position to extend the number to $n + 1$ bits and then perform the addition. Overflow is a problem in computers because the number of bits that hold a number is finite, and a result that exceeds the finite value by 1 cannot be accommodated.

The complement form of representing negative numbers is unfamiliar to those used to the signed-magnitude system. To determine the value of a negative number in signed-2's complement, it is necessary to convert the number to a positive number to place it in a more familiar form. For example, the signed binary number 11111001 is negative because the leftmost bit is 1. Its 2's complement is 00000111, which is the binary equivalent of +7. We therefore recognize the original negative number to be equal to −7.

Arithmetic Subtraction

Subtraction of two signed binary numbers when negative numbers are in 2's-complement form is simple and can be stated as follows:

> Take the 2's complement of the subtrahend (including the sign bit) and add it to the minuend (including the sign bit). A carry out of the sign-bit position is discarded.

This procedure is adopted because a subtraction operation can be changed to an addition operation if the sign of the subtrahend is changed, as is demonstrated by the following relationship:

$$(\pm A) - (+B) = (\pm A) + (-B);$$
$$(\pm A) - (-B) = (\pm A) + (+B).$$

But changing a positive number to a negative number is easily done by taking the 2's complement of the positive number. The reverse is also true, because the complement of a negative number in complement form produces the equivalent positive number. To see this, consider the subtraction $(-6) - (-13) = +7$. In binary with eight bits, this operation is written as $(11111010 - 11110011)$. The subtraction is changed to addition by taking the 2's complement of the subtrahend (−13), giving (+13). In binary, this is $11111010 + 00001101 = 100000111$. Removing the end carry, we obtain the correct answer: 00000111 (+7).

It is worth noting that binary numbers in the signed-complement system are added and subtracted by the same basic addition and subtraction rules as unsigned numbers. Therefore, **computers need only one common hardware circuit to handle both types of arithmetic.** This consideration has resulted in the signed-complement system being used in virtually all arithmetic units of computer systems. The user or programmer must interpret the results of such addition or subtraction differently, depending on whether it is assumed that the numbers are signed or unsigned.

1.7 BINARY CODES

Digital systems use signals that have two distinct values and circuit elements that have two stable states. There is a direct analogy among binary signals, binary circuit elements, and binary digits. A binary number of n digits, for example, may be represented by n binary circuit elements, each having an output signal equivalent to 0 or 1. Digital systems represent and manipulate not only binary numbers, but also many other discrete elements of information. Any discrete element of information that is distinct among a group of quantities can be represented with a binary code (i.e., a pattern of 0's and 1's). The codes must be in binary because, in today's technology, only circuits that represent and manipulate patterns of 0's and 1's can be manufactured economically for use in computers. However, it must be realized that binary codes merely change the symbols, not the meaning of the elements of information that they represent. If we inspect the bits of a computer at random, we will find that most of the time they represent some type of coded information rather than binary numbers.

An n-bit binary code is a group of n bits that assumes up to 2^n distinct combinations of 1's and 0's, with each combination representing one element of the set that is being coded. A set of four elements can be coded with two bits, with each element assigned one of the following bit combinations: 00, 01, 10, 11. A set of eight elements requires a three-bit code and a set of 16 elements requires a four-bit code. The bit combination of an n-bit code is determined from the count in binary from 0 to $2^n - 1$. Each element must be assigned a unique binary bit combination, and no two elements can have the same value; otherwise, the code assignment will be ambiguous.

Although the *minimum* number of bits required to code 2^n distinct quantities is n, there is no *maximum* number of bits that may be used for a binary code. For example, the 10 decimal digits can be coded with 10 bits, and each decimal digit can be assigned a bit combination of nine 0's and a 1. In this particular binary code, the digit 6 is assigned the bit combination 0001000000.

Binary-Coded Decimal Code

Although the binary number system is the most natural system for a computer because it is readily represented in today's electronic technology, most people are more accustomed to the decimal system. One way to resolve this difference is to convert decimal numbers to binary, perform all arithmetic calculations in binary, and then convert the binary results back to decimal. This method requires that we store decimal numbers in the computer so that they can be converted to binary. Since the computer can accept only binary values, we must represent the decimal digits by means of a code that contains 1's and 0's. It is also possible to perform the arithmetic operations directly on decimal numbers when they are stored in the computer in coded form.

A binary code will have some unassigned bit combinations if the number of elements in the set is not a multiple power of 2. The 10 decimal digits form such a set. A binary code that distinguishes among 10 elements must contain at least four bits, but 6 out of the 16 possible combinations remain unassigned. Different binary codes can be obtained by arranging four bits into 10 distinct combinations. The code most commonly used for the decimal digits is the straight binary assignment listed in Table 1.4. This scheme is called *binary-coded decimal* and is commonly referred to as BCD. Other decimal codes are possible and a few of them are presented later in this section.

Table 1.4 gives the four-bit code for one decimal digit. A number with k decimal digits will require $4k$ bits in BCD. Decimal 396 is represented in BCD with 12 bits as 0011 1001 0110, with **each group of 4 bits representing one decimal digit.** A decimal number in BCD is the same as its equivalent binary number only when the number is between 0 and 9. A BCD number greater than 10 looks different from its equivalent binary number, even though both contain 1's and 0's. Moreover, **the binary combinations 1010 through 1111 are not used and have no meaning in BCD.** Consider decimal 185 and its corresponding value in BCD and binary:

$$(185)_{10} = (0001\ 1000\ 0101)_{BCD} = (10111001)_2$$

Table 1.4
Binary-Coded Decimal (BCD)

Decimal Symbol	BCD Digit
0	0000
1	0001
2	0010
3	0011
4	0100
5	0101
6	0110
7	0111
8	1000
9	1001

The BCD value has 12 bits to encode the characters of the decimal value, but the equivalent binary number needs only 8 bits. It is obvious that the representation of a BCD number needs more bits than its equivalent binary value. However, there is an advantage in the use of decimal numbers, because computer input and output data are generated by people who use the decimal system.

It is important to realize that BCD numbers are decimal numbers and not binary numbers, although they use bits in their representation. The only difference between a decimal number and BCD is that decimals are written with the symbols 0, 1, 2, ... , 9 and BCD numbers use the binary code 0000, 0001, 0010, ... , 1001. The decimal value is exactly the same. Decimal 10 is represented in BCD with eight bits as 0001 0000 and decimal 15 as 0001 0101. The corresponding binary values are 1010 and 1111 and have only four bits.

BCD Addition

Consider the addition of two decimal digits in BCD, together with a possible carry from a previous less significant pair of digits. Since each digit does not exceed 9, the sum cannot be greater than $9 + 9 + 1 = 19$, with the 1 being a previous carry. Suppose we add the BCD digits as if they were binary numbers. Then the binary sum will produce a result in the range from 0 to 19. In binary, this range will be from 0000 to 10011, but in BCD, it is from 0000 to 1 1001, with the first (i.e., leftmost) 1 being a carry and the next four bits being the BCD sum. When the binary sum is equal to or less than 1001 (without a carry), the corresponding BCD digit is correct. However, when the binary sum is greater than or equal to 1010, the result is an invalid BCD digit. The addition of $6 = (0110)_2$ to the binary sum converts it to the correct digit and also produces a carry as required. This is because a carry in the most significant bit position of the binary sum and a decimal carry differ by $16 - 10 = 6$. Consider the following three BCD additions:

$$
\begin{array}{rrrrrr}
4 & 0100 & 4 & 0100 & 8 & 1000 \\
\underline{+5} & \underline{+0101} & \underline{+8} & \underline{+1000} & \underline{+9} & \underline{1001} \\
9 & 1001 & 12 & 1100 & 17 & 10001 \\
 & & & \underline{+0110} & & \underline{+0110} \\
 & & & 10010 & & 10111
\end{array}
$$

In each case, the two BCD digits are added as if they were two binary numbers. If the binary sum is greater than or equal to 1010, we add 0110 to obtain the correct BCD sum and a carry. In the first example, the sum is equal to 9 and is the correct BCD sum. In the second example, the binary sum produces an invalid BCD digit. The addition of 0110 produces the correct BCD sum, 0010 (i.e., the number 2), and a carry. In the third example, the binary sum produces a carry. This condition occurs when the sum is greater than or equal to 16. Although the other four bits are less than 1001, the binary sum requires a correction because of the carry. Adding 0110, we obtain the required BCD sum 0111 (i.e., the number 7) and a BCD carry.

The addition of two n-digit unsigned BCD numbers follows the same procedure. Consider the addition of 184 + 576 = 760 in BCD:

BCD	1	1		
	0001	1000	0100	184
	+0101	0111	0110	+576
Binary sum	0111	10000	1010	
Add 6		0110	0110	
BCD sum	0111	0110	0000	760

The first, least significant pair of BCD digits produces a BCD digit sum of 0000 and a carry for the next pair of digits. The second pair of BCD digits plus a previous carry produces a digit sum of 0110 and a carry for the next pair of digits. The third pair of digits plus a carry produces a binary sum of 0111 and does not require a correction.

Decimal Arithmetic

The representation of signed decimal numbers in BCD is similar to the representation of signed numbers in binary. We can use either the familiar signed-magnitude system or the signed-complement system. The sign of a decimal number is usually represented with four bits to conform to the four-bit code of the decimal digits. It is customary to designate a plus with four 0's and a minus with the BCD equivalent of 9, which is 1001.

The signed-magnitude system is seldom used in computers. The signed-complement system can be either the 9's or the 10's complement, but the 10's complement is the one most often used. To obtain the 10's complement of a BCD number, we first take the 9's complement and then add 1 to the least significant digit. The 9's complement is calculated from the subtraction of each digit from 9.

The procedures developed for the signed-2's-complement system in the previous section also apply to the signed-10's-complement system for decimal numbers. Addition is done by summing all digits, including the sign digit, and discarding the end carry. This operation assumes that all negative numbers are in 10's-complement form. Consider the addition $(+375) + (-240) = +135$, done in the signed-complement system:

$$
\begin{array}{rr}
0 & 375 \\
+9 & 760 \\
\hline
0 & 135
\end{array}
$$

The 9 in the leftmost position of the second number represents a minus, and 9760 is the 10's complement of 0240. The two numbers are added and the end carry is discarded to obtain +135. Of course, the decimal numbers inside the computer, including the sign digits, must be in BCD. The addition is done with BCD digits as described previously.

The subtraction of decimal numbers, either unsigned or in the signed-10's-complement system, is the same as in the binary case: Take the 10's complement of the subtrahend and add it to the minuend. Many computers have special hardware to perform arithmetic calculations directly with decimal numbers in BCD. The user of the computer can specify programmed instructions to perform the arithmetic operation with decimal numbers directly, without having to convert them to binary.

Other Decimal Codes

Binary codes for decimal digits require a minimum of four bits per digit. Many different codes can be formulated by arranging four bits into 10 distinct combinations. BCD and three other representative codes are shown in Table 1.5. Each code uses only 10 out of a possible 16 bit combinations that can be arranged with four bits. The other six unused combinations have no meaning and should be avoided.

BCD and the 2421 code are examples of weighted codes. In a weighted code, each bit position is assigned a weighting factor in such a way that each digit can be evaluated by adding the weights of all the 1's in the coded combination. The BCD code has weights of 8, 4, 2, and 1, which correspond to the power-of-two values of each bit. The bit assignment 0110, for example, is interpreted by the weights to represent decimal 6 because $8 \times 0 + 4 \times 1 + 2 \times 1 + 1 \times 0 = 6$. The bit combination 1101, when weighted by the respective digits 2421, gives the decimal equivalent of $2 \times 1 + 4 \times 1 + 2 \times 0 + 1 \times 1 = 7$. Note that some digits can be coded in two possible ways in the 2421 code. For instance, decimal 4 can be assigned to bit combination 0100 or 1010, since both combinations add up to a total weight of 4.

Table 1.5
Four Different Binary Codes for the Decimal Digits

Decimal Digit	BCD 8421	2421	Excess-3	8, 4, −2, −1
0	0000	0000	0011	0000
1	0001	0001	0100	0111
2	0010	0010	0101	0110
3	0011	0011	0110	0101
4	0100	0100	0111	0100
5	0101	1011	1000	1011
6	0110	1100	1001	1010
7	0111	1101	1010	1001
8	1000	1110	1011	1000
9	1001	1111	1100	1111
	1010	0101	0000	0001
Unused	1011	0110	0001	0010
bit	1100	0111	0010	0011
combi-	1101	1000	1101	1100
nations	1110	1001	1110	1101
	1111	1010	1111	1110

BCD adders add BCD values directly, digit by digit, without converting the numbers to binary. However, it is necessary to add 6 to the result if it is greater than 9. BCD adders require significantly more hardware and no longer have a speed advantage of conventional binary adders [5].

The 2421 and the excess-3 codes are examples of self-complementing codes. Such codes have the property that the 9's complement of a decimal number is obtained directly by changing 1's to 0's and 0's to 1's (i.e., by complementing each bit in the pattern). For example, decimal 395 is represented in the excess-3 code as 0110 1100 1000. The 9's complement of 604 is represented as 1001 0011 0111, which is obtained simply by complementing each bit of the code (as with the 1's complement of binary numbers).

The excess-3 code has been used in some older computers because of its self-complementing property. **Excess-3 is an unweighted code in which each coded combination is obtained from the corresponding binary value plus 3.** Note that the BCD code is not self-complementing.

The 8, 4, −2, −1 code is an example of assigning both positive and negative weights to a decimal code. In this case, the bit combination 0110 is interpreted as decimal 2 and is calculated from $8 \times 0 + 4 \times 1 + (-2) \times 1 + (-1) \times 0 = 2$.

Gray Code

The output data of many physical systems are quantities that are continuous. These data must be converted into digital form before they are applied to a digital system. Continuous or analog information is converted into digital form by means of an analog-to-digital converter. It is sometimes convenient to use the Gray code shown in Table 1.6 to represent digital data that have been converted from analog data. The advantage of the Gray code over the straight binary number sequence is that only one bit in the code group changes in going from one number to the next. For example, in going from 7 to 8, the Gray code changes from 0100 to 1100. Only the first bit changes, from 0 to 1; the other three bits remain the same. By contrast, with binary numbers the change from 7 to 8 will be from 0111 to 1000, which causes all four bits to change values.

The Gray code is used in applications in which the normal sequence of binary numbers generated by the hardware may produce an error or ambiguity during the transition from one number to the next. If binary numbers are used, a change, for example, from 0111 to 1000 may produce an intermediate erroneous number 1001 if the value of the rightmost bit takes longer to change than do the values of the other three bits. This could have serious consequences for the machine using the information. The Gray code eliminates this problem, since only one bit changes its value during any transition between two numbers.

A typical application of the Gray code is the representation of analog data by a continuous change in the angular position of a shaft. The shaft is partitioned into segments, and each segment is assigned a number. If adjacent segments are made to correspond with the Gray-code sequence, ambiguity is eliminated between the angle of the shaft and the value encoded by the sensor.

Table 1.6
Gray Code

Gray Code	Decimal Equivalent
0000	0
0001	1
0011	2
0010	3
0110	4
0111	5
0101	6
0100	7
1100	8
1101	9
1111	10
1110	11
1010	12
1011	13
1001	14
1000	15

ASCII Character Code

Many applications of digital computers require the handling not only of numbers, but also of other characters or symbols, such as the letters of the alphabet. For instance, consider a high-tech company with thousands of employees. To represent the names and other pertinent information, it is necessary to formulate a binary code for the letters of the alphabet. In addition, the same binary code must represent numerals and special characters (such as $). An alphanumeric character set is a set of elements that includes the 10 decimal digits, the 26 letters of the alphabet, and a number of special characters. Such a set contains between 36 and 64 elements if only capital letters are included, or between 64 and 128 elements if both uppercase and lowercase letters are included. In the first case, we need a binary code of six bits, and in the second, we need a binary code of seven bits.

The standard binary code for the alphanumeric characters is the American Standard Code for Information Interchange (ASCII), which uses seven bits to code 128 characters, as shown in Table 1.7. The seven bits of the code are designated by b_1 through b_7, with b_7 the most significant bit. The letter A, for example, is represented in ASCII as 1000001 (column 100, row 0001). The ASCII code also contains 94 graphic characters that can be printed and 34 nonprinting characters used for various control functions. The graphic characters consist of the 26 uppercase letters (A through Z), the 26 lowercase letters (a through z), the 10 numerals (0 through 9), and 32 special printable characters, such as %, *, and $.

Table 1.7
American Standard Code for Information Interchange (ASCII)

	$b_7b_6b_5$							
$b_4b_3b_2b_1$	**000**	**001**	**010**	**011**	**100**	**101**	**110**	**111**
0000	NUL	DLE	SP	0	@	P	`	p
0001	SOH	DC1	!	1	A	Q	a	q
0010	STX	DC2	"	2	B	R	b	r
0011	ETX	DC3	#	3	C	S	c	s
0100	EOT	DC4	$	4	D	T	d	t
0101	ENQ	NAK	%	5	E	U	e	u
0110	ACK	SYN	&	6	F	V	f	v
0111	BEL	ETB	'	7	G	W	g	w
1000	BS	CAN	(	8	H	X	h	x
1001	HT	EM	)	9	I	Y	i	y
1010	LF	SUB	*	:	J	Z	j	z
1011	VT	ESC	+	;	K	[	k	{
1100	FF	FS	,	<	L	\	l	|
1101	CR	GS	-	=	M	]	m	}
1110	SO	RS	.	>	N	∧	n	~
1111	SI	US	/	?	O	_	o	DEL

Control Characters

NUL	Null	DLE	Data-link escape
SOH	Start of heading	DC1	Device control 1
STX	Start of text	DC2	Device control 2
ETX	End of text	DC3	Device control 3
EOT	End of transmission	DC4	Device control 4
ENQ	Enquiry	NAK	Negative acknowledge
ACK	Acknowledge	SYN	Synchronous idle
BEL	Bell	ETB	End-of-transmission block
BS	Backspace	CAN	Cancel
HT	Horizontal tab	EM	End of medium
LF	Line feed	SUB	Substitute
VT	Vertical tab	ESC	Escape
FF	Form feed	FS	File separator
CR	Carriage return	GS	Group separator
SO	Shift out	RS	Record separator
SI	Shift in	US	Unit separator
SP	Space	DEL	Delete

The 34 control characters are designated in the ASCII table with abbreviated names. They are listed again below the table with their functional names. The control characters are used for routing data and arranging the printed text into a prescribed format. There are three types of control characters: format effectors, information separators, and communication-control characters. Format effectors are characters that control the layout of printing. They include the familiar word processor and typewriter controls such as backspace (BS), horizontal tabulation (HT), and carriage return (CR). Information separators are used to separate the data into divisions such as paragraphs and pages. They include characters such as record separator (RS) and file separator (FS). The communication-control characters are useful during the transmission of text between remote devices so that it can be distinguished from other messages using the same communication channel before it and after it. Examples of communication-control characters are STX (start of text) and ETX (end of text), which are used to frame a text message transmitted through a communication channel.

ASCII is a seven-bit code, but most computers manipulate an eight-bit quantity as a single unit called a *byte*. Therefore, ASCII characters most often are stored one per byte. The extra bit is sometimes used for other purposes, depending on the application. For example, some printers recognize eight-bit ASCII characters with the most significant bit set to 0. An additional 128 eight-bit characters with the most significant bit set to 1 are used for other symbols, such as the Greek alphabet or italic type font.

Error-Detecting Code

To detect errors in data communication and processing, an eighth bit is sometimes added to the ASCII character to indicate its parity. A *parity bit* is an extra bit included with a message to make the total number of 1's either even or odd. Consider the following two characters and their even and odd parity:

	With even parity	With odd parity
ASCII A = 1000001	01000001	11000001
ASCII T = 1010100	11010100	01010100

In each case, we insert an extra bit in the leftmost position of the code to produce an even number of 1's in the character for even parity or an odd number of 1's in the character for odd parity. In general, one or the other parity is adopted, with even parity being more common.

The parity bit is helpful in detecting errors during the transmission of information from one location to another. This function is handled by generating an even parity bit at the sending end for each character. The eight-bit characters that include parity bits are transmitted to their destination. The parity of each character is then checked at the receiving end. If the parity of the received character is not even, then at least one bit has changed value during the transmission. This method detects one, three, or any odd combination of errors in each character that is transmitted. An even combination of errors, however, goes undetected, and additional error detection codes may be needed to take care of that possibility.

What is done after an error is detected depends on the particular application. One possibility is to request retransmission of the message on the assumption that the error was random and will not occur again. Thus, if the receiver detects a parity error, it sends back the ASCII NAK (negative acknowledge) control character consisting of an even-parity eight bits 10010101. If no error is detected, the receiver sends back an ACK (acknowledge) control character, namely, 00000110. The sending end will respond to an NAK by transmitting the message again until the correct parity is received. If, after a number of attempts, the transmission is still in error, a message can be sent to the operator to check for malfunctions in the transmission path.

1.8 BINARY STORAGE AND REGISTERS

The binary information in a digital computer must have a physical existence in some medium for storing individual bits. A *binary cell* is a device that possesses two stable states and is capable of storing one bit (0 or 1) of information. The input to the cell receives excitation signals that set it to one of the two states. The output of the cell is a physical quantity that distinguishes between the two states. The information stored in a cell is 1 when the cell is in one stable state and 0 when the cell is in the other stable state.

Registers

A *register* is a group of binary cells. A register with n cells can store any discrete quantity of information that contains n bits. The state of a register is an n-tuple of 1's and 0's, with each bit designating the state of one cell in the register. The content of a register is a function of the interpretation given to the information stored in it. Consider, for example, a 16-bit register with the following binary content:

$$1100001111001001$$

A register with 16 cells can be in one of 2^{16} possible states. If one assumes that the content of the register represents a binary integer, then the register can store any binary number from 0 to $2^{16} - 1$. For the particular example shown, the content of the register is the binary equivalent of the decimal number 50,121. If one assumes instead that the register stores alphanumeric characters of an eight-bit code, then the content of the register is any two meaningful characters. For the ASCII code with an even parity placed in the eighth most significant bit position, the register contains the two characters C (the leftmost eight bits) and I (the rightmost eight bits). If, however, one interprets the content of the register to be four decimal digits represented by a four-bit code, then the content of the register is a four-digit decimal number. In the excess-3 code, the register holds the decimal number 9,096. The content of the register is meaningless in BCD, because the bit combination 1100 is not assigned to any decimal digit. From this example, it is clear that a register can store discrete elements of information and that the same bit configuration may be interpreted differently for different types of data depending on the application.

Register Transfer

A digital system is characterized by its registers and the components that perform data processing. In digital systems, a *register transfer* operation is a basic operation that consists of a transfer of binary information from one set of registers into another set of registers. The transfer may be direct, from one register to another, or may pass through data-processing circuits to perform an operation. Figure 1.1 illustrates the transfer of information among registers and demonstrates pictorially the transfer of binary information from a keyboard into a register in the memory unit. The input unit is assumed to have a keyboard, a control circuit, and an input register. Each time a key is struck, the control circuit enters an equivalent eight-bit alphanumeric character code into the input register. We shall assume that the code used is the ASCII code with an odd-parity bit. The information from the input register is transferred into the eight least significant cells of a processor register. After every transfer, the input register is cleared to enable the control to insert a new eight-bit code when the keyboard is struck again. Each eight-bit character transferred to the processor register is preceded by a shift of the previous character to the next eight cells on its left. When a transfer of four characters is completed, the processor register is full, and its contents are transferred into a memory register. The content stored in the

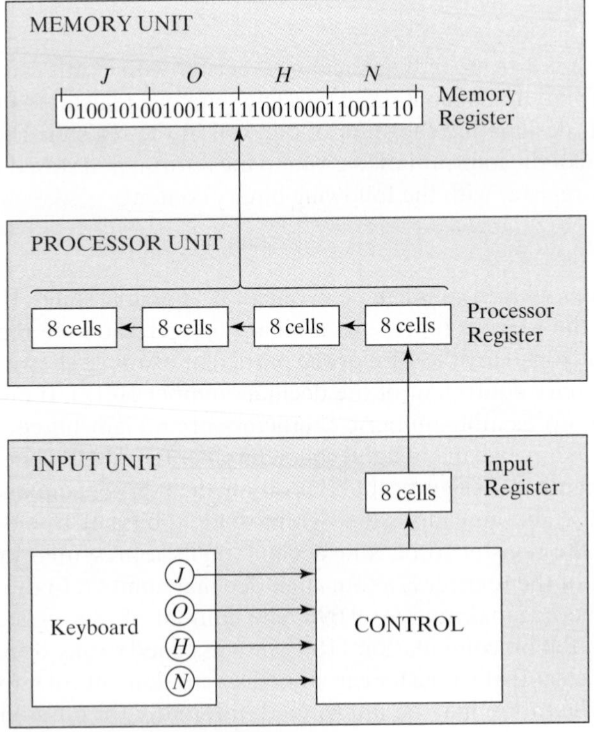

FIGURE 1.1
Transfer of information among registers

memory register shown in Fig. 1.1 came from the transfer of the characters "J," "O," "H," and "N" after the four appropriate keys were struck.

To process discrete quantities of information in binary form, a computer must be provided with devices that hold the data to be processed and with circuit elements that manipulate individual bits of information. **The device most commonly used for holding data is a register.** Binary variables are manipulated by means of digital logic circuits. Figure 1.2 illustrates the process of adding two 10-bit binary numbers. The memory unit, which normally consists of millions of registers, is shown with only three of its registers. The part of the processor unit shown consists of three registers—*R1*, *R2*, and *R3*—together with digital logic circuits that manipulate the bits of *R1* and *R2* and transfer into *R3* a binary number equal to their arithmetic sum. Memory registers store information and are incapable of processing the two operands. However, the information stored in memory can be transferred to processor registers, and the results obtained in processor registers can be transferred back into a memory register for storage until needed again. The diagram shows the contents of two operands transferred from two memory registers

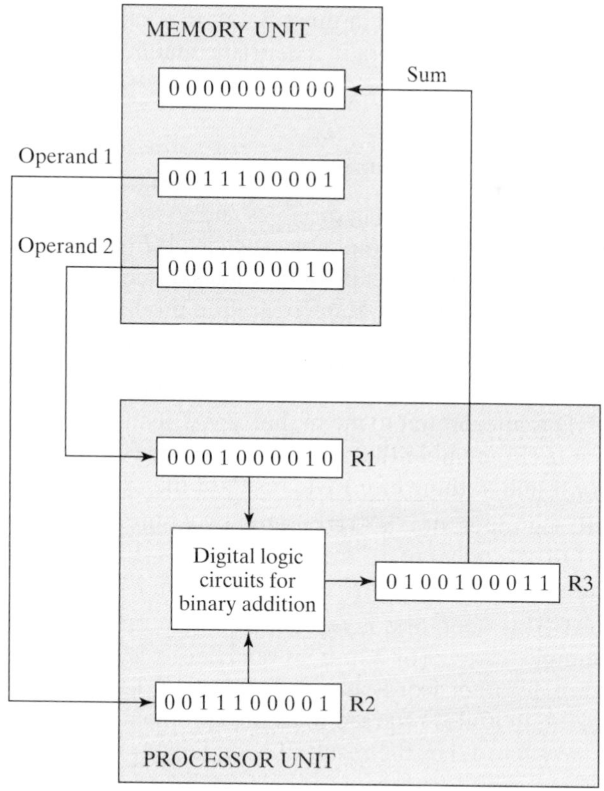

FIGURE 1.2
Example of binary information processing

into *R1* and *R2*. The digital logic circuits produce the sum, which is transferred to register *R3*. The contents of *R3* can now be transferred back to one of the memory registers.

The last two examples demonstrated the information-flow capabilities of a digital system in a simple manner. The registers of the system are the basic elements for storing and holding the binary information. Digital logic circuits process the binary information stored in the registers. Digital logic circuits and registers are covered in Chapters 2 through 6. The memory unit is explained in Chapter 7. The description of register operations at the register transfer level and the design of digital systems are covered in Chapter 8.

1.9 BINARY LOGIC

Binary logic deals with variables that take on two discrete values and with operations that assume logical meaning. The two values the variables assume may be called by different names (*true* and *false*, *yes* and *no*, etc.), but for our purpose, it is convenient to think in terms of bits and assign the values 1 and 0. The binary logic introduced in this section is equivalent to an algebra called Boolean algebra. The formal presentation of Boolean algebra is covered in more detail in Chapter 2. The purpose of this section is to introduce Boolean algebra in a heuristic manner and relate it to digital logic circuits and binary signals.

Definition of Binary Logic

Binary logic consists of binary variables and a set of logical operations. The variables are designated by letters of the alphabet, such as A, B, C, x, y, z, etc., with each variable having two and only two distinct possible values: 1 and 0. There are three basic logical operations: AND, OR, and NOT. Each operation produces a binary result, denoted by z.

1. AND: This operation is represented by a dot or by the absence of an operator. For example, $x \cdot y = z$ or $xy = z$ is read "x AND y is equal to z." The logical operation AND is interpreted to mean that $z = 1$ if and only if $x = 1$ and $y = 1$; otherwise $z = 0$. (Remember that x, y, and z are binary variables and can be equal either to 1 or 0, and nothing else.) The result of the operation $x \cdot y$ is z.
2. OR: This operation is represented by a plus sign. For example, $x + y = z$ is read "x OR y is equal to z," meaning that $z = 1$ if $x = 1$ or if $y = 1$ or if both $x = 1$ and $y = 1$. If both $x = 0$ and $y = 0$, then $z = 0$.
3. NOT: This operation is represented by a prime (sometimes by an overbar). For example, $x' = z$ (or $\bar{x} = z$) is read "not x is equal to z," meaning that z is what x is not. In other words, if $x = 1$, then $z = 0$, but if $x = 0$, then $z = 1$. The NOT operation is also referred to as the complement operation, since it changes a 1 to 0 and a 0 to 1, i.e., the result of complementing 1 is 0, and vice versa.

Binary logic resembles binary arithmetic, and the operations AND and OR have similarities to multiplication and addition, respectively. In fact, the symbols used for

Table 1.8
Truth Tables of Logical Operations

AND			OR			NOT	
x	y	$x \cdot y$	x	y	$x + y$	x	x'
0	0	0	0	0	0	0	1
0	1	0	0	1	1	1	0
1	0	0	1	0	1		
1	1	1	1	1	1		

AND and OR are the same as those used for multiplication and addition. However, **binary logic should not be confused with binary arithmetic.** One should realize that an arithmetic variable designates a number that may consist of many digits. A logic variable is always either 1 or 0. For example, in binary arithmetic, we have $1 + 1 = 10$ (read "one plus one is equal to 2"), whereas in binary logic, we have $1 + 1 = 1$ (read "one OR one is equal to one").

For each combination of the values of x and y, there is a value of z specified by the definition of the logical operation. Definitions of logical operations may be listed in a compact form called *truth tables*. A truth table is a table of all possible combinations of the variables, showing the relation between the values that the variables may take and the result of the operation. The truth tables for the operations AND and OR with variables x and y are obtained by listing all possible values that the variables may have when combined in pairs. For each combination, the result of the operation is then listed in a separate row. The truth tables for AND, OR, and NOT are given in Table 1.8. These tables clearly demonstrate the definition of the operations.

Logic Gates

Logic gates are electronic circuits that operate on one or more input signals to produce an output signal. Electrical signals such as voltages or currents exist as analog signals having values over a given continuous range, say, 0 to 3 V, but in a digital system these voltages are interpreted to be either of two recognizable values, 0 or 1. Voltage-operated logic circuits respond to two separate voltage levels that represent a binary variable equal to logic 1 or logic 0. For example, a particular digital system may define logic 0 as a signal equal to 0 V and logic 1 as a signal equal to 3 V. In practice, each voltage level has an acceptable range, as shown in Fig. 1.3. The input terminals of digital circuits accept binary signals within the allowable range and respond at the output terminals with binary signals that fall within the specified range. The intermediate region between the allowed regions is crossed only during a state transition. Any desired information for computing or control can be operated on by passing binary signals through various combinations of logic gates, with each signal representing a particular binary variable. When the physical signal is in a particular range it is interpreted to be either a 0 or a 1.

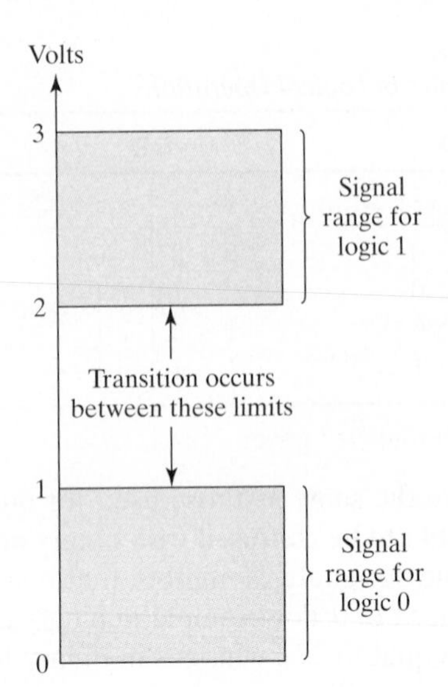

FIGURE 1.3
Signal levels for binary logic values

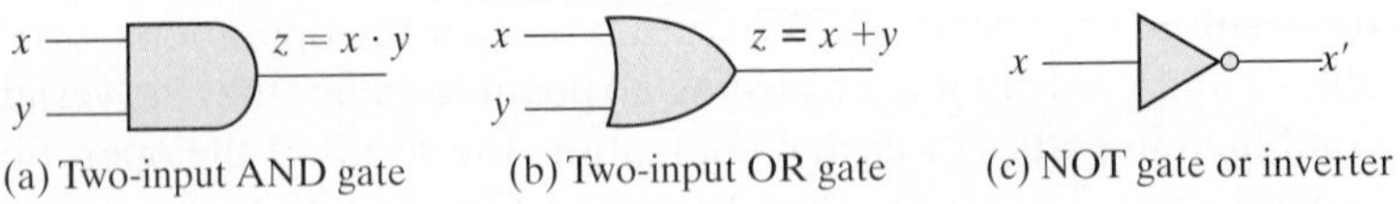

x, y, $z = x \cdot y$ (a) Two-input AND gate

x, y, $z = x + y$ (b) Two-input OR gate

x, x' (c) NOT gate or inverter

FIGURE 1.4
Symbols for digital logic circuits

The graphic symbols used to designate the three types of gates are shown in Fig. 1.4. The gates are blocks of hardware that produce the equivalent of logic-1 or logic-0 output signals if input logic requirements are satisfied. The input signals x and y in the AND and OR gates may exist in one of four possible states: 00, 10, 11, or 01. These input signals are shown in Fig. 1.5 together with the corresponding output signal for each gate. The timing diagrams illustrate the idealized response of each gate to the four input signal combinations. The horizontal axis of the timing diagram represents the time, and the vertical axis shows the signal as it changes between the two possible voltage levels. In reality, the transitions between logic values occur quickly, but not instantaneously. The low level represents logic 0, the high level logic 1. The AND gate responds with a logic 1 output signal when both input signals are logic 1. The OR gate responds with a logic 1 output signal if any input signal is logic 1. The NOT gate is commonly referred to as an inverter. The reason for this name is apparent from the signal response in the timing diagram, which shows that the output signal inverts the logic sense of the input signal.

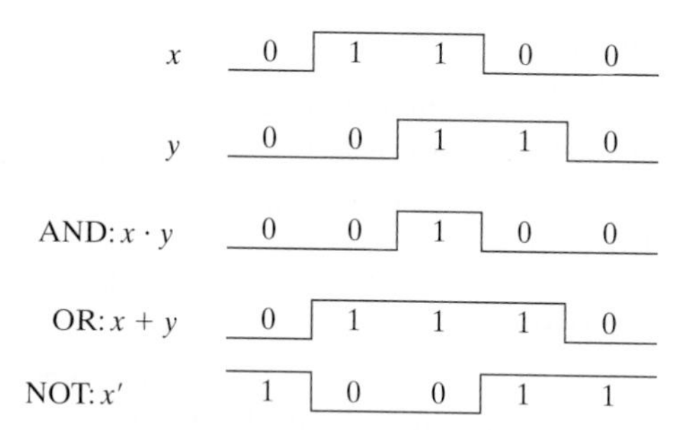

FIGURE 1.5
Input–output signals for gates

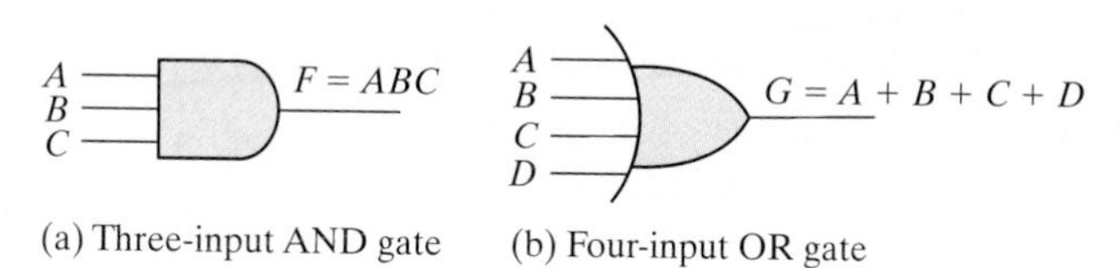

FIGURE 1.6
Gates with multiple inputs

AND and OR gates may have more than two inputs. An AND gate with three inputs and an OR gate with four inputs are shown in Fig. 1.6. The three-input AND gate responds with logic 1 output if all three inputs are logic 1. The output produces logic 0 if any input is logic 0. The four-input OR gate responds with logic 1 if any input is logic 1; its output becomes logic 0 only when all inputs are logic 0.

PROBLEMS

(Answers to problems marked with * appear at the end of the text.)

1.1 List the octal and hexadecimal numbers from 10 to 32. Using A and B for the last two digits, list the numbers from 10 to 32 in base 12.

1.2* What is the exact number of bytes in a system that contains (a) 16K bytes, (b) 32M bytes, and (c) 3.2G bytes?

1.3 Convert the following numbers with the indicated bases to decimal:
(a)* $(432)_5$ (b)* $(A98)_{12}$
(c) $(475)_8$ (d) $(2345)_6$

1.4 What is the largest binary number that can be expressed with 12 bits? What are the equivalent decimal and hexadecimal numbers?

1.5* Determine the base of the numbers in each case for the following operations to be correct:
(a) $12 \times 4 = 52$ (b) $75/3 = 26$ (c) $24 + 17 = 40$.

1.6* The solutions to the quadratic equation $x^2 - 13x + 32 = 0$ are $x = 5$ and $x = 4$. What is the base of the numbers?

1.7* Convert the hexadecimal number ABCD to binary, and then convert it from binary to octal.

1.8 Convert the decimal number 512 to binary in two ways: (a) convert directly to binary; (b) convert first to hexadecimal and then from hexadecimal to binary. Which method is faster?

1.9 Express the following numbers in decimal:
(a)* $(11010.0101)_2$ (b)* $(A6.5)_{16}$
(c)* $(276.24)_8$ (d) $(BABA.B)_{16}$
(e) $(10110.1101)_2$

1.10 Convert the following binary numbers to hexadecimal and to decimal: (a) 1.10010, (b) 1100.010. Explain why the decimal answer in (b) is 8 times that in (a).

1.11 Perform the following division in binary: 111111 ÷ 101.

1.12* Add and multiply the following numbers without converting them to decimal.
(a) Binary numbers 1100 and 110.
(b) Hexadecimal numbers AB and 1C.

1.13 Do the following conversion problems:
(a) Convert decimal 35.125 to binary.
(b) Calculate the binary equivalent of 1/3 out to eight places. Then convert from binary to decimal. How close is the result to 1/3?
(c) Convert the binary result in (b) into hexadecimal. Then convert the result to decimal. Is the answer the same?

1.14 Obtain the 1's and 2's complements of the following binary numbers:
(a) 11110000 (b) 00000000
(c) 11011000 (d) 01010101
(e) 10000000 (f) 11111111.

1.15 Find the 9's and the 10's complement of the following decimal numbers:
(a) 25,918,036 (b) 99,999,999
(c) 25,000,000 (d) 00,000,000.

1.16 (a) Find the 16's complement of CAD9.
(b) Convert CAD9 to binary.
(c) Find the 2's complement of the result in (b).
(d) Convert the answer in (c) to hexadecimal and compare with the answer in (a).

1.17 Perform subtraction on the given unsigned numbers using the 10's complement of the subtrahend. Where the result should be negative, find its 10's complement and affix a minus sign. Verify your answers.
(a) 3,699 − 2,579 (b) 974 − 1,800
(c) 2,943 − 4,361 (d) 7,631 − 745

1.18 Perform subtraction on the given unsigned binary numbers using the 2's complement of the subtrahend. Where the result should be negative, find its 2's complement and affix a minus sign.
(a) 10101 − 10010 (b) 10010 − 100110
(c) 10011 − 110101 (d) 101000 − 101101

1.19* The following decimal numbers are shown in sign-magnitude form: +9,081 and +954. Convert them to signed-10's-complement form and perform the following operations (note that the sum is +10,035 and requires five digits and a sign).
(a) (+9,081) + (+801) (b) (+9,286) + (−954)
(c) (−9,081) + (+801) (d) (−9,286) + (−954)

1.20 Convert decimal +56 and +35 to binary, using the signed-2's-complement representation and enough digits to accommodate the numbers. Then perform the binary equivalent of (+56) + (+35), (+56) + (−35), and (−56) + (+35). Convert the answers back to decimal and verify that they are correct.

1.21 If the numbers $(+9{,}542)_{10}$ and $(+641)_{10}$ are in signed magnitude format, their sum is $(+10{,}183)_{10}$ and requires five digits and a sign. Convert the numbers to signed-10's-complement form and find the following sums:
(a) (+9,742) + (+641) (b) (+9,742) + (−641)
(c) (−9,742) + (+641) (d) (−9,742) + (−641)

1.22 Convert decimal 7,654 to both BCD and ASCII codes. For ASCII, an even parity bit is to be appended at the left.

1.23 Represent the unsigned decimal numbers 694 and 538 in BCD, and then show the steps necessary to form their sum.

1.24 Formulate a weighted binary code for the octal digits, using the following weights:
(a)* 6, 3, 1, 1
(b) 6, 4, 2, 1

1.25 Represent the decimal number 6,514 in (a) BCD, (b) excess-3 code, (c) 2421 code, and (d) a 6311 code.

1.26 Find the 9's complement of decimal 6,514 and express it in 2421 code. Show that the result is the 1's complement of the answer to (c) in CR_PROBlem 1.25. This demonstrates that the 2421 code is self-complementing.

1.27 Assign a binary code in some orderly manner to the 52 playing cards. Use the minimum number of bits.

1.28 Write the expression "George B." in ASCII, using an eight-bit code. Include the period and the space. Treat the leftmost bit of each character as a parity bit. Each eight-bit code should have odd parity. (George Boole was a 19th-century mathematician. Boolean algebra, introduced in the next chapter, bears his name.)

1.29* Decode the following ASCII code:
1000100 1101001 1100111 1101001 1110100 1100001 1101100 1010011 1111001 1110011 1110100 1100101 1101101.

1.30 The following is a string of ASCII characters whose bit patterns have been converted into hexadecimal for compactness: C9 EE F3 74 69 74 F5 74 65. Of the eight bits in each pair of digits, the leftmost is a parity bit. The remaining bits are the ASCII code.
(a) Convert the string to bit form and decode the ASCII.
(b) Determine the parity used: odd or even?

1.31* How many printing characters are there in ASCII? How many of them are special characters (not letters or numerals)?

1.32* What bit must be complemented to change an ASCII letter from capital to lowercase and vice versa?

1.33* The state of a 12-bit register is 010101100100. What is its content if it represents
(a) Three decimal digits in BCD?
(b) Three decimal digits in the excess-3 code?
(c) Three decimal digits in the 84-2-1 code?
(d) A binary number?

1.34 List the ASCII code for the 8 octal digits with an odd parity bit in the leftmost position.

1.35 By means of a timing diagram similar to Fig. 1.5, show the signals of the outputs f and g in Fig. P1.35 as functions of the three inputs a, b, and c. Use all eight possible combinations of a, b, and c.

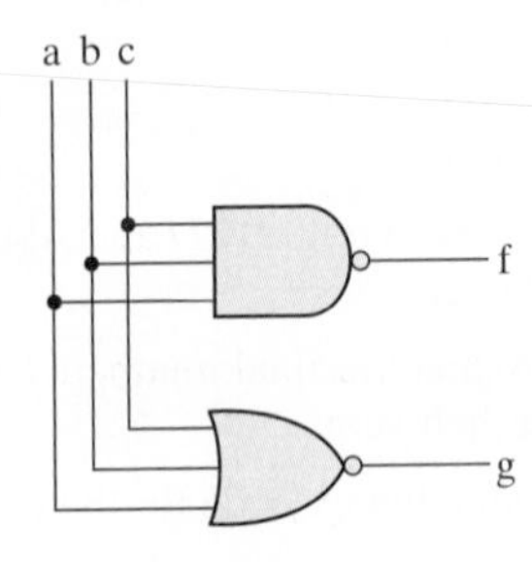

FIGURE P1.35

1.36 By means of a timing diagram similar to Fig. 1.5, show the signals of the outputs f and g in Fig. P1.36 as functions of the two inputs a and b. Use all four possible combinations of a and b.

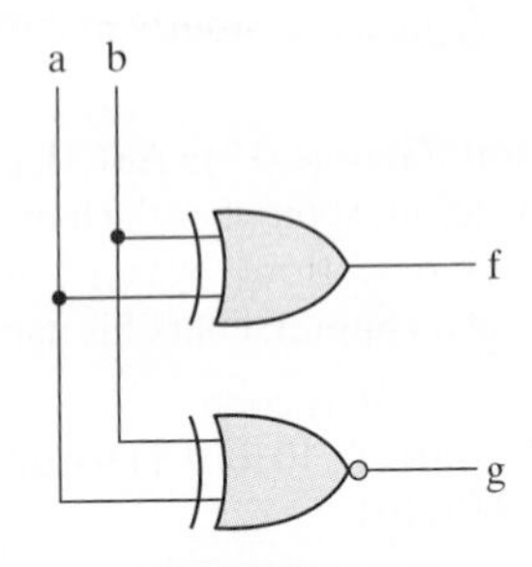

FIGURE P1.36

REFERENCES

1. Cavanagh, J. J. 1984. *Digital Computer Arithmetic.* New York: McGraw-Hill.
2. Mano, M. M. 1988. *Computer Engineering: Hardware Design*. Englewood Cliffs, NJ: Prentice-Hall.
3. Nelson, V. P., H. T. Nagle, J. D. Irwin, and B. D. Carroll. 1997. *Digital Logic Circuit Analysis and Design.* Upper Saddle River, NJ: Prentice Hall.
4. Schmid, H. 1974. *Decimal Computation*. New York: John Wiley.
5. Katz, R. H. and Borriello, G. 2004. *Contemporary Logic Design*, 2nd ed. Upper Saddle River, NJ: Prentice-Hall.

WEB SEARCH TOPICS

BCD code
ASCII
Storage register
Binary logic
BCD addition
Binary codes
Binary numbers
Excess-3 code

Chapter 2

Boolean Algebra and Logic Gates

2.1 INTRODUCTION

Because binary logic is used in all of today's digital computers and devices, the cost of the circuits that implement it is an important factor addressed by designers—be they computer engineers, electrical engineers, or computer scientists. Finding simpler and cheaper, but equivalent, realizations of a circuit can reap huge payoffs in reducing the overall cost of the design. Mathematical methods that simplify circuits rely primarily on Boolean algebra. Therefore, this chapter provides a basic vocabulary and a brief foundation in Boolean algebra that will enable you to optimize simple circuits and to understand the purpose of algorithms used by software tools to optimize complex circuits involving millions of logic gates.

2.2 BASIC DEFINITIONS

Boolean algebra, like any other deductive mathematical system, may be defined with a set of elements, a set of operators, and a number of unproved axioms or postulates. A *set* of elements is any collection of objects, usually having a common property. If S is a set, and x and y are certain objects, then the notation $x \in S$ means that x is a member of the set S and $y \notin S$ means that y is not an element of S. A set with a denumerable number of elements is specified by braces: $A = \{1, 2, 3, 4\}$ indicates that the elements of set A are the numbers 1, 2, 3, and 4. A *binary operator* defined on a set S of elements is a rule that assigns, to each pair of elements from S, a unique element from S. As an example, consider the relation $a * b = c$. We say that $*$ is a binary operator if it specifies a rule for finding c from the pair (a, b) and also if $a, b, c \in S$. However, $*$ is not a binary operator if $a, b \in S$, and if $c \notin S$.

The postulates of a mathematical system form the basic assumptions from which it is possible to deduce the rules, theorems, and properties of the system. The most common postulates used to formulate various algebraic structures are as follows:

1. *Closure.* A set S is closed with respect to a binary operator if, for every pair of elements of S, the binary operator specifies a rule for obtaining a unique element of S. For example, the set of natural numbers $N = \{1, 2, 3, 4, \ldots\}$ is closed with respect to the binary operator $+$ by the rules of arithmetic addition, since, for any $a, b \in N$, there is a unique $c \in N$ such that $a + b = c$. The set of natural numbers is *not* closed with respect to the binary operator $-$ by the rules of arithmetic subtraction, because $2 - 3 = -1$ and $2, 3 \in N$, but $(-1) \notin N$.
2. *Associative law.* A binary operator $*$ on a set S is said to be associative whenever

$$(x * y) * z = x * (y * z) \text{ for all } x, y, z, \in S$$

3. *Commutative law.* A binary operator $*$ on a set S is said to be commutative whenever

$$x * y = y * x \text{ for all } x, y \in S$$

4. *Identity element.* A set S is said to have an identity element with respect to a binary operation $*$ on S if there exists an element $e \in S$ with the property that

$$e * x = x * e = x \text{ for every } x \in S$$

Example: The element 0 is an identity element with respect to the binary operator $+$ on the set of integers $I = \{\ldots, -3, -2, -1, 0, 1, 2, 3, \ldots\}$, since

$$x + 0 = 0 + x = x \text{ for any } x \in I$$

The set of natural numbers, N, has no identity element, since 0 is excluded from the set.

5. *Inverse.* A set S having the identity element e with respect to a binary operator $*$ is said to have an inverse whenever, for every $x \in S$, there exists an element $y \in S$ such that

$$x * y = e$$

Example: In the set of integers, I, and the operator $+$, with $e = 0$, the inverse of an element a is $(-a)$, since $a + (-a) = 0$.

6. *Distributive law.* If $*$ and $\cdot$ are two binary operators on a set S, $*$ is said to be distributive over $\cdot$ whenever

$$x * (y \cdot z) = (x * y) \cdot (x * z)$$

A *field* is an example of an algebraic structure. A field is a set of elements, together with two binary operators, each having properties 1 through 5 and both operators combining to give property 6. The set of real numbers, together with the binary operators $+$ and $\cdot$, forms the field of real numbers. The field of real numbers is the basis for arithmetic and ordinary algebra. The operators and postulates have the following meanings:

The binary operator $+$ defines addition.
The additive identity is 0.
The additive inverse defines subtraction.
The binary operator $\cdot$ defines multiplication.
The multiplicative identity is 1.
For $a \neq 0$, the multiplicative inverse of $a = 1/a$ defines division (i.e., $a \cdot 1/a = 1$).
The only distributive law applicable is that of $\cdot$ over $+$:

$$a \cdot (b + c) = (a \cdot b) + (a \cdot c)$$

2.3 AXIOMATIC DEFINITION OF BOOLEAN ALGEBRA

In 1854, George Boole developed an algebraic system now called *Boolean algebra*. In 1938, Claude E. Shannon introduced a two-valued Boolean algebra called *switching algebra* that represented the properties of bistable electrical switching circuits. For the formal definition of Boolean algebra, we shall employ the postulates formulated by E. V. Huntington in 1904.

Boolean algebra is an algebraic structure defined by a set of elements, B, together with two binary operators, $+$ and $\cdot$, provided that the following (Huntington) postulates are satisfied:

1. (a) The structure is closed with respect to the operator $+$.
(b) The structure is closed with respect to the operator $\cdot$.

2. (a) The element 0 is an identity element with respect to $+$; that is, $x + 0 = 0 + x = x$.
(b) The element 1 is an identity element with respect to $\cdot$; that is, $x \cdot 1 = 1 \cdot x = x$.

3. (a) The structure is commutative with respect to $+$; that is, $x + y = y + x$.
(b) The structure is commutative with respect to $\cdot$; that is, $x \cdot y = y \cdot x$.

4. (a) The operator $\cdot$ is distributive over $+$; that is, $x \cdot (y + z) = (x \cdot y) + (x \cdot z)$.
(b) The operator $+$ is distributive over $\cdot$; that is, $x + (y \cdot z) = (x + y) \cdot (x + z)$.

5. For every element $x \in B$, there exists an element $x' \in B$ (called the *complement* of x) such that (a) $x + x' = 1$ and (b) $x \cdot x' = 0$.

6. There exist at least two elements $x, y \in B$ such that $x \neq y$.

Comparing Boolean algebra with arithmetic and ordinary algebra (the field of real numbers), we note the following differences:

1. Huntington postulates do not include the associative law. However, this law holds for Boolean algebra and can be derived (for both operators) from the other postulates.
2. The distributive law of $+$ over $\cdot$ (i.e., $x + (y \cdot z) = (x + y) \cdot (x + z)$) is valid for Boolean algebra, but not for ordinary algebra.

3. Boolean algebra does not have additive or multiplicative inverses; therefore, there are no subtraction or division operations.
4. Postulate 5 defines an operator called the *complement* that is not available in ordinary algebra.
5. Ordinary algebra deals with the real numbers, which constitute an infinite set of elements. Boolean algebra deals with the as yet undefined set of elements, B, but in the two-valued Boolean algebra defined next (and of interest in our subsequent use of that algebra), B is defined as a set with only two elements, 0 and 1.

Boolean algebra resembles ordinary algebra in some respects. The choice of the symbols $+$ and $\cdot$ is intentional, to facilitate Boolean algebraic manipulations by persons already familiar with ordinary algebra. Although one can use some knowledge from ordinary algebra to deal with Boolean algebra, the beginner must be careful not to substitute the rules of ordinary algebra where they are not applicable.

It is important to distinguish between the elements of the set of an algebraic structure and the variables of an algebraic system. For example, the elements of the field of real numbers are numbers, whereas variables such as a, b, c, etc., used in ordinary algebra, are symbols that *stand for* real numbers. Similarly, in Boolean algebra, one defines the elements of the set B, and variables such as x, y, and z are merely symbols that *represent* the elements. At this point, it is important to realize that, in order to have a Boolean algebra, one must show that

1. the elements of the set B,
2. the rules of operation for the two binary operators, and
3. the set of elements, B, together with the two operators, satisfy the six Huntington postulates.

One can formulate many Boolean algebras, depending on the choice of elements of B and the rules of operation. In our subsequent work, **we deal only with a two-valued Boolean algebra** (i.e., a Boolean algebra with only two elements). Two-valued Boolean algebra has applications in set theory (the algebra of classes) and in propositional logic. Our interest here is in the application of Boolean algebra to gate-type circuits commonly used in digital devices and computers.

Two-Valued Boolean Algebra

A two-valued Boolean algebra is defined on a set of two elements, $B = \{0, 1\}$, with rules for the two binary operators $+$ and $\cdot$ as shown in the following operator tables (the rule for the complement operator is for verification of postulate 5):

x	y	$x \cdot y$
0	0	0
0	1	0
1	0	0
1	1	1

x	y	$x + y$
0	0	0
0	1	1
1	0	1
1	1	1

x	x'
0	1
1	0

These rules are exactly the same as the AND, OR, and NOT operations, respectively, defined in Table 1.8. We must now show that the Huntington postulates are valid for the set $B = \{0, 1\}$ and the two binary operators $+$ and $\cdot$.

1. That the structure is *closed* with respect to the two operators is obvious from the tables, since the result of each operation is either 1 or 0 and $1, 0 \in B$.
2. From the tables, we see that

   (a) $0 + 0 = 0 \qquad 0 + 1 = 1 + 0 = 1;$
   (b) $1 \cdot 1 = 1 \qquad 1 \cdot 0 = 0 \cdot 1 = 0.$

   This establishes the two *identity elements,* 0 for $+$ and 1 for $\cdot$, as defined by postulate 2.
3. The *commutative* laws are obvious from the symmetry of the binary operator tables.
4. (a) The *distributive* law $x \cdot (y + z) = (x \cdot y) + (x \cdot z)$ can be shown to hold from the operator tables by forming a truth table of all possible values of x, y, and z. For each combination, we derive $x \cdot (y + z)$ and show that the value is the same as the value of $(x \cdot y) + (x \cdot z)$:

x	y	z
0	0	0
0	0	1
0	1	0
0	1	1
1	0	0
1	0	1
1	1	0
1	1	1

y + z	x · (y + z)
0	0
1	0
1	0
1	0
0	0
1	1
1	1
1	1

x · y	x · z	(x · y) + (x · z)
0	0	0
0	0	0
0	0	0
0	0	0
0	0	0
0	1	1
1	0	1
1	1	1

   (b) The *distributive* law of $+$ over $\cdot$ can be shown to hold by means of a truth table similar to the one in part (a).
5. From the complement table, it is easily shown that

   (a) $x + x' = 1$, since $0 + 0' = 0 + 1 = 1$ and $1 + 1' = 1 + 0 = 1$.
   (b) $x \cdot x' = 0$, since $0 \cdot 0' = 0 \cdot 1 = 0$ and $1 \cdot 1' = 1 \cdot 0 = 0$.

   Thus, postulate 1 is verified.
6. Postulate 6 is satisfied because the two-valued Boolean algebra has two elements, 1 and 0, with $1 \neq 0$.

We have just established a two-valued Boolean algebra having a set of two elements, 1 and 0, two binary operators with rules equivalent to the AND and OR operations, and a complement operator equivalent to the NOT operator. Thus, Boolean algebra has been defined in a formal mathematical manner and has been shown to be equivalent to the binary logic presented heuristically in Section 1.9. The heuristic presentation is helpful in understanding the application of Boolean algebra to gate-type circuits. The formal presentation is necessary for developing the theorems and properties of the algebraic system. The two-valued Boolean algebra defined in this section is also called "switching algebra" by engineers. To emphasize the similarities between two-valued Boolean algebra and other binary systems, that algebra was called "binary logic" in Section 1.9. From here on, we shall drop the adjective "two-valued" from Boolean algebra in subsequent discussions.

2.4 BASIC THEOREMS AND PROPERTIES OF BOOLEAN ALGEBRA

Duality

In Section 2.3, the Huntington postulates were listed in pairs and designated by part (a) and part (b). One part may be obtained from the other if the binary operators and the identity elements are interchanged. This important property of Boolean algebra is called the *duality principle* and states that every algebraic expression deducible from the postulates of Boolean algebra remains valid if the operators and identity elements are interchanged. In a two-valued Boolean algebra, the identity elements and the elements of the set B are the same: 1 and 0. The duality principle has many applications. If the *dual* of an algebraic expression is desired, we simply interchange OR and AND operators and replace 1's by 0's and 0's by 1's.

Basic Theorems

Table 2.1 lists six theorems of Boolean algebra and four of its postulates. The notation is simplified by omitting the binary operator whenever doing so does not lead to confusion. The theorems and postulates listed are the most basic relationships in Boolean

Table 2.1
Postulates and Theorems of Boolean Algebra

Postulate 2	(a)	$x + 0 = x$	(b)	$x \cdot 1 = x$
Postulate 5	(a)	$x + x' = 1$	(b)	$x \cdot x' = 0$
Theorem 1	(a)	$x + x = x$	(b)	$x \cdot x = x$
Theorem 2	(a)	$x + 1 = 1$	(b)	$x \cdot 0 = 0$
Theorem 3, involution		$(x')' = x$		
Postulate 3, commutative	(a)	$x + y = y + x$	(b)	$xy = yx$
Theorem 4, associative	(a)	$x + (y + z) = (x + y) + z$	(b)	$x(yz) = (xy)z$
Postulate 4, distributive	(a)	$x(y + z) = xy + xz$	(b)	$x + yz = (x + y)(x + z)$
Theorem 5, DeMorgan	(a)	$(x + y)' = x'y'$	(b)	$(xy)' = x' + y'$
Theorem 6, absorption	(a)	$x + xy = x$	(b)	$x(x + y) = x$

algebra. The theorems, like the postulates, are listed in pairs; each relation is the dual of the one paired with it. The postulates are basic axioms of the algebraic structure and need no proof. The theorems must be proven from the postulates. Proofs of the theorems with one variable are presented next. At the right is listed the number of the postulate which justifies that particular step of the proof.

THEOREM 1(a): $x + x = x$.

Statement	Justification
$x + x = (x + x) \cdot 1$	postulate 2(b)
$= (x + x)(x + x')$	5(a)
$= x + xx'$	4(b)
$= x + 0$	5(b)
$= x$	2(a)

THEOREM 1(b): $x \cdot x = x$.

Statement	Justification
$x \cdot x = xx + 0$	postulate 2(a)
$= xx + xx'$	5(b)
$= x(x + x')$	4(a)
$= x \cdot 1$	5(a)
$= x$	2(b)

Note that theorem 1(b) is the dual of theorem 1(a) and that each step of the proof in part (b) is the dual of its counterpart in part (a). Any dual theorem can be similarly derived from the proof of its corresponding theorem.

THEOREM 2(a): $x + 1 = 1$.

Statement	Justification
$x + 1 = 1 \cdot (x + 1)$	postulate 2(b)
$= (x + x')(x + 1)$	5(a)
$= x + x' \cdot 1$	4(b)
$= x + x'$	2(b)
$= 1$	5(a)

THEOREM 2(b): $x \cdot 0 = 0$ by duality.

THEOREM 3: $(x')' = x$. From postulate 5, we have $x + x' = 1$ and $x \cdot x' = 0$, which together define the complement of x. The complement of x' is x and is also $(x')'$.

Therefore, since the complement is unique, we have $(x')' = x$. The theorems involving two or three variables may be proven algebraically from the postulates and the theorems that have already been proven. Take, for example, the absorption theorem:

THEOREM 6(a): $x + xy = x$.

Statement	Justification
$x + xy = x \cdot 1 + xy$	postulate 2(b)
$= x(1 + y)$	4(a)
$= x(y + 1)$	3(a)
$= x \cdot 1$	2(a)
$= x$	2(b)

THEOREM 6(b): $x(x + y) = x$ by duality.

The theorems of Boolean algebra can be proven by means of truth tables. In truth tables, both sides of the relation are checked to see whether they yield identical results for all possible combinations of the variables involved. The following truth table verifies the first absorption theorem:

x	y	xy	x + xy
0	0	0	0
0	1	0	0
1	0	0	1
1	1	1	1

The algebraic proofs of the associative law and DeMorgan's theorem are long and will not be shown here. However, their validity is easily shown with truth tables. For example, the truth table for the first DeMorgan's theorem, $(x + y)' = x'y'$, is as follows:

x	y	x + y	(x + y)'	x'	y'	x'y'
0	0	0	1	1	1	1
0	1	1	0	1	0	0
1	0	1	0	0	1	0
1	1	1	0	0	0	0

Operator Precedence

The operator precedence for evaluating Boolean expressions is (1) parentheses, (2) NOT, (3) AND, and (4) OR. In other words, expressions inside parentheses must be evaluated before all other operations. The next operation that holds precedence is the complement, and then follows the AND and, finally, the OR. As an example, consider the truth table for one of DeMorgan's theorems. The left side of the expression is $(x + y)'$. Therefore, the expression inside the parentheses is evaluated first and the result then complemented. The right side of the expression is $x'y'$, so the complement of x and the complement of y are both evaluated first and the result is then ANDed. Note that in ordinary arithmetic, the same precedence holds (except for the complement) when multiplication and addition are replaced by AND and OR, respectively.

2.5 BOOLEAN FUNCTIONS

Boolean algebra is an algebra that deals with binary variables and logic operations. A Boolean function described by an algebraic expression consists of binary variables, the constants 0 and 1, and the logic operation symbols. For a given value of the binary variables, the function can be equal to either 1 or 0. As an example, consider the Boolean function

$$F_1 = x + y'z$$

The function F_1 is equal to 1 if x is equal to 1 or if both y' and z are equal to 1. F_1 is equal to 0 otherwise. The complement operation dictates that when $y' = 1$, $y = 0$. Therefore, $F_1 = 1$ if $x = 1$ or if $y = 0$ and $z = 1$. A Boolean function expresses the logical relationship between binary variables and is evaluated by determining the binary value of the expression for all possible values of the variables.

A Boolean function can be represented in a truth table. The number of rows in the truth table is 2^n, where n is the number of variables in the function. The binary combinations for the truth table are obtained from the binary numbers by counting from 0 through $2^n - 1$. Table 2.2 shows the truth table for the function F_1. There are eight possible binary combinations for assigning bits to the three variables x, y, and z. The column labeled F_1 contains either 0 or 1 for each of these combinations. The table shows that the function is equal to 1 when $x = 1$ or when $yz = 01$ and is equal to 0 otherwise.

A Boolean function can be transformed from an algebraic expression into a circuit diagram composed of logic gates connected in a particular structure. The logic-circuit diagram (also called a schematic) for F_1 is shown in Fig. 2.1. There is an inverter for input y to generate its complement. There is an AND gate for the term $y'z$ and an OR gate

Table 2.2
Truth Tables for F_1 and F_2

x	y	z	F_1	F_2
0	0	0	0	0
0	0	1	1	1
0	1	0	0	0
0	1	1	0	1
1	0	0	1	1
1	0	1	1	1
1	1	0	1	0
1	1	1	1	0

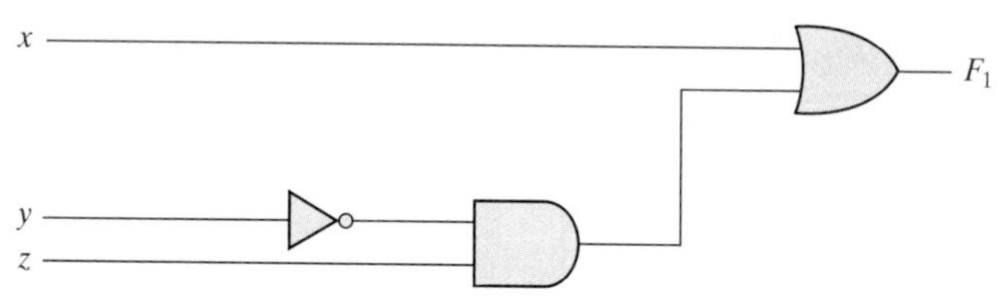

FIGURE 2.1
Gate implementation of $F_1 = x + y'z$

that combines x with $y'z$. In logic-circuit diagrams, the variables of the function are taken as the inputs of the circuit and the binary variable F_1 is taken as the output of the circuit. The schematic expresses the relationship between the output of the circuit and its inputs. Rather than listing each combination of inputs and outputs, it indicates how to compute the logic value of each output from the logic values of the inputs.

There is only one way that a Boolean function can be represented in a truth table. However, when the function is in algebraic form, it can be expressed in a variety of ways, all of which have equivalent logic. The particular expression used to represent the function will dictate the interconnection of gates in the logic-circuit diagram. Conversely, the interconnection of gates will dictate the logic expression. Here is a key fact that motivates our use of Boolean algebra: By manipulating a Boolean expression according to the rules of Boolean algebra, it is sometimes possible to obtain a simpler expression for the same function and thus reduce the number of gates in the circuit and the number of inputs to the gate. Designers are motivated to reduce the complexity and number of gates because their effort can significantly reduce the cost of a circuit. Consider, for example, the following Boolean function:

$$F_2 = x'y'z + x'yz + xy'$$

A schematic of an implementation of this function with logic gates is shown in Fig. 2.2(a). Input variables x and y are complemented with inverters to obtain x' and y'. The three terms in the expression are implemented with three AND gates. The OR gate forms the logical OR of the three terms. The truth table for F_2 is listed in Table 2.2. The function is equal to 1 when $xyz = 001$ or 011 or when $xy = 10$ (irrespective of the value of z) and is equal to 0 otherwise. This set of conditions produces four 1's and four 0's for F_2.

Now consider the possible simplification of the function by applying some of the identities of Boolean algebra:

$$F_2 = x'y'z + x'yz + xy' = x'z(y' + y) + xy' = x'z + xy'$$

The function is reduced to only two terms and can be implemented with gates as shown in Fig. 2.2(b). It is obvious that the circuit in (b) is simpler than the one in (a), yet both implement the same function. By means of a truth table, it is possible to verify that the two expressions are equivalent. The simplified expression is equal to 1 when $xz = 01$ or when $xy = 10$. This produces the same four 1's in the truth table. Since both expressions

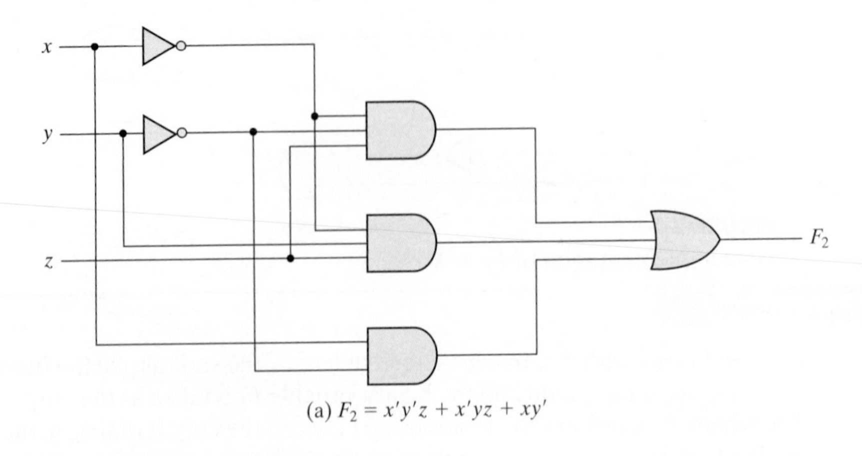

(a) $F_2 = x'y'z + x'yz + xy'$

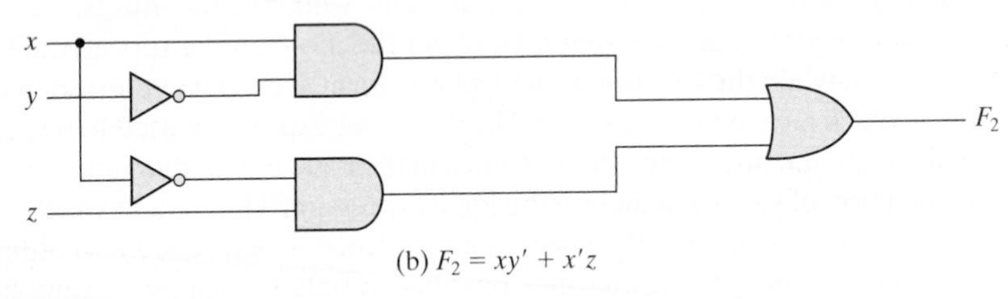

(b) $F_2 = xy' + x'z$

FIGURE 2.2
Implementation of Boolean function F_2 with gates

produce the same truth table, they are equivalent. Therefore, the two circuits have the same outputs for all possible binary combinations of inputs of the three variables. Each circuit implements the same identical function, but the one with fewer gates and fewer inputs to gates is preferable because it requires fewer wires and components. In general, there are many equivalent representations of a logic function. Finding the most economic representation of the logic is an important design task.

Algebraic Manipulation

When a Boolean expression is implemented with logic gates, each term requires a gate and each variable within the term designates an input to the gate. We define a *literal* to be a single variable within a term, in complemented or uncomplemented form. The function of Fig. 2.2(a) has three terms and eight literals, and the one in Fig. 2.2(b) has two terms and four literals. By reducing the number of terms, the number of literals, or both in a Boolean expression, it is often possible to obtain a simpler circuit. The manipulation of Boolean algebra consists mostly of reducing an expression for the purpose of obtaining a simpler circuit. Functions of up to five variables can be simplified by the map method described in the next chapter. For complex Boolean functions and many different outputs, designers of digital circuits use computer minimization programs that are capable of producing optimal circuits with millions of logic gates. The concepts introduced in this chapter provide the framework for those tools. The only manual method available is a cut-and-try procedure employing the basic relations and other manipulation techniques that become familiar with use, but remain, nevertheless, subject to human error. The examples that follow illustrate the algebraic manipulation of Boolean algebra to acquaint the reader with this important design task.

EXAMPLE 2.1

Simplify the following Boolean functions to a minimum number of literals.

1. $x(x' + y) = xx' + xy = 0 + xy = xy.$
2. $x + x'y = (x + x')(x + y) = 1(x + y) = x + y.$
3. $(x + y)(x + y') = x + xy + xy' + yy' = x(1 + y + y') = x.$
4. $$\begin{aligned} xy + x'z + yz &= xy + x'z + yz(x + x') \\ &= xy + x'z + xyz + x'yz \\ &= xy(1 + z) + x'z(1 + y) \\ &= xy + x'z. \end{aligned}$$
5. $(x + y)(x' + z)(y + z) = (x + y)(x' + z)$, by duality from function 4.

■

Functions 1 and 2 are the dual of each other and use dual expressions in corresponding steps. An easier way to simplify function 3 is by means of postulate 4(b) from Table 2.1: $(x + y)(x + y') = x + yy' = x$. The fourth function illustrates the fact that an increase in the number of literals sometimes leads to a simpler final expression. Function 5 is not minimized directly, but can be derived from the dual of the steps used to derive function 4. Functions 4 and 5 are together known as the *consensus theorem.*

Complement of a Function

The complement of a function F is F' and is obtained from an interchange of 0's for 1's and 1's for 0's in the value of F. The complement of a function may be derived algebraically through DeMorgan's theorems, listed in Table 2.1 for two variables. DeMorgan's theorems can be extended to three or more variables. The three-variable form of the first DeMorgan's theorem is derived as follows, from postulates and theorems listed in Table 2.1:

$$\begin{aligned} (A + B + C)' &= (A + x)' && \text{let } B + C = x \\ &= A'x' && \text{by theorem 5(a) (DeMorgan)} \\ &= A'(B + C)' && \text{substitute } B + C = x \\ &= A'(B'C') && \text{by theorem 5(a) (DeMorgan)} \\ &= A'B'C' && \text{by theorem 4(b) (associative)} \end{aligned}$$

DeMorgan's theorems for any number of variables resemble the two-variable case in form and can be derived by successive substitutions similar to the method used in the preceding derivation. These theorems can be generalized as follows:

$$(A + B + C + D + \cdots + F)' = A'B'C'D' \ldots F'$$
$$(ABCD \ldots F)' = A' + B' + C' + D' + \cdots + F'$$

The generalized form of DeMorgan's theorems states that the complement of a function is obtained by interchanging AND and OR operators and complementing each literal.

EXAMPLE 2.2

Find the complement of the functions $F_1 = x'yz' + x'y'z$ and $F_2 = x(y'z' + yz)$. By applying DeMorgan's theorems as many times as necessary, the complements are obtained as follows:

$$F_1' = (x'yz' + x'y'z)' = (x'yz')'(x'y'z)' = (x + y' + z)(x + y + z')$$
$$F_2' = [x(y'z' + yz)]' = x' + (y'z' + yz)' = x' + (y'z')'(yz)'$$
$$= x' + (y + z)(y' + z')$$
$$= x' + yz' + y'z$$

■

A simpler procedure for deriving the complement of a function is to take the dual of the function and complement each literal. This method follows from the generalized forms of DeMorgan's theorems. Remember that the dual of a function is obtained from the interchange of AND and OR operators and 1's and 0's.

EXAMPLE 2.3

Find the complement of the functions F_1 and F_2 of Example 2.2 by taking their duals and complementing each literal.

1. $F_1 = x'yz' + x'y'z$.
The dual of F_1 is $(x' + y + z')(x' + y' + z)$.
Complement each literal: $(x + y' + z)(x + y + z') = F_1'$.

2. $F_2 = x(y'z' + yz)$.
The dual of F_2 is $x + (y' + z')(y + z)$.
Complement each literal: $x' + (y + z)(y' + z') = F_2'$.

■

2.6 CANONICAL AND STANDARD FORMS

Minterms and Maxterms

A binary variable may appear either in its normal form (x) or in its complement form (x'). Now consider two binary variables x and y combined with an AND operation. Since each variable may appear in either form, there are four possible combinations: $x'y'$, $x'y$, xy', and xy. Each of these four AND terms is called a *minterm*, or a *standard product*. In a similar manner, n variables can be combined to form 2^n minterms. The 2^n different minterms may be determined by a method similar to the one shown in Table 2.3 for three variables. The binary numbers from 0 to $2^n - 1$ are listed under the n variables. Each minterm is obtained from an AND term of the n variables, with each variable being primed if the corresponding bit of the binary number is a 0 and unprimed if a 1. A symbol for each minterm is also shown in the table and is of the form m_j, where the subscript j denotes the decimal equivalent of the binary number of the minterm designated.

In a similar fashion, n variables forming an OR term, with each variable being primed or unprimed, provide 2^n possible combinations, called *maxterms*, or *standard sums*. The eight maxterms for three variables, together with their symbolic designations, are listed in Table 2.3. Any 2^n maxterms for n variables may be determined similarly. It is important to note that (1) each maxterm is obtained from an OR term of the n variables, with each variable being unprimed if the corresponding bit is a 0 and primed if a 1, and (2) each maxterm is the complement of its corresponding minterm and vice versa.

A Boolean function can be expressed algebraically from a given truth table by forming a minterm for each combination of the variables that produces a 1 in the function and then taking the OR of all those terms. For example, the function f_1 in Table 2.4 is determined by expressing the combinations 001, 100, and 111 as $x'y'z$, $xy'z'$, and xyz, respectively. Since each one of these minterms results in $f_1 = 1$, we have

$$f_1 = x'y'z + xy'z' + xyz = m_1 + m_4 + m_7$$

Table 2.3
Minterms and Maxterms for Three Binary Variables

			Minterms		Maxterms	
x	**y**	**z**	**Term**	**Designation**	**Term**	**Designation**
0	0	0	$x'y'z'$	m_0	$x + y + z$	M_0
0	0	1	$x'y'z$	m_1	$x + y + z'$	M_1
0	1	0	$x'yz'$	m_2	$x + y' + z$	M_2
0	1	1	$x'yz$	m_3	$x + y' + z'$	M_3
1	0	0	$xy'z'$	m_4	$x' + y + z$	M_4
1	0	1	$xy'z$	m_5	$x' + y + z'$	M_5
1	1	0	xyz'	m_6	$x' + y' + z$	M_6
1	1	1	xyz	m_7	$x' + y' + z'$	M_7

Table 2.4
Functions of Three Variables

x	y	z	Function f_1	Function f_2
0	0	0	0	0
0	0	1	1	0
0	1	0	0	0
0	1	1	0	1
1	0	0	1	0
1	0	1	0	1
1	1	0	0	1
1	1	1	1	1

Similarly, it may be easily verified that

$$f_2 = x'yz + xy'z + xyz' + xyz = m_3 + m_5 + m_6 + m_7$$

These examples demonstrate an important property of Boolean algebra: Any Boolean function can be expressed as a sum of minterms (with "sum" meaning the ORing of terms).

Now consider the complement of a Boolean function. It may be read from the truth table by forming a minterm for each combination that produces a 0 in the function and then ORing those terms. The complement of f_1 is read as

$$f_1' = x'y'z' + x'yz' + x'yz + xy'z + xyz'$$

If we take the complement of f_1', we obtain the function f_1:

$$\begin{aligned} f_1 &= (x + y + z)(x + y' + z)(x' + y + z')(x' + y' + z) \\ &= M_0 \cdot M_2 \cdot M_3 \cdot M_5 \cdot M_6 \end{aligned}$$

Similarly, it is possible to read the expression for f_2 from the table:

$$\begin{aligned} f_2 &= (x + y + z)(x + y + z')(x + y' + z)(x' + y + z) \\ &= M_0 M_1 M_2 M_4 \end{aligned}$$

These examples demonstrate a second property of Boolean algebra: Any Boolean function can be expressed as a product of maxterms (with "product" meaning the ANDing of terms). The procedure for obtaining the product of maxterms directly from the truth table is as follows: Form a maxterm for each combination of the variables that produces a 0 in the function, and then form the AND of all those maxterms. **Boolean functions expressed as a sum of minterms or product of maxterms are said to be in *canonical form*.**

Sum of Minterms

Previously, we stated that, for n binary variables, one can obtain 2^n distinct minterms and that any Boolean function can be expressed as a sum of minterms. **The minterms whose sum defines the Boolean function are those which give the 1's of the function in a**

truth table. Since the function can be either 1 or 0 for each minterm, and since there are 2^n minterms, one can calculate all the functions that can be formed with n variables to be 2^{2n}. It is sometimes convenient to express a Boolean function in its sum-of-minterms form. If the function is not in this form, it can be made so by first expanding the expression into a sum of AND terms. Each term is then inspected to see if it contains all the variables. If it misses one or more variables, it is ANDed with an expression such as $x + x'$, where x is one of the missing variables. The next example clarifies this procedure.

EXAMPLE 2.4

Express the Boolean function $F = A + B'C$ as a sum of minterms. The function has three variables: A, B, and C. The first term A is missing two variables; therefore,

$$A = A(B + B') = AB + AB'$$

This function is still missing one variable, so

$$\begin{aligned} A &= AB(C + C') + AB'(C + C') \\ &= ABC + ABC' + AB'C + AB'C' \end{aligned}$$

The second term $B'C$ is missing one variable; hence,

$$B'C = B'C(A + A') = AB'C + A'B'C$$

Combining all terms, we have

$$\begin{aligned} F &= A + B'C \\ &= ABC + ABC' + AB'C + AB'C' + A'B'C \end{aligned}$$

But $AB'C$ appears twice, and according to theorem 1 ($x + x = x$), it is possible to remove one of those occurrences. Rearranging the minterms in ascending order, we finally obtain

$$\begin{aligned} F &= A'B'C + AB'C' + AB'C + ABC' + ABC \\ &= m_1 + m_4 + m_5 + m_6 + m_7 \end{aligned}$$

■

When a Boolean function is in its sum-of-minterms form, it is sometimes convenient to express the function in the following brief notation:

$$F(A, B, C) = \Sigma(1, 4, 5, 6, 7)$$

The summation symbol Σ stands for the ORing of terms; the numbers following it are the indices of the minterms of the function. The letters in parentheses following F form a list of the variables in the order taken when the minterm is converted to an AND term.

An alternative procedure for deriving the minterms of a Boolean function is to obtain the truth table of the function directly from the algebraic expression and then read the minterms from the truth table. Consider the Boolean function given in Example 2.4:

$$F = A + B'C$$

The truth table shown in Table 2.5 can be derived directly from the algebraic expression by listing the eight binary combinations under variables A, B, and C and inserting

Table 2.5
Truth Table for F = A + B′C

A	B	C	F
0	0	0	0
0	0	1	1
0	1	0	0
0	1	1	0
1	0	0	1
1	0	1	1
1	1	0	1
1	1	1	1

1's under F for those combinations for which $A = 1$ and $BC = 01$. From the truth table, we can then read the five minterms of the function to be 1, 4, 5, 6, and 7.

Product of Maxterms

Each of the 2^{2^n} functions of n binary variables can be also expressed as a product of maxterms. To express a Boolean function as a product of maxterms, it must first be brought into a form of OR terms. This may be done by using the distributive law, $x + yz = (x + y)(x + z)$. Then any missing variable x in each OR term is ORed with xx'. The procedure is clarified in the following example.

EXAMPLE 2.5

Express the Boolean function $F = xy + x'z$ as a product of maxterms. First, convert the function into OR terms by using the distributive law:

$$\begin{aligned} F = xy + x'z &= (xy + x')(xy + z) \\ &= (x + x')(y + x')(x + z)(y + z) \\ &= (x' + y)(x + z)(y + z) \end{aligned}$$

The function has three variables: x, y, and z. Each OR term is missing one variable; therefore,

$$\begin{aligned} x' + y &= x' + y + zz' = (x' + y + z)(x' + y + z') \\ x + z &= x + z + yy' = (x + y + z)(x + y' + z) \\ y + z &= y + z + xx' = (x + y + z)(x' + y + z) \end{aligned}$$

Combining all the terms and removing those which appear more than once, we finally obtain

$$\begin{aligned} F &= (x + y + z)(x + y' + z)(x' + y + z)(x' + y + z') \\ &= M_0 M_2 M_4 M_5 \end{aligned}$$

A convenient way to express this function is as follows:

$$F(x, y, z) = \Pi(0, 2, 4, 5)$$

The product symbol, Π, denotes the ANDing of maxterms; the numbers are the indices of the maxterms of the function.

■

Conversion between Canonical Forms

The complement of a function expressed as the sum of minterms equals the sum of minterms missing from the original function. This is because the original function is expressed by those minterms which make the function equal to 1, whereas its complement is a 1 for those minterms for which the function is a 0. As an example, consider the function

$$F(A, B, C) = \Sigma(1, 4, 5, 6, 7)$$

This function has a complement that can be expressed as

$$F'(A, B, C) = \Sigma(0, 2, 3) = m_0 + m_2 + m_3$$

Now, if we take the complement of F' by DeMorgan's theorem, we obtain F in a different form:

$$F = (m_0 + m_2 + m_3)' = m'_0 \cdot m'_2 \cdot m'_3 = M_0 M_2 M_3 = \Pi(0, 2, 3)$$

The last conversion follows from the definition of minterms and maxterms as shown in Table 2.3. From the table, it is clear that the following relation holds:

$$m'_j = M_j$$

That is, the **maxterm with subscript j is a complement of the minterm with the same subscript j and vice versa.**

The last example demonstrates the conversion between a function expressed in sum-of-minterms form and its equivalent in product-of-maxterms form. A similar argument will show that the conversion between the product of maxterms and the sum of minterms is similar. We now state a general conversion procedure: To convert from one canonical form to another, interchange the symbols Σ and Π and list those numbers missing from the original form. In order to find the missing terms, one must realize that the total number of minterms or maxterms is 2^n, where n is the number of binary variables in the function.

A Boolean function can be converted from an algebraic expression to a product of maxterms by means of a truth table and the canonical conversion procedure. Consider, for example, the Boolean expression

$$F = xy + x'z$$

First, we derive the truth table of the function, as shown in Table 2.6. The 1's under F in the table are determined from the combination of the variables for which $xy = 11$ or $xz = 01$. The minterms of the function are read from the truth table to be 1, 3, 6, and 7. The function expressed as a sum of minterms is

$$F(x, y, z) = \Sigma(1, 3, 6, 7)$$

Table 2.6
Truth Table for $F = xy + x'z$

x	y	z	F
0	0	0	0
0	0	1	1
0	1	0	0
0	1	1	1
1	0	0	0
1	0	1	0
1	1	0	1
1	1	1	1

Minterms

Maxterms

Since there is a total of eight minterms or maxterms in a function of three variables, we determine the missing terms to be 0, 2, 4, and 5. The function expressed as a product of maxterms is

$$F(x, y, z) = \Pi(0, 2, 4, 5)$$

the same answer as obtained in Example 2.5

Standard Forms

The two canonical forms of Boolean algebra are basic forms that one obtains from reading a given function from the truth table. These forms are very seldom the ones with the least number of literals, because each minterm or maxterm must contain, by definition, *all* the variables, either complemented or uncomplemented.

Another way to express Boolean functions is in *standard* form. In this configuration, the terms that form the function may contain one, two, or any number of literals. There are two types of standard forms: the sum of products and products of sums.

The *sum of products* is a Boolean expression containing AND terms, called *product terms,* with one or more literals each. The *sum* denotes the ORing of these terms. An example of a function expressed as a sum of products is

$$F_1 = y' + xy + x'yz'$$

The expression has three product terms, with one, two, and three literals. Their sum is, in effect, an OR operation.

The logic diagram of a sum-of-products expression consists of a group of AND gates followed by a single OR gate. This configuration pattern is shown in Fig. 2.3(a). Each product term requires an AND gate, except for a term with a single literal. The logic sum is formed with an OR gate whose inputs are the outputs of the AND gates and the single literal. It is assumed that the input variables are directly available in their complements, so inverters are not included in the diagram. This circuit configuration is referred to as a *two-level implementation.*

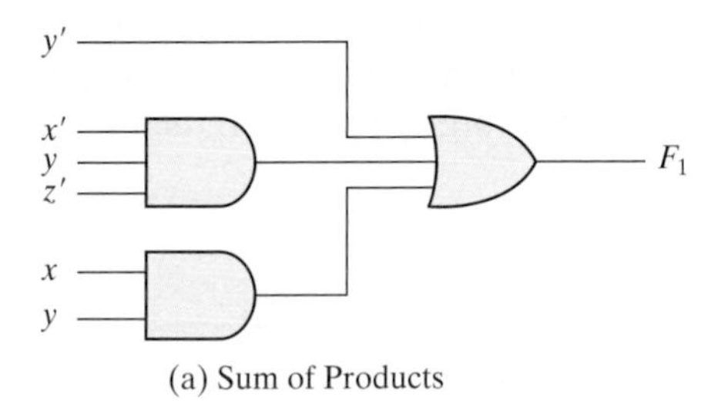

(a) Sum of Products

(b) Product of Sums

FIGURE 2.3
Two-level implementation

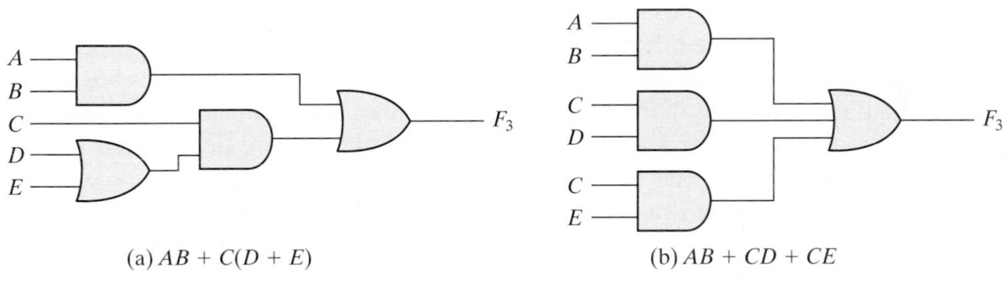

(a) $AB + C(D + E)$

(b) $AB + CD + CE$

FIGURE 2.4
Three- and two-level implementation

A *product of sums* is a Boolean expression containing OR terms, called *sum terms*. Each term may have any number of literals. The *product* denotes the ANDing of these terms. An example of a function expressed as a product of sums is

$$F_2 = x(y' + z)(x' + y + z')$$

This expression has three sum terms, with one, two, and three literals. The product is an AND operation. The use of the words *product* and *sum* stems from the similarity of the AND operation to the arithmetic product (multiplication) and the similarity of the OR operation to the arithmetic sum (addition). The gate structure of the product-of-sums expression consists of a group of OR gates for the sum terms (except for a single literal), followed by an AND gate, as shown in Fig. 2.3(b). **This standard type of expression results in a two-level structure of gates.**

A Boolean function may be expressed in a nonstandard form. For example, the function

$$F_3 = AB + C(D + E)$$

is neither in sum-of-products nor in product-of-sums form. The implementation of this expression is shown in Fig. 2.4(a) and requires two AND gates and two OR gates. There are three levels of gating in this circuit. It can be changed to a standard form by using the distributive law to remove the parentheses:

$$F_3 = AB + C(D + E) = AB + CD + CE$$

The sum-of-products expression is implemented in Fig. 2.4(b). In general, a two-level implementation is preferred because it produces the least amount of delay through the gates when the signal propagates from the inputs to the output. However, the number of inputs to a given gate might not be practical.

2.7 OTHER LOGIC OPERATIONS

When the binary operators AND and OR are placed between two variables, x and y, they form two Boolean functions, $x \cdot y$ and $x + y$, respectively. Previously we stated that there are 2^{2n} functions for n binary variables. Thus, for two variables, $n = 2$, and the number of possible Boolean functions is 16. Therefore, the AND and OR functions are only 2 of a total of 16 possible functions formed with two binary variables. It would be instructive to find the other 14 functions and investigate their properties.

The truth tables for the 16 functions formed with two binary variables are listed in Table 2.7. Each of the 16 columns, F_0 to F_{15}, represents a truth table of one possible function for the two variables, x and y. Note that the functions are determined from the 16 binary combinations that can be assigned to F. The 16 functions can be expressed algebraically by means of Boolean functions, as is shown in the first column of Table 2.8. The Boolean expressions listed are simplified to their minimum number of literals.

Although each function can be expressed in terms of the Boolean operators AND, OR, and NOT, there is no reason one cannot assign special operator symbols for expressing the other functions. Such operator symbols are listed in the second column of Table 2.8. However, of all the new symbols shown, only the exclusive-OR symbol, $\oplus$, is in common use by digital designers.

Each of the functions in Table 2.8 is listed with an accompanying name and a comment that explains the function in some way.[1] The 16 functions listed can be subdivided into three categories:

1. Two functions that produce a constant 0 or 1.
2. Four functions with unary operations: complement and transfer.
3. Ten functions with binary operators that define eight different operations: AND, OR, NAND, NOR, exclusive-OR, equivalence, inhibition, and implication.

Table 2.7
Truth Tables for the 16 Functions of Two Binary Variables

x	y	F_0	F_1	F_2	F_3	F_4	F_5	F_6	F_7	F_8	F_9	F_{10}	F_{11}	F_{12}	F_{13}	F_{14}	F_{15}
0	0	0	0	0	0	0	0	0	0	1	1	1	1	1	1	1	1
0	1	0	0	0	0	1	1	1	1	0	0	0	0	1	1	1	1
1	0	0	0	1	1	0	0	1	1	0	0	1	1	0	0	1	1
1	1	0	1	0	1	0	1	0	1	0	1	0	1	0	1	0	1

[1]The symbol ^ is also used to indicate the exclusive or operator, e.g., $x\hat{}y$. The symbol for the AND function is sometimes omitted from the product of two variables, e.g., xy.

Table 2.8
Boolean Expressions for the 16 Functions of Two Variables

Boolean Functions	Operator Symbol	Name	Comments
$F_0 = 0$		Null	Binary constant 0
$F_1 = xy$	$x \cdot y$	AND	x and y
$F_2 = xy'$	x/y	Inhibition	x, but not y
$F_3 = x$		Transfer	x
$F_4 = x'y$	y/x	Inhibition	y, but not x
$F_5 = y$		Transfer	y
$F_6 = xy' + x'y$	$x \oplus y$	Exclusive-OR	x or y, but not both
$F_7 = x + y$	$x + y$	OR	x or y
$F_8 = (x + y)'$	$x \downarrow y$	NOR	Not-OR
$F_9 = xy + x'y'$	$(x \oplus y)'$	Equivalence	x equals y
$F_{10} = y'$	y'	Complement	Not y
$F_{11} = x + y'$	$x \subset y$	Implication	If y, then x
$F_{12} = x'$	x'	Complement	Not x
$F_{13} = x' + y$	$x \supset y$	Implication	If x, then y
$F_{14} = (xy)'$	$x \uparrow y$	NAND	Not-AND
$F_{15} = 1$		Identity	Binary constant 1

Constants for binary functions can be equal to only 1 or 0. The complement function produces the complement of each of the binary variables. A function that is equal to an input variable has been given the name *transfer*, because the variable x or y is transferred through the gate that forms the function without changing its value. Of the eight binary operators, two (inhibition and implication) are used by logicians, but are seldom used in computer logic. The AND and OR operators have been mentioned in conjunction with Boolean algebra. The other four functions are used extensively in the design of digital systems.

The NOR function is the complement of the OR function, and its name is an abbreviation of *not-OR*. Similarly, NAND is the complement of AND and is an abbreviation of *not-AND*. The exclusive-OR, abbreviated XOR, is similar to OR, but excludes the combination of *both* x and y being equal to 1; it holds only when x and y differ in value. (It is sometimes referred to as the binary difference operator.) Equivalence is a function that is 1 when the two binary variables are equal (i.e., when both are 0 or both are 1). The exclusive-OR and equivalence functions are the complements of each other. This can be easily verified by inspecting Table 2.7: The truth table for exclusive-OR is F_6 and for equivalence is F_9, and these two functions are the complements of each other. For this reason, the equivalence function is called exclusive-NOR, abbreviated XNOR.

Boolean algebra, as defined in Section 2.2, has two binary operators, which we have called AND and OR, and a unary operator, NOT (complement). From the definitions, we have deduced a number of properties of these operators and now have defined other binary operators in terms of them. There is nothing unique about this procedure. We could have just as well started with the operator NOR ($\downarrow$), for example, and later defined AND, OR, and NOT in terms of it. There are, nevertheless, good reasons for introducing Boolean algebra in the way it has been introduced. The concepts of "and," "or," and "not" are familiar and are used by people to express everyday logical ideas. Moreover, the Huntington postulates reflect the dual nature of the algebra, emphasizing the symmetry of $+$ and $\cdot$ with respect to each other.

2.8 DIGITAL LOGIC GATES

Since Boolean functions are expressed in terms of AND, OR, and NOT operations, it is easier to implement a Boolean function with these type of gates. Still, the possibility of constructing gates for the other logic operations is of practical interest. Factors to be weighed in considering the construction of other types of logic gates are (1) the feasibility and economy of producing the gate with physical components, (2) the possibility of extending the gate to more than two inputs, (3) the basic properties of the binary operator, such as commutativity and associativity, and (4) the ability of the gate to implement Boolean functions alone or in conjunction with other gates.

Of the 16 functions defined in Table 2.8, two are equal to a constant and four are repeated. There are only 10 functions left to be considered as candidates for logic gates. Two—inhibition and implication—are not commutative or associative and thus are impractical to use as standard logic gates. The other eight—complement, transfer, AND, OR, NAND, NOR, exclusive-OR, and equivalence—are used as standard gates in digital design.

The graphic symbols and truth tables of the eight gates are shown in Fig. 2.5. Each gate has one or two binary input variables, designated by x and y, and one binary output variable, designated by F. The AND, OR, and inverter circuits were defined in Fig. 1.6. The inverter circuit inverts the logic sense of a binary variable, producing the NOT, or complement, function. The small circle in the output of the graphic symbol of an inverter (referred to as a *bubble*) designates the logic complement. The triangle symbol by itself designates a buffer circuit. A buffer produces the *transfer* function, but does not produce a logic operation, since the binary value of the output is equal to the binary value of the input. This circuit is used for power amplification of the signal and is equivalent to two inverters connected in cascade.

The NAND function is the complement of the AND function, as indicated by a graphic symbol that consists of an AND graphic symbol followed by a small circle. The NOR function is the complement of the OR function and uses an OR graphic symbol followed by a small circle. NAND and NOR gates are used extensively as standard logic gates and are in fact far more popular than the AND and OR gates. This is because NAND and NOR gates are easily constructed with transistor circuits and because digital circuits can be easily implemented with them.

Name	Graphic symbol	Algebraic function	Truth table
AND	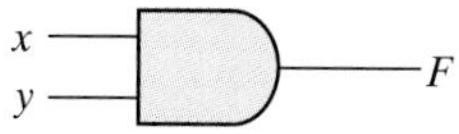	$F = x \cdot y$	x y | F: 0 0 | 0; 0 1 | 0; 1 0 | 0; 1 1 | 1
OR	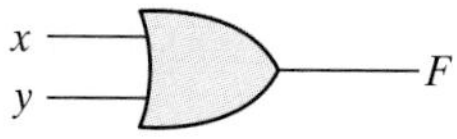	$F = x + y$	x y | F: 0 0 | 0; 0 1 | 1; 1 0 | 1; 1 1 | 1
Inverter	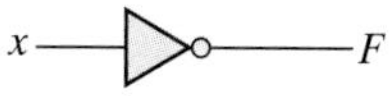	$F = x'$	x | F: 0 | 1; 1 | 0
Buffer	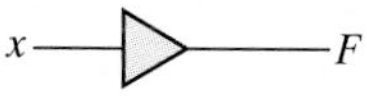	$F = x$	x | F: 0 | 0; 1 | 1
NAND	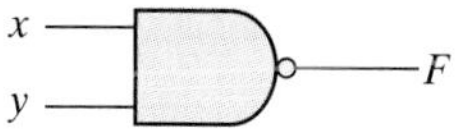	$F = (xy)'$	x y | F: 0 0 | 1; 0 1 | 1; 1 0 | 1; 1 1 | 0
NOR	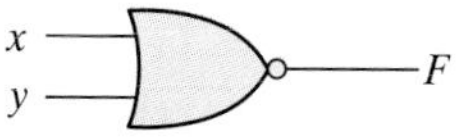	$F = (x + y)'$	x y | F: 0 0 | 1; 0 1 | 0; 1 0 | 0; 1 1 | 0
Exclusive-OR (XOR)	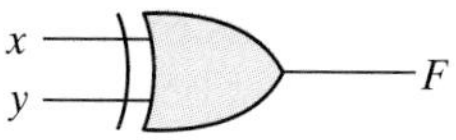	$F = xy' + x'y$ $= x \oplus y$	x y | F: 0 0 | 0; 0 1 | 1; 1 0 | 1; 1 1 | 0
Exclusive-NOR or equivalence	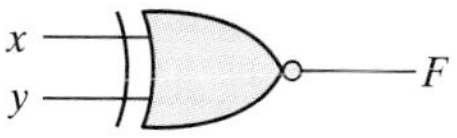	$F = xy + x'y'$ $= (x \oplus y)'$	x y | F: 0 0 | 1; 0 1 | 0; 1 0 | 0; 1 1 | 1

FIGURE 2.5
Digital logic gates

The exclusive-OR gate has a graphic symbol similar to that of the OR gate, except for the additional curved line on the input side. The equivalence, or exclusive-NOR, gate is the complement of the exclusive-OR, as indicated by the small circle on the output side of the graphic symbol.

Extension to Multiple Inputs

The gates shown in Fig. 2.5—except for the inverter and buffer—can be extended to have more than two inputs. A gate can be extended to have multiple inputs if the binary operation it represents is commutative and associative. The AND and OR operations, defined in Boolean algebra, possess these two properties. For the OR function, we have

$$x + y = y + x \quad \text{(commutative)}$$

and

$$(x + y) + z = x + (y + z) = x + y + z \quad \text{(associative)}$$

which indicates that the gate inputs can be interchanged and that the OR function can be extended to three or more variables.

The NAND and NOR functions are commutative, and their gates can be extended to have more than two inputs, provided that the definition of the operation is modified slightly. The difficulty is that the NAND and NOR operators are not associative (i.e., $(x \downarrow y) \downarrow z \neq x \downarrow (y \downarrow z)$), as shown in Fig. 2.6 and the following equations:

$$(x \downarrow y) \downarrow z = [(x + y)' + z]' = (x + y)z' = xz' + yz'$$
$$x \downarrow (y \downarrow z) = [x + (y + z)']' = x'(y + z) = x'y + x'z$$

To overcome this difficulty, we define the multiple NOR (or NAND) gate as a complemented OR (or AND) gate. Thus, by definition, we have

$$x \downarrow y \downarrow z = (x + y + z)'$$
$$x \uparrow y \uparrow z = (xyz)'$$

The graphic symbols for the three-input gates are shown in Fig. 2.7. In writing cascaded NOR and NAND operations, one must use the correct parentheses to signify the proper sequence of the gates. To demonstrate this principle, consider the circuit of Fig. 2.7(c). The Boolean function for the circuit must be written as

$$F = [(ABC)'(DE)']' = ABC + DE$$

The second expression is obtained from one of DeMorgan's theorems. It also shows that an expression in sum-of-products form can be implemented with NAND gates. (NAND and NOR gates are discussed further in Section 3.7)

The exclusive-OR and equivalence gates are both commutative and associative and can be extended to more than two inputs. However, multiple-input exclusive-OR gates are uncommon from the hardware standpoint. In fact, even a two-input function is usually constructed with other types of gates. Moreover, the definition of the function must be modified when extended to more than two variables. Exclusive-OR is an *odd* function (i.e., it is equal to 1 if the input variables have an odd number of 1's). The construction

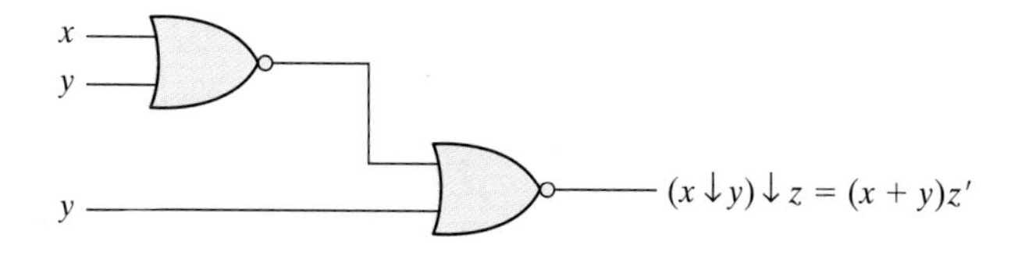

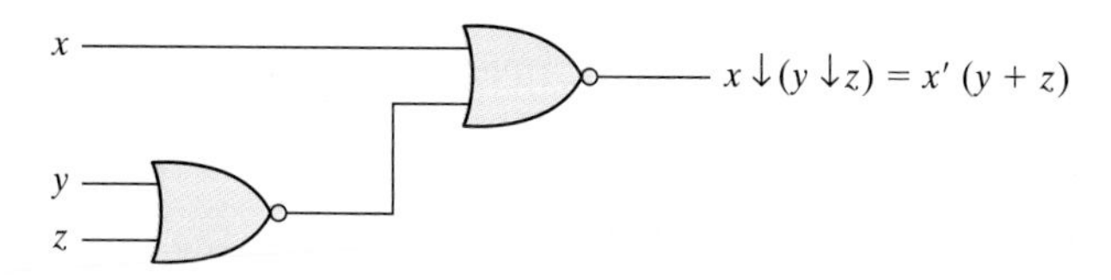

FIGURE 2.6
Demonstrating the nonassociativity of the NOR operator: $(x \downarrow y) \downarrow z \neq x \downarrow (y \downarrow z)$

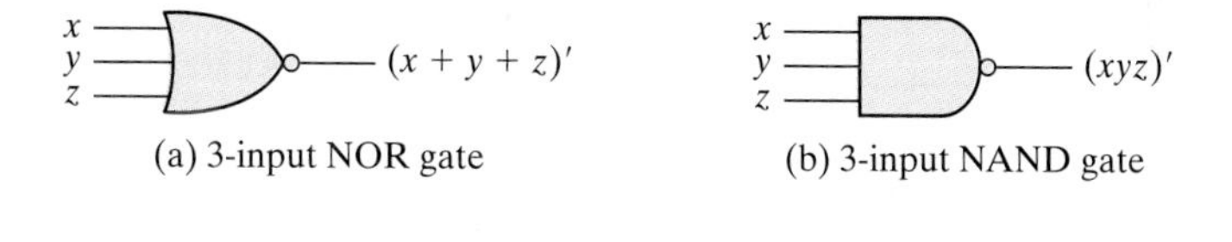

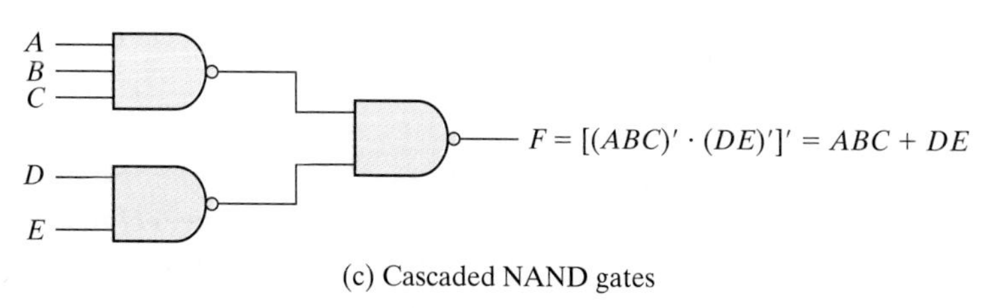

FIGURE 2.7
Multiple-input and cascaded NOR and NAND gates

of a three-input exclusive-OR function is shown in Fig. 2.8. This function is normally implemented by cascading two-input gates, as shown in (a). Graphically, it can be represented with a single three-input gate, as shown in (b). The truth table in (c) clearly indicates that the output F is equal to 1 if only one input is equal to 1 or if all three inputs are equal to 1 (i.e., when the total number of 1's in the input variables is *odd*). (Exclusive-OR gates are discussed further in Section 3.9.)

Positive and Negative Logic

The binary signal at the inputs and outputs of any gate has one of two values, except during transition. One signal value represents logic 1 and the other logic 0. Since two signal values are assigned to two logic values, there exist two different assignments of

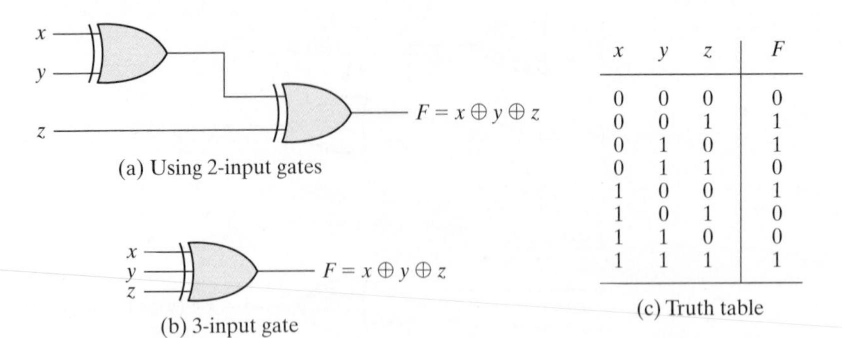

x	y	z	F
0	0	0	0
0	0	1	1
0	1	0	1
0	1	1	0
1	0	0	1
1	0	1	0
1	1	0	0
1	1	1	1

FIGURE 2.8
Three-input exclusive-OR gate

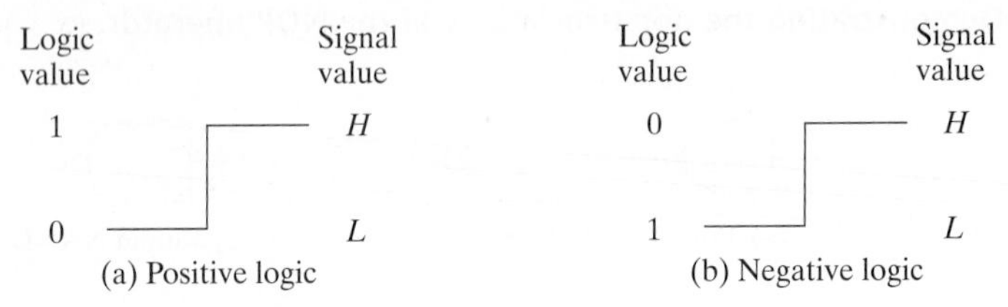

FIGURE 2.9
Signal assignment and logic polarity

signal level to logic value, as shown in Fig. 2.9. The higher signal level is designated by H and the lower signal level by L. **Choosing the high-level H to represent logic 1 defines a positive logic system. Choosing the low-level L to represent logic 1 defines a negative logic system.** The terms *positive* and *negative* are somewhat misleading, since both signals may be positive or both may be negative. It is not the actual values of the signals that determine the type of logic, but rather the assignment of logic values to the relative amplitudes of the two signal levels.

Hardware digital gates are defined in terms of signal values such as H and L. It is up to the user to decide on a positive or negative logic polarity. Consider, for example, the electronic gate shown in Fig. 2.10(b). The truth table for this gate is listed in Fig. 2.10(a). It specifies the physical behavior of the gate when H is 3 V and L is 0 V. The truth table of Fig. 2.10(c) assumes a positive logic assignment, with $H = 1$ and $L = 0$. This truth table is the same as the one for the AND operation. The graphic symbol for a positive logic AND gate is shown in Fig. 2.10(d).

Now consider the negative logic assignment for the same physical gate with $L = 1$ and $H = 0$. The result is the truth table of Fig. 2.10(e). This table represents the OR operation, even though the entries are reversed. The graphic symbol for the negative-logic OR gate is shown in Fig. 2.10(f). The small triangles in the inputs and output

x	y	z
L	L	L
L	H	L
H	L	L
H	H	H

(a) Truth table with H and L

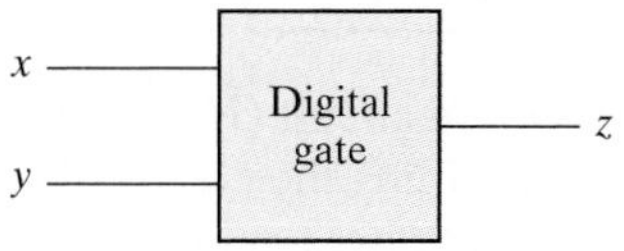

(b) Gate block diagram

x	y	z
0	0	0
0	1	0
1	0	0
1	1	1

(c) Truth table for positive logic

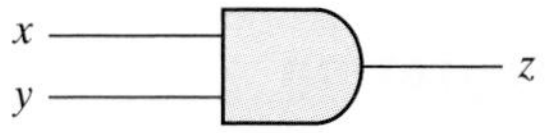

(d) Positive logic AND gate

x	y	z
1	1	1
1	0	1
0	1	1
0	0	0

(e) Truth table for negative logic

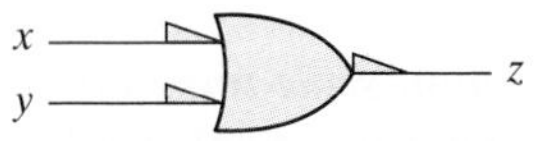

(f) Negative logic OR gate

FIGURE 2.10
Demonstration of positive and negative logic

designate a *polarity indicator*, the presence of which along a terminal signifies that negative logic is assumed for the signal. Thus, the same physical gate can operate either as a positive-logic AND gate or as a negative-logic OR gate.

The conversion from positive logic to negative logic and vice versa is essentially an operation that changes 1's to 0's and 0's to 1's in both the inputs and the output of a gate. Since this operation produces the dual of a function, the change of all terminals from one polarity to the other results in taking the dual of the function. The upshot is that all AND operations are converted to OR operations (or graphic symbols) and vice versa. In addition, one must not forget to include the polarity-indicator triangle in the graphic symbols when negative logic is assumed. In this book, we will not use negative logic gates and will assume that all gates operate with a positive logic assignment.

2.9 INTEGRATED CIRCUITS

An integrated circuit (IC) is fabricated on a die of a silicon semiconductor crystal, called a *chip,* containing the electronic components for constructing digital gates. The complex chemical and physical processes used to form a semiconductor circuit are not a subject of this book. The various gates are interconnected inside the chip to form the required circuit. The chip is mounted in a ceramic or plastic container, and connections are welded to external pins to form the integrated circuit. The number of pins may range from 14 on a small IC package to several thousand on a larger package. Each IC has a numeric designation printed on the surface of the package for identification. Vendors provide data books, catalogs, and Internet websites that contain descriptions and information about the ICs that they manufacture.

Levels of Integration

Digital ICs are often categorized according to the complexity of their circuits, as measured by the number of logic gates in a single package. The differentiation between those chips which have a few internal gates and those having hundreds of thousands of gates is made by customary reference to a package as being either a small-, medium-, large-, or very large-scale integration device.

Small-scale integration (SSI) devices contain several independent gates in a single package. The inputs and outputs of the gates are connected directly to the pins in the package. The number of gates is usually fewer than 10 and is limited by the number of pins available in the IC.

Medium-scale integration (MSI) devices have a complexity of approximately 10 to 1,000 gates in a single package. They usually perform specific elementary digital operations. MSI digital functions are introduced in Chapter 4 as decoders, adders, and multiplexers and in Chapter 6 as registers and counters.

Large-scale integration (LSI) devices contain thousands of gates in a single package. They include digital systems such as processors, memory chips, and programmable logic devices. Some LSI components are presented in Chapter 7.

Very large-scale integration (VLSI) devices now contain millions of gates within a single package. Examples are large memory arrays and complex microcomputer chips. Because of their small size and low cost, VLSI devices have revolutionized the computer system design technology, giving the designer the capability to create structures that were previously uneconomical to build.

Digital Logic Families

Digital integrated circuits are classified not only by their complexity or logical operation, but also by the specific circuit technology to which they belong. The circuit technology is referred to as a *digital logic family*. Each logic family has its own basic electronic circuit upon which more complex digital circuits and components are developed. The basic circuit in each technology is a NAND, NOR, or inverter gate. The electronic components employed in the construction of the basic circuit are usually used to name the technology. Many different logic families of digital integrated circuits have been introduced commercially. The following are the most popular:

TTL	transistor–transistor logic;
ECL	emitter-coupled logic;
MOS	metal-oxide semiconductor;
CMOS	complementary metal-oxide semiconductor.

TTL is a logic family that has been in use for 50 years and is considered to be standard. ECL has an advantage in systems requiring high-speed operation. MOS is suitable for circuits that need high component density, and CMOS is preferable in systems requiring low power consumption, such as digital cameras, personal media players, and other handheld portable devices. Low power consumption is essential for VLSI design; therefore, CMOS has become the dominant logic family, while TTL and ECL continue to decline in use. The most important parameters distinguishing logic families are listed below; CMOS integrated circuits are discussed briefly in the appendix.

Fan-out specifies the number of standard loads that the output of a typical gate can drive without impairing its normal operation. A standard load is usually defined as the amount of current needed by an input of another similar gate in the same family.

Fan-in is the number of inputs available in a gate.

Power dissipation is the power consumed by the gate that must be available from the power supply.

Propagation delay is the average transition delay time for a signal to propagate from input to output. For example, if the input of an inverter switches from 0 to 1, the output will switch from 1 to 0, but after a time determined by the propagation delay of the device. The operating speed is inversely proportional to the propagation delay.

Noise margin is the maximum external noise voltage added to an input signal that does not cause an undesirable change in the circuit output.

Computer-Aided Design of VLSI Circuits

Integrated circuits having submicron geometric features are manufactured by optically projecting patterns of light onto silicon wafers. Prior to exposure, the wafers are coated with a photoresistive material that either hardens or softens when exposed to light. Removing extraneous photoresist leaves patterns of exposed silicon. The exposed regions are then implanted with dopant atoms to create a semiconductor material having the electrical properties of transistors and the logical properties of gates. The design process translates a functional specification or description of the circuit (i.e., what it must do) into a physical specification or description (how it must be implemented in silicon).

The design of digital systems with VLSI circuits containing millions of transistors and gates is an enormous and formidable task. Systems of this complexity are usually impossible to develop and verify without the assistance of computer-aided design (CAD) tools, which consist of software programs that support computer-based representations of circuits and aid in the development of digital hardware by automating the design process. Electronic design automation (EDA) covers all phases of the design of integrated circuits. A typical design flow for creating VLSI circuits consists of a sequence of steps beginning with design entry (e.g., entering a schematic) and culminating with the generation of the database that contains the photomask used to fabricate the IC. There are a variety of options available for creating the physical realization of a digital circuit in silicon. The designer can choose between an application-specific integrated circuit (ASIC), a field-programmable gate array (FPGA), a programmable logic device (PLD), and a full-custom IC. With each of these devices comes a set of CAD tools that provide the necessary software to facilitate the hardware fabrication of the unit. Each of these technologies has a market niche determined by the size of the market and the unit cost of the devices that are required to implement a design.

Some CAD systems include an editing program for creating and modifying schematic diagrams on a computer screen. This process is called *schematic capture* or *schematic entry*. With the aid of menus, keyboard commands, and a mouse, a schematic editor can draw circuit diagrams of digital circuits on the computer screen. Components can be placed on the screen from a list in an internal library and can then be connected with lines that represent wires. The schematic entry software creates and manages a database containing the information produced with the schematic. Primitive gates and functional blocks have associated models that allow the functionality (i.e., logical behavior) and timing of the circuit to be verified. Verification is performed by applying inputs to the circuit and using a logic simulator to determine and display the outputs in text or waveform format.

An important development in the design of digital systems is the use of a hardware description language (HDL). Such a language resembles a computer programming language, but is specifically oriented to describing digital hardware. It represents logic diagrams and other digital information in textual form to describe the functionality and structure of a circuit. Moreover, the HDL description of a circuit's functionality can be abstract, without reference to specific hardware, thereby freeing a designer to devote attention to higher level functional detail (e.g., under certain conditions the circuit must detect a particular pattern of 1's and 0's in a serial bit stream of data) rather than transistor-level detail. HDL-based models of a circuit or system are simulated to check and verify its functionality before it is submitted to fabrication, thereby reducing the risk and waste of manufacturing a circuit that fails to operate correctly. In tandem with the emergence of HDL-based design languages, tools have been developed to automatically and optimally synthesize the logic described by an HDL model of a circuit. These two advances in technology have led to an **almost total reliance by industry on HDL-based synthesis tools and methodologies for the design of the circuits of complex digital systems.** Two HDLs—Verilog and VHDL—have been approved as standards by the Institute of Electronics and Electrical Engineers (IEEE) and are in use by design teams worldwide. The Verilog HDL is introduced in Section 3.10, and because of its importance, we include several exercises and design problems based on Verilog throughout the book.

PROBLEMS

(Answers to problems marked with * appear at the end of the text.)

2.1 Demonstrate the validity of the following identities by means of truth tables:
(a) DeMorgan's theorem for three variables: $(x + y + z)' = x'y'z'$ and $(xyz)' = x' + y' + z'$
(b) The distributive law: $xy + z = (x + z)(y + z)$
(c) The distributive law: $x(y + z) = xy + xz$
(d) The associative law: $x + (y + z) = (x + y) + z$
(e) The associative law and $x(yz) = (xy)z$

2.2 Simplify the following Boolean expressions to a minimum number of literals:
(a)* $x'y' + x'y$
(b)* $(x' + y)(x' + y')$
(c)* $x'y'z + xy'z + yz$
(d)* $(A + B)'(A' + B')'$
(e) $(a + b + c')(a'b' + c)$
(f) $a'b'c + ab'c + abc + a'bc$

2.3 Simplify the following Boolean expressions to a minimum number of literals:
(a)* $A'B'C + AB'C + BC$
(b)* $x'y'z' + y'z$
(c)* $(x + y)'(x' + y')$
(d)* $x'y'z' + w'x'yz' + wx'yz'$
(e)* $(BC' + A'D)(AB' + CD')$
(f) $(a + c)(a' + b + c)(a' + b' + c)$

2.4 Reduce the following Boolean expressions to the indicated number of literals:
(a)* $A'C' + ABC + AC$ to two literals
(b)* $(x'y' + z)' + z + xy + wz$ to three literals
(c)* $A'B(D' + C'D) + B(A + A'CD)$ to one literal
(d)* $(A' + C)(A' + C')(A + B + C'D)$ to four literals
(e) $A'BD' + ABC'D' + ABCD'$ to two literals

2.5 Draw logic diagrams of the circuits that implement the original and simplified expressions in Problem 2.2.

2.6 Draw logic diagrams of the circuits that implement the original and simplified expressions in Problem 2.3.

2.7 Draw logic diagrams of the circuits that implement the original and simplified expressions in Problem 2.4.

2.8 Find the complement of $F = AC + BD$; then show that $FF' = 0$ and $F + F' = 1$.

2.9 Find the complement of the following expressions:
(a)* $x'y' + xy$
(b) $ac + ab' + a'bc'$
(c) $z + z'(v'w + xy)$
(d) $(B + C')(A + C')(B' + C')$

2.10 Given the Boolean functions F_1 and F_2, show that
(a) The Boolean function $E = F_1 + F_2$ contains the sum of the minterms of F_1 and F_2.
(b) The Boolean function $G = F_1F_2$ contains only the minterms that are common to F_1 and F_2.

2.11 List the truth table of the function:
(a)* $F = y'z' + x'y' + xz + x'z$
(b) $F = ac + bc'$

2.12 We can perform logical operations on strings of bits by considering each pair of corresponding bits separately (called bitwise operation). Given two eight-bit strings $A = 10110001$ and $B = 00001110$, evaluate the eight-bit result after the following logical operations:
(a)* AND (b) OR (c)* XOR (d)* NOT A (e) NOT B (f) NAND
(g) NOR

2.13 Draw logic diagrams to implement the following Boolean expressions:

(a) $F = (u + x')(y' + z')'$
(b) $F = (u \oplus y)' + x$
(c) $F = (u' + x')(y + z')$
(d) $F = u(x \oplus z) + y'$
(e) $F = u + (yz + uxy)'$
(f) $F = u + x + x'(u + y') + x'y'$

2.14 Implement the Boolean function

$$F = x'y' + x'z + xy$$

(a) With AND, OR, and inverter gates
(b)* With OR and inverter gates
(c) With AND and inverter gates
(d) With NAND and inverter gates
(e) With NOR and inverter gates

2.15* Simplify the following Boolean functions T_1 and T_2 to a minimum number of literals:

A	B	C	T_1	T_2
0	0	0	0	1
0	0	1	0	1
0	1	0	1	0
0	1	1	1	0
1	0	0	1	0
1	0	1	1	0
1	1	0	0	1
1	1	1	0	1

2.16 The logical product of all maxterms of a Boolean function of n variables is 1.

(a) Prove the previous statement for $n = 3$.
(b) Suggest a procedure for a general proof.

2.17 Obtain the truth table of the following functions, and express each function in sum-of-minterms and product-of-maxterms form:

(a)* $(b + c'd)(a' + cd')$
(b) $(ad + b'c + bd')(b + d)$
(c) $(b + d)(b + d')(a + c)$
(d) $ad + bcd + ab'c' + b'c'd'$

2.18 For the Boolean function

$$F = xy'z' + x'y'z' + xy + x'y$$

(a) Obtain the truth table of F.
(b) Draw the logic diagram, using the original Boolean expression.
(c)* Use Boolean algebra to simplify the function to a minimum number of literals.
(d) Obtain the truth table of the function from the simplified expression and show that it is the same as the one in part (a).
(e) Draw the logic diagram from the simplified expression, and compare the total number of gates with the diagram of part (b).

2.19* Express the following function as a sum of minterms and as a product of maxterms:

$$F(A, B, C, D) = B'D + A'D + BD + BCD$$

2.20 Express the complement of the following functions in sum-of-minterms form:

(a) $F(A,B,C,D) = \Sigma(0, 3, 5, 7, 9, 11, 13)$

(b) $F(x, y, z) = \Pi(2, 4, 6, 8)$

2.21 Convert each of the following to the other canonical form:

(a) $F(x, y, z) = \Sigma(1, 3, 5, 7, 9)$

(b) $F(A, B, C, D) = \Pi(3, 5, 8, 11, 13, 15)$

2.22* Convert each of the following expressions into sum of products and product of sums:

(a) $(u + x'w)(x + u'v)$

(b) $x' + z(x + y')(y + z')$

2.23 Draw the logic diagram corresponding to the following Boolean expressions without simplifying them:

(a) $BC' + AB + ACD + CD$

(b) $(A + B)(C + BD)(A' + B + D)$

(c) $(AB + A'B')(CD' + C'D)(A'B + AB')$

(d) $A + CD + (A + D')(C' + D) + C'D'$

2.24 Show that the dual of the exclusive-OR is equal to its complement.

2.25 By substituting the Boolean expression equivalent of the binary operations as defined in Table 2.8, show the following:

(a) The inhibition operation is neither commutative nor associative.

(b) The exclusive-OR operation is commutative and associative.

2.26 Show that a positive logic NOR gate is a negative logic NAND gate and vice versa.

2.27 Write the Boolean equations and draw the logic diagram of the circuit whose outputs are defined by the following truth table:

Table P2.27

a	*b*	*c*	f_1
0	0	0	1
0	0	1	0
0	1	0	1
0	1	1	1
1	0	0	0
1	0	1	1
1	1	0	1
1	1	1	0

2.28 Write Boolean expressions and construct the truth tables describing the outputs of the circuits described by the logic diagrams in Fig. P2.28.

2.29 Determine whether the following Boolean equation is true or false.

(a) $y'z' + yz' + x'z = x' + xz'$

(b) $x'y' + xz' + yz = y'z' + xy + x'z$

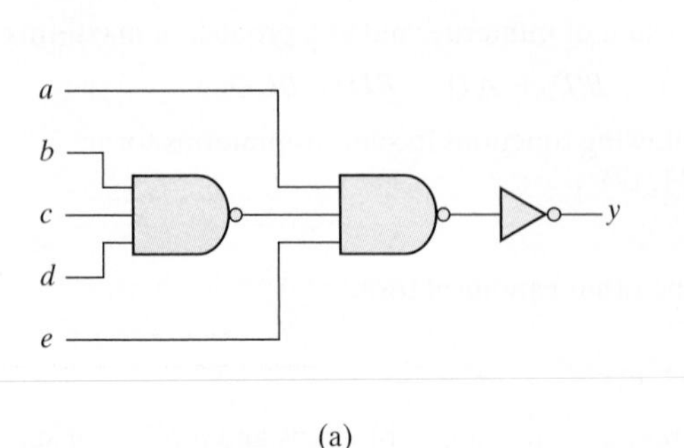

(a)

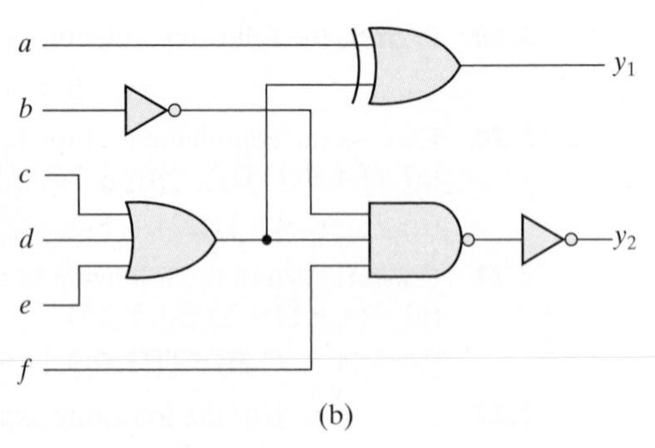

(b)

FIGURE P2.28

2.30 Write the following Boolean expressions in sum of products form:

$$(b + d)(a' + b' + c)(a + c)$$

2.31 Write the following Boolean expression in product of sums form:

$$a'b + a'c' + bc$$

REFERENCES

1. Boole, G. 1854. *An Investigation of the Laws of Thought*. New York: Dover.
2. Dietmeyer, D. L. 1988. *Logic Design of Digital Systems*, 3rd ed. Boston: Allyn and Bacon.
3. Huntington, E. V. Sets of independent postulates for the algebra of logic. *Trans. Am. Math. Soc.*, 5 (1904): 288–309.
4. *IEEE Standard Hardware Description Language Based on the Verilog Hardware Description Language*, Language Reference Manual (LRM), IEEE Std.1364-1995, 1996, 2001, 2005, The Institute of Electrical and Electronics Engineers, Piscataway, NJ.
5. *IEEE Standard VHDL Language Reference Manual* (LRM), IEEE Std. 1076-1987, 1988, The Institute of Electrical and Electronics Engineers, Piscataway, NJ.
6. Mano, M. M. and C. R. Kime. 2000. *Logic and Computer Design Fundamentals*, 2nd ed. Upper Saddle River, NJ: Prentice Hall.
7. Shannon, C. E. A symbolic analysis of relay and switching circuits. *Trans. AIEE*, 57 (1938): 713–723.

WEB SEARCH TOPICS

Algebraic field
Boolean logic
Boolean gates
Bipolar transistor
Field-effect transistor
Emitter-coupled logic
TTL logic
CMOS logic
CMOS process

Chapter 3

Gate-Level Minimization

3.1 INTRODUCTION

Gate-level minimization is the design task of finding an optimal gate-level implementation of the Boolean functions describing a digital circuit. This task is well understood, but is difficult to execute by manual methods when the logic has more than a few inputs. Fortunately, computer-based logic synthesis tools can minimize a large set of Boolean equations efficiently and quickly. Nevertheless, it is important that a designer understand the underlying mathematical description and solution of the problem. This chapter serves as a foundation for your understanding of that important topic and will enable you to execute a manual design of simple circuits, preparing you for skilled use of modern design tools. The chapter will also introduce a hardware description language that is used by modern design tools.

3.2 THE MAP METHOD

The complexity of the digital logic gates that implement a Boolean function is directly related to the complexity of the algebraic expression from which the function is implemented. Although the truth table representation of a function is unique, when it is expressed algebraically it can appear in many different, but equivalent, forms. Boolean expressions may be simplified by algebraic means as discussed in Section 2.4. However, this procedure of minimization is awkward because it lacks specific rules to predict each succeeding step in the manipulative process. The map method presented here provides a simple, straightforward procedure for minimizing Boolean functions. This method may be regarded as a pictorial form of a truth table. The map method is also known as the *Karnaugh map* or *K-map*.

A K-map is a diagram made up of squares, with each square representing one minterm of the function that is to be minimized. Since any Boolean function can be expressed as a sum of minterms, it follows that a Boolean function is recognized graphically in the map from the area enclosed by those squares whose minterms are included in the function. In fact, the map presents a visual diagram of all possible ways a function may be expressed in standard form. By recognizing various patterns, the user can derive alternative algebraic expressions for the same function, from which the simplest can be selected.

The simplified expressions produced by the map are always in one of the two standard forms: sum of products or product of sums. It will be assumed that the simplest algebraic expression is an algebraic expression with a minimum number of terms and with the smallest possible number of literals in each term. This expression produces a circuit diagram with a minimum number of gates and the minimum number of inputs to each gate. We will see subsequently that the simplest expression is not unique: It is sometimes possible to find two or more expressions that satisfy the minimization criteria. In that case, either solution is satisfactory.

Two-Variable K-Map

The two-variable map is shown in Fig. 3.1(a). There are four minterms for two variables; hence, the map consists of four squares, one for each minterm. The map is redrawn in (b) to show the relationship between the squares and the two variables x and y. The 0 and 1 marked in each row and column designate the values of variables. Variable x appears primed in row 0 and unprimed in row 1. Similarly, y appears primed in column 0 and unprimed in column 1.

If we mark the squares whose minterms belong to a given function, the two-variable map becomes another useful way to represent any one of the 16 Boolean functions of two variables. As an example, the function xy is shown in Fig. 3.2(a). Since xy is equal to m_3, a 1 is placed inside the square that belongs to m_3. Similarly, the function $x + y$ is represented in the map of Fig. 3.2(b) by three squares marked with 1's. These squares are found from the minterms of the function:

$$m_1 + m_2 + m_3 = x'y + xy' + xy = x + y$$

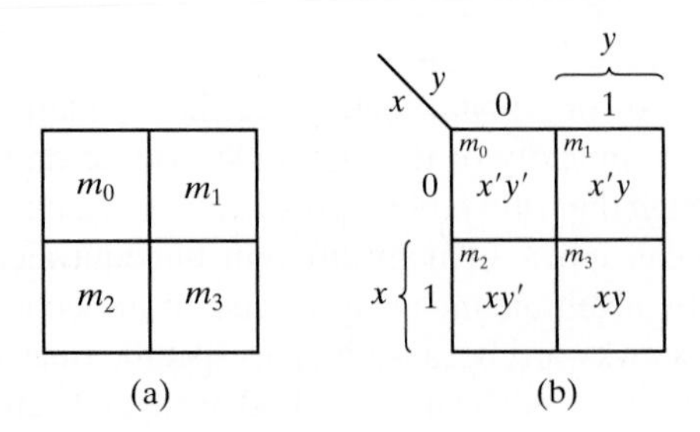

FIGURE 3.1
Two-variable K-map

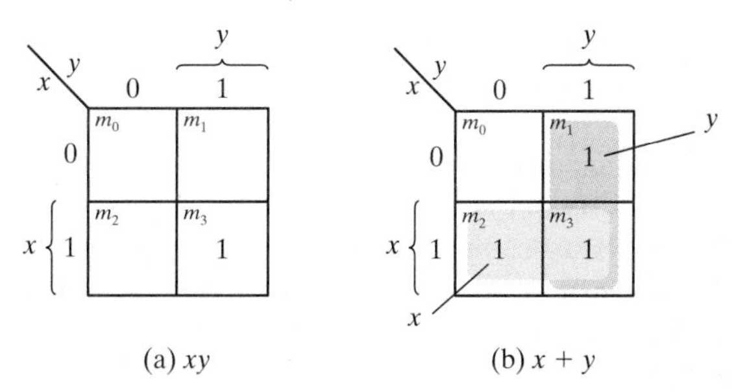

FIGURE 3.2
Representation of functions in the map

The three squares could also have been determined from the intersection of variable x in the second row and variable y in the second column, which encloses the area belonging to x or y. In each example, the minterms at which the function is asserted are marked with a 1.

Three-Variable K-Map

A three-variable K-map is shown in Fig. 3.3. There are eight minterms for three binary variables; therefore, the map consists of eight squares. Note that the minterms are arranged, not in a binary sequence, but in a sequence similar to the Gray code (Table 1.6). The characteristic of this sequence is that **only one bit changes in value from one adjacent column to the next.** The map drawn in part (b) is marked with numbers in each row and each column to show the relationship between the squares and the three variables. For example, the square assigned to m_5 corresponds to row 1 and column 01. When these two numbers are concatenated, they give the binary number 101, whose decimal equivalent is 5. Each cell of the map corresponds to a unique minterm, so another way of looking at square $m_5 = xy'z$ is to consider it to be in the row marked x and the column belonging to $y'z$ (column 01). Note that there are four squares in which each variable is equal to 1 and four in which each is equal to 0. The variable appears unprimed in the former four

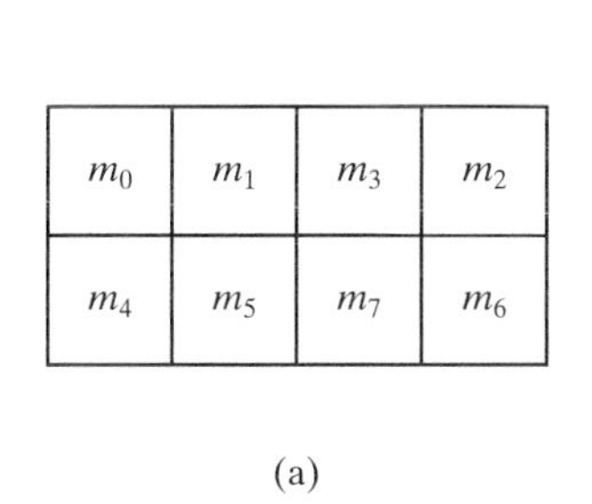

(a)

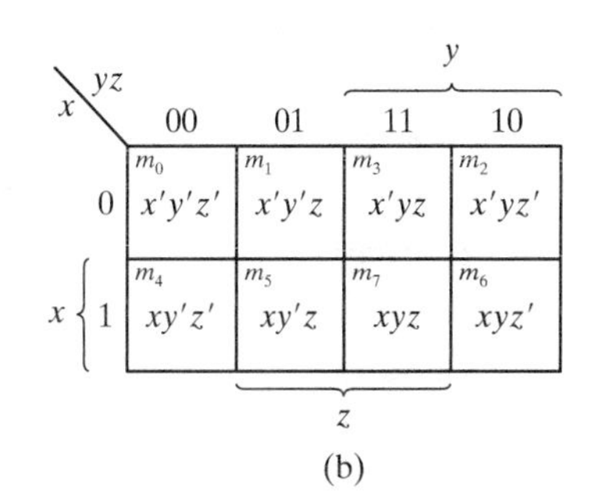

(b)

FIGURE 3.3
Three-variable K-map

squares and primed in the latter. For convenience, we write the variable with its letter symbol under the four squares in which it is unprimed.

To understand the usefulness of the map in simplifying Boolean functions, we must recognize the basic property possessed by adjacent squares: **Any two adjacent squares in the map differ by only one variable,** which is primed in one square and unprimed in the other. For example, m_5 and m_7 lie in two adjacent squares. Variable y is primed in m_5 and unprimed in m_7, whereas the other two variables are the same in both squares. From the postulates of Boolean algebra, it follows that the sum of two minterms in adjacent squares can be simplified to a single product term consisting of only two literals. To clarify this concept, consider the sum of two adjacent squares such as m_5 and m_7:

$$m_5 + m_7 = xy'z + xyz = xz(y' + y) = xz$$

Here, the two squares differ by the variable y, which can be removed when the sum of the two minterms is formed. Thus, any two minterms in adjacent squares (vertically or horizontally, but not diagonally, adjacent) that are ORed together will cause a removal of the dissimilar variable. The next four examples explain the procedure for minimizing a Boolean function with a K-map.

EXAMPLE 3.1

Simplify the Boolean function

$$F(x, y, z) = \Sigma(2, 3, 4, 5)$$

First, a 1 is marked in each minterm square that represents the function. This is shown in Fig. 3.4, in which the squares for minterms 010, 011, 100, and 101 are marked with 1's. The next step is to find possible adjacent squares. These are indicated in the map by two shaded rectangles, each enclosing two 1's. The upper right rectangle represents the area enclosed by $x'y$. This area is determined by observing that the two-square area is in row 0, corresponding to x', and the last two columns, corresponding to y. Similarly, the lower left rectangle represents the product term xy'. (The second row represents x and the two left columns represent y'.) The sum of four minterms can be replaced by a sum of

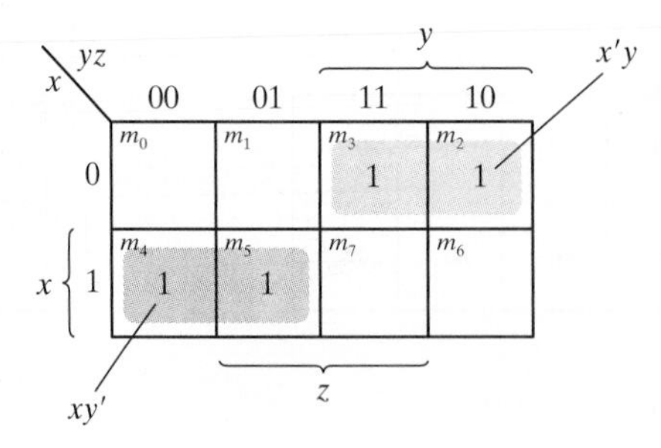

FIGURE 3.4
Map for Example 3.1, $F(x, y, z) = \Sigma(2, 3, 4, 5) = x'y + xy'$

only two product terms. The logical sum of these two product terms gives the simplified expression

$$F = x'y + xy'$$

■

In certain cases, two squares in the map are considered to be adjacent even though they do not touch each other. In Fig. 3.3(b), m_0 is adjacent to m_2 and m_4 is adjacent to m_6 because their minterms differ by one variable. This difference can be readily verified algebraically:

$$m_0 + m_2 = x'y'z' + x'yz' = x'z'(y' + y) = x'z'$$
$$m_4 + m_6 = xy'z' + xyz' = xz' + (y' + y) = xz'$$

Consequently, we must modify the definition of adjacent squares to include this and other similar cases. We do so by considering the map as being drawn on a surface in which the right and left edges touch each other to form adjacent squares.

EXAMPLE 3.2

Simplify the Boolean function

$$F\ (x, y, z) = \Sigma(3, 4, 6, 7)$$

The map for this function is shown in Fig. 3.5. There are four squares marked with 1's, one for each minterm of the function. Two adjacent squares are combined in the third column to give a two-literal term yz. The remaining two squares with 1's are also adjacent by the new definition. These two squares, when combined, give the two-literal term xz'. The simplified function then becomes

$$F = yz + xz'$$

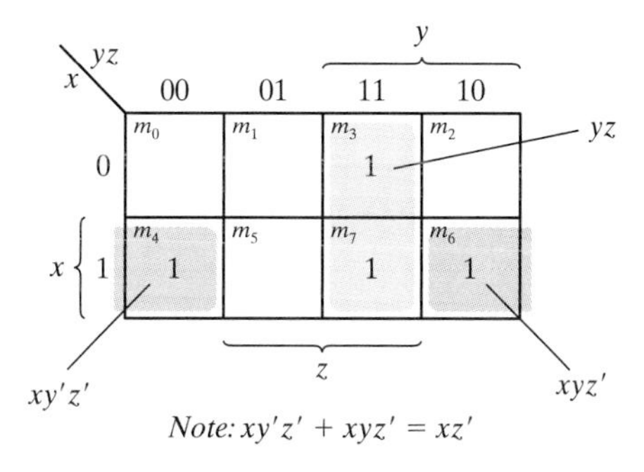

FIGURE 3.5
Map for Example 3.2, $F\ (x, y, z) = \Sigma(3, 4, 6, 7) = yz + xz'$

■

Consider now any combination of four adjacent squares in the three-variable map. Any such combination represents the logical sum of four minterms and results in an expression with only one literal. As an example, the logical sum of the four adjacent minterms 0, 2, 4, and 6 reduces to the single literal term z':

$$\begin{aligned} m_0 + m_2 + m_4 + m_6 &= x'y'z' + x'yz' + xy'z' + xyz' \\ &= x'z'(y' + y) + xz'(y' + y) \\ &= x'z' + xz' = z'(x' + x) = z' \end{aligned}$$

The number of adjacent squares that may be combined must always represent a number that is a power of two, such as 1, 2, 4, and 8. As more adjacent squares are combined, we obtain a product term with fewer literals.

One square represents one minterm, giving a term with three literals.
Two adjacent squares represent a term with two literals.
Four adjacent squares represent a term with one literal.
Eight adjacent squares encompass the entire map and produce a function that is always equal to 1.

EXAMPLE 3.3

Simplify the Boolean function

$$F(x, y, z) = \Sigma(0, 2, 4, 5, 6)$$

The map for F is shown in Fig. 3.6. First, we combine the four adjacent squares in the first and last columns to give the single literal term z'. The remaining single square, representing minterm 5, is combined with an adjacent square that has already been used once. This is not only permissible, but rather desirable, because the two adjacent squares give the two-literal term xy' and the single square represents the three-literal minterm $xy'z$. The simplified function is

$$F = z' + xy'$$

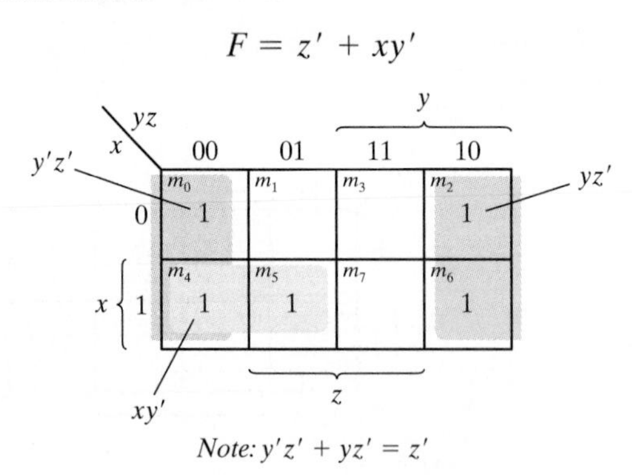

FIGURE 3.6
Map for Example 3.3, $F(x, y, z) = \Sigma(0, 2, 4, 5, 6) = z' + xy'$

■

If a function is not expressed in sum-of-minterms form, it is possible to use the map to obtain the minterms of the function and then simplify the function to an expression with a minimum number of terms. It is necessary, however, to make sure that the algebraic expression is in sum-of-products form. Each product term can be plotted in the map in one, two, or more squares. The minterms of the function are then read directly from the map.

EXAMPLE 3.4

For the Boolean function

$$F = A'C + A'B + AB'C + BC$$

(a) Express this function as a sum of minterms.
(b) Find the minimal sum-of-products expression.

Note that F is a sum of products. Three product terms in the expression have two literals and are represented in a three-variable map by two squares each. The two squares corresponding to the first term, $A'C$, are found in Fig. 3.7 from the coincidence of A' (first row) and C (two middle columns) to give squares 001 and 011. Note that, in marking 1's in the squares, it is possible to find a 1 already placed there from a preceding term. This happens with the second term, $A'B$, which has 1's in squares 011 and 010. Square 011 is common with the first term, $A'C$, though, so only one 1 is marked in it. Continuing in this fashion, we determine that the term $AB'C$ belongs in square 101, corresponding to minterm 5, and the term BC has two 1's in squares 011 and 111. The function has a total of five minterms, as indicated by the five 1's in the map of Fig. 3.7. The minterms are read directly from the map to be 1, 2, 3, 5, and 7. The function can be expressed in sum-of-minterms form as

$$F(A, B, C) = \Sigma(1, 2, 3, 5, 7)$$

The sum-of-products expression, as originally given, has too many terms. It can be simplified, as shown in the map, to an expression with only two terms:

$$F = C + A'B$$

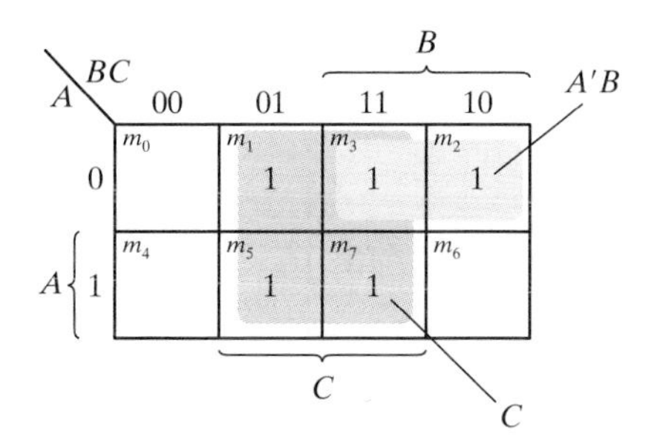

FIGURE 3.7
Map of Example 3.4, $A'C + A'B + AB'C + BC = C + A'B$

■

3.3 FOUR-VARIABLE K-MAP

The map for Boolean functions of four binary variables (w, x, y, z) is shown in Fig. 3.8. In Fig. 3.8(a) are listed the 16 minterms and the squares assigned to each. In Fig. 3.8(b), the map is redrawn to show the relationship between the squares and the four variables. The rows and columns are numbered in a Gray code sequence, with only one digit changing value between two adjacent rows or columns. The minterm corresponding to each square can be obtained from the concatenation of the row number with the column number. For example, the numbers of the third row (11) and the second column (01), when concatenated, give the binary number 1101, the binary equivalent of decimal 13. Thus, the square in the third row and second column represents minterm m_{13}.

The map minimization of four-variable Boolean functions is similar to the method used to minimize three-variable functions. Adjacent squares are defined to be squares next to each other. In addition, the map is considered to lie on a surface with the top and bottom edges, as well as the right and left edges, touching each other to form adjacent squares. For example, m_0 and m_2 form adjacent squares, as do m_3 and m_{11}. The combination of adjacent squares that is useful during the simplification process is easily determined from inspection of the four-variable map:

One square represents one minterm, giving a term with four literals.
Two adjacent squares represent a term with three literals.
Four adjacent squares represent a term with two literals.
Eight adjacent squares represent a term with one literal.
Sixteen adjacent squares produce a function that is always equal to 1.

No other combination of squares can simplify the function. The next two examples show the procedure used to simplify four-variable Boolean functions.

m_0	m_1	m_3	m_2
m_4	m_5	m_7	m_6
m_{12}	m_{13}	m_{15}	m_{14}
m_8	m_9	m_{11}	m_{10}

(a)

wx \ yz	00	01	11	10
00	m_0 $w'x'y'z'$	m_1 $w'x'y'z$	m_3 $w'x'yz$	m_2 $w'x'yz'$
01	m_4 $w'xy'z'$	m_5 $w'xy'z$	m_7 $w'xyz$	m_6 $w'xyz'$
11	m_{12} $wxy'z'$	m_{13} $wxy'z$	m_{15} $wxyz$	m_{14} $wxyz'$
10	m_8 $wx'y'z'$	m_9 $wx'y'z$	m_{11} $wx'yz$	m_{10} $wx'yz'$

(y spans columns 11 and 10; z spans columns 01 and 11; x spans rows 01 and 11; w spans rows 11 and 10)

(b)

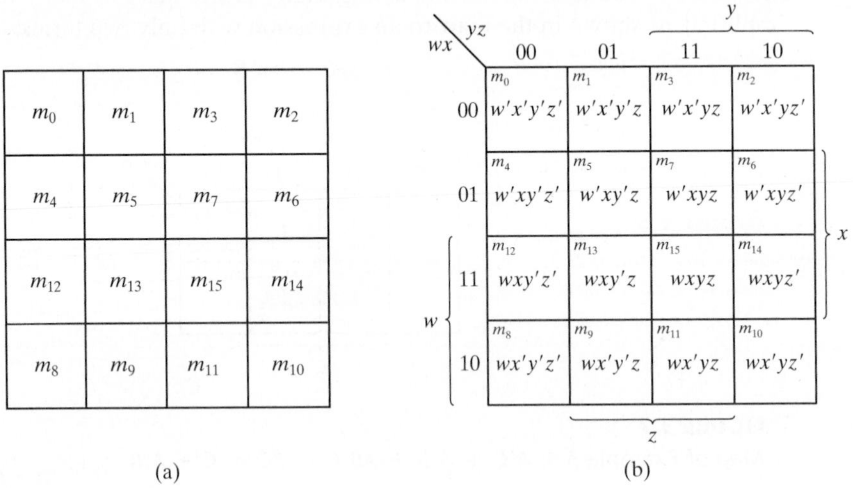

FIGURE 3.8
Four-variable map

EXAMPLE 3.5

Simplify the Boolean function

$$F(w, x, y, z) = \Sigma(0, 1, 2, 4, 5, 6, 8, 9, 12, 13, 14)$$

Since the function has four variables, a four-variable map must be used. The minterms listed in the sum are marked by 1's in the map of Fig. 3.9. Eight adjacent squares marked with 1's can be combined to form the one literal term y'. The remaining three 1's on the right cannot be combined to give a simplified term; they must be combined as two or four adjacent squares. The larger the number of squares combined, the smaller is the number of literals in the term. In this example, the top two 1's on the right are combined with the top two 1's on the left to give the term $w'z'$. Note that it is permissible to use the same square more than once. We are now left with a square marked by 1 in the third row and fourth column (square 1110). Instead of taking this square alone (which will give a term with four literals), we combine it with squares already used to form an area of four adjacent squares. These squares make up the two middle rows and the two end columns, giving the term xz'. The simplified function is

$$F = y' + w'z' + xz'$$

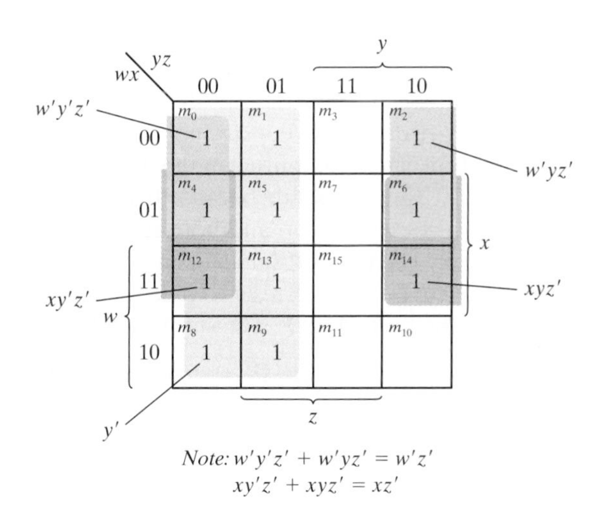

FIGURE 3.9
Map for Example 3.5, $F(w, x, y, z) = \Sigma(0,1, 2, 4, 5, 6, 8, 9, 12, 13, 14) = y' + w'z' + xz'$

■

EXAMPLE 3.6

Simplify the Boolean function

$$F = A'B'C' + B'CD' + A'BCD' + AB'C'$$

The area in the map covered by this function consists of the squares marked with 1's in Fig. 3.10. The function has four variables and, as expressed, consists of three terms with

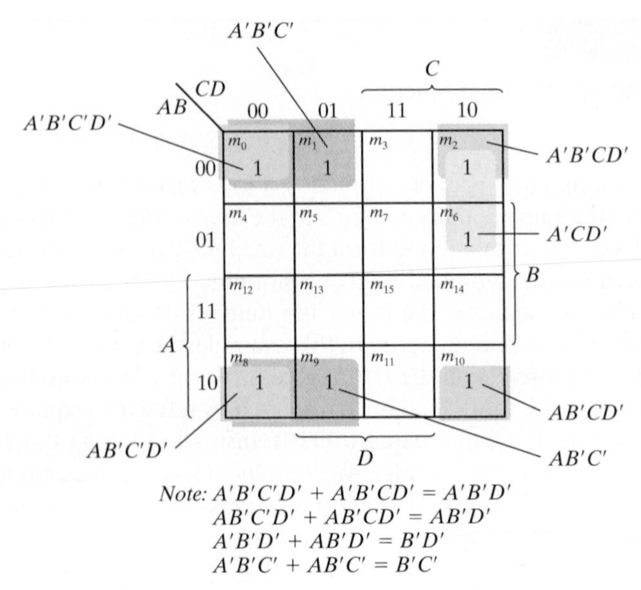

FIGURE 3.10
Map for Example 3.6, $A'B'C' + B'CD' + A'BCD' + AB'C' = B'D' + B'C' + A'CD'$

three literals each and one term with four literals. Each term with three literals is represented in the map by two squares. For example, $A'B'C'$ is represented in squares 0000 and 0001. The function can be simplified in the map by taking the 1's in the four corners to give the term $B'D'$. This is possible because these four squares are adjacent when the map is drawn in a surface with top and bottom edges, as well as left and right edges, touching one another. The two left-hand 1's in the top row are combined with the two 1's in the bottom row to give the term $B'C'$. The remaining 1 may be combined in a two-square area to give the term $A'CD'$. The simplified function is

$$F = B'D' + B'C' + A'CD'$$

■

Prime Implicants

In choosing adjacent squares in a map, we must ensure that (1) all the minterms of the function are covered when we combine the squares, (2) the number of terms in the expression is minimized, and (3) there are no redundant terms (i.e., minterms already covered by other terms). Sometimes there may be two or more expressions that satisfy the simplification criteria. The procedure for combining squares in the map may be made more systematic if we understand the meaning of two special types of terms. **A *prime implicant* is a product term obtained by combining the maximum possible number of adjacent squares in the map.** If a minterm in a square is covered by only one prime implicant, that prime implicant is said to be *essential.*

The prime implicants of a function can be obtained from the map by combining all possible maximum numbers of squares. This means that a single 1 on a map represents a prime implicant if it is not adjacent to any other 1's. Two adjacent 1's form a prime implicant, provided that they are not within a group of four adjacent squares. Four adjacent 1's form a prime implicant if they are not within a group of eight adjacent squares, and so on. The essential prime implicants are found by looking at each square marked with a 1 and checking the number of prime implicants that cover it. The prime implicant is essential if it is the only prime implicant that covers the minterm.

Consider the following four-variable Boolean function:

$$F(A, B, C, D) = \Sigma(0, 2, 3, 5, 7, 8, 9, 10, 11, 13, 15)$$

The minterms of the function are marked with 1's in the maps of Fig. 3.11. The partial map (Fig. 3.11(a)) shows two essential prime implicants, each formed by collapsing four cells into a term having only two literals. One term is essential because there is only one way to include minterm m_0 within four adjacent squares. These four squares define the term $B'D'$. Similarly, there is only one way that minterm m_5 can be combined with four adjacent squares, and this gives the second term BD. The two essential prime implicants cover eight minterms. The three minterms that were omitted from the partial map (m_3, m_9, and m_{11}) must be considered next.

Figure 3.11(b) shows all possible ways that the three minterms can be covered with prime implicants. Minterm m_3 can be covered with either prime implicant CD or prime implicant $B'C$. Minterm m_9 can be covered with either AD or AB'. Minterm m_{11} is covered with any one of the four prime implicants. The simplified expression is obtained from the logical sum of the two essential prime implicants and any two prime implicants

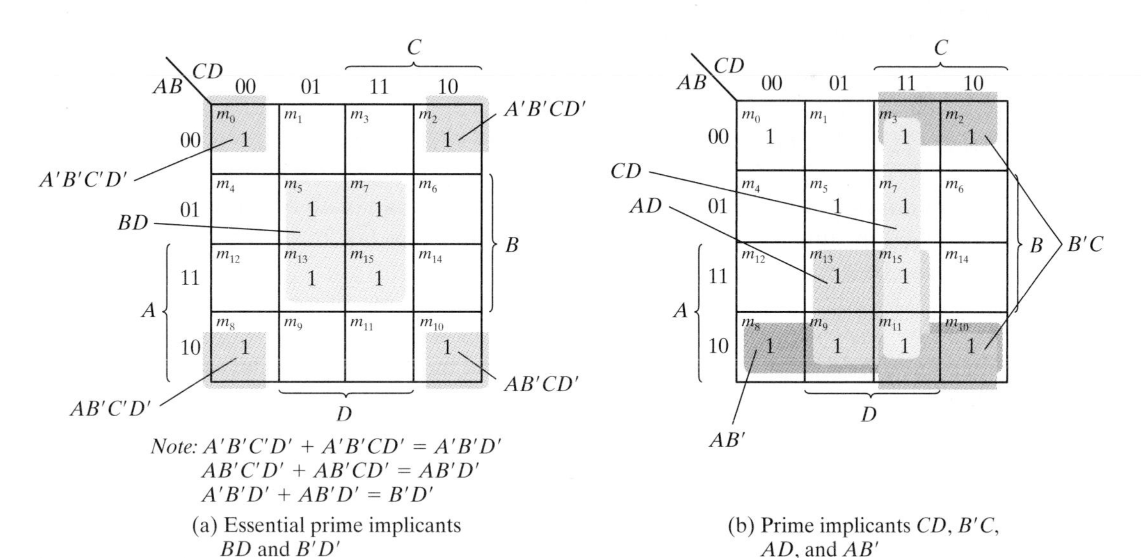

(a) Essential prime implicants BD and $B'D'$

(b) Prime implicants CD, $B'C$, AD, and AB'

FIGURE 3.11
Simplification using prime implicants

that cover minterms m_3, m_9, and m_{11}. There are four possible ways that the function can be expressed with four product terms of two literals each:

$$\begin{aligned} F &= BD + B'D' + CD + AD \\ &= BD + B'D' + CD + AB' \\ &= BD + B'D' + B'C + AD \\ &= BD + B'D' + B'C + AB' \end{aligned}$$

The previous example has demonstrated that the identification of the prime implicants in the map helps in determining the alternatives that are available for obtaining a simplified expression.

The procedure for finding the simplified expression from the map requires that we first determine all the essential prime implicants. The simplified expression is obtained from the logical sum of all the essential prime implicants, plus other prime implicants that may be needed to cover any remaining minterms not covered by the essential prime implicants. Occasionally, there may be more than one way of combining squares, and each combination may produce an equally simplified expression.

Five-Variable Map

Maps for more than four variables are not as simple to use as maps for four or fewer variables. A five-variable map needs 32 squares and a six-variable map needs 64 squares. When the number of variables becomes large, the number of squares becomes excessive and the geometry for combining adjacent squares becomes more involved.

Maps for more than four variables are difficult to use and will not be considered here.

3.4 PRODUCT-OF-SUMS SIMPLIFICATION

The minimized Boolean functions derived from the map in all previous examples were expressed in sum-of-products form. With a minor modification, the product-of-sums form can be obtained.

The procedure for obtaining a minimized function in product-of-sums form follows from the basic properties of Boolean functions. The 1's placed in the squares of the map represent the minterms of the function. The minterms not included in the standard sum-of-products form of a function denote the complement of the function. From this observation, we see that the complement of a function is represented in the map by the squares not marked by 1's. If we mark the empty squares by 0's and combine them into valid adjacent squares, we obtain a simplified sum-of-products expression of the complement of the function (i.e., of F'). The complement of F' gives us back the function F in product-of-sums form (a consequence of DeMorgan's theorem). Because of the generalized DeMorgan's theorem, the function so obtained is automatically in product-of-sums form. The best way to show this is by example.

EXAMPLE 3.7

Simplify the following Boolean function into (a) sum-of-products form and (b) product-of-sums form:

$$F(A, B, C, D) = \Sigma(0, 1, 2, 5, 8, 9, 10)$$

The 1's marked in the map of Fig. 3.12 represent all the minterms of the function. The squares marked with 0's represent the minterms not included in F and therefore denote the complement of F. Combining the squares with 1's gives the simplified function in sum-of-products form:

(a) $F = B'D' + B'C' + A'C'D$

If the squares marked with 0's are combined, as shown in the diagram, we obtain the simplified complemented function:

$$F' = AB + CD + BD'$$

Applying DeMorgan's theorem (by taking the dual and complementing each literal as described in Section 2.4), we obtain the simplified function in product-of-sums form:

(b) $F = (A' + B')(C' + D')(B' + D)$

■

The gate-level implementation of the simplified expressions obtained in Example 3.7 is shown in Fig. 3.13. The sum-of-products expression is implemented in (a) with a group of AND gates, one for each AND term. The outputs of the AND gates are connected to the inputs of a single OR gate. The same function is implemented in (b) in its product-of-sums

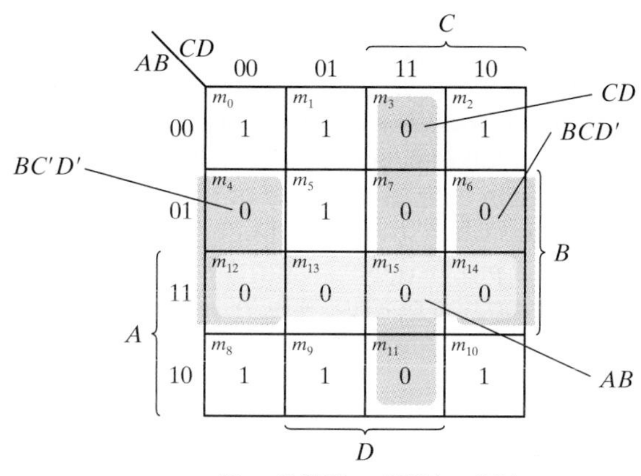

FIGURE 3.12
Map for Example 3.7, $F(A, B, C, D) = \Sigma(0,1, 2, 5, 8, 9,10) = B'D' + B'C' + A'C'D = (A' + B')(C' + D')(B' + D)$

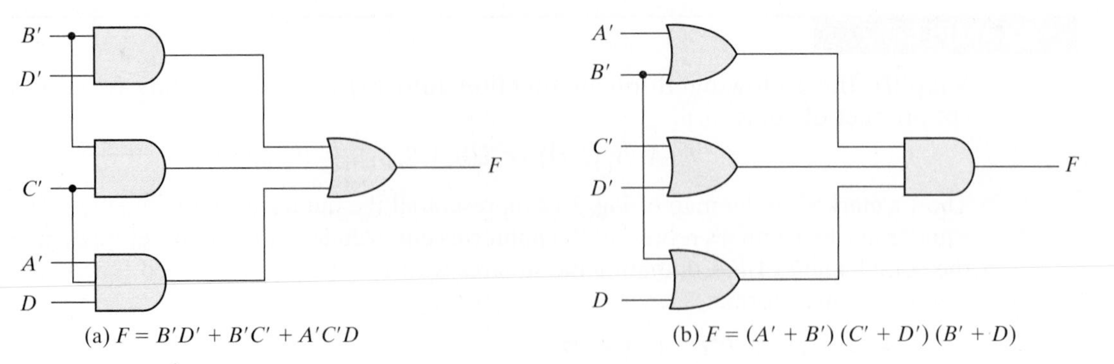

(a) $F = B'D' + B'C' + A'C'D$

(b) $F = (A' + B')(C' + D')(B' + D)$

FIGURE 3.13
Gate implementations of the function of Example 3.7

Table 3.1
Truth Table of Function F

x	*y*	*z*	*F*
0	0	0	0
0	0	1	1
0	1	0	0
0	1	1	1
1	0	0	1
1	0	1	0
1	1	0	1
1	1	1	0

form with a group of OR gates, one for each OR term. The outputs of the OR gates are connected to the inputs of a single AND gate. In each case, it is assumed that the input variables are directly available in their complement, so inverters are not needed. The configuration pattern established in Fig. 3.13 is the general form by which any Boolean function is implemented when expressed in one of the standard forms. AND gates are connected to a single OR gate when in sum-of-products form; OR gates are connected to a single AND gate when in product-of-sums form. Either configuration forms two levels of gates. Thus, the implementation of a function in a standard form is said to be a two-level implementation. The two-level implementation may not be practical, depending on the number of inputs to the gates.

Example 3.7 showed the procedure for obtaining the product-of-sums simplification when the function is originally expressed in the sum-of-minterms canonical form. The procedure is also valid when the function is originally expressed in the product-of-maxterms canonical form. Consider, for example, the truth table that defines the function F in Table 3.1. In sum-of-minterms form, this function is expressed as

$$F(x, y, z) = \Sigma(1, 3, 4, 6)$$

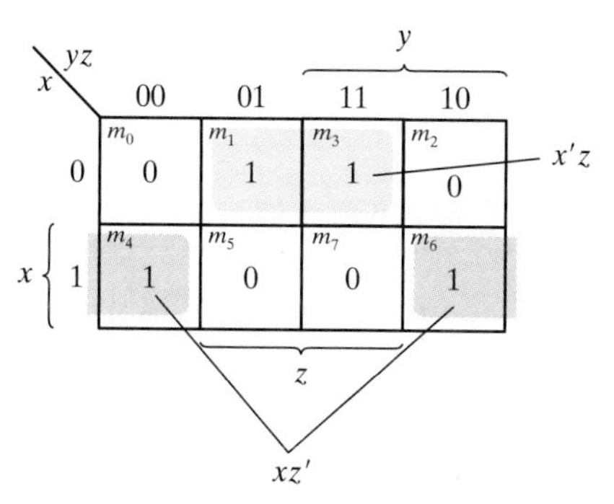

FIGURE 3.14
Map for the function of Table 3.1

In product-of-maxterms form, it is expressed as

$$F(x, y, z) = \Pi(0, 2, 5, 7)$$

In other words, the 1's of the function represent the minterms and the 0's represent the maxterms. The map for this function is shown in Fig. 3.14. One can start simplifying the function by first marking the 1's for each minterm that the function is a 1. The remaining squares are marked by 0's. If, instead, the product of maxterms is initially given, one can start marking 0's in those squares listed in the function; the remaining squares are then marked by 1's. Once the 1's and 0's are marked, the function can be simplified in either one of the standard forms. For the sum of products, we combine the 1's to obtain

$$F = x'z + xz'$$

For the product of sums, we combine the 0's to obtain the simplified complemented function

$$F' = xz + x'z'$$

which shows that the exclusive-OR function is the complement of the equivalence function (Section 2.6). Taking the complement of F', we obtain the simplified function in product-of-sums form:

$$F = (x' + z')(x + z)$$

To enter a function expressed in product-of-sums form into the map, use the complement of the function to find the squares that are to be marked by 0's. For example, the function

$$F = (A' + B' + C')(B + D)$$

can be entered into the map by first taking its complement, namely,

$$F' = ABC + B'D'$$

and then marking 0's in the squares representing the minterms of F'. The remaining squares are marked with 1's.

3.5 DON'T-CARE CONDITIONS

The logical sum of the minterms associated with a Boolean function specifies the conditions under which the function is equal to 1. The function is equal to 0 for the rest of the minterms. This pair of conditions assumes that all the combinations of the values for the variables of the function are valid. In practice, in some applications the function is not specified for certain combinations of the variables. As an example, the four-bit binary code for the decimal digits has six combinations that are not used and consequently are considered to be unspecified. Functions that have unspecified outputs for some input combinations are called *incompletely specified functions*. In most applications, we simply don't care what value is assumed by the function for the unspecified minterms. For this reason, it is customary to call the unspecified minterms of a function *don't-care conditions*. These don't-care conditions can be used on a map to provide further simplification of the Boolean expression.

A don't-care minterm is a combination of variables whose logical value is not specified. Such a minterm cannot be marked with a 1 in the map, because it would require that the function always be a 1 for such a combination. Likewise, putting a 0 on the square requires the function to be 0. To distinguish the don't-care condition from 1's and 0's, an X is used. Thus, an X inside a square in the map indicates that we don't care whether the value of 0 or 1 is assigned to F for the particular minterm.

In choosing adjacent squares to simplify the function in a map, the don't-care minterms may be assumed to be either 0 or 1. When simplifying the function, we can choose to include each don't-care minterm with either the 1's or the 0's, depending on which combination gives the simplest expression.

EXAMPLE 3.8

Simplify the Boolean function

$$F(w, x, y, z) = \Sigma(1, 3, 7, 11, 15)$$

which has the don't-care conditions

$$d(w, x, y, z) = \Sigma(0, 2, 5)$$

The minterms of F are the variable combinations that make the function equal to 1. The minterms of d are the don't-care minterms that may be assigned either 0 or 1. The map simplification is shown in Fig. 3.15. The minterms of F are marked by 1's, those of d are marked by X's, and the remaining squares are filled with 0's. To get the simplified expression in sum-of-products form, we must include all five 1's in the map, but we may or may not include any of the X's, depending on the way the function is simplified. The term yz covers the four minterms in the third column. The remaining minterm, m_1, can be combined

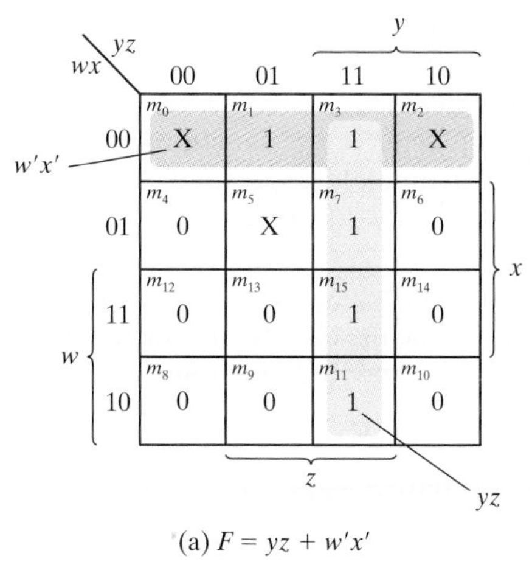

(a) $F = yz + w'x'$

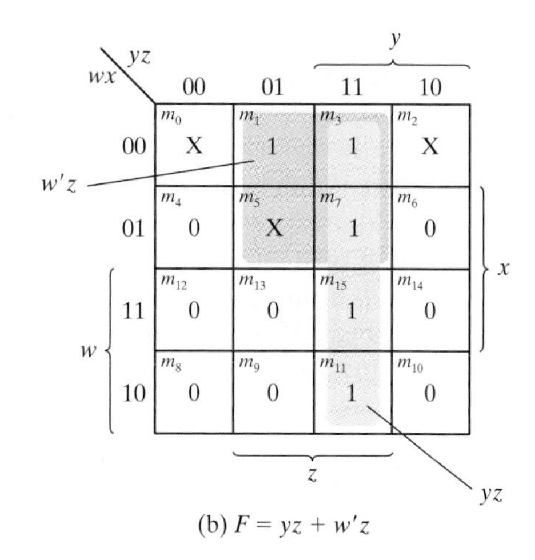

(b) $F = yz + w'z$

FIGURE 3.15
Example with don't-care conditions

with minterm m_3 to give the three-literal term $w'x'z$. However, by including one or two adjacent X's we can combine four adjacent squares to give a two-literal term. In Fig. 3.15(a), don't-care minterms 0 and 2 are included with the 1's, resulting in the simplified function

$$F = yz + w'x'$$

In Fig. 3.15(b), don't-care minterm 5 is included with the 1's, and the simplified function is now

$$F = yz + w'z$$

Either one of the preceding two expressions satisfies the conditions stated for this example.

■

The previous example has shown that the don't-care minterms in the map are initially marked with X's and are considered as being either 0 or 1. The choice between 0 and 1 is made depending on the way the incompletely specified function is simplified. Once the choice is made, the simplified function obtained will consist of a sum of minterms that includes those minterms which were initially unspecified and have been chosen to be included with the 1's. Consider the two simplified expressions obtained in Example 3.8:

$$F(w, x, y, z) = yz + w'x' = \Sigma(0, 1, 2, 3, 7, 11, 15)$$
$$F(w, x, y, z) = yz + w'z = \Sigma(1, 3, 5, 7, 11, 15)$$

Both expressions include minterms 1, 3, 7, 11, and 15 that make the function F equal to 1. The don't-care minterms 0, 2, and 5 are treated differently in each expression.

The first expression includes minterms 0 and 2 with the 1's and leaves minterm 5 with the 0's. The second expression includes minterm 5 with the 1's and leaves minterms 0 and 2 with the 0's. The two expressions represent two functions that are not algebraically equal. Both cover the specified minterms of the function, but each covers different don't-care minterms. As far as the incompletely specified function is concerned, either expression is acceptable because the only difference is in the value of F for the don't-care minterms.

It is also possible to obtain a simplified product-of-sums expression for the function of Fig. 3.15. In this case, the only way to combine the 0's is to include don't-care minterms 0 and 2 with the 0's to give a simplified complemented function:

$$F' = z' + wy'$$

Taking the complement of F' gives the simplified expression in product-of-sums form:

$$F(w, x, y, z) = z(w' + y) = \Sigma(1, 3, 5, 7, 11, 15)$$

In this case, we include minterms 0 and 2 with the 0's and minterm 5 with the 1's.

3.6 NAND AND NOR IMPLEMENTATION

Digital circuits are frequently constructed with NAND or NOR gates rather than with AND and OR gates. NAND and NOR gates are easier to fabricate with electronic components and are the basic gates used in all IC digital logic families. Because of the prominence of NAND and NOR gates in the design of digital circuits, rules and procedures have been developed for the conversion from Boolean functions given in terms of AND, OR, and NOT into equivalent NAND and NOR logic diagrams.

NAND Circuits

The NAND gate is said to be a *universal* gate because any logic circuit can be implemented with it. To show that any Boolean function can be implemented with NAND gates, we need only show that the logical operations of AND, OR, and complement can be obtained with NAND gates alone. This is indeed shown in Fig. 3.16. The complement operation is obtained from a one-input NAND gate that behaves exactly like an inverter. The AND operation requires two NAND gates. The first produces the NAND operation and the second inverts the logical sense of the signal. The OR operation is achieved through a NAND gate with additional inverters in each input.

A convenient way to implement a Boolean function with NAND gates is to obtain the simplified Boolean function in terms of Boolean operators and then convert the function to NAND logic. The conversion of an algebraic expression from AND, OR, and complement to NAND can be done by simple circuit manipulation techniques that change AND–OR diagrams to NAND diagrams.

To facilitate the conversion to NAND logic, it is convenient to define an alternative graphic symbol for the gate. Two equivalent graphic symbols for the NAND gate are shown in Fig. 3.17. The AND-invert symbol has been defined previously and consists

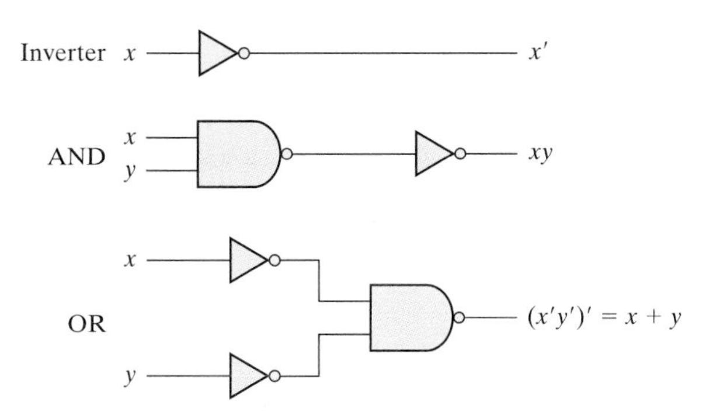

FIGURE 3.16
Logic operations with NAND gates

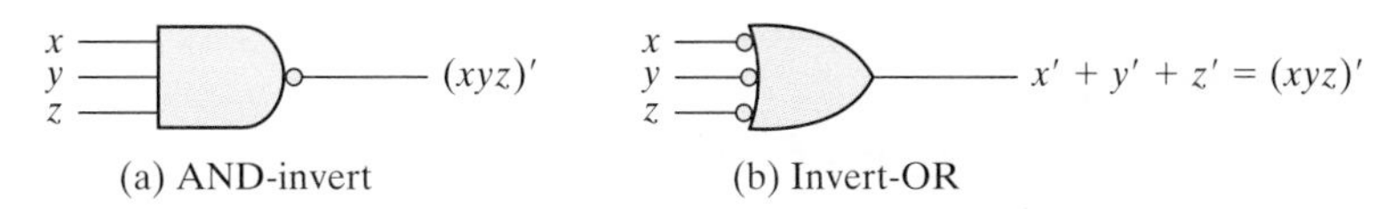

FIGURE 3.17
Two graphic symbols for a three-input NAND gate

of an AND graphic symbol followed by a small circle negation indicator referred to as a bubble. Alternatively, it is possible to represent a NAND gate by an OR graphic symbol that is preceded by a bubble in each input. The invert-OR symbol for the NAND gate follows DeMorgan's theorem and the convention that the negation indicator (bubble) denotes complementation. The two graphic symbols' representations are useful in the analysis and design of NAND circuits. When both symbols are mixed in the same diagram, the circuit is said to be in mixed notation.

Two-Level Implementation

The implementation of Boolean functions with NAND gates requires that the functions be in sum-of-products form. To see the relationship between a sum-of-products expression and its equivalent NAND implementation, consider the logic diagrams drawn in Fig. 3.18. All three diagrams are equivalent and implement the function

$$F = AB + CD$$

The function is implemented in Fig. 3.18(a) with AND and OR gates. In Fig. 3.18(b), the AND gates are replaced by NAND gates and the OR gate is replaced by a NAND gate with an OR-invert graphic symbol. Remember that a bubble denotes complementation and two bubbles along the same line represent double complementation, so both can be removed. Removing the bubbles on the gates of (b) produces the circuit of (a). Therefore, the two diagrams implement the same function and are equivalent.

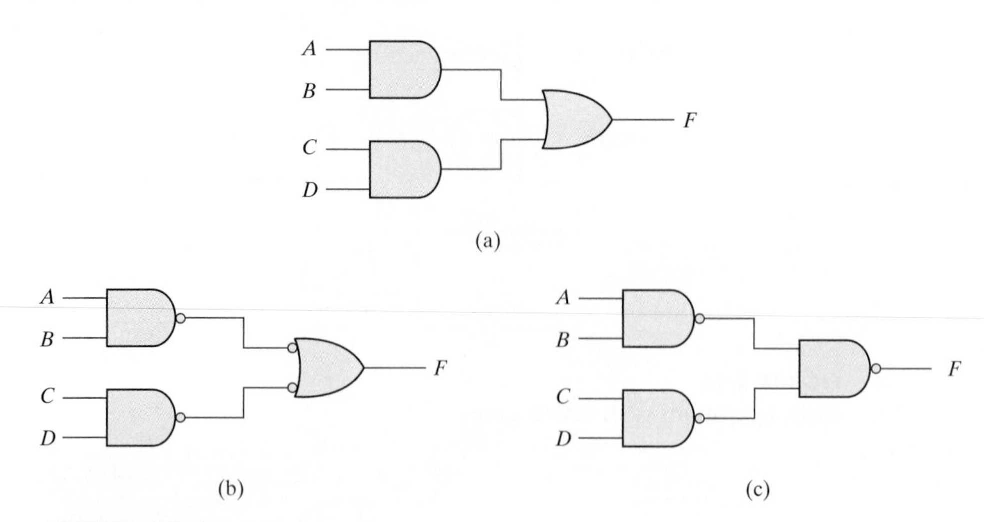

FIGURE 3.18
Three ways to implement $F = AB + CD$

In Fig. 3.18(c), the output NAND gate is redrawn with the AND-invert graphic symbol. In drawing NAND logic diagrams, the circuit shown in either Fig. 3.18(b) or (c) is acceptable. The one in Fig. 3.18(b) is in mixed notation and represents a more direct relationship to the Boolean expression it implements. The NAND implementation in Fig. 3.18(c) can be verified algebraically. The function it implements can easily be converted to sum-of-products form by DeMorgan's theorem:

$$F = ((AB)'(CD)')' = AB + CD$$

EXAMPLE 3.9

Implement the following Boolean function with NAND gates:

$$F(x, y, z) = (1, 2, 3, 4, 5, 7)$$

The first step is to simplify the function into sum-of-products form. This is done by means of the map of Fig. 3.19(a), from which the simplified function is obtained:

$$F = xy' + x'y + z$$

The two-level NAND implementation is shown in Fig. 3.19(b) in mixed notation. Note that input z must have a one-input NAND gate (an inverter) to compensate for the bubble in the second-level gate. An alternative way of drawing the logic diagram is given in Fig. 3.19(c). Here, all the NAND gates are drawn with the same graphic symbol. The inverter with input z has been removed, but the input variable is complemented and denoted by z'.

■

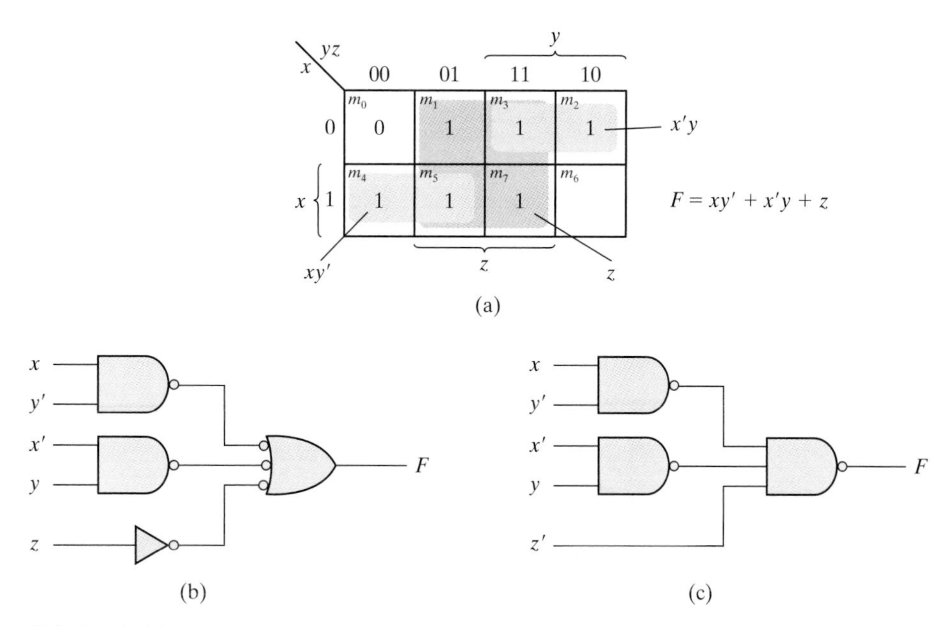

FIGURE 3.19
Solution to Example 3.9

The procedure described in the previous example indicates that a Boolean function can be implemented with two levels of NAND gates. The procedure for obtaining the logic diagram from a Boolean function is as follows:

1. Simplify the function and express it in sum-of-products form.
2. Draw a NAND gate for each product term of the expression that has at least two literals. The inputs to each NAND gate are the literals of the term. This procedure produces a group of first-level gates.
3. Draw a single gate using the AND-invert or the invert-OR graphic symbol in the second level, with inputs coming from outputs of first-level gates.
4. A term with a single literal requires an inverter in the first level. However, if the single literal is complemented, it can be connected directly to an input of the second-level NAND gate.

Multilevel NAND Circuits

The standard form of expressing Boolean functions results in a two-level implementation. There are occasions, however, when the design of digital systems results in gating structures with three or more levels. The most common procedure in the design of multilevel circuits is to express the Boolean function in terms of AND, OR, and complement operations. The function can then be implemented with AND and OR gates. After that, if necessary, it can be converted into an all-NAND circuit. Consider, for example, the Boolean function

$$F = A(CD + B) + BC'$$

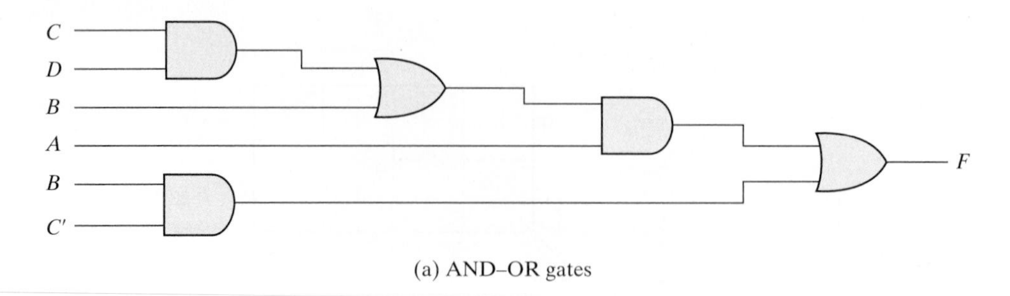

(a) AND–OR gates

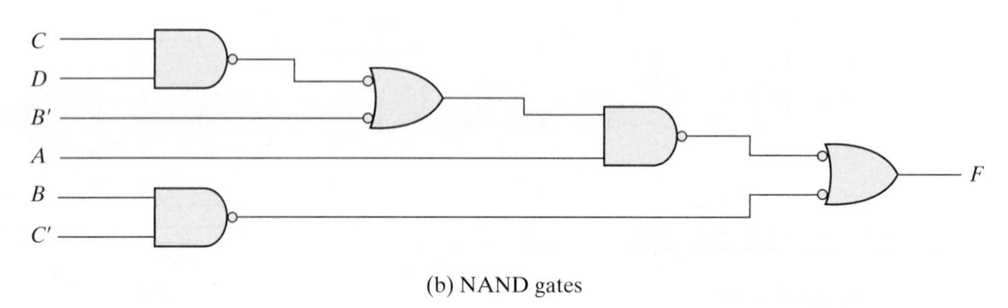

(b) NAND gates

FIGURE 3.20
Implementing $F = A(CD + B) + BC'$

Although it is possible to remove the parentheses and reduce the expression into a standard sum-of-products form, we choose to implement it as a multilevel circuit for illustration. The AND–OR implementation is shown in Fig. 3.20(a). There are four levels of gating in the circuit. The first level has two AND gates. The second level has an OR gate followed by an AND gate in the third level and an OR gate in the fourth level. A logic diagram with a pattern of alternating levels of AND and OR gates can easily be converted into a NAND circuit with the use of mixed notation, shown in Fig. 3.20(b). The procedure is to change every AND gate to an AND-invert graphic symbol and every OR gate to an invert-OR graphic symbol. The NAND circuit performs the same logic as the AND–OR diagram as long as there are two bubbles along the same line. The bubble associated with input B causes an extra complementation, which must be compensated for by changing the input literal to B'.

The general procedure for converting a multilevel AND–OR diagram into an all-NAND diagram using mixed notation is as follows:

1. Convert all AND gates to NAND gates with AND-invert graphic symbols.
2. Convert all OR gates to NAND gates with invert-OR graphic symbols.
3. Check all the bubbles in the diagram. For every bubble that is not compensated by another small circle along the same line, insert an inverter (a one-input NAND gate) or complement the input literal.

As another example, consider the multilevel Boolean function

$$F = (AB' + A'B)(C + D')$$

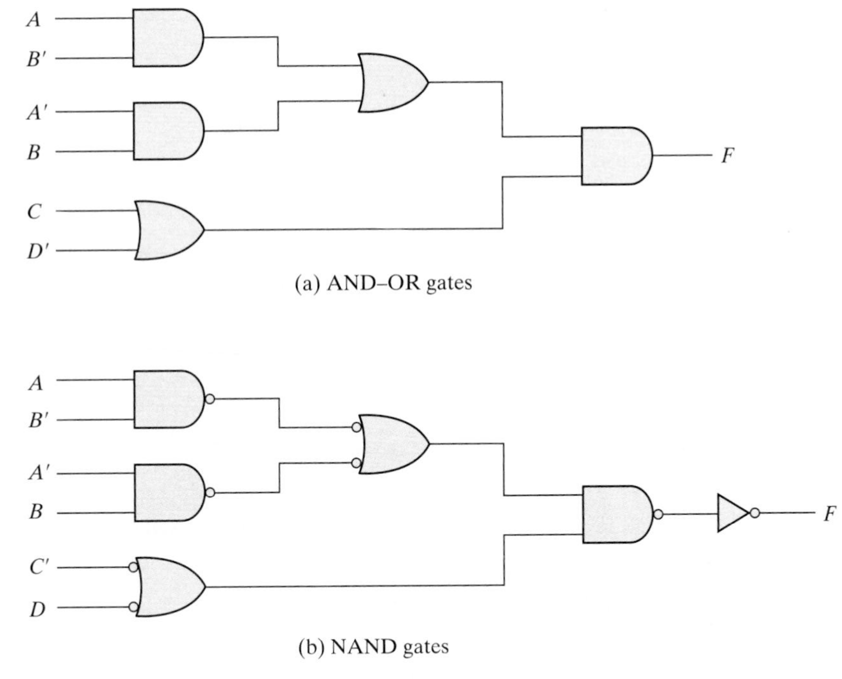

(a) AND–OR gates

(b) NAND gates

FIGURE 3.21
Implementing $F = (AB' + A'B)\ (C + D')$

The AND–OR implementation of this function is shown in Fig. 3.21(a) with three levels of gating. The conversion to NAND with mixed notation is presented in Fig. 3.21(b) of the diagram. The two additional bubbles associated with inputs C and D' cause these two literals to be complemented to C' and D. The bubble in the output NAND gate complements the output value, so we need to insert an inverter gate at the output in order to complement the signal again and get the original value back.

NOR Implementation

The NOR operation is the dual of the NAND operation. Therefore, all procedures and rules for NOR logic are the duals of the corresponding procedures and rules developed for NAND logic. The NOR gate is another universal gate that can be used to implement any Boolean function. The implementation of the complement, OR, and AND operations with NOR gates is shown in Fig. 3.22. The complement operation is obtained from a one-input NOR gate that behaves exactly like an inverter. The OR operation requires two NOR gates, and the AND operation is obtained with a NOR gate that has inverters in each input.

The two graphic symbols for the mixed notation are shown in Fig. 3.23. The OR-invert symbol defines the NOR operation as an OR followed by a complement. The invert-AND symbol complements each input and then performs an AND operation. The two symbols designate the same NOR operation and are logically identical because of DeMorgan's theorem.

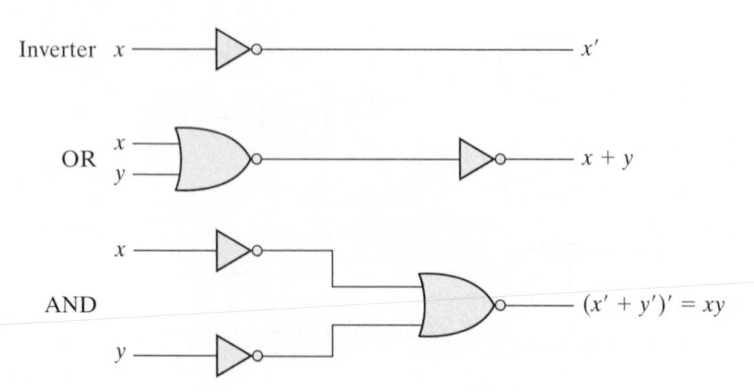

FIGURE 3.22
Logic operations with NOR gates

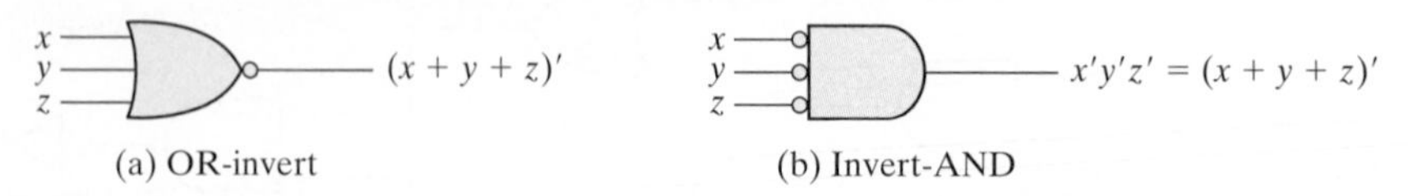

FIGURE 3.23
Two graphic symbols for the NOR gate

A two-level implementation with NOR gates requires that the function be simplified into product-of-sums form. Remember that the simplified product-of-sums expression is obtained from the map by combining the 0's and complementing. A product-of-sums expression is implemented with a first level of OR gates that produce the sum terms followed by a second-level AND gate to produce the product. The transformation from the OR–AND diagram to a NOR diagram is achieved by changing the OR gates to NOR gates with OR-invert graphic symbols and the AND gate to a NOR gate with an invert-AND graphic symbol. A single literal term going into the second-level gate must be complemented. Figure 3.24 shows the NOR implementation of a function expressed as a product of sums:

$$F = (A + B)(C + D)E$$

The OR–AND pattern can easily be detected by the removal of the bubbles along the same line. Variable E is complemented to compensate for the third bubble at the input of the second-level gate.

The procedure for converting a multilevel AND–OR diagram to an all-NOR diagram is similar to the one presented for NAND gates. For the NOR case, we must convert each OR gate to an OR-invert symbol and each AND gate to an invert-AND symbol. Any bubble that is not compensated by another bubble along the same line needs an inverter, or the complementation of the input literal.

The transformation of the AND–OR diagram of Fig. 3.21(a) into a NOR diagram is shown in Fig. 3.25. The Boolean function for this circuit is

$$F = (AB' + A'B)(C + D')$$

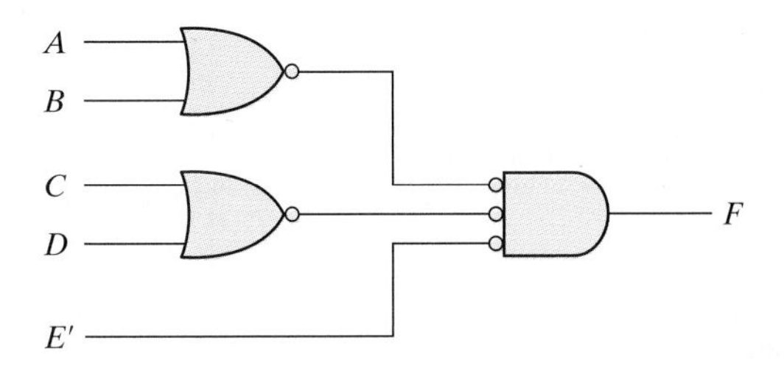

FIGURE 3.24
Implementing $F = (A + B)(C + D)E$

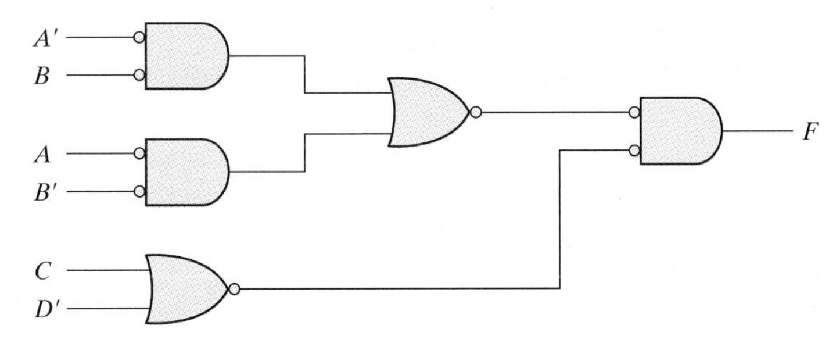

FIGURE 3.25
Implementing $F = (AB' + A'B)(C + D')$ with NOR gates

The equivalent AND–OR diagram can be recognized from the NOR diagram by removing all the bubbles. To compensate for the bubbles in four inputs, it is necessary to complement the corresponding input literals.

3.7 OTHER TWO-LEVEL IMPLEMENTATIONS

The types of gates most often found in integrated circuits are NAND and NOR gates. For this reason, NAND and NOR logic implementations are the most important from a practical point of view. Some (but not all) NAND or NOR gates allow the possibility of a wire connection between the outputs of two gates to provide a specific logic function. This type of logic is called *wired logic.* For example, open-collector TTL NAND gates, when tied together, perform wired-AND logic. The wired-AND logic performed with two NAND gates is depicted in Fig. 3.26(a). The AND gate is drawn with the lines going through the center of the gate to distinguish it from a conventional gate. The wired-AND gate is not a physical gate, but only a symbol to designate the function obtained from the indicated wired connection. The logic function implemented by the circuit of Fig. 3.26(a) is

$$F = (AB)' \cdots (CD)' = (AB + CD)' = (A' + B')(C' + D')$$

and is called an AND–OR–INVERT function.

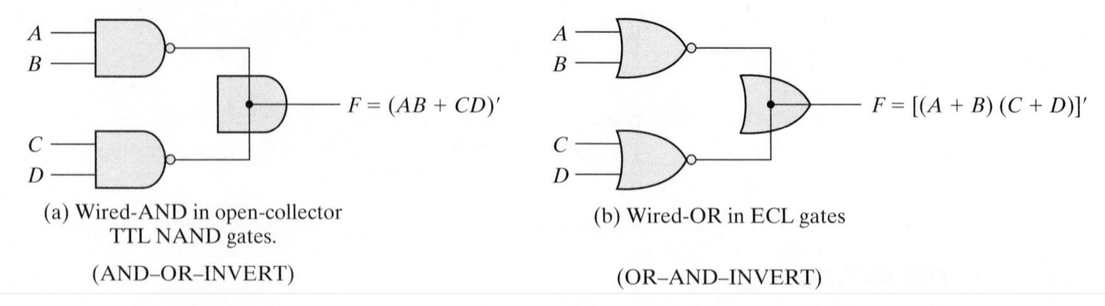

FIGURE 3.26
Wired logic
(a) Wired-AND logic with two NAND gates
(b) Wired-OR in emitter-coupled logic (ECL) gates

Similarly, the NOR outputs of ECL gates can be tied together to perform a wired-OR function. The logic function implemented by the circuit of Fig. 3.26(b) is

$$F = (A + B)' + (C + D)' = [(A + B)(C + D)]'$$

and is called an OR–AND–INVERT function.

A wired-logic gate does not produce a physical second-level gate, since it is just a wire connection. Nevertheless, for discussion purposes, we will consider the circuits of Fig. 3.26 as two-level implementations. The first level consists of NAND (or NOR) gates and the second level has a single AND (or OR) gate. The wired connection in the graphic symbol will be omitted in subsequent discussions.

Nondegenerate Forms

It will be instructive from a theoretical point of view to find out how many two-level combinations of gates are possible. We consider four types of gates: AND, OR, NAND, and NOR. If we assign one type of gate for the first level and one type for the second level, we find that there are 16 possible combinations of two-level forms. (The same type of gate can be in the first and second levels, as in a NAND–NAND implementation.) Eight of these combinations are said to be *degenerate* forms because they degenerate to a single operation. This can be seen from a circuit with AND gates in the first level and an AND gate in the second level. The output of the circuit is merely the AND function of all input variables. The remaining eight *nondegenerate* forms produce an implementation in sum-of-products form or product-of-sums form. The eight *nondegenerate* forms are as follows:

AND–OR	OR–AND
NAND–NAND	NOR–NOR
NOR–OR	NAND–AND
OR–NAND	AND–NOR

The first gate listed in each of the forms constitutes a first level in the implementation. The second gate listed is a single gate placed in the second level. Note that any two forms listed on the same line are duals of each other.

The AND–OR and OR–AND forms are the basic two-level forms discussed in Section 3.4. The NAND–NAND and NOR–NOR forms were presented in Section 3.5. The remaining four forms are investigated in this section.

AND–OR–INVERT Implementation

The two forms, NAND–AND and AND–NOR, are equivalent and can be treated together. Both perform the AND–OR–INVERT function, as shown in Fig. 3.27. The AND–NOR form resembles the AND–OR form, but with an inversion done by the bubble in the output of the NOR gate. It implements the function

$$F = (AB + CD + E)'$$

By using the alternative graphic symbol for the NOR gate, we obtain the diagram of Fig. 3.27(b). Note that the single variable E is *not* complemented, because the only change made is in the graphic symbol of the NOR gate. Now we move the bubble from the input terminal of the second-level gate to the output terminals of the first-level gates. An inverter is needed for the single variable in order to compensate for the bubble. Alternatively, the inverter can be removed, provided that input E is complemented. The circuit of Fig. 3.27(c) is a NAND–AND form and was shown in Fig. 3.26 to implement the AND–OR–INVERT function.

An AND–OR implementation requires an expression in sum-of-products form. The AND–OR–INVERT implementation is similar, except for the inversion. Therefore, if the *complement* of the function is simplified into sum-of-products form (by combining the 0's in the map), it will be possible to implement F' with the AND–OR part of the function. When F' passes through the always present output inversion (the INVERT part), it will

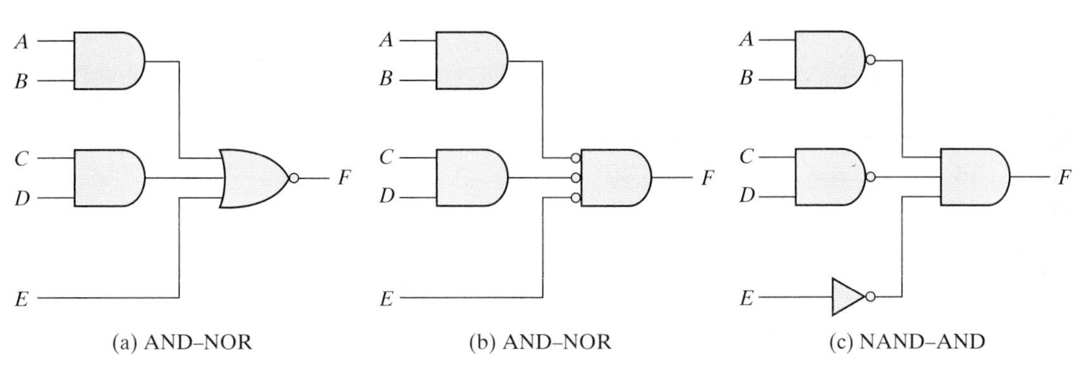

FIGURE 3.27
AND–OR–INVERT circuits, $F = (AB + CD + E)'$

generate the output F of the function. An example for the AND–OR–INVERT implementation will be shown subsequently.

OR–AND–INVERT Implementation

The OR–NAND and NOR–OR forms perform the OR–AND–INVERT function, as shown in Fig. 3.28. The OR–NAND form resembles the OR–AND form, except for the inversion done by the bubble in the NAND gate. It implements the function

$$F = [(A + B)(C + D)E]'$$

By using the alternative graphic symbol for the NAND gate, we obtain the diagram of Fig. 3.28(b). The circuit in Fig. 3.28(c) is obtained by moving the small circles from the inputs of the second-level gate to the outputs of the first-level gates. The circuit of Fig. 3.28(c) is a NOR–OR form and was shown in Fig. 3.26 to implement the OR–AND–INVERT function.

The OR–AND–INVERT implementation requires an expression in product-of-sums form. If the complement of the function is simplified into that form, we can implement F' with the OR–AND part of the function. When F' passes through the INVERT part, we obtain the complement of F', or F, in the output.

Tabular Summary and Example

Table 3.2 summarizes the procedures for implementing a Boolean function in any one of the four 2-level forms. Because of the INVERT part in each case, it is convenient to use the simplification of F' (the complement) of the function. When F' is implemented in one of these forms, we obtain the complement of the function in the AND–OR or OR–AND form. The four 2-level forms invert this function, giving an output that is the complement of F'. This is the normal output F.

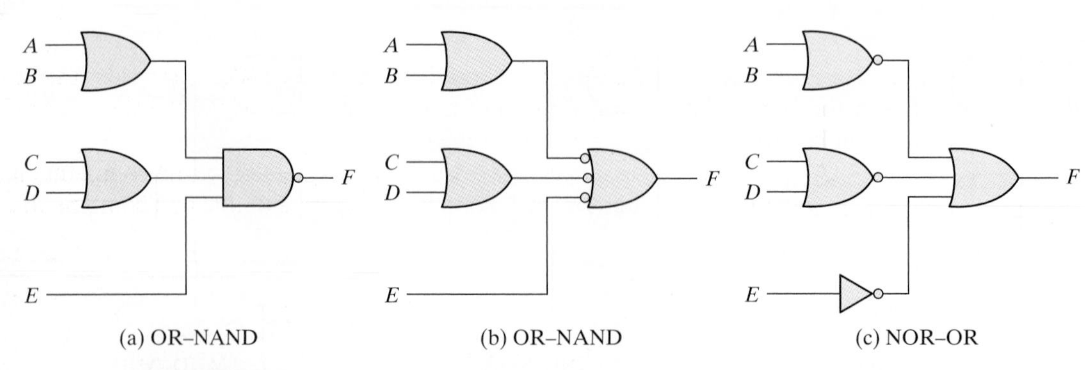

FIGURE 3.28
OR–AND–INVERT circuits, $F = [(A + B)(C + D)E]'$

Table 3.2
Implementation with Other Two-Level Forms

Equivalent Nondegenerate Form (a)	**(b)***	**Implements the Function**	**Simplify F' into**	**To Get an Output of**
AND–NOR	NAND–AND	AND–OR–INVERT	Sum-of-products form by combining 0's in the map.	F
OR–NAND	NOR–OR	OR–AND–INVERT	Product-of-sums form by combining 1's in the map and then complementing.	F

*Form (b) requires an inverter for a single literal term.

EXAMPLE 3.10

Implement the function of Fig. 3.29(a) with the four 2-level forms listed in Table 3.2. The complement of the function is simplified into sum-of-products form by combining the 0's in the map:

$$F' = x'y + xy' + z$$

The normal output for this function can be expressed as

$$F = (x'y + xy' + z)'$$

which is in the AND–OR–INVERT form. The AND–NOR and NAND–AND implementations are shown in Fig. 3.29(b). Note that a one-input NAND, or inverter, gate is needed in the NAND–AND implementation, but not in the AND–NOR case. The inverter can be removed if we apply the input variable z' instead of z.

The OR–AND–INVERT forms require a simplified expression of the complement of the function in product-of-sums form. To obtain this expression, we first combine the 1's in the map:

$$F = x'y'z' + xyz'$$

Then we take the complement of the function:

$$F' = (x + y + z)(x' + y' + z)$$

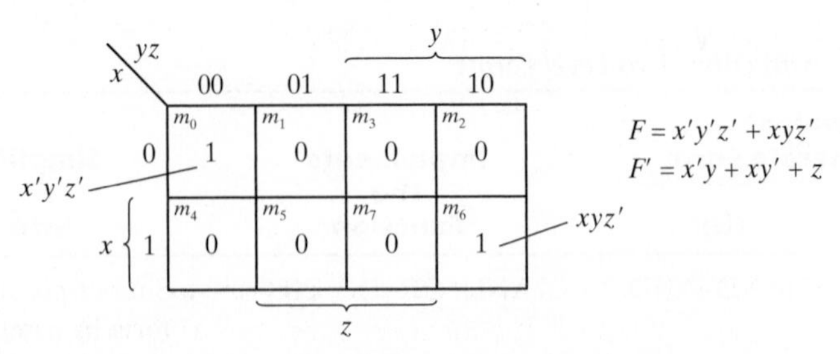

(a) Map simplification in sum of products

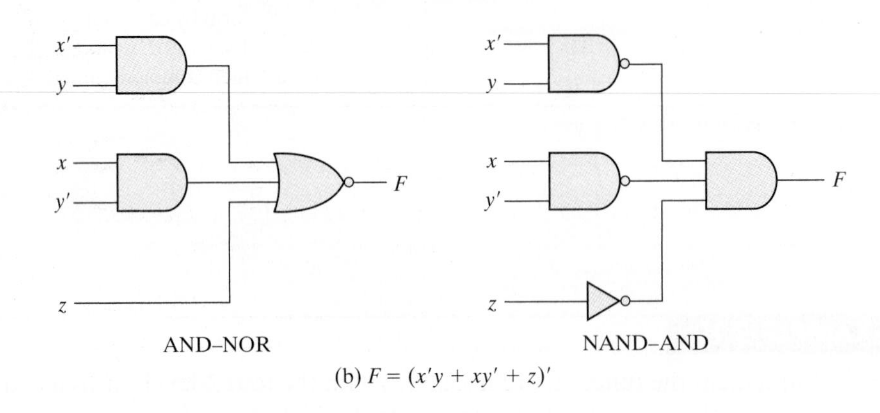

(b) $F = (x'y + xy' + z)'$

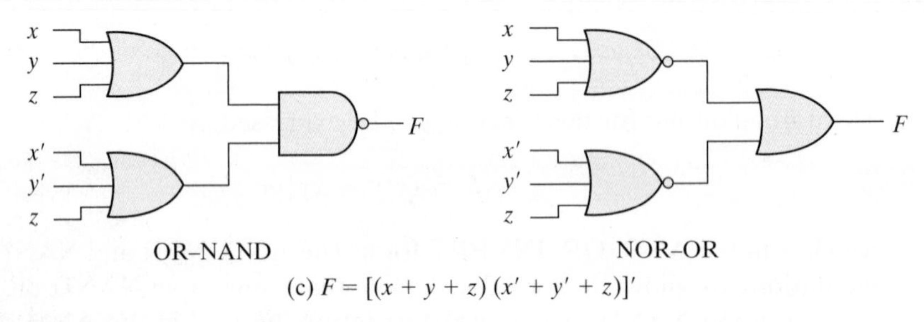

(c) $F = [(x + y + z)(x' + y' + z)]'$

FIGURE 3.29
Other two-level implementations

The normal output F can now be expressed in the form

$$F = [(x + y + z)(x' + y' + z)]'$$

which is the OR–AND–INVERT form. From this expression, we can implement the function in the OR–NAND and NOR–OR forms, as shown in Fig. 3.29(c).

■

3.8 EXCLUSIVE-OR FUNCTION

The exclusive-OR (XOR), denoted by the symbol $\oplus$, is a logical operation that performs the following Boolean operation:

$$x \oplus y = xy' + x'y$$

The exclusive-OR is equal to 1 if only x is equal to 1 or if only y is equal to 1 (i.e., x and y differ in value), but not when both are equal to 1 or when both are equal to 0. The exclusive-NOR, also known as equivalence, performs the following Boolean operation:

$$(x \oplus y)' = xy + x'y'$$

The exclusive-NOR is equal to 1 if both x and y are equal to 1 or if both are equal to 0. The exclusive-NOR can be shown to be the complement of the exclusive-OR by means of a truth table or by algebraic manipulation:

$$(x \oplus y)' = (xy' + x'y)' = (x' + y)(x + y') = xy + x'y'$$

The following identities apply to the exclusive-OR operation:

$$x \oplus 0 = x$$
$$x \oplus 1 = x'$$
$$x \oplus x = 0$$
$$x \oplus x' = 1$$
$$x \oplus y' = x' \oplus y = (x \oplus y)'$$

Any of these identities can be proven with a truth table or by replacing the $\oplus$ operation by its equivalent Boolean expression. Also, it can be shown that the exclusive-OR operation is both commutative and associative; that is,

$$A \oplus B = B \oplus A$$

and

$$(A \oplus B) \oplus C = A \oplus (B \oplus C) = A \oplus B \oplus C$$

This means that the two inputs to an exclusive-OR gate can be interchanged without affecting the operation. It also means that we can evaluate a three-variable exclusive-OR operation in any order, and for this reason, three or more variables can be expressed without parentheses. This would imply the possibility of using exclusive-OR gates with three or more inputs. However, multiple-input exclusive-OR gates are difficult to fabricate with hardware. In fact, even a two-input function is usually constructed with other types of gates. A two-input exclusive-OR function is constructed with conventional gates using two inverters, two AND gates, and an OR gate, as shown in Fig. 3.30(a). Figure 3.30(b) shows the implementation of the exclusive-OR with four NAND gates. The first NAND gate performs the operation $(xy)' = (x' + y')$. The other two-level NAND circuit produces the sum of products of its inputs:

$$(x' + y')x + (x' + y')y = xy' + x'y = x \oplus y$$

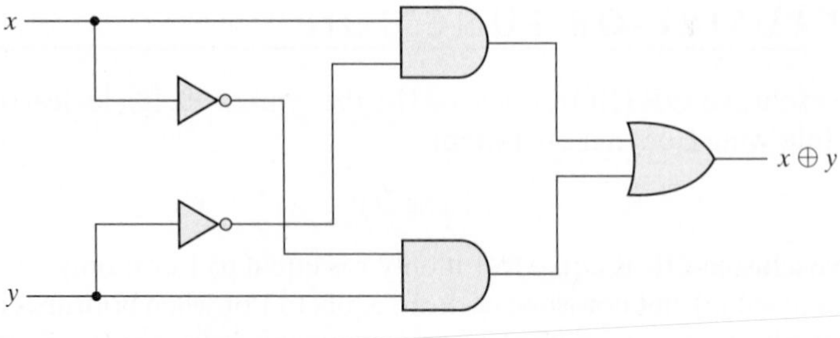

(a) Exclusive-OR with AND–OR–NOT gates

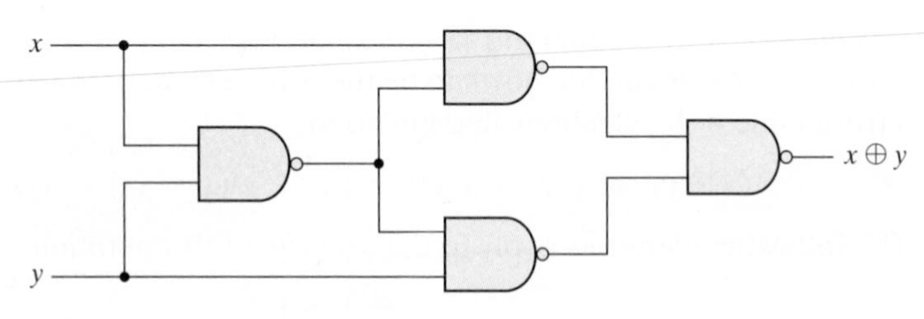

(b) Exclusive-OR with NAND gates

FIGURE 3.30
Exclusive-OR implementations

Only a limited number of Boolean functions can be expressed in terms of exclusive-OR operations. Nevertheless, this function emerges quite often during the design of digital systems. It is particularly useful in arithmetic operations and error detection and correction circuits.

Odd Function

The exclusive-OR operation with three or more variables can be converted into an ordinary Boolean function by replacing the $\oplus$ symbol with its equivalent Boolean expression. In particular, the three-variable case can be converted to a Boolean expression as follows:

$$\begin{aligned} A \oplus B \oplus C &= (AB' + A'B)C' + (AB + A'B')C \\ &= AB'C' + A'BC' + ABC + A'B'C \\ &= \Sigma(1, 2, 4, 7) \end{aligned}$$

The Boolean expression clearly indicates that the three-variable exclusive-OR function is equal to 1 if only one variable is equal to 1 or if all three variables are equal to 1. Contrary to the two-variable case, in which only one variable must be equal to 1, in the case of three or more variables the requirement is that an odd number of variables be equal to 1. As a consequence, the multiple-variable exclusive-OR operation is defined as an *odd function.*

The Boolean function derived from the three-variable exclusive-OR operation is expressed as the logical sum of four minterms whose binary numerical values are 001, 010, 100, and 111. Each of these binary numbers has an odd number of 1's. The remaining four minterms not included in the function are 000, 011, 101, and 110, and they have an even number of 1's in their binary numerical values. In general, an n-variable exclusive-OR function is an odd function defined as the logical sum of the $2^n/2$ minterms whose binary numerical values have an odd number of 1's.

The definition of an odd function can be clarified by plotting it in a map. Figure 3.31(a) shows the map for the three-variable exclusive-OR function. The four minterms of the function are a unit distance apart from each other. The odd function is identified from the four minterms whose binary values have an odd number of 1's. The complement of an odd function is an even function. As shown in Fig. 3.31(b), the three-variable even function is equal to 1 when an even number of its variables is equal to 1 (including the condition that none of the variables is equal to 1).

The three-input odd function is implemented by means of two-input exclusive-OR gates, as shown in Fig. 3.32(a). The complement of an odd function is obtained by replacing the output gate with an exclusive-NOR gate, as shown in Fig. 3.32(b).

Consider now the four-variable exclusive-OR operation. By algebraic manipulation, we can obtain the sum of minterms for this function:

$$\begin{aligned} A \oplus B \oplus C \oplus D &= (AB' + A'B) \oplus (CD' + C'D) \\ &= (AB' + A'B)(CD + C'D') + (AB + A'B')(CD' + C'D) \\ &= \Sigma(1, 2, 4, 7, 8, 11, 13, 14) \end{aligned}$$

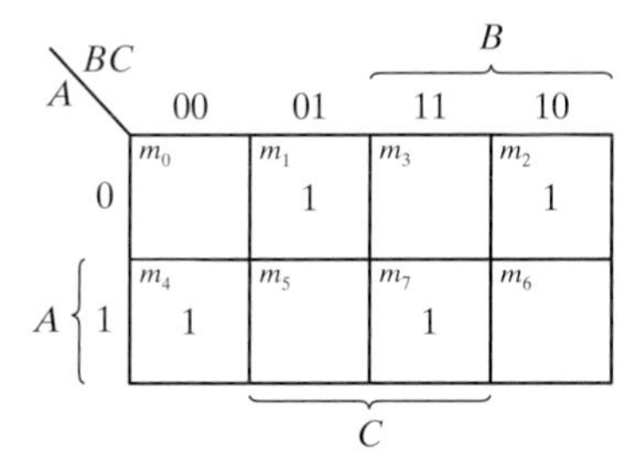

(a) Odd function $F = A \oplus B \oplus C$

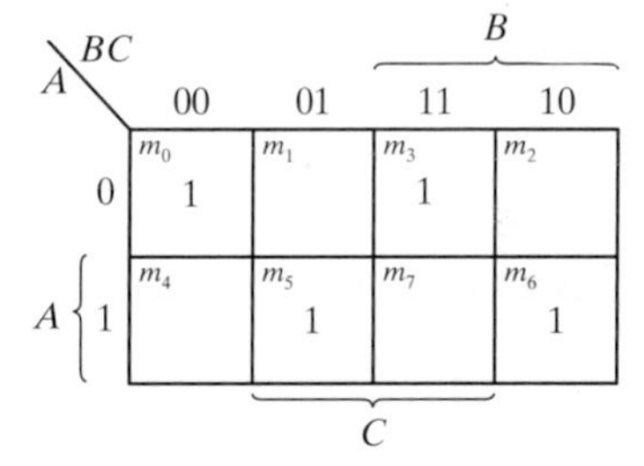

(b) Even function $F = (A \oplus B \oplus C)'$

FIGURE 3.31
Map for a three-variable exclusive-OR function

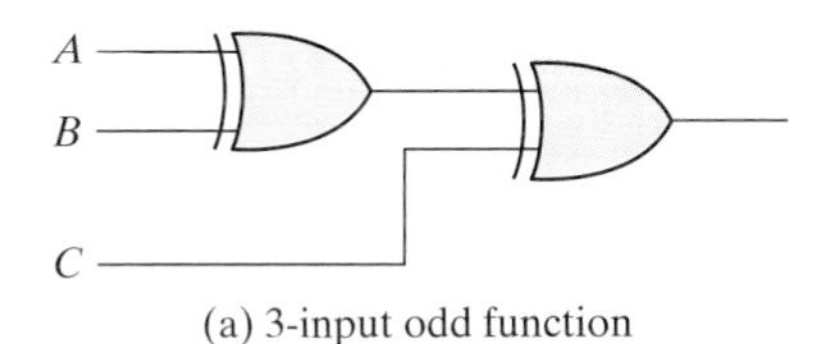

(a) 3-input odd function

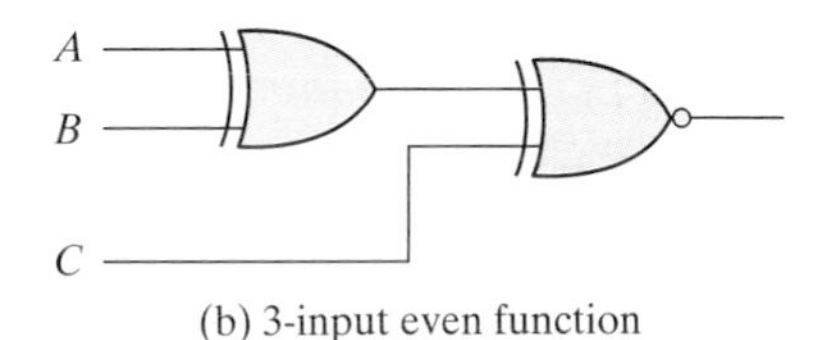

(b) 3-input even function

FIGURE 3.32
Logic diagram of odd and even functions

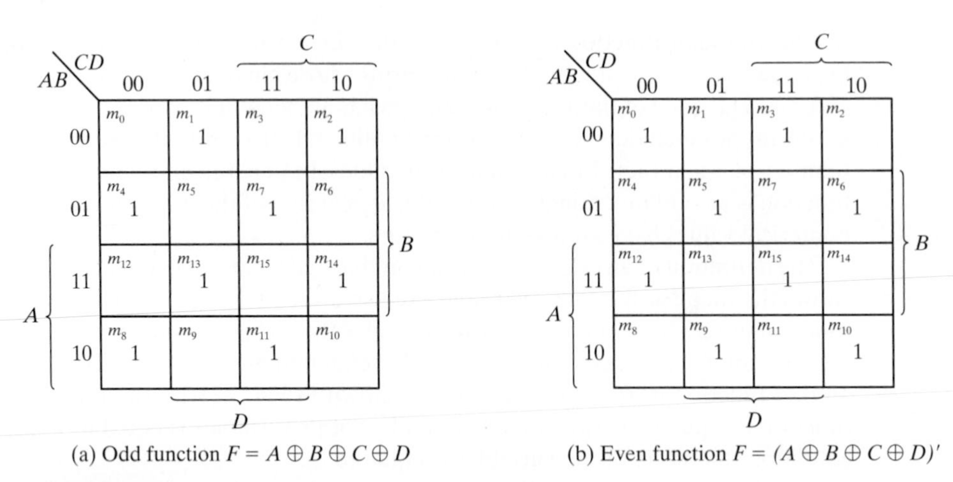

(a) Odd function $F = A \oplus B \oplus C \oplus D$

(b) Even function $F = (A \oplus B \oplus C \oplus D)'$

FIGURE 3.33
Map for a four-variable exclusive-OR function

There are 16 minterms for a four-variable Boolean function. Half of the minterms have binary numerical values with an odd number of 1's; the other half of the minterms have binary numerical values with an even number of 1's. In plotting the function in the map, the binary numerical value for a minterm is determined from the row and column numbers of the square that represents the minterm. The map of Fig. 3.33(a) is a plot of the four-variable exclusive-OR function. This is an odd function because the binary values of all the minterms have an odd number of 1's. The complement of an odd function is an even function. As shown in Fig. 3.33(b), the four-variable even function is equal to 1 when an even number of its variables is equal to 1.

Parity Generation and Checking

Exclusive-OR functions are very useful in systems requiring error detection and correction codes. As discussed in Section 1.6, a parity bit is used for the purpose of detecting errors during the transmission of binary information. A parity bit is an extra bit included with a binary message to make the number of 1's either odd or even. The message, including the parity bit, is transmitted and then checked at the receiving end for errors. An error is detected if the checked parity does not correspond with the one transmitted. The circuit that generates the parity bit in the transmitter is called a *parity generator*. The circuit that checks the parity in the receiver is called a *parity checker*.

As an example, consider a three-bit message to be transmitted together with an even-parity bit. Table 3.3 shows the truth table for the parity generator. The three bits—x, y, and z—constitute the message and are the inputs to the circuit. The parity bit P is the output. For even parity, the bit P must be generated to make the total number of 1's (including P) even. From the truth table, we see that P constitutes an

Table 3.3
Even-Parity-Generator Truth Table

Three-Bit Message			Parity Bit
x	***y***	***z***	***P***
0	0	0	0
0	0	1	1
0	1	0	1
0	1	1	0
1	0	0	1
1	0	1	0
1	1	0	0
1	1	1	1

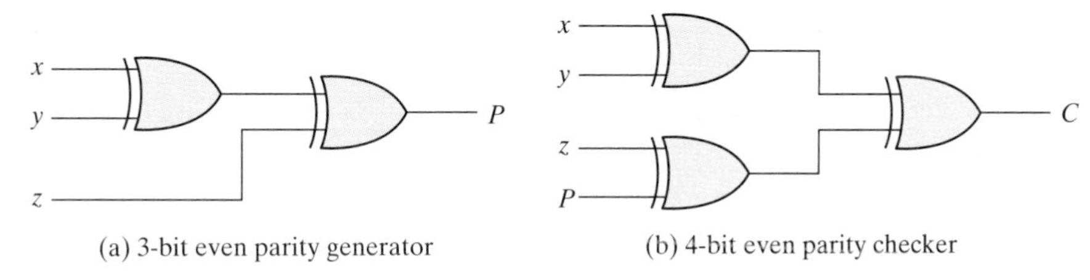

FIGURE 3.34
Logic diagram of a parity generator and checker

odd function because it is equal to 1 for those minterms whose numerical values have an odd number of 1's. Therefore, P can be expressed as a three-variable exclusive-OR function:

$$P = x \oplus y \oplus z$$

The logic diagram for the parity generator is shown in Fig. 3.34(a).

The three bits in the message, together with the parity bit, are transmitted to their destination, where they are applied to a parity-checker circuit to check for possible errors in the transmission. Since the information was transmitted with even parity, the four bits received must have an even number of 1's. An error occurs during the transmission if the four bits received have an odd number of 1's, indicating that one bit has changed in value during transmission. The output of the parity checker, denoted by C, will be equal to 1 if an error occurs—that is, if the four bits received have an odd number of 1's. Table 3.4 is the truth table for the even-parity checker. From it, we see that the function C consists of the eight minterms with binary numerical values having an odd number of 1's. The table corresponds to the map of Fig. 3.33(a), which

Table 3.4
Even-Parity-Checker Truth Table

Four Bits Received				Parity Error Check
x	*y*	*z*	*P*	*C*
0	0	0	0	0
0	0	0	1	1
0	0	1	0	1
0	0	1	1	0
0	1	0	0	1
0	1	0	1	0
0	1	1	0	0
0	1	1	1	1
1	0	0	0	1
1	0	0	1	0
1	0	1	0	0
1	0	1	1	1
1	1	0	0	0
1	1	0	1	1
1	1	1	0	1
1	1	1	1	0

represents an odd function. The parity checker can be implemented with exclusive-OR gates:

$$C = x \oplus y \oplus z \oplus P$$

The logic diagram of the parity checker is shown in Fig. 3.34(b).

It is worth noting that the parity generator can be implemented with the circuit of Fig. 3.34(b) if the input P is connected to logic 0 and the output is marked with P. This is because $z \oplus 0 = z$, causing the value of z to pass through the gate unchanged. The advantage of this strategy is that the same circuit can be used for both parity generation and checking.

It is obvious from the foregoing example that parity generation and checking circuits always have an output function that includes half of the minterms whose numerical values have either an odd or even number of 1's. As a consequence, they can be implemented with exclusive-OR gates. A function with an even number of 1's is the complement of an odd function. It is implemented with exclusive-OR gates, except that the gate associated with the output must be an exclusive-NOR to provide the required complementation.

3.9 HARDWARE DESCRIPTION LANGUAGE

Manual methods for designing logic circuits are feasible only when the circuit is small. For anything else (i.e., a practical circuit), designers use computer-based design tools. Coupled with the correct-by-construction methodology, computer-based design tools leverage the creativity and the effort of a designer and reduce the risk of producing a flawed design. Prototype integrated circuits are too expensive and time consuming to build, so all modern design tools rely on a hardware description language to describe, design, and test a circuit in software before it is ever manufactured.

A *hardware description language* (HDL) is a computer-based language that describes the hardware of digital systems in a textual form. It resembles an ordinary computer programming language, such as C, but is specifically oriented to describing hardware structures and the behavior of logic circuits. It can be used to represent logic diagrams, truth tables, Boolean expressions, and complex abstractions of the behavior of a digital system. One way to view an HDL is to observe that it describes a relationship between signals that are the inputs to a circuit and the signals that are the outputs of the circuit. For example, an HDL description of an AND gate describes how the logic value of the gate's output is determined by the logic values of its inputs.

As a *documentation* language, an HDL is used to represent and document digital systems in a form that can be read by both humans and computers and is suitable as an exchange language between designers. The language content can be stored, retrieved, edited, and transmitted easily and processed by computer software in an efficient manner.

HDLs are used in several major steps in the design flow of an integrated circuit: design entry, functional simulation or verification, logic synthesis, timing verification, and fault simulation.

Design entry creates an HDL-based description of the functionality that is to be implemented in hardware. Depending on the HDL, the description can be in a variety of forms: Boolean logic equations, truth tables, a netlist of interconnected gates, or an abstract behavioral model. The HDL model may also represent a partition of a larger circuit into smaller interconnected and interacting functional units.

Logic simulation displays the behavior of a digital system through the use of a computer. A simulator interprets the HDL description and either produces readable output, such as a time-ordered sequence of input and output signal values, or displays waveforms of the signals. The simulation of a circuit predicts how the hardware will behave before it is actually fabricated. Simulation detects functional errors in a design without having to physically create and operate the circuit. Errors that are detected during a simulation can be corrected by modifying the appropriate HDL statements. The stimulus (i.e., the logic values of the inputs to a circuit) that tests the functionality of the design is called a *test bench*. Thus, to simulate a digital system, the design is first described in an HDL and then verified by simulating the design and checking it with a test bench, which is also written in the HDL. An alternative and more complex approach relies on formal mathematical methods to prove that a circuit is functionally correct. We will focus exclusively on simulation.

Logic synthesis is the process of deriving a list of physical components and their interconnections (called a *netlist*) from the model of a digital system described in an HDL. The netlist can be used to fabricate an integrated circuit or to lay out a printed circuit board with the hardware counterparts of the gates in the list. Logic synthesis is similar to compiling a program in a conventional high-level language. The difference is that, instead of producing an object code, logic synthesis produces a database describing the elements and structure of a circuit. The database specifies how to fabricate a physical integrated circuit that implements in silicon the functionality described by statements made in an HDL. Logic synthesis is based on formal exact procedures that implement digital circuits and addresses that part of a digital design which can be automated with computer software. The design of today's large, complex circuits is made possible by logic synthesis software.

Timing verification confirms that the fabricated, integrated circuit will operate at a specified speed. Because each logic gate in a circuit has a propagation delay, a signal transition at the input of a circuit cannot immediately cause a change in the logic value of the output of a circuit. Propagation delays ultimately limit the speed at which a circuit can operate. Timing verification checks each signal path to verify that it is not compromised by propagation delay. This step is done after logic synthesis specifies the actual devices that will compose a circuit and before the circuit is released for production.

In VLSI circuit design, *fault simulation* compares the behavior of an ideal circuit with the behavior of a circuit that contains a process-induced flaw. Dust and other particulates in the atmosphere of the clean room can cause a circuit to be fabricated with a fault. A circuit with a fault will not exhibit the same functionality as a fault-free circuit. Fault simulation is used to identify input stimuli that can be used to reveal the difference between the faulty circuit and the fault-free circuit. These test patterns will be used to test fabricated devices to ensure that only good devices are shipped to the customer. Test generation and fault simulation may occur at different steps in the design process, but they are always done before production in order to avoid the disaster of producing a circuit whose internal logic cannot be tested.

Companies that design integrated circuits use proprietary and public HDLs. In the public domain, there are two standard HDLs that are supported by the IEEE: VHDL and Verilog. VHDL is a Department of Defense–mandated language. (The *V* in VHDL stands for the first letter in VHSIC, an acronym for very high-speed integrated circuit.) Verilog began as a proprietary HDL of Cadence Design Systems, but Cadence transferred control of Verilog to a consortium of companies and universities known as Open Verilog International (OVI) as a step leading to its adoption as an IEEE standard. VHDL is more difficult to learn than Verilog. Because Verilog is an easier language than VHDL to describe, learn, and use, we have chosen it for this book. However, the Verilog HDL descriptions listed throughout the book are not just about Verilog, but also serve to introduce a design methodology based on the concept of computer-aided modeling of digital systems by means of a typical hardware description language. Our emphasis will be on the modeling, verification, and synthesis (both manual and automated) of Verilog models of circuits having specified behavior. The Verilog HDL was initially approved as a standard HDL in 1995; revised and enhanced versions of the language were approved in 2001 and 2005. We will address only those features of Verilog, including the latest standard, that support our discussion of HDL-based design methodology for integrated circuits.

Module Declaration

The language reference manual for the Verilog HDL presents a syntax that describes precisely the constructs that can be used in the language. In particular, a Verilog model is composed of text using keywords, of which there are about 100. Keywords are predefined lowercase identifiers that define the language constructs. Examples of keywords are **module**, **endmodule**, **input**, **output**, **wire**, **and**, **or**, and **not**. For clarity, keywords will be displayed in boldface in the text in all examples of code and wherever it is appropriate to call attention to their use. Any text between two forward slashes (//) and the end of the line is interpreted as a comment and will have no effect on a simulation using the model. Multiline comments begin with /* and terminate with */. Blank spaces are ignored, but they may not appear within the text of a keyword, a user-specified identifier, an operator, or the representation of a number. Verilog is case sensitive, which means that uppercase and lowercase letters are distinguishable (e.g., **not** is not the same as NOT). The term *module* refers to the text enclosed by the keyword pair **module** ... **endmodule**. A module is the fundamental descriptive unit in the Verilog language. It is declared by the keyword **module** and must always be terminated by the keyword **endmodule**.

Combinational logic can be described by a schematic connection of gates, by a set of Boolean equations, or by a truth table. Each type of description can be developed in Verilog. We will demonstrate each style, beginning with a simple example of a Verilog gate-level description to illustrate some aspects of the language.

The HDL description of the circuit of Fig. 3.35 is shown in HDL Example 3.1. The first line of text is a comment (optional) providing useful information to the reader. The second line begins with the keyword **module** and starts the declaration (description) of the module; the last line completes the declaration with the keyword **endmodule**. The keyword **module** is followed by a name and a list of ports. The name (*Simple_Circuit* in this example) is an identifier. Identifiers are names given to modules, variables (e.g., a signal), and other elements of the language so that they can be referenced in the design. In general, we choose meaningful names for modules. Identifiers are composed of alphanumeric characters and the underscore (_), and are case sensitive. Identifiers must start with an alphabetic character or an underscore, but they cannot start with a number.

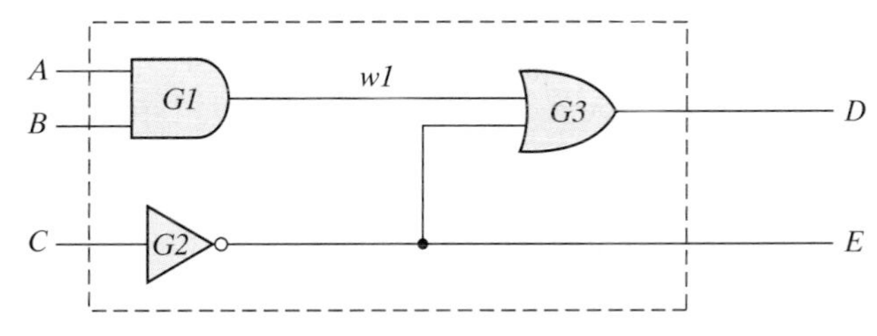

FIGURE 3.35
Circuit to demonstrate an HDL

HDL Example 3.1 (Combinational Logic Modeled with Primitives)

```
// Verilog model of circuit of Figure 3.35. IEEE 1364–1995 Syntax

module Simple_Circuit (A, B, C, D, E);
 output       D, E;
 input        A, B, C;
 wire         w1;

 and          G1 (w1, A, B); // Optional gate instance name
 not          G2 (E, C);
 or           G3 (D, w1, E);
endmodule
```

The *port list* of a module is the interface between the module and its environment. In this example, the ports are the inputs and outputs of the circuit. The logic values of the inputs to a circuit are determined by the environment; the logic values of the outputs are determined within the circuit and result from the action of the inputs on the circuit. The port list is enclosed in parentheses, and commas are used to separate elements of the list. The statement is terminated with a semicolon (;). In our examples, all keywords (which must be in lowercase) are printed in bold for clarity, but that is not a requirement of the language. Next, the keywords **input** and **output** specify which of the ports are inputs and which are outputs. Internal connections are declared as wires. The circuit in this example has one internal connection, at terminal *w1*, and is declared with the keyword **wire.** The structure of the circuit is specified by a list of (predefined) *primitive* gates, each identified by a descriptive keyword (**and, not, or**). The elements of the list are referred to as *instantiations* of a gate, each of which is referred to as a *gate instance*. Each *gate instantiation* consists of an optional name (such as *G1, G2*, etc.) followed by the gate output and inputs separated by commas and enclosed within parentheses. The output of a primitive gate is always listed first, followed by the inputs. For example, the OR gate of the schematic is represented by the **or** primitive, is named *G3*, and has output *D* and inputs *w1* and *E*. (*Note*: The output of a primitive must be listed first, but the inputs and outputs of a module may be listed in any order.) The module description ends with the keyword **endmodule.** Each statement must be terminated with a semicolon, but there is no semicolon after **endmodule.**

It is important to understand the distinction between the terms *declaration* and *instantiation*. A Verilog module is declared. Its declaration specifies the input–output behavior of the hardware that it represents. Predefined primitives are not declared, because their definition is specified by the language and is not subject to change by the user. Primitives are used (i.e., instantiated), just as gates are used to populate a printed circuit board. We'll see that once a module has been declared, it may be used (instantiated) within a design. Note that *Simple_Circuit* is not a computational model like those developed in an ordinary programming language: The sequential ordering of the statements instantiating gates in the model has no significance and does not specify a sequence of computations. A Verilog model is a *descriptive* model. *Simple_Circuit* describes what primitives form a circuit and how they are connected. The input–output behavior of the circuit is

Table 3.5
Output of Gates after Delay

	Time Units (ns)	Input ABC	Output E w1 D
Initial	—	000	1 0 1
Change	—	111	1 0 1
	10	111	0 0 1
	20	111	0 0 1
	30	111	0 1 0
	40	111	0 1 0
	50	111	0 1 1

implicitly specified by the description because the behavior of each logic gate is defined. Thus, an HDL-based model can be used to simulate the circuit that it represents.

Gate Delays

All physical circuits exhibit a propagation delay between the transition of an input and a resulting transition of an output. When an HDL model of a circuit is simulated, it is sometimes necessary to specify the amount of delay from the input to the output of its gates. In Verilog, the propagation delay of a gate is specified in terms of *time units* and by the symbol #. The numbers associated with time delays in Verilog are dimensionless. The association of a time unit with physical time is made with the '**timescale** compiler directive. (Compiler directives start with the (') back quote, or grave accent, symbol.) Such a directive is specified before the declaration of a module and applies to all numerical values of time in the code that follows. An example of a timescale directive is

```
` timescale 1ns/100ps
```

The first number specifies the unit of measurement for time delays. The second number specifies the precision for which the delays are rounded off, in this case to 0.1 ns. If no timescale is specified, a simulator may display dimensionless values or default to a certain time unit, usually 1 ns $(=10^{-9}\text{ s})$. Our examples will use only the default time unit.

HDL Example 3.2 repeats the description of the simple circuit of Example 3.1, but with propagation delays specified for each gate. The **and**, **or**, and **not** gates have a time delay of 30, 20, and 10 ns, respectively. If the circuit is simulated and the inputs change from $A, B, C = 0$ to $A, B, C = 1$, the outputs change as shown in Table 3.5 (calculated by hand or generated by a simulator). The output of the inverter at E changes from 1 to 0 after a 10-ns delay. The output of the AND gate at $w1$ changes from 0 to 1 after a 30-ns delay. The output of the OR gate at D changes from 1 to 0 at $t = 30$ ns and then changes back to 1 at $t = 50$ ns. In both cases, the change in the output of the OR gate results from a change in its inputs 20 ns earlier. It is clear from this result that although output D eventually returns to a final value of 1 after the input changes, the gate delays produce a negative spike that lasts 20 ns before the final value is reached.

HDL Example 3.2 (Gate-Level Model with Propagation Delays)

```
// Verilog model of simple circuit with propagation delay

module Simple_Circuit_prop_delay (A, B, C, D, E);
 output D, E;
 input  A, B, C;
 wire   w1;

 and          #(30) G1 (w1, A, B);
 not          #(10) G2 (E, C);
 or           #(20) G3 (D, w1, E);
endmodule
```

In order to simulate a circuit with an HDL, it is necessary to apply inputs to the circuit so that the simulator will generate an output response. An HDL description that provides the stimulus to a design is called a *test bench.* The writing of test benches is explained in more detail at the end of Section 4.12. Here, we demonstrate the procedure with a simple example without dwelling on too many details. HDL Example 3.3 shows a test bench for simulating the circuit with delay. (Note the distinguishing name *Simple_Circuit_prop_delay*.) In its simplest form, a test bench is a module containing a signal generator and an instantiation of the model that is to be verified. Note that the test bench (*t_Simple_Circuit_prop_delay*) has no input or output ports, because it does not interact with its environment. In general, we prefer to name the test bench with the prefix *t_* concatenated with the name of the module that is to be tested by the test bench, but that choice is left to the designer. Within the test bench, the inputs to the circuit are declared with keyword **reg** and the outputs are declared with the keyword **wire**. The module *Simple_Circuit_prop_delay* is instantiated with the instance name M1. Every instantiation of a module must include a unique instance name. Note that using a test bench is similar to testing actual hardware by attaching signal generators to the inputs of a circuit and attaching

HDL Example 3.3 (Test Bench)

```
// Test bench for Simple_Circuit_prop_delay

module t_Simple_Circuit_prop_delay;
 wire   D, E;
 reg    A, B, C;

Simple_Circuit_prop_delay M1 (A, B, C, D, E); // Instance name required

initial
  begin
    A = 1'b0; B = 1'b0; C = 1'b0;
    #100 A = 1'b1; B = 1'b1; C = 1'b1;
  end

  initial #200 $finish;
endmodule
```

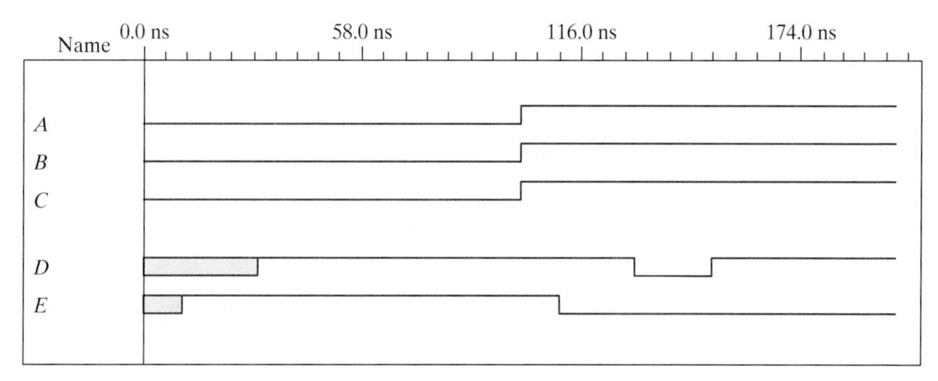

FIGURE 3.36
Simulation output of HDL Example 3.3

probes (wires) to the outputs of the circuit. (The interaction between the signal generators of the stimulus module and the instantiated circuit module is illustrated in Fig. 4.36.)

Hardware signal generators are not used to verify an HDL model: The entire simulation exercise is done with software models executing on a digital computer under the direction of an HDL simulator. The waveforms of the input signals are abstractly modeled (generated) by Verilog statements specifying waveform values and transitions. The **initial** keyword is used with a set of statements that begin executing when the simulation is initialized; the signal activity associated with **initial** terminates execution when the last statement has finished executing. The **initial** statements are commonly used to describe waveforms in a test bench. The set of statements to be executed is called a *block statement* and consists of several statements enclosed by the keywords **begin** and **end**. The action specified by the statements begins when the simulation is launched, and the statements are executed in sequence, left to right, from top to bottom, by a simulator in order to provide the input to the circuit. Initially, $A, B, C = 0$. (A, B, and C are each set to 1'b0, which signifies one binary digit with a value of 0.) After 100 ns, the inputs change to $A, B, C = 1$. After another 100 ns, the simulation terminates at time 200 ns. A second **initial** statement uses the **$finish** system task to specify termination of the simulation. If a statement is preceded by a delay value (e.g., #100), the simulator postpones executing the statement until the specified time delay has elapsed. The timing diagram of waveforms that result from the simulation is shown in Figure 3.36. The total simulation generates waveforms over an interval of 200 ns. The inputs A, B, and C change from 0 to 1 after 100 ns. Output E is unknown for the first 10 ns (denoted by shading), and output D is unknown for the first 30 ns. Output E goes from 1 to 0 at 110 ns. Output D goes from 1 to 0 at 130 ns and back to 1 at 150 ns, just as we predicted in Table 3.5.

Boolean Expressions

Boolean equations describing combinational logic are specified in Verilog with a continuous assignment statement consisting of the keyword **assign** followed by a Boolean expression. To distinguish arithmetic operators from logical operators, Verilog uses the symbols (&), (/), and (~) for AND, OR, and NOT (complement), respectively. Thus, to describe the simple circuit of Fig. 3.35 with a Boolean expression, we use the statement

```
assign D = (A && B)||(!C);
```

HDL Example 3.4 describes a circuit that is specified with the following two Boolean expressions:

$$E = A + BC + B'D$$
$$F = B'C + BC'D'$$

The equations specify how the logic values E and F are determined by the values of A, B, C, and D.

HDL Example 3.4 (Combinational Logic Modeled with Boolean Equations)

```
// Verilog model: Circuit with Boolean expressions

module Circuit_Boolean_CA (E, F, A, B, C, D);
  output      E, F;
  input       A, B, C, D;

  assign E = A || (B && C) || ((!B) && D);
  assign F = ((!B) && C) || (B && (!C) && (!D));
endmodule
```

The circuit has two outputs E and F and four inputs A, B, C, and D. The two **assign** statements describe the Boolean equations. The values of E and F during simulation are determined dynamically by the values of A, B, C, and D. The simulator detects when the test bench changes a value of one or more of the inputs. When this happens, the simulator updates the values of E and F. The continuous assignment mechanism is so named because the relationship between the assigned value and the variables is permanent. The mechanism acts just like combinational logic, has a gate-level equivalent circuit, and is referred to as *implicit combinational logic*.

We have shown that a digital circuit can be described with HDL statements, just as it can be drawn in a circuit diagram or specified with a Boolean expression. A third alternative is to describe combinational logic with a truth table.

User-Defined Primitives

The logic gates used in Verilog descriptions with keywords **and, or,** etc., are defined by the system and are referred to as *system primitives. (Caution: Other languages may use these words differently.)* The user can create additional primitives by defining them in tabular form. These types of circuits are referred to as *user-defined primitives* (UDPs). One way of specifying a digital circuit in tabular form is by means of a truth table. UDP descriptions do not use the keyword pair **module . . . endmodule.** Instead, they are declared with the keyword pair **primitive . . . endprimitive.** The best way to demonstrate a UDP declaration is by means of an example.

HDL Example 3.5 defines a UDP with a truth table. It proceeds according to the following general rules:

- It is declared with the keyword **primitive**, followed by a name and port list.
- There can be only one output, and it must be listed first in the port list and declared with keyword **output**.
- There can be any number of inputs. The order in which they are listed in the **input** declaration must conform to the order in which they are given values in the table that follows.
- The truth table is enclosed within the keywords **table** and **endtable**.
- The values of the inputs are listed in order, ending with a colon (:). The output is always the last entry in a row and is followed by a semicolon (;).
- The declaration of a UDP ends with the keyword **endprimitive**.

HDL Example 3.5 (User-Defined Primitive)

```
// Verilog model: User-defined Primitive
primitive UDP_02467 (D, A, B, C);
  output D;
  input   A, B, C;
//Truth table for D 5 f (A, B, C) 5 Σ(0, 2, 4, 6, 7);
  table
//      A     B     C     :     D       // Column header comment
        0     0     0     :     1;
        0     0     1     :     0;
        0     1     0     :     1;
        0     1     1     :     0;
        1     0     0     :     1;
        1     0     1     :     0;
        1     1     0     :     1;
        1     1     1     :     1;
  endtable
endprimitive

// Instantiate primitive

// Verilog model: Circuit instantiation of Circuit_UDP_02467

module Circuit_with_UDP_02467 (e, f, a, b, c, d);
  output        e, f;
  input         a, b, c, d

  UDP_02467             (e, a, b, c);
  and                   (f, e, d);          // Option gate instance name omitted
endmodule
```

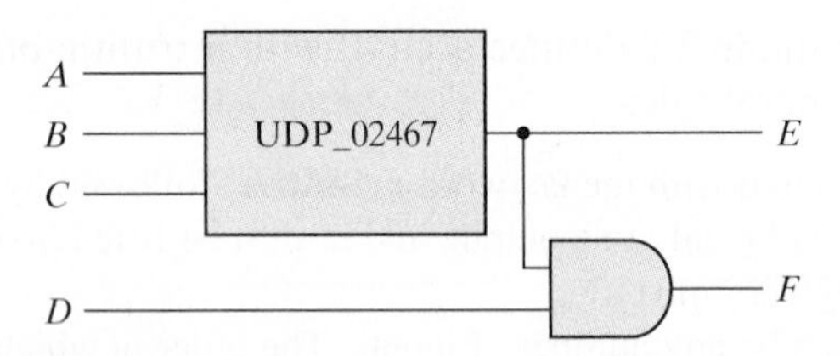

FIGURE 3.37
Schematic for *Circuit with_UDP_02467*

Note that the variables listed on top of the table are part of a comment and are shown only for clarity. The system recognizes the variables by the order in which they are listed in the input declaration. A user-defined primitive can be instantiated in the construction of other modules (digital circuits), just as the system primitives are used. For example, the declaration

```
Circuit_with_UDP_02467 (E, F, A, B, C, D);
```

will produce a circuit that implements the hardware shown in Figure 3.37.

Although Verilog HDL uses this kind of description for UDPs only, other HDLs and computer-aided design (CAD) systems use other procedures to specify digital circuits in tabular form. The tables can be processed by CAD software to derive an efficient gate structure of the design. None of Verilog's predefined primitives describes sequential logic. The model of a sequential UDP requires that its output be declared as a **reg** data type, and that a column be added to the truth table to describe the next state. So the columns are organized as inputs : state : next state.

In this section, we introduced the Verilog HDL and presented simple examples to illustrate alternatives for modeling combinational logic. A more detailed presentation of Verilog HDL can be found in the next chapter. The reader familiar with combinational circuits can go directly to Section 4.12 to continue with this subject.

PROBLEMS

(Answers to problems marked with * appear at the end of the text.)

3.1* Simplify the following Boolean functions, using three-variable maps:
(a) $F(x, y, z) = \Sigma(0, 2, 4, 6)$ (b) $F(x, y, z) = \Sigma(0, 2, 4, 5, 6, 7)$
(c) $F(x, y, z) = \Sigma(0, 1, 2, 3, 4, 6)$ (d) $F(x, y, z) = \Sigma(1, 4, 5, 7)$

3.2 Simplify the following Boolean functions, using three-variable maps:
(a)* $F(x, y, z) = \Sigma(0, 1, 4, 7)$ (b)* $F(x, y, z) = \Sigma(1, 2, 3, 6, 7)$
(c) $F(x, y, z) = \Sigma(2, 3, 5, 6)$ (d) $F(x, y, z) = \Sigma(1, 2, 4, 7)$
(e) $F(x, y, z) = \Sigma(0, 2, 4, 6)$ (f) $F(x, y, z) = \Sigma(3, 4, 5, 6, 7)$

3.3* Simplify the following Boolean expressions, using three-variable maps:
(a)* $F(x, y, z) = xyz + x'y + xyz'$ (b)* $F(x, y, z) = x'yz + xyz' + xyz + x'yz' + xy'z'$
(c)* $F(x, y, z) = x'yz + xz$ (d) $F(x, y, z) = xyz + x'y + xyz' + x'y'z'$

3.4 Simplify the following Boolean functions, using *Karnaugh* maps:

(a)* $F(x, y, z) = \Sigma(0, 1, 4, 5)$ (b)* $F(A, B, C) = \Sigma(0, 2, 3, 7)$

(c)* $F(A, B, C, D) = \Sigma(1, 5, 9, 12, 13, 15)$ (d)* $F(w, x, y, z) = \Sigma(0, 2, 3, 8, 10, 11)$

(e) $F(w, x, y, z) = \Sigma(11, 12, 13, 14, 15)$ (f) $F(w, x, y, z) = \Sigma(8, 10, 12, 13, 14)$

3.5 Simplify the following Boolean functions, using four-variable maps:

(a)* $F(w, x, y, z) = \Sigma(0, 4, 6, 8, 14, 15)$

(b) $F(A, B, C, D) = \Sigma(2, 3, 6, 7, 12, 13, 14)$

(c) $F(w, x, y, z) = \Sigma(1, 3, 4, 5, 6, 7, 9, 11, 13, 15)$

(d)* $F(A, B, C, D) = \Sigma(0, 2, 4, 6, 8, 10, 12, 14)$

3.6 Simplify the following Boolean expressions, using four-variable maps:

(a)* $B'D'(A'C' + C) + AC'D' + BD(A'C + C')$

(b)* $x'z + w'xy' + w(x'y + xy')$

(c) $A'BCD + ABC + CD + B'D$

(d) $A'B'C'D' + BC'D + A'C'D + A'BCD + ACD'$

3.7 Simplify the following Boolean expressions, using four-variable maps:

(a)* $w'z + xz + x'y + wx'z$

(b) $ACD' + B'C'D + BCD + BC'$

(c)* $AB'C + B'C' + A'BCD + ACD' + AB'C' + A'C'D$

(d) $wxy + xz + w'xz + y'z + wy'$

3.8 Find the minterms of the following Boolean expressions by first plotting each function in a map:

(a)* $wxy + yz + xy'z + wz'$ (b)* $AC'D + BC'D + ACD' + A'B'D + A'D'$

(c) $wyz + w'x' + wx'z' + x'z'$ (d) $A'B + A'CD + B'CD + BC'D'$

3.9 Find all the prime implicants for the following Boolean functions, and determine which are essential:

(a)* $F(w, x, y, z) = \Sigma(0, 2, 4, 6, 8, 10, 12, 14)$

(b)* $F(A, B, C, D) = \Sigma(0, 2, 3, 5, 7, 8, 10, 11, 14, 15)$

(c) $F(A, B, C, D) = \Sigma(2, 3, 4, 5, 6, 7, 9, 11, 12, 13)$

(d) $F(w, x, y, z) = \Sigma(1, 3, 5, 7, 9, 11, 13, 15)$

(e) $F(A, B, C, D) = \Sigma(0, 1, 2, 5, 7, 8, 9, 10, 13, 15)$

(f) $F(w, x, y, z) = \Sigma(1, 3, 4, 6, 7, 9, 10, 12, 13, 15)$

3.10 Simplify the following Boolean functions by first finding the essential prime implicants:

(a) $F(w, x, y, z) = \Sigma(0, 2, 5, 7, 8, 10, 13, 14, 15)$

(b) $F(A, B, C, D) = \Sigma(0, 2, 3, 5, 7, 8, 10, 11, 14, 15)$

(c)* $F(A, B, C, D) = \Sigma(2, 3, 6, 7, 10, 11, 14, 15)$

(d) $F(w, x, y, z) = \Sigma(0, 1, 4, 5, 6, 7, 9, 11, 14, 15)$

(e) $F(A, B, C, D) = \Sigma(0, 1, 3, 7, 8, 9, 10, 13, 15)$

(f) $F(w, x, y, z) = \Sigma(0, 1, 2, 4, 5, 6, 7, 10, 12, 15)$

3.11 Convert the following Boolean function from a sum-of-products form to a simplified product-of-sums form.

$$F(w, x, y, z) = \Sigma(0, 1, 3, 5, 7, 9, 10, 13, 15)$$

3.12 Simplify the following Boolean functions:

(a)* $F(A, B, C, D) = \Pi(0, 2, 4, 6, 8, 10, 12, 14)$

(b) $F(A, B, C, D) = \Pi(1, 3, 5, 7, 9, 11, 13, 15)$

3.13 Simplify the following expressions to (1) sum-of-products and (2) products-of-sums:

(a)* $xz' + y'z' + yz' + xy'$

(b) $AC'D' + C'D + AB' + AB'CD$

(c) $(A' + B + D')(A' + B' + C')(A' + B' + C)(B' + C + D')$

(d) $BD' + AB'CD + AC'D + BC'$

3.14 Give three possible ways to express the following Boolean function with eight or fewer literals:

$$F(A, B, C, D) = A'B'D' + AB'D' + BD + ABCD'$$

3.15 Simplify the following Boolean function F, together with the don't-care conditions d, and then express the simplified function in sum-of-minterms form:

(a) $F(x, y, z) = \Sigma(0, 1, 3, 5, 7)$
$d(x, y, z) = \Sigma(2, 4, 6)$

(b)* $F(A, B, C, D) = \Sigma(0, 4, 8, 10, 14)$
$d(A, B, C, D) = \Sigma(2, 6, 12)$

(c) $F(A, B, C, D) = \Sigma(5, 6, 7, 11, 14, 15)$
$d(A, B, C, D) = \Sigma(3, 9, 13)$

(d) $F(A, B, C, D) = \Sigma(2, 4, 7, 10, 12, 14)$
$d(A, B, C, D) = \Sigma(0, 3, 6, 8, 13)$

3.16 Simplify the following functions, and implement them with two-level NAND gate circuits:

(a) $F(A, B, C, D) = AD + BC'D' + ABC + A'BC'D$

(b) $F(A, B, C, D) = A'B'C'D + CD' + AC'D$

(c) $F(A, B, C, D) = (A' + C' + D')(A' + C')(C' + D')$

(d) $F(A, B, C, D) = A' + AB + B'C + ACD$

3.17* Draw a NAND and a NOR logic diagram that implements the complement of the following function:

$$F(A, B, C, D) = \Sigma(4, 5, 6, 7, 9, 13, 15)$$

3.18 Draw a logic diagram using only two-input NAND gates to implement the following function:

$$F(A, B, C, D) = (A \oplus B)'(C \oplus D)$$

3.19 Simplify the following functions, and implement them with two-level NOR gate circuits:

(a)* $F(w, x, y, z) = wx'y + xy'z' + w'yz' + xy$

(b) $F(w, x, y, z) = \Sigma(4, 7, 11, 12, 15)$

(c) $F(x, y, z) = [(x + y)(x + z)]'$

3.20 Draw the multiple-level NOR circuit for the following expression:

$$F = BC(D + C)A + (BC' + DE') + BD'$$

3.21 Draw the multiple-level NAND circuit for the following expression:

$$F = w(x' + y' + z) + xy'z$$

3.22 Convert the logic diagram of the circuit shown in Fig. 4.4 into a multiple-level NOR circuit.

3.23 Implement the following Boolean function F, together with the don't-care conditions d, using no more than two NOR gates:

$$F(A, B, C, D) = \Sigma(2, 4, 10, 12, 14)$$
$$d(A, B, C, D) = \Sigma(1, 5, 8, 10)$$

Assume that both the normal and complement inputs are available.

3.24 Implement the following Boolean function F, using the two-level forms of logic (a) AND-OR, (b) ORNAND, (c) NOR-OR, (d) NAND-NAND, (e) OR-AND, (f) NOR-NOR, and (g) NAND-AND:

$$F(A, B, C, D) = \Sigma(1, 5, 8, 9, 10, 11, 12, 13, 15)$$

3.25 List the eight degenerate two-level forms and show that they reduce to a single operation. Explain how the degenerate two-level forms can be used to extend the number of inputs to a gate.

3.26 With the use of maps, find the simplest sum-of-products form of the function $F = fg$, where

$$f = abc' + b'd' + a'd' + b'cd'$$

and

$$g = (a + b + c' + d')(a' + b' + d)(a' + d')$$

3.27 Show that the dual of the exclusive-NOR is also its complement.

3.28 Derive the circuits for a four-bit parity generator and three-bit parity checker using an odd parity bit.

3.29 Implement the following four Boolean expressions with three half adders:

$$D = (A \oplus B) \oplus (C)$$
$$E = (A \oplus B) \oplus (AB)$$
$$F = (A \oplus B)(AB)$$
$$G = (A \oplus B)C$$

3.30* Implement the following Boolean expression with exclusive-NOR and AND gates:

$$F = A'B'C'D' + ABC'D' + A'B'CD + ABCD$$

3.31 Write a Verilog gate-level description of the circuit shown in
(a) Fig. 3.20(a) (b) Fig. 3.20(b) (c) Fig. 3.21(a)
(d) Fig. 3.21(b) (e) Fig. 3.24 (f) Fig. 3.25

3.32 Using continuous assignment statements, write a Verilog description of the circuit shown in
(a) Fig. 3.20(a) (b) Fig. 3.20(b) (c) Fig. 3.21(a)
(d) Fig. 3.21(b) (e) Fig. 3.24 (f) Fig. 3.25

3.33 The exclusive-OR circuit of Fig. 3.30(a) has gates with a delay of 3 ns for an inverter, a 6 ns delay for an AND gate, and a 8 ns delay for an OR gate. The input of the circuit goes from $xy = 00$ to $xy = 01$.
(a) Determine the signals at the output of each gate from $t = 0$ to $t = 50$ ns.
(b) Write a Verilog gate-level description of the circuit, including the delays.
(c) Write a stimulus module (i.e., a test bench similar to HDL Example 3.3), and simulate the circuit to verify the answer in part (a).

3.34 Using continuous assignments, write a Verilog description of the circuit specified by the following Boolean functions:

$$Out\_1 = (A + B')C'(C + D)$$
$$Out\_2 = (C'D + BCD + CD')(A' + B)$$
$$Out\_3 = (AB + C)D + B'C$$

Write a test bench and simulate the circuit's behavior.

3.35* Find the syntax errors in the following declarations (note that names for primitive gates are optional):

```
module Exmpl-3(A, B, C, D, F)          // Line 1
  inputs      A, B, C, Output D, F,    // Line 2
  output      B                        // Line 3
  and         g1(A, B, D);             // Line 4
  not         (D, A, C),               // Line 5
  OR          (F, B; C);               // Line 6
endmodule;                             // Line 7
```

3.36 Draw the logic diagram of the digital circuit specified by the following Verilog description:

(a)
```
module Circuit_A (A, B, C, D, F);
  input     A, B, C, D;
  output    F;
  wire      w, x, y, z, a, d;
  or        (x, B, C, d);
  and       (y, a ,C);
  and       (w, z ,B);
  and       (z, y, A);
  or        (F, x, w);
  not       (a, A);
  not       (d, D);
endmodule
```

(b)
```
module Circuit_B (F1, F2, F3, A0, A1, B0, B1);
  output    F1, F2, F3;
  input     A0, A1, B0, B1;
  nor       (F1, F2, F3);
  or        (F2, w1, w2, w3);
  and       (F3, w4, w5);
  and       (w1, w6, B1);
  or        (w2, w6, w7, B0);
  and       (w3, w7, B0, B1);
  not       (w6, A1);
  not       (w7, A0);
  xor       (w4, A1, B1);
  xnor      (w5, A0, B0);
endmodule
```

(c)
```
module Circuit_C (y1, y2, y3, a, b);
  output y1, y2, y3;
  input a, b;

  assign y1 = a || b;
  and (y2, a, b);
  assign y3 = a && b;
endmodule
```

3.37 A majority logic function is a Boolean function that is equal to 1 if the majority of the variables are equal to 1, equal to 0 otherwise.
(a) Write a truth table for a four-bit majority function.
(b) Write a Verilog user-defined primitive for a four-bit majority function.

3.38 Simulate the behavior of *Circuit_with_UDP_02467*, using the stimulus waveforms shown in Fig. P3.38.

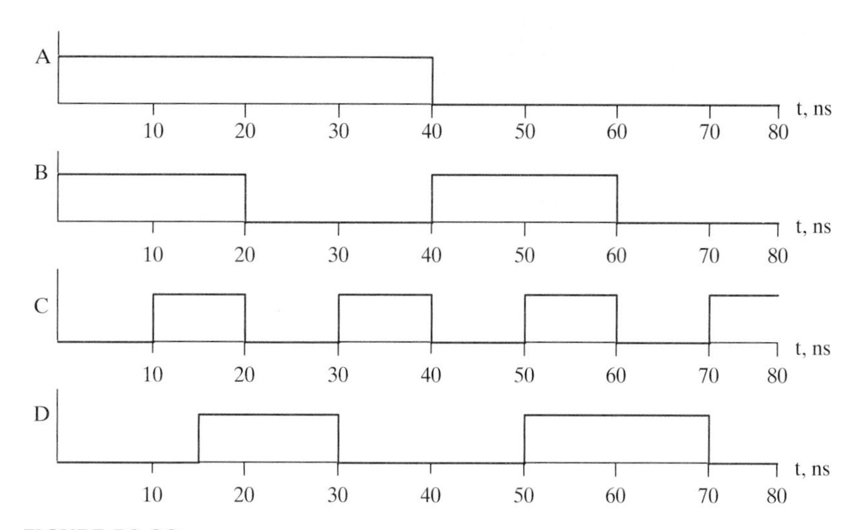

FIGURE P3.38
Stimulus waveforms for Problem 3.38

3.39 Using primitive gates, write a Verilog model of a circuit that will produce two outputs, s and c, equal to the sum and carry produced by adding two binary input bits a and b (e.g., $s = 1$ and $c = 0$ if $a = 0$ and $b = 1$). (*Hint:* Begin by developing a truth table for s and c.)

REFERENCES

1. BHASKER, J. 1997. *A Verilog HDL Primer*. Allentown, PA: Star Galaxy Press.
2. CILETTI, M. D. 1999. *Modeling, Synthesis and Rapid Prototyping with the Verilog HDL*. Upper Saddle River, NJ: Prentice Hall.
3. HILL, F. J., and G. R. PETERSON. 1981. *Introduction to Switching Theory and Logical Design*, 3rd ed. New York: John Wiley.
4. *IEEE Standard Hardware Description Language Based on the Verilog Hardware Description Language* (IEEE Std. 1364-1995). 1995. New York: The Institute of Electrical and Electronics Engineers.
5. KARNAUGH, M. A Map Method for Synthesis of Combinational Logic Circuits. *Transactions of AIEE, Communication and Electronics*. 72, part I (Nov. 1953): 593–99.
6. KOHAVI, Z. 1978. *Switching and Automata Theory*, 2nd ed. New York: McGraw-Hill.

7. Mano, M. M. and C. R. Kime. 2004. *Logic and Computer Design Fundamentals*, 3rd ed. Upper Saddle River, NJ: Prentice Hall.
8. McCluskey, E. J. 1986. *Logic Design Principles*. Englewood Cliffs, NJ: Prentice-Hall.
9. Palnitkar, S. 1996. *Verilog HDL: A Guide to Digital Design and Synthesis*. Mountain View, CA: SunSoft Press (a Prentice Hall title).

WEB SEARCH TOPICS

Boolean minimization
Karnaugh map
Wired logic
Emitter-coupled logic
Open-collector logic
Quine McCluskey method
Expresso software
Consensus theorem
Don't-care conditions

Chapter 4

Combinational Logic

4.1 INTRODUCTION

Logic circuits for digital systems may be combinational or sequential. A combinational circuit consists of logic gates whose outputs at any time are determined from only the present combination of inputs. A combinational circuit performs an operation that can be specified logically by a set of Boolean functions. In contrast, sequential circuits employ storage elements in addition to logic gates. Their outputs are a function of the inputs and the state of the storage elements. Because the state of the storage elements is a function of previous inputs, the outputs of a sequential circuit depend not only on present values of inputs, but also on past inputs, and the circuit behavior must be specified by a time sequence of inputs and internal states. Sequential circuits are the building blocks of digital systems and are discussed in Chapters 5 and 8.

4.2 COMBINATIONAL CIRCUITS

A combinational circuit consists of an interconnection of logic gates. Combinational logic gates react to the values of the signals at their inputs and produce the value of the output signal, transforming binary information from the given input data to a required output data. A block diagram of a combinational circuit is shown in Fig. 4.1. The n input binary variables come from an external source; the m output variables are produced by the internal combinational logic circuit and go to an external destination. Each input and output variable exists physically as an analog signal whose values are interpreted to be a binary signal that represents logic 1 and logic 0. (*Note*: Logic simulators show only 0's and 1's, not the actual analog signals.) In many applications, the source and

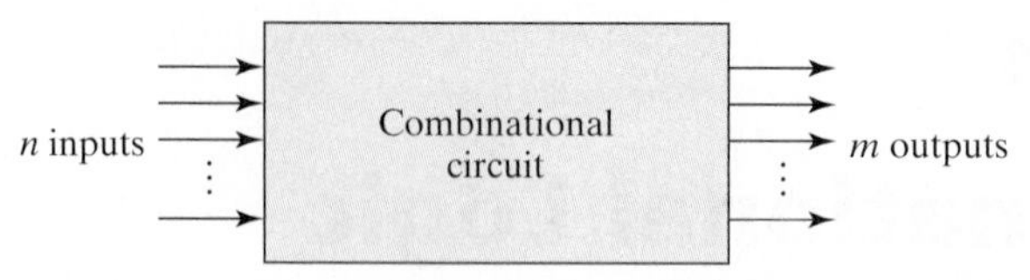

FIGURE 4.1
Block diagram of combinational circuit

destination are storage registers. If the registers are included with the combinational gates, then the total circuit must be considered to be a sequential circuit.

For n input variables, there are 2^n possible combinations of the binary inputs. For each possible input combination, there is one possible value for each output variable. Thus, a combinational circuit can be specified with a truth table that lists the output values for each combination of input variables. A combinational circuit also can be described by m Boolean functions, one for each output variable. Each output function is expressed in terms of the n input variables.

In Chapter 1, we learned about binary numbers and binary codes that represent discrete quantities of information. The binary variables are represented physically by electric voltages or some other type of signal. The signals can be manipulated in digital logic gates to perform required functions. In Chapter 2, we introduced Boolean algebra as a way to express logic functions algebraically. In Chapter 3, we learned how to simplify Boolean functions to achieve economical (simpler) gate implementations. The purpose of the current chapter is to use the knowledge acquired in previous chapters to formulate systematic analysis and design procedures for combinational circuits. The solution of some typical examples will provide a useful catalog of elementary functions that are important for the understanding of digital systems. We'll address three tasks: (1) Analyze the behavior of a given logic circuit, (2) synthesize a circuit that will have a given behavior, and (3) write hardware description language (HDL) models for some common circuits.

There are several combinational circuits that are employed extensively in the design of digital systems. These circuits are available in integrated circuits and are classified as standard components. They perform specific digital functions commonly needed in the design of digital systems. In this chapter, we introduce the most important standard combinational circuits, such as adders, subtractors, comparators, decoders, encoders, and multiplexers. These components are available in integrated circuits as medium-scale integration (MSI) circuits. They are also used as *standard cells* in complex very large-scale integrated (VLSI) circuits such as application-specific integrated circuits (ASICs). The standard cell functions are interconnected within the VLSI circuit in the same way that they are used in multiple-IC MSI design.

4.3 ANALYSIS PROCEDURE

The analysis of a combinational circuit requires that we determine the function that the circuit implements. This task starts with a given logic diagram and culminates with a set of Boolean functions, a truth table, or, possibly, an explanation of the circuit operation.

If the logic diagram to be analyzed is accompanied by a function name or an explanation of what it is assumed to accomplish, then the analysis problem reduces to a verification of the stated function. The analysis can be performed manually by finding the Boolean functions or truth table or by using a computer simulation program.

The first step in the analysis is to make sure that the given circuit is combinational and not sequential. **The diagram of a combinational circuit has logic gates with no feedback paths** or **memory elements.** A feedback path is a connection from the output of one gate to the input of a second gate whose output forms part of the input to the first gate. Feedback paths in a digital circuit define a sequential circuit and must be analyzed by special methods and will not be considered here.

Once the logic diagram is verified to be that of a combinational circuit, one can proceed to obtain the output Boolean functions or the truth table. If the function of the circuit is under investigation, then it is necessary to interpret the operation of the circuit from the derived Boolean functions or truth table. The success of such an investigation is enhanced if one has previous experience and familiarity with a wide variety of digital circuits.

To obtain the output Boolean functions from a logic diagram, we proceed as follows:

1. Label all gate outputs that are a function of input variables with arbitrary symbols—but with meaningful names. Determine the Boolean functions for each gate output.
2. Label the gates that are a function of input variables and previously labeled gates with other arbitrary symbols. Find the Boolean functions for these gates.
3. Repeat the process outlined in step 2 until the outputs of the circuit are obtained.
4. By repeated substitution of previously defined functions, obtain the output Boolean functions in terms of input variables.

The analysis of the combinational circuit of Fig. 4.2 illustrates the proposed procedure. We note that the circuit has three binary inputs—A, B, and C—and two binary outputs—F_1 and F_2. The outputs of various gates are labeled with intermediate symbols. The outputs of gates that are a function only of input variables are T_1 and T_2. Output F_2 can easily be derived from the input variables. The Boolean functions for these three outputs are

$$F_2 = AB + AC + BC$$
$$T_1 = A + B + C$$
$$T_2 = ABC$$

Next, we consider outputs of gates that are a function of already defined symbols:

$$T_3 = F'_2T_1$$
$$F_1 = T_3 + T_2$$

To obtain F_1 as a function of A, B, and C, we form a series of substitutions as follows:

$$\begin{aligned}F_1 &= T_3 + T_2 = F'_2T_1 + ABC = (AB + AC + BC)'(A + B + C) + ABC \\ &= (A' + B')(A' + C')(B' + C')(A + B + C) + ABC \\ &= (A' + B'C')(AB' + AC' + BC' + B'C) + ABC \\ &= A'BC' + A'B'C + AB'C' + ABC\end{aligned}$$

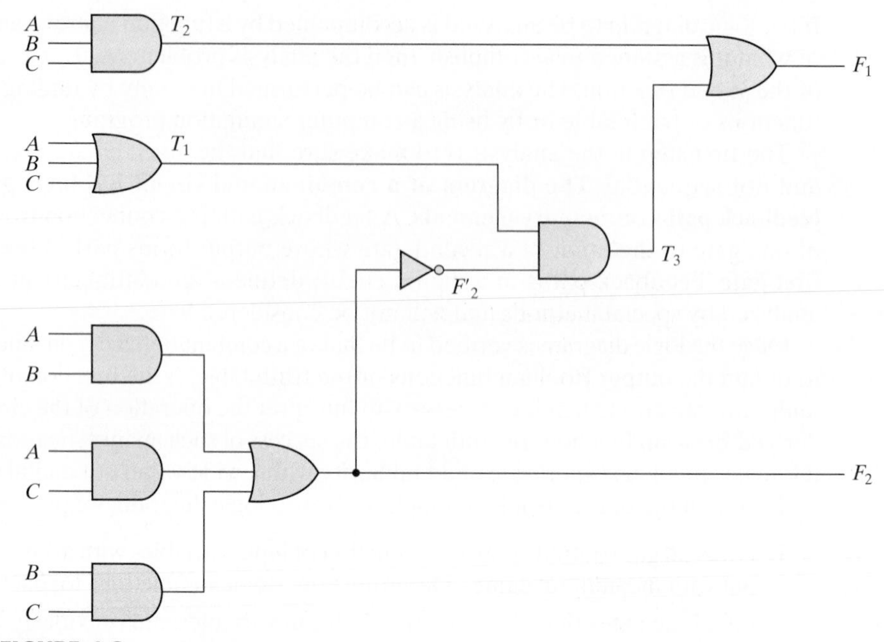

FIGURE 4.2
Logic diagram for analysis example

If we want to pursue the investigation and determine the information transformation task achieved by this circuit, we can draw the circuit from the derived Boolean expressions and try to recognize a familiar operation. The Boolean functions for F_1 and F_2 implement a circuit discussed in Section 4.5. Merely finding a Boolean representation of a circuit doesn't provide insight into its behavior, but in this example we will observe that the Boolean equations and truth table for F_1 and F_2 match those describing the functionality of what we call a full adder.

The derivation of the truth table for a circuit is a straightforward process once the output Boolean functions are known. To obtain the truth table directly from the logic diagram without going through the derivations of the Boolean functions, we proceed as follows:

1. Determine the number of input variables in the circuit. For n inputs, form the 2^n possible input combinations and list the binary numbers from 0 to $(2^n - 1)$ in a table.
2. Label the outputs of selected gates with arbitrary symbols.
3. Obtain the truth table for the outputs of those gates which are a function of the input variables only.
4. Proceed to obtain the truth table for the outputs of those gates which are a function of previously defined values until the columns for all outputs are determined.

Table 4.1
Truth Table for the Logic Diagram of Fig. 4.2

A	B	C	F_2	F'_2	T_1	T_2	T_3	F_1
0	0	0	0	1	0	0	0	0
0	0	1	0	1	1	0	1	1
0	1	0	0	1	1	0	1	1
0	1	1	1	0	1	0	0	0
1	0	0	0	1	1	0	1	1
1	0	1	1	0	1	0	0	0
1	1	0	1	0	1	0	0	0
1	1	1	1	0	1	1	0	1

This process is illustrated with the circuit of Fig. 4.2. In Table 4.1, we form the eight possible combinations for the three input variables. The truth table for F_2 is determined directly from the values of A, B, and C, with F_2 equal to 1 for any combination that has two or three inputs equal to 1. The truth table for F'_2 is the complement of F_2. The truth tables for T_1 and T_2 are the OR and AND functions of the input variables, respectively. The values for T_3 are derived from T_1 and F'_2:T_3 is equal to 1 when both T_1 and F'_2 are equal to 1, and T_3 is equal to 0 otherwise. Finally, F_1 is equal to 1 for those combinations in which either T_2 or T_3 or both are equal to 1. Inspection of the truth table combinations for A, B, C, F_1, and F_2 shows that it is identical to the truth table of the full adder given in Section 4.5 for x, y, z, S, and C, respectively.

Another way of analyzing a combinational circuit is by means of logic simulation. This is not practical, however, because the number of input patterns that might be needed to generate meaningful outputs could be very large. But simulation has a very practical application in verifying that the functionality of a circuit actually matches its specification. In Section 4.12, we demonstrate the logic simulation and verification of the circuit of Fig. 4.2, using Verilog HDL.

4.4 DESIGN PROCEDURE

The design of combinational circuits starts from the specification of the design objective and culminates in a logic circuit diagram or a set of Boolean functions from which the logic diagram can be obtained. The procedure involves the following steps:

1. From the specifications of the circuit, determine the required number of inputs and outputs and assign a symbol to each.
2. Derive the truth table that defines the required relationship between inputs and outputs.

3. Obtain the simplified Boolean functions for each output as a function of the input variables.
4. Draw the logic diagram and verify the correctness of the design (manually or by simulation).

A truth table for a combinational circuit consists of input columns and output columns. The input columns are obtained from the 2^n binary numbers for the n input variables. The binary values for the outputs are determined from the stated specifications. The output functions specified in the truth table give the exact definition of the combinational circuit. It is important that the verbal specifications be interpreted correctly in the truth table, as they are often incomplete, and any wrong interpretation may result in an incorrect truth table.

The output binary functions listed in the truth table are simplified by any available method, such as algebraic manipulation, the map method, or a computer-based simplification program. Frequently, there is a variety of simplified expressions from which to choose. In a particular application, certain criteria will serve as a guide in the process of choosing an implementation. A practical design must consider such constraints as the number of gates, number of inputs to a gate, propagation time of the signal through the gates, number of interconnections, limitations of the driving capability of each gate (i.e., the number of gates to which the output of the circuit may be connected), and various other criteria that must be taken into consideration when designing integrated circuits. Since the importance of each constraint is dictated by the particular application, it is difficult to make a general statement about what constitutes an acceptable implementation. In most cases, the simplification begins by satisfying an elementary objective, such as producing the simplified Boolean functions in a standard form. Then the simplification proceeds with further steps to meet other performance criteria.

Code Conversion Example

The availability of a large variety of codes for the same discrete elements of information results in the use of different codes by different digital systems. It is sometimes necessary to use the output of one system as the input to another. A conversion circuit must be inserted between the two systems if each uses different codes for the same information. Thus, a code converter is a circuit that makes the two systems compatible even though each uses a different binary code.

To convert from binary code A to binary code B, the input lines must supply the bit combination of elements as specified by code A and the output lines must generate the corresponding bit combination of code B. A combinational circuit performs this transformation by means of logic gates. The design procedure will be illustrated by an example that converts binary coded decimal (BCD) to the excess-3 code for the decimal digits.

The bit combinations assigned to the BCD and excess-3 codes are listed in Table 1.5 (Section 1.7). Since each code uses four bits to represent a decimal digit, there must be

Table 4.2
Truth Table for Code Conversion Example

Input BCD				Output Excess-3 Code			
A	***B***	***C***	***D***	***w***	***x***	***y***	***z***
0	0	0	0	0	0	1	1
0	0	0	1	0	1	0	0
0	0	1	0	0	1	0	1
0	0	1	1	0	1	1	0
0	1	0	0	0	1	1	1
0	1	0	1	1	0	0	0
0	1	1	0	1	0	0	1
0	1	1	1	1	0	1	0
1	0	0	0	1	0	1	1
1	0	0	1	1	1	0	0

four input variables and four output variables. We designate the four input binary variables by the symbols A, B, C, and D, and the four output variables by w, x, y, and z. The truth table relating the input and output variables is shown in Table 4.2. The bit combinations for the inputs and their corresponding outputs are obtained directly from Section 1.7. Note that four binary variables may have 16 bit combinations, but only 10 are listed in the truth table. The six bit combinations not listed for the input variables are don't-care combinations. These values have no meaning in BCD and we assume that they will never occur in actual operation of the circuit. Therefore, we are at liberty to assign to the output variables either a 1 or a 0, whichever gives a simpler circuit.

The maps in Fig. 4.3 are plotted to obtain simplified Boolean functions for the outputs. Each one of the four maps represents one of the four outputs of the circuit as a function of the four input variables. The 1's marked inside the squares are obtained from the minterms that make the output equal to 1. The 1's are obtained from the truth table by going over the output columns one at a time. For example, the column under output z has five 1's; therefore, the map for z has five 1's, each being in a square corresponding to the minterm that makes z equal to 1. The six don't-care minterms 10 through 15 are marked with an X. One possible way to simplify the functions into sum-of-products form is listed under the map of each variable. (See Chapter 3.)

A two-level logic diagram for each output may be obtained directly from the Boolean expressions derived from the maps. There are various other possibilities for a logic diagram that implements this circuit. The expressions obtained in Fig. 4.3 may be manipulated algebraically for the purpose of using common gates for two or more outputs. This manipulation, shown next, illustrates the flexibility obtained with multiple-output systems when

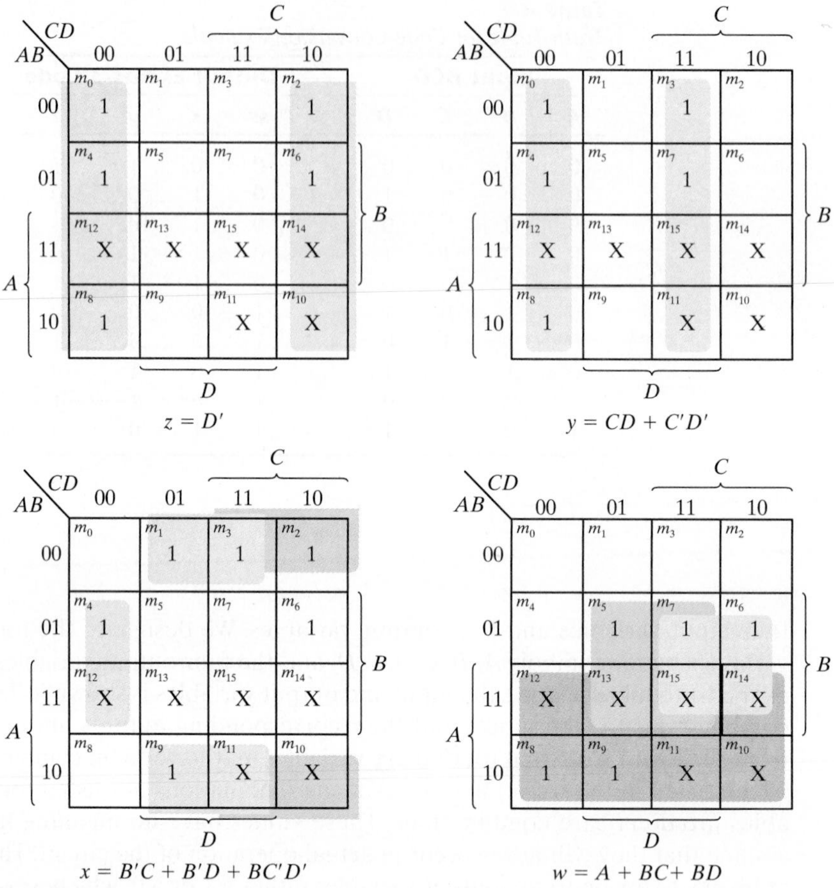

AB \ CD	00	01	11	10
00	m_0 1	m_1	m_3	m_2 1
01	m_4 1	m_5	m_7	m_6 1
11	m_{12} X	m_{13} X	m_{15} X	m_{14} X
10	m_8 1	m_9	m_{11} X	m_{10} X

$z = D'$

AB \ CD	00	01	11	10
00	m_0 1	m_1	m_3 1	m_2
01	m_4 1	m_5	m_7 1	m_6
11	m_{12} X	m_{13} X	m_{15} X	m_{14} X
10	m_8 1	m_9	m_{11} X	m_{10} X

$y = CD + C'D'$

AB \ CD	00	01	11	10
00	m_0	m_1 1	m_3 1	m_2 1
01	m_4 1	m_5	m_7	m_6 1
11	m_{12} X	m_{13} X	m_{15} X	m_{14} X
10	m_8	m_9 1	m_{11} X	m_{10} X

$x = B'C + B'D + BC'D'$

AB \ CD	00	01	11	10
00	m_0	m_1	m_3	m_2
01	m_4	m_5 1	m_7 1	m_6 1
11	m_{12} X	m_{13} X	m_{15} X	m_{14} X
10	m_8 1	m_9 1	m_{11} X	m_{10} X

$w = A + BC + BD$

FIGURE 4.3
Maps for BCD-to-excess-3 code converter

implemented with three or more levels of gates:

$$z = D'$$
$$y = CD + C'D' = CD + (C + D)'$$
$$x = B'C + B'D + BC'D' = B'(C + D) + BC'D'$$
$$= B'(C + D) + B(C + D)'$$
$$w = A + BC + BD = A + B(C + D)$$

The logic diagram that implements these expressions is shown in Fig. 4.4. Note that the OR gate whose output is $C + D$ has been used to implement partially each of three outputs.

Not counting input inverters, the implementation in sum-of-products form requires seven AND gates and three OR gates. The implementation of Fig. 4.4 requires four AND gates, four OR gates, and one inverter. If only the normal inputs are available, the first

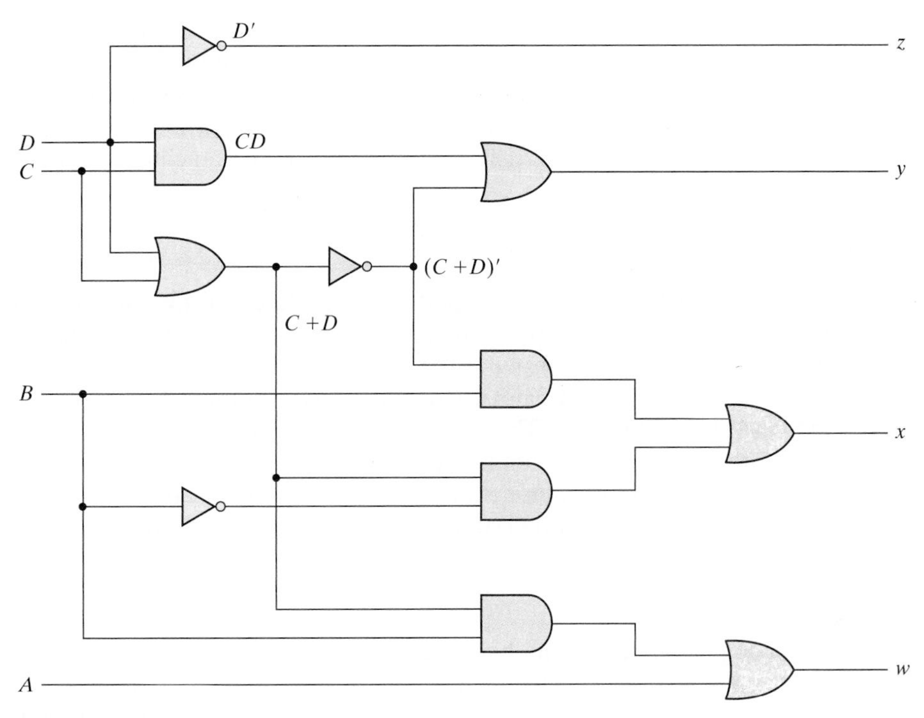

FIGURE 4.4
Logic diagram for BCD-to-excess-3 code converter

implementation will require inverters for variables B, C, and D, and the second implementation will require inverters for variables B and D. Thus, the three-level logic circuit requires fewer gates, all of which in turn require no more than two inputs.

4.5 BINARY ADDER–SUBTRACTOR

Digital computers perform a variety of information-processing tasks. Among the functions encountered are the various arithmetic operations. The most basic arithmetic operation is the addition of two binary digits. This simple addition consists of four possible elementary operations: $0 + 0 = 0$, $0 + 1 = 1$, $1 + 0 = 1$, and $1 + 1 = 10$. The first three operations produce a sum of one digit, but when both augend and addend bits are equal to 1, the binary sum consists of two digits. The higher significant bit of this result is called a *carry*. When the augend and addend numbers contain more significant digits, the carry obtained from the addition of two bits is added to the next higher order pair of significant bits. A combinational circuit that performs the addition of two bits is called a *half adder*. One that performs the addition of three bits (two significant bits and a previous carry) is a *full adder*. The names of the circuits stem from the fact that two half adders can be employed to implement a full adder.

A binary adder–subtractor is a combinational circuit that performs the arithmetic operations of addition and subtraction with binary numbers. We will develop this circuit by means of a hierarchical design. The half adder design is carried out first, from which we develop the full adder. Connecting n full adders in cascade produces a binary adder for two n-bit numbers. The subtraction circuit is included in a complementing circuit.

Half Adder

From the verbal explanation of a half adder, we find that this circuit needs two binary inputs and two binary outputs. The input variables designate the augend and addend bits; the output variables produce the sum and carry. We assign symbols x and y to the two inputs and S (for sum) and C (for carry) to the outputs. The truth table for the half adder is listed in Table 4.3. The C output is 1 only when both inputs are 1. The S output represents the least significant bit of the sum.

The simplified Boolean functions for the two outputs can be obtained directly from the truth table. The simplified sum-of-products expressions are

$$S = x'y + xy'$$
$$C = xy$$

The logic diagram of the half adder implemented in sum of products is shown in Fig. 4.5(a). It can be also implemented with an exclusive-OR and an AND gate as shown in Fig. 4.5(b). This form is used to show that two half adders can be used to construct a full adder.

Table 4.3
Half Adder

x	*y*	*C*	*S*
0	0	0	0
0	1	0	1
1	0	0	1
1	1	1	0

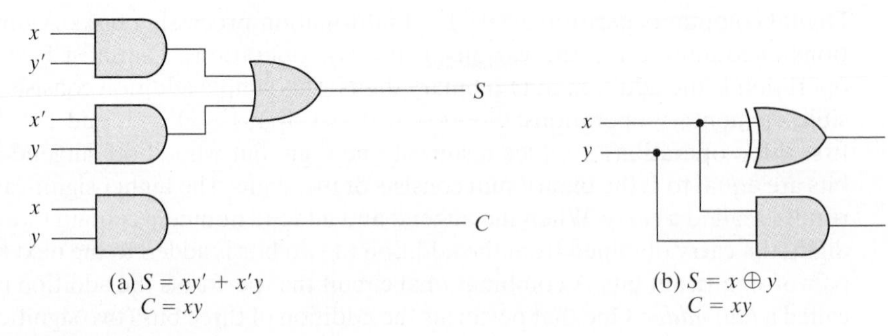

FIGURE 4.5
Implementation of half adder

Full Adder

Addition of n-bit binary numbers requires the use of a full adder, and the process of addition proceeds on a bit-by-bit basis, right to left, beginning with the least significant bit. After the least significant bit, addition at each position adds not only the respective bits of the words, but must also consider a possible carry bit from addition at the previous position.

A full adder is a combinational circuit that forms the arithmetic sum of three bits. It consists of three inputs and two outputs. Two of the input variables, denoted by x and y, represent the two significant bits to be added. The third input, z, represents the carry from the previous lower significant position. Two outputs are necessary because the arithmetic sum of three binary digits ranges in value from 0 to 3, and binary representation of 2 or 3 needs two bits. The two outputs are designated by the symbols S for sum and C for carry. The binary variable S gives the value of the least significant bit of the sum. The binary variable C gives the output carry formed by adding the input carry and the bits of the words. The truth table of the full adder is listed in Table 4.4. The eight rows under the input variables designate all possible combinations of the three variables. The output variables are determined from the arithmetic sum of the input bits. When all input bits are 0, the output is 0. The S output is equal to 1 when only one input is equal to 1 or when all three inputs are equal to 1. The C output has a carry of 1 if two or three inputs are equal to 1.

The input and output bits of the combinational circuit have different interpretations at various stages of the problem. On the one hand, physically, the binary signals of the inputs are considered binary digits to be added arithmetically to form a two-digit sum at the output. On the other hand, the same binary values are considered as variables of Boolean functions when expressed in the truth table or when the circuit is implemented with logic gates. The maps for the outputs of the full adder are shown in Fig. 4.6. The simplified expressions are

$$S = x'y'z + x'yz' + xy'z' + xyz$$
$$C = xy + xz + yz$$

The logic diagram for the full adder implemented in sum-of-products form is shown in Fig. 4.7. It can also be implemented with two half adders and one OR gate, as shown

Table 4.4
Full Adder

x	y	z	C	S
0	0	0	0	0
0	0	1	0	1
0	1	0	0	1
0	1	1	1	0
1	0	0	0	1
1	0	1	1	0
1	1	0	1	0
1	1	1	1	1

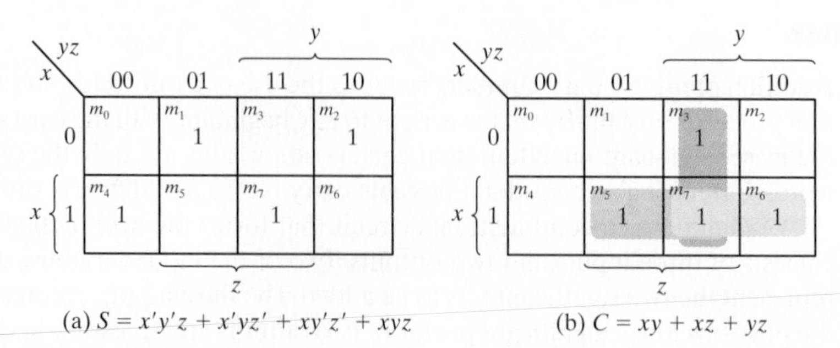

(a) $S = x'y'z + x'yz' + xy'z' + xyz$ (b) $C = xy + xz + yz$

FIGURE 4.6
K-Maps for full adder

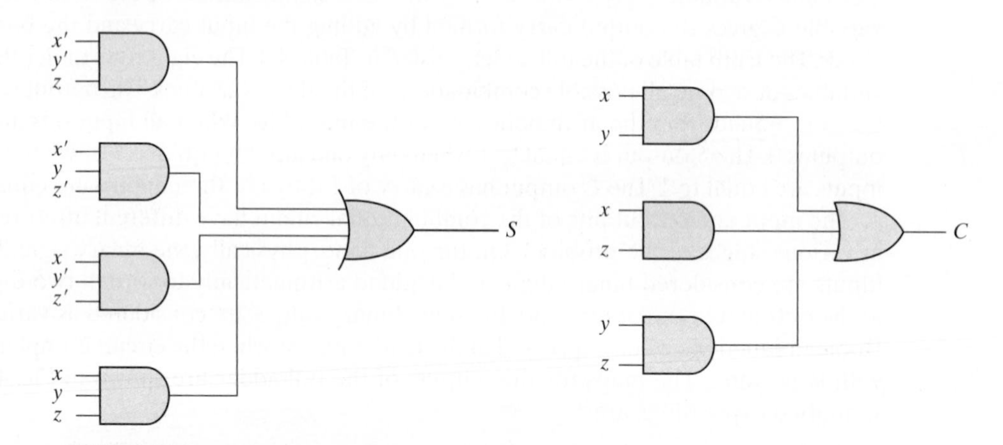

FIGURE 4.7
Implementation of full adder in sum-of-products form

in Fig. 4.8. The S output from the second half adder is the exclusive-OR of z and the output of the first half adder, giving

$$\begin{aligned} S &= z \oplus (x \oplus y) \\ &= z'(xy' + x'y) + z(xy' + x'y)' \\ &= z'(xy' + x'y) + z(xy + x'y') \\ &= xy'z' + x'yz' + xyz + x'y'z \end{aligned}$$

The carry output is

$$C = z(xy' + x'y) + xy = xy'z + x'yz + xy$$

Binary Adder

A binary adder is a digital circuit that produces the arithmetic sum of two binary numbers. It can be constructed with full adders connected in cascade, with the output carry from each full adder connected to the input carry of the next full adder in the chain.

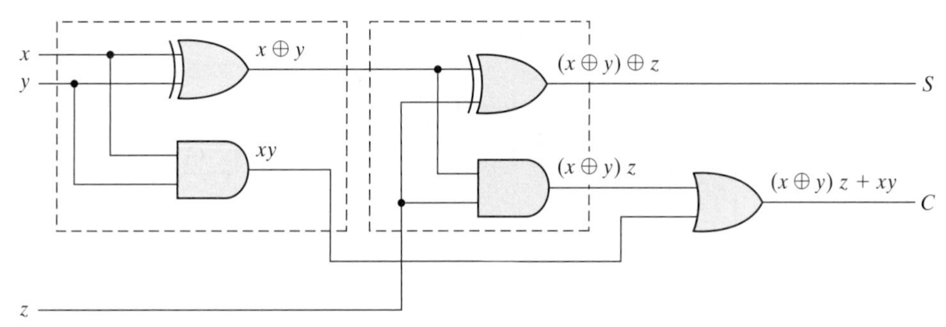

FIGURE 4.8
Implementation of full adder with two half adders and an OR gate

Addition of n-bit numbers requires a chain of n full adders or a chain of one-half adder and $n-1$ full adders. In the former case, the input carry to the least significant position is fixed at 0. Figure 4.9 shows the interconnection of four full-adder (FA) circuits to provide a four-bit binary ripple carry adder. The augend bits of A and the addend bits of B are designated by subscript numbers from right to left, with subscript 0 denoting the least significant bit. The carries are connected in a chain through the full adders. The input carry to the adder is C_0, and it ripples through the full adders to the output carry C_4. The S outputs generate the required sum bits. An n-bit adder requires n full adders, with each output carry connected to the input carry of the next higher order full adder.

To demonstrate with a specific example, consider the two binary numbers $A = 1011$ and $B = 0011$. Their sum $S = 1110$ is formed with the four-bit adder as follows:

Subscript *i*:	**3**	**2**	**1**	**0**	
Input carry	0	1	1	0	C_i
Augend	1	0	1	1	A_i
Addend	0	0	1	1	B_i
Sum	1	1	1	0	S_i
Output carry	0	0	1	1	C_{i+1}

The bits are added with full adders, starting from the least significant position (subscript 0), to form the sum bit and carry bit. The input carry C_0 in the least significant position must be 0. The value of C_{i+1} in a given significant position is the output carry of the full adder. This value is transferred into the input carry of the full adder that adds the bits one higher significant position to the left. The sum bits are thus generated starting from the rightmost position and are available as soon as the corresponding previous carry bit is generated. All the carries must be generated for the correct sum bits to appear at the outputs.

The four-bit adder is a typical example of a standard component. It can be used in many applications involving arithmetic operations. Observe that the design of this circuit

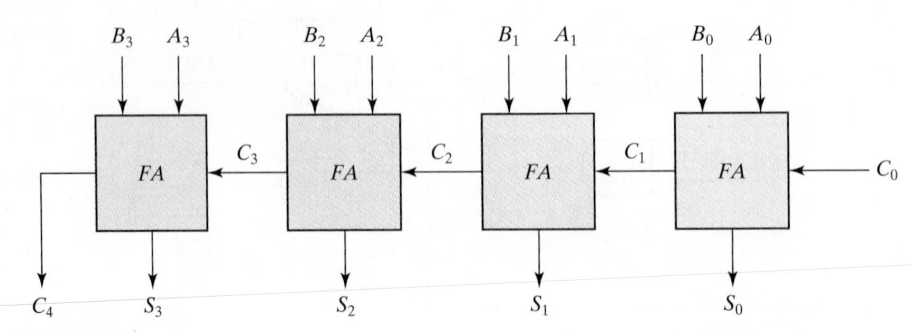

FIGURE 4.9
Four-bit adder

by the classical method would require a truth table with $2^9 = 512$ entries, since there are nine inputs to the circuit. By using an iterative method of cascading a standard function, it is possible to obtain a simple and straightforward implementation.

Carry Propagation

The addition of two binary numbers in parallel implies that all the bits of the augend and addend are available for computation at the same time. As in any combinational circuit, the signal must propagate through the gates before the correct output sum is available in the output terminals. The total propagation time is equal to the propagation delay of a typical gate, times the number of gate levels in the circuit. The longest propagation delay time in an adder is the time it takes the carry to propagate through the full adders. Since each bit of the sum output depends on the value of the input carry, the value of S_i at any given stage in the adder will be in its steady-state final value only after the input carry to that stage has been propagated. In this regard, consider output S_3 in Fig. 4.9. Inputs A_3 and B_3 are available as soon as input signals are applied to the adder. However, input carry C_3 does not settle to its final value until C_2 is available from the previous stage. Similarly, C_2 has to wait for C_1 and so on down to C_0. Thus, only after the carry propagates and ripples through all stages will the last output S_3 and carry C_4 settle to their final correct value.

The number of gate levels for the carry propagation can be found from the circuit of the full adder. The circuit is redrawn with different labels in Fig. 4.10 for convenience. The input and output variables use the subscript i to denote a typical stage of the adder. The signals at P_i and G_i settle to their steady-state values after they propagate through their respective gates. These two signals are common to all half adders and depend on only the input augend and addend bits. The signal from the input carry C_i to the output carry C_{i+1} propagates through an AND gate and an OR gate, which constitute two gate levels. If there are four full adders in the adder, the output carry C_4 would have $2 \times 4 = 8$ gate levels from C_0 to C_4. For an n-bit adder, there are $2n$ gate levels for the carry to propagate from input to output.

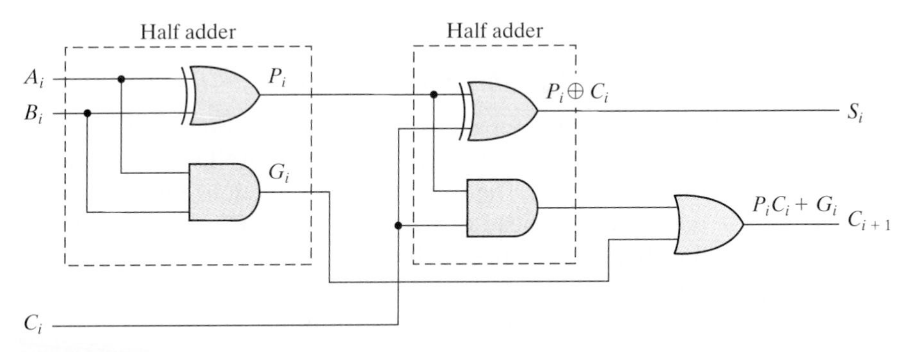

FIGURE 4.10
Full adder with *P* and *G* shown

The carry propagation time is an important attribute of the adder because it limits the speed with which two numbers are added. Although the adder—or, for that matter, any combinational circuit—will always have some value at its output terminals, the outputs will not be correct unless the signals are given enough time to propagate through the gates connected from the inputs to the outputs. Since all other arithmetic operations are implemented by successive additions, the time consumed during the addition process is critical. An obvious solution for reducing the carry propagation delay time is to employ faster gates with reduced delays. However, physical circuits have a limit to their capability. Another solution is to increase the complexity of the equipment in such a way that the carry delay time is reduced. There are several techniques for reducing the carry propagation time in a parallel adder. The most widely used technique employs the principle of *carry lookahead logic*.

Consider the circuit of the full adder shown in Fig. 4.10. If we define two new binary variables

$$P_i = A_i \oplus B_i$$
$$G_i = A_iB_i$$

the output sum and carry can respectively be expressed as

$$S_i = P_i \oplus C_i$$
$$C_{i+1} = G_i + P_iC_i$$

G_i is called a *carry generate*, and it produces a carry of 1 when both A_i and B_i are 1, regardless of the input carry C_i. P_i is called a *carry propagate*, because it determines whether a carry into stage i will propagate into stage $i + 1$ (i.e., whether an assertion of C_i will propagate to an assertion of C_{i+1}).

We now write the Boolean functions for the carry outputs of each stage and substitute the value of each C_i from the previous equations:

$$C_0 = \text{input carry}$$
$$C_1 = G_0 + P_0C_0$$

$$C_2 = G_1 + P_1C_1 = G_1 + P_1(G_0 + P_0C_0) = G_1 + P_1G_0 + P_1P_0C_0$$

$$C_3 = G_2 + P_2C_2 = G_2 + P_2G_1 + P_2P_1G_0 = P_2P_1P_0C_0$$

Since the Boolean function for each output carry is expressed in sum-of-products form, each function can be implemented with one level of AND gates followed by an OR gate (or by a two-level NAND). The three Boolean functions for C_1, C_2, and C_3 are implemented in the carry lookahead generator shown in Fig. 4.11. Note that this circuit can add in less time because C_3 does not have to wait for C_2 and C_1 to propagate; in fact, C_3 is propagated at the same time as C_1 and C_2. This gain in speed of operation is achieved at the expense of additional complexity (hardware).

The construction of a four-bit adder with a carry lookahead scheme is shown in Fig. 4.12. Each sum output requires two exclusive-OR gates. The output of the first exclusive-OR gate generates the P_i variable, and the AND gate generates the G_i variable. The carries are propagated through the carry lookahead generator (similar to that in Fig. 4.11) and applied as inputs to the second exclusive-OR gate. All output carries are generated after

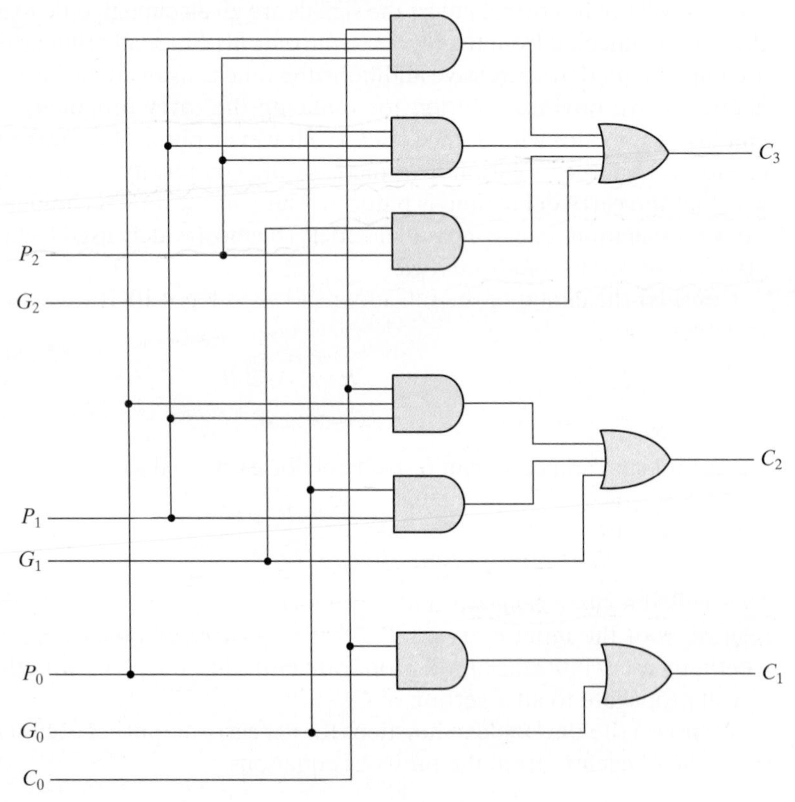

FIGURE 4.11
Logic diagram of carry lookahead generator

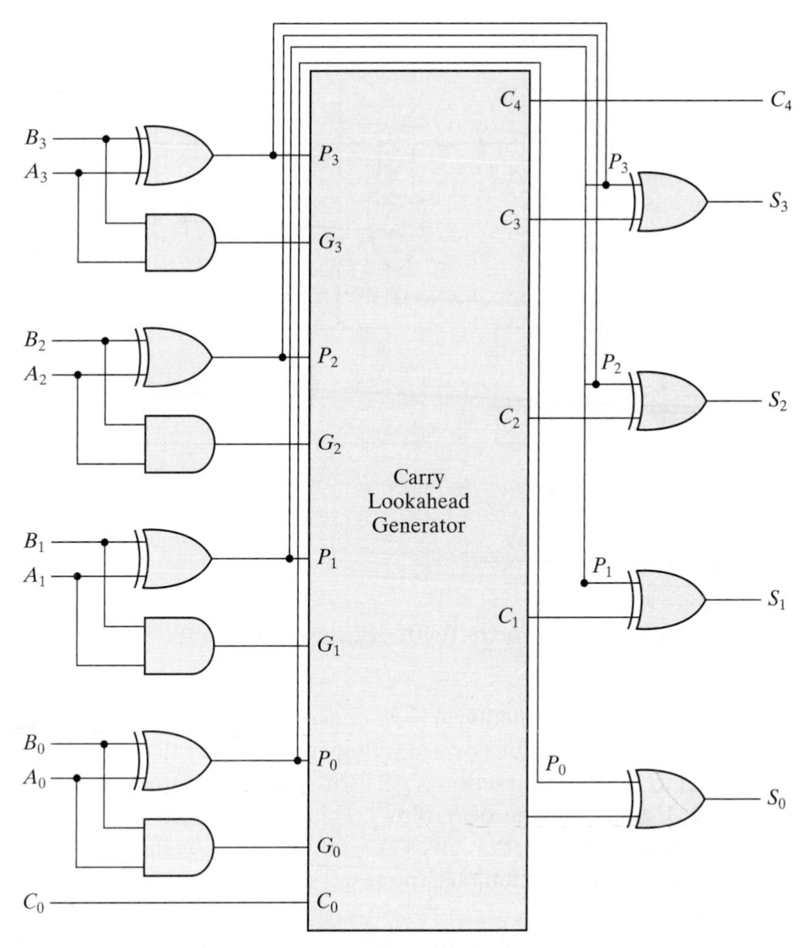

FIGURE 4.12
Four-bit adder with carry lookahead

a delay through two levels of gates. Thus, outputs S_1 through S_3 have equal propagation delay times. The two-level circuit for the output carry C_4 is not shown. This circuit can easily be derived by the equation-substitution method.

Binary Subtractor

The subtraction of unsigned binary numbers can be done most conveniently by means of complements, as discussed in Section 1.5. Remember that the subtraction $A - B$ can be done by taking the 2's complement of B and adding it to A. The 2's complement can be obtained by taking the 1's complement and adding 1 to the least significant pair of bits. The 1's complement can be implemented with inverters, and a 1 can be added to the sum through the input carry.

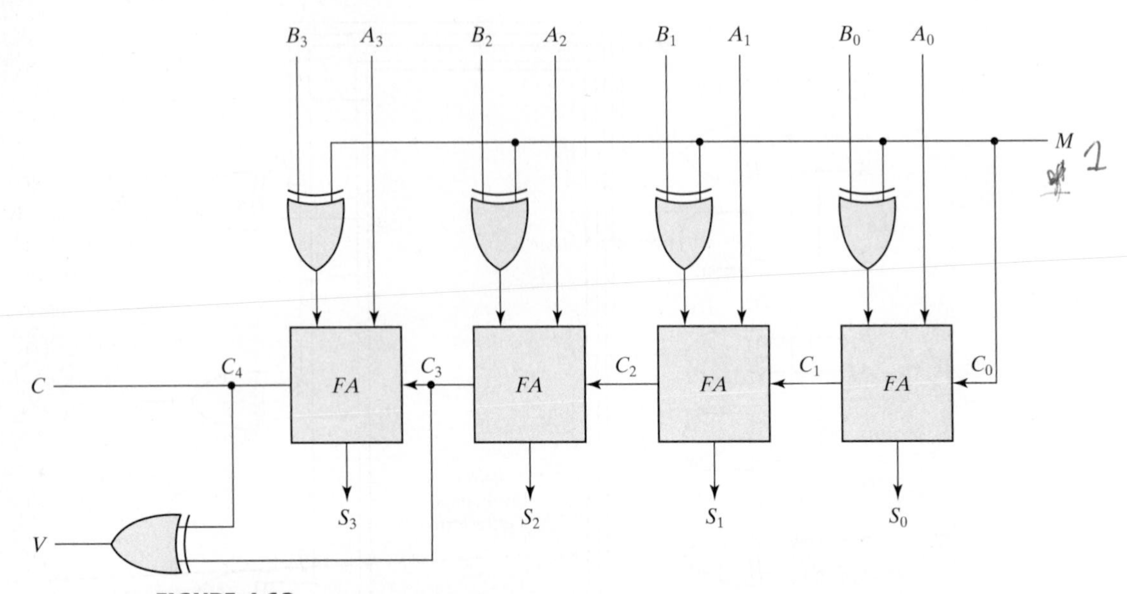

FIGURE 4.13
Four-bit adder–subtractor (with overflow detection)

The circuit for subtracting $A - B$ consists of an adder with inverters placed between each data input B and the corresponding input of the full adder. The input carry C_0 must be equal to 1 when subtraction is performed. The operation thus performed becomes A, plus the 1's complement of B, plus 1. This is equal to A plus the 2's complement of B. For unsigned numbers, that gives $A - B$ if $A \geq B$ or the 2's complement of $(B - A)$ if $A < B$. For signed numbers, the result is $A - B$, provided that there is no overflow. (See Section 1.6.)

The addition and subtraction operations can be combined into one circuit with one common binary adder by including an exclusive-OR gate with each full adder. A four-bit adder–subtractor circuit is shown in Fig. 4.13. The mode input M controls the operation. When $M = 0$, the circuit is an adder, and when $M = 1$, the circuit becomes a subtractor. Each exclusive-OR gate receives input M and one of the inputs of B. When $M = 0$, we have $B \oplus 0 = B$. The full adders receive the value of B, the input carry is 0, and the circuit performs A plus B. When $M = 1$, we have $B \oplus 1 = B'$ and $C_0 = 1$. The B inputs are all complemented and a 1 is added through the input carry. The circuit performs the operation A plus the 2's complement of B. (The exclusive-OR with output V is for detecting an overflow.)

It is worth noting that binary numbers in the signed-complement system are added and subtracted by the same basic addition and subtraction rules as are unsigned numbers. Therefore, computers need only one common hardware circuit to handle both types of arithmetic. The user or programmer must interpret the results of such addition or subtraction differently, depending on whether it is assumed that the numbers are signed or unsigned.

Overflow

When two numbers with n digits each are added and the sum is a number occupying $n + 1$ digits, we say that an overflow occurred. This is true for binary or decimal numbers, signed or unsigned. When the addition is performed with paper and pencil, an overflow is not a problem, since there is no limit by the width of the page to write down the sum. Overflow is a problem in digital computers because the number of bits that hold the number is finite and a result that contains $n + 1$ bits cannot be accommodated by an n-bit word. For this reason, many computers detect the occurrence of an overflow, and when it occurs, a corresponding flip-flop is set that can then be checked by the user.

The detection of an overflow after the addition of two binary numbers depends on whether the numbers are considered to be signed or unsigned. When two unsigned numbers are added, an overflow is detected from the end carry out of the most significant position. In the case of signed numbers, two details are important: the leftmost bit always represents the sign, and negative numbers are in 2's-complement form. When two signed numbers are added, the sign bit is treated as part of the number and the end carry does not indicate an overflow.

An overflow cannot occur after an addition if one number is positive and the other is negative, since adding a positive number to a negative number produces a result whose magnitude is smaller than the larger of the two original numbers. An overflow may occur if the two numbers added are both positive or both negative. To see how this can happen, consider the following example: Two signed binary numbers, +70 and +80, are stored in two eight-bit registers. The range of numbers that each register can accommodate is from binary +127 to binary −128. Since the sum of the two numbers is +150, it exceeds the capacity of an eight-bit register. This is also true for −70 and −80. The two additions in binary are shown next, together with the last two carries:

carries:	0 1	carries:	1 0
+70	0 1000110	−70	1 0111010
+80	0 1010000	−80	1 0110000
+150	1 0010110	−150	0 1101010

Note that the eight-bit result that should have been positive has a negative sign bit (i.e., the eighth bit) and the eight-bit result that should have been negative has a positive sign bit. If, however, the carry out of the sign bit position is taken as the sign bit of the result, then the nine-bit answer so obtained will be correct. But since the answer cannot be accommodated within eight bits, we say that an overflow has occurred.

An overflow condition can be detected by observing the carry into the sign bit position and the carry out of the sign bit position. If these two carries are not equal, an overflow has occurred. This is indicated in the examples in which the two carries are explicitly shown. If the two carries are applied to an exclusive-OR gate, an overflow is detected when the output of the gate is equal to 1. For this method to work correctly, the 2's complement of a negative number must be computed by taking the 1's complement and adding 1. This takes care of the condition when the maximum negative number is complemented.

The binary adder–subtractor circuit with outputs C and V is shown in Fig. 4.13. If the two binary numbers are considered to be unsigned, then the C bit detects a carry after addition or a borrow after subtraction. If the numbers are considered to be signed, then the V bit detects an overflow. If $V = 0$ after an addition or subtraction, then no overflow occurred and the n-bit result is correct. If $V = 1$, then the result of the operation contains $n + 1$ bits, but only the rightmost n bits of the number fit in the space available, so an overflow has occurred. The $(n + 1)$ th bit is the actual sign and has been shifted out of position.

4.6 DECIMAL ADDER

Computers or calculators that perform arithmetic operations directly in the decimal number system represent decimal numbers in binary coded form. An adder for such a computer must employ arithmetic circuits that accept coded decimal numbers and present results in the same code. For binary addition, it is sufficient to consider a pair of significant bits together with a previous carry. A decimal adder requires a minimum of nine inputs and five outputs, since four bits are required to code each decimal digit and the circuit must have an input and output carry. There is a wide variety of possible decimal adder circuits, depending upon the code used to represent the decimal digits. Here we examine a decimal adder for the BCD code. (See Section 1.7.)

BCD Adder

Consider the arithmetic addition of two decimal digits in BCD, together with an input carry from a previous stage. Since each input digit does not exceed 9, the output sum cannot be greater than $9 + 9 + 1 = 19$, the 1 in the sum being an input carry. Suppose we apply two BCD digits to a four-bit binary adder. The adder will form the sum in *binary* and produce a result that ranges from 0 through 19. These binary numbers are listed in Table 4.5 and are labeled by symbols K, Z_8, Z_4, Z_2, and Z_1. K is the carry, and the subscripts under the letter Z represent the weights 8, 4, 2, and 1 that can be assigned to the four bits in the BCD code. The columns under the binary sum list the binary value that appears in the outputs of the four-bit binary adder. The output sum of two decimal digits must be represented in BCD and should appear in the form listed in the columns under "BCD Sum." The problem is to find a rule by which the binary sum is converted to the correct BCD digit representation of the number in the BCD sum.

In examining the contents of the table, it becomes apparent that when the binary sum is equal to or less than 1001, the corresponding BCD number is identical, and therefore no conversion is needed. When the binary sum is greater than 1001, we obtain an invalid BCD representation. The addition of binary 6 (0110) to the binary sum converts it to the correct BCD representation and also produces an output carry as required.

Table 4.5
Derivation of BCD Adder

Binary Sum					BCD Sum					Decimal
K	Z_8	Z_4	Z_2	Z_1	***C***	S_8	S_4	S_2	S_1	
0	0	0	0	0	0	0	0	0	0	0
0	0	0	0	1	0	0	0	0	1	1
0	0	0	1	0	0	0	0	1	0	2
0	0	0	1	1	0	0	0	1	1	3
0	0	1	0	0	0	0	1	0	0	4
0	0	1	0	1	0	0	1	0	1	5
0	0	1	1	0	0	0	1	1	0	6
0	0	1	1	1	0	0	1	1	1	7
0	1	0	0	0	0	1	0	0	0	8
0	1	0	0	1	0	1	0	0	1	9
0	1	0	1	0	1	0	0	0	0	10
0	1	0	1	1	1	0	0	0	1	11
0	1	1	0	0	1	0	0	1	0	12
0	1	1	0	1	1	0	0	1	1	13
0	1	1	1	0	1	0	1	0	0	14
0	1	1	1	1	1	0	1	0	1	15
1	0	0	0	0	1	0	1	1	0	16
1	0	0	0	1	1	0	1	1	1	17
1	0	0	1	0	1	1	0	0	0	18
1	0	0	1	1	1	1	0	0	1	19

The logic circuit that detects the necessary correction can be derived from the entries in the table. It is obvious that a correction is needed when the binary sum has an output carry $K = 1$. The other six combinations from 1010 through 1111 that need a correction have a 1 in position Z_8. To distinguish them from binary 1000 and 1001, which also have a 1 in position Z_8, we specify further that either Z_4 or Z_2 must have a 1. The condition for a correction and an output carry can be expressed by the Boolean function

$$C = K + Z_8Z_4 + Z_8Z_2$$

When $C = 1$, it is necessary to add 0110 to the binary sum and provide an output carry for the next stage.

A BCD adder that adds two BCD digits and produces a sum digit in BCD is shown in Fig. 4.14. The two decimal digits, together with the input carry, are first added in the top four-bit adder to produce the binary sum. When the output carry is equal to 0, nothing is added to the binary sum. When it is equal to 1, binary 0110 is added to the binary sum through the bottom four-bit adder. The output carry generated from the bottom

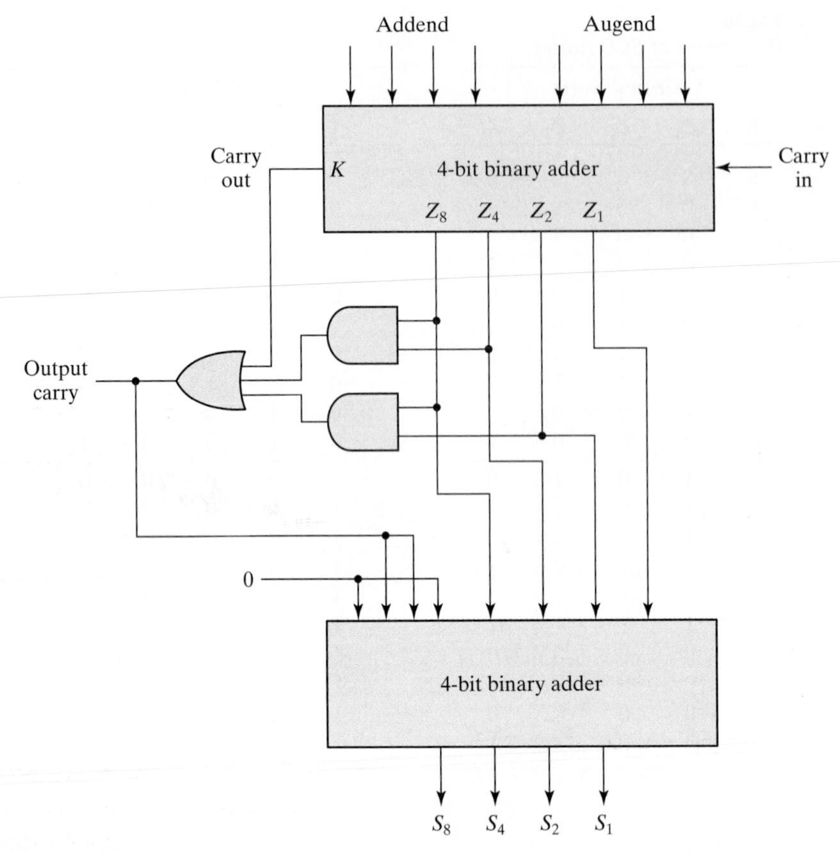

FIGURE 4.14
Block diagram of a BCD adder

adder can be ignored, since it supplies information already available at the output carry terminal. A decimal parallel adder that adds n decimal digits needs n BCD adder stages. The output carry from one stage must be connected to the input carry of the next higher order stage.

4.7 BINARY MULTIPLIER

Multiplication of binary numbers is performed in the same way as multiplication of decimal numbers. The multiplicand is multiplied by each bit of the multiplier, starting from the least significant bit. Each such multiplication forms a partial product. Successive partial products are shifted one position to the left. The final product is obtained from the sum of the partial products.

To see how a binary multiplier can be implemented with a combinational circuit, consider the multiplication of two 2-bit numbers as shown in Fig. 4.15. The multiplicand

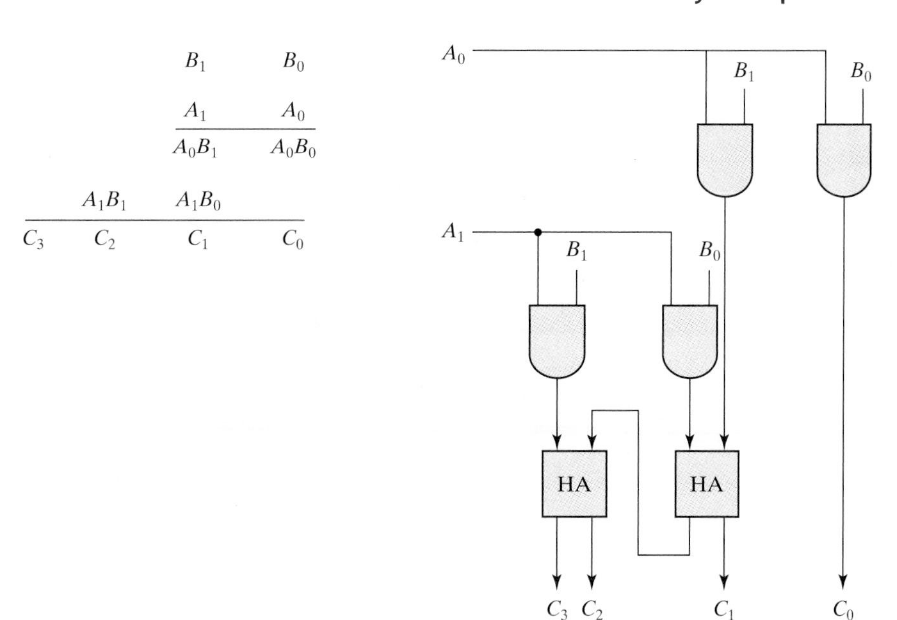

FIGURE 4.15
Two-bit by two-bit binary multiplier

bits are B_1 and B_0, the multiplier bits are A_1 and A_0, and the product is $C_3C_2C_1C_0$. The first partial product is formed by multiplying B_1B_0 by A_0. The multiplication of two bits such as A_0 and B_0 produces a 1 if both bits are 1; otherwise, it produces a 0. This is identical to an AND operation. Therefore, the partial product can be implemented with AND gates as shown in the diagram. The second partial product is formed by multiplying B_1B_0 by A_1 and shifting one position to the left. The two partial products are added with two half-adder (HA) circuits. Usually, there are more bits in the partial products and it is necessary to use full adders to produce the sum of the partial products. Note that the least significant bit of the product does not have to go through an adder, since it is formed by the output of the first AND gate.

A combinational circuit binary multiplier with more bits can be constructed in a similar fashion. A bit of the multiplier is ANDed with each bit of the multiplicand in as many levels as there are bits in the multiplier. The binary output in each level of AND gates is added with the partial product of the previous level to form a new partial product. The last level produces the product. For J multiplier bits and K multiplicand bits, we need $(J \times K)$ AND gates and $(J - 1)$ K-bit adders to produce a product of $(J + K)$ bits.

As a second example, consider a multiplier circuit that multiplies a binary number represented by four bits by a number represented by three bits. Let the multiplicand be represented by $B_3B_2B_1B_0$ and the multiplier by $A_2A_1A_0$. Since $K = 4$ and $J = 3$, we need 12 AND gates and two 4-bit adders to produce a product of seven bits. The logic diagram of the multiplier is shown in Fig. 4.16.

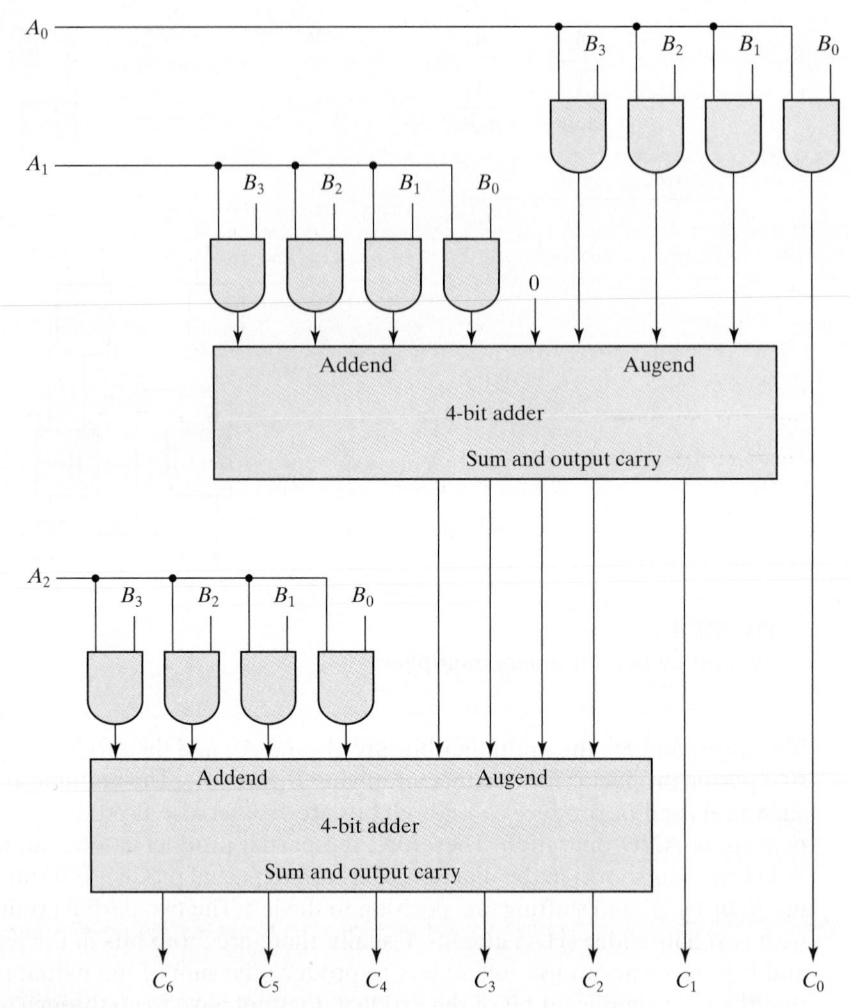

FIGURE 4.16
Four-bit by three-bit binary multiplier

4.8 MAGNITUDE COMPARATOR

The comparison of two numbers is an operation that determines whether one number is greater than, less than, or equal to the other number. A *magnitude comparator* is a combinational circuit that compares two numbers A and B and determines their relative magnitudes. The outcome of the comparison is specified by three binary variables that indicate whether $A > B$, $A = B$, or $A < B$.

On the one hand, the circuit for comparing two n-bit numbers has 2^{2n} entries in the truth table and becomes too cumbersome, even with $n = 3$. On the other hand, as one may suspect, a comparator circuit possesses a certain amount of regularity. Digital functions that possess an inherent well-defined regularity can usually be designed by means of an algorithm—a procedure which specifies a finite set of steps that, if followed, give the solution to a problem. We illustrate this method here by deriving an algorithm for the design of a four-bit magnitude comparator.

The algorithm is a direct application of the procedure a person uses to compare the relative magnitudes of two numbers. Consider two numbers, A and B, with four digits each. Write the coefficients of the numbers in descending order of significance:

$$A = A_3 \, A_2 \, A_1 \, A_0$$
$$B = B_3 \, B_2 \, B_1 \, B_0$$

Each subscripted letter represents one of the digits in the number. The two numbers are equal if all pairs of significant digits are equal: $A_3 = B_3, A_2 = B_2, A_1 = B_1$, and $A_0 = B_0$. When the numbers are binary, the digits are either 1 or 0, and the equality of each pair of bits can be expressed logically with an exclusive-NOR function as

$$x_i = A_iB_i + A_i'B_i' \quad \text{for } i = 0, 1, 2, 3$$

where $x_i = 1$ only if the pair of bits in position i are equal (i.e., if both are 1 or both are 0).

The equality of the two numbers A and B is displayed in a combinational circuit by an output binary variable that we designate by the symbol $(A = B)$. This binary variable is equal to 1 if the input numbers, A and B, are equal, and is equal to 0 otherwise. For equality to exist, all x_i variables must be equal to 1, a condition that dictates an AND operation of all variables:

$$(A = B) = x_3x_2x_1x_0$$

The *binary* variable $(A = B)$ is equal to 1 only if all pairs of digits of the two numbers are equal.

To determine whether A is greater or less than B, we inspect the relative magnitudes of pairs of significant digits, starting from the most significant position. If the two digits of a pair are equal, we compare the next lower significant pair of digits. The comparison continues until a pair of unequal digits is reached. If the corresponding digit of A is 1 and that of B is 0, we conclude that $A > B$. If the corresponding digit of A is 0 and that of B is 1, we have $A < B$. The sequential comparison can be expressed logically by the two Boolean functions

$$(A > B) = A_3B_3' + x_3A_2B_2' + x_3x_2A_1B_1' + x_3x_2x_1A_0B_0'$$
$$(A < B) = A_3'B_3 + x_3A_2'B_2 + x_3x_2A_1'B_1' + x_3x_2x_1A'n_0B_0'$$

The symbols $(A > B)$ and $(A < B)$ are *binary* output variables that are equal to 1 when $A > B$ and $A < B$, respectively.

The gate implementation of the three output variables just derived is simpler than it seems because it involves a certain amount of repetition. The unequal outputs can use the same gates that are needed to generate the equal output. The logic diagram of the four-bit magnitude comparator is shown in Fig. 4.17. The four x outputs are generated

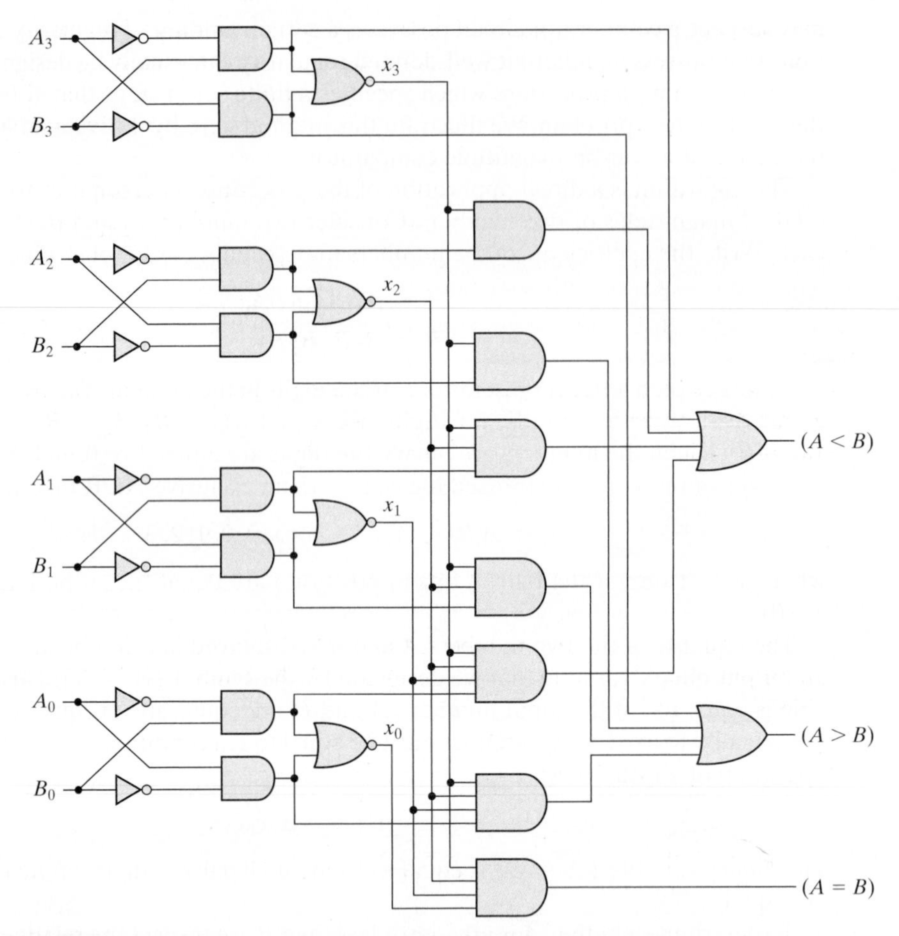

FIGURE 4.17
Four-bit magnitude comparator

with exclusive-NOR circuits and are applied to an AND gate to give the output binary variable $(A = B)$. The other two outputs use the x variables to generate the Boolean functions listed previously. This is a multilevel implementation and has a regular pattern. The procedure for obtaining magnitude comparator circuits for binary numbers with more than four bits is obvious from this example.

4.9 DECODERS

Discrete quantities of information are represented in digital systems by binary codes. A binary code of n bits is capable of representing up to 2^n distinct elements of coded information. A *decoder* is a combinational circuit that converts binary information from

n input lines to a maximum of 2^n unique output lines. If the n-bit coded information has unused combinations, the decoder may have fewer than 2^n outputs.

The decoders presented here are called n-to-m-line decoders, where $m \leq 2^n$. Their purpose is to generate the 2^n (or fewer) minterms of n input variables. Each combination of inputs will assert a unique output. The name *decoder* is also used in conjunction with other code converters, such as a BCD-to-seven-segment decoder.

As an example, consider the three-to-eight-line decoder circuit of Fig. 4.18. The three inputs are decoded into eight outputs, each representing one of the minterms of the three input variables. The three inverters provide the complement of the inputs, and each one of the eight AND gates generates one of the minterms. A particular application of this decoder is binary-to-octal conversion. The input variables represent a binary number, and the outputs represent the eight digits of a number in the octal number system. However, a three-to-eight-line decoder can be used for decoding *any* three-bit code to provide eight outputs, one for each element of the code.

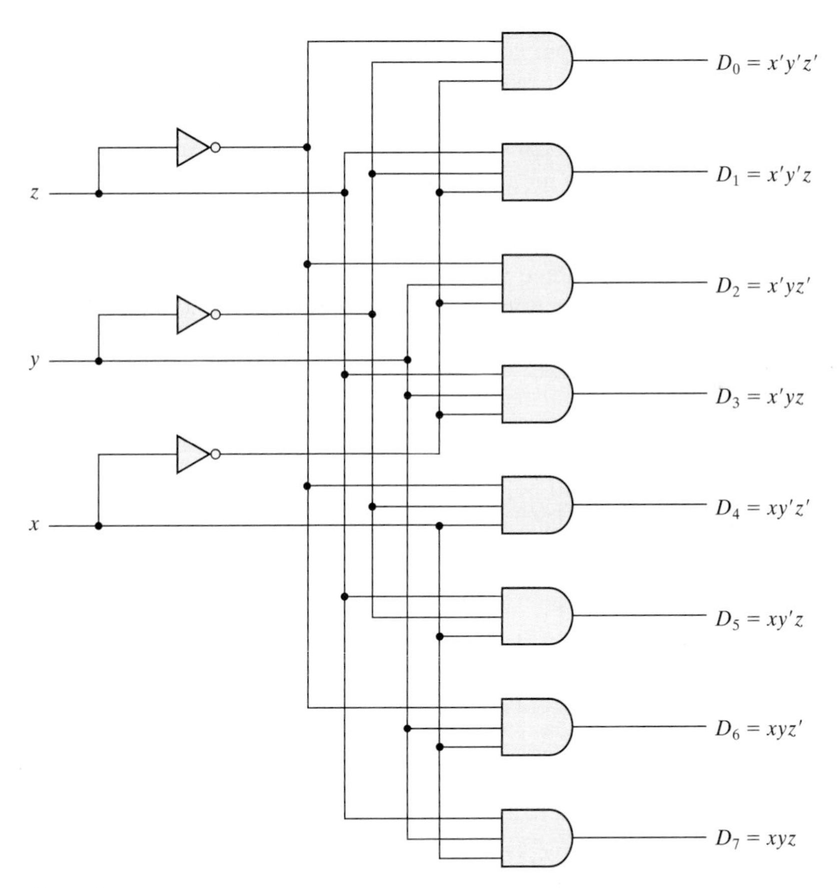

FIGURE 4.18
Three-to-eight-line decoder

Table 4.6
Truth Table of a Three-to-Eight-Line Decoder

Inputs			Outputs							
x	y	z	D_0	D_1	D_2	D_3	D_4	D_5	D_6	D_7
0	0	0	1	0	0	0	0	0	0	0
0	0	1	0	1	0	0	0	0	0	0
0	1	0	0	0	1	0	0	0	0	0
0	1	1	0	0	0	1	0	0	0	0
1	0	0	0	0	0	0	1	0	0	0
1	0	1	0	0	0	0	0	1	0	0
1	1	0	0	0	0	0	0	0	1	0
1	1	1	0	0	0	0	0	0	0	1

The operation of the decoder may be clarified by the truth table listed in Table 4.6. For each possible input combination, there are seven outputs that are equal to 0 and only one that is equal to 1. The output whose value is equal to 1 represents the minterm equivalent of the binary number currently available in the input lines.

Some decoders are constructed with NAND gates. Since a NAND gate produces the AND operation with an inverted output, it becomes more economical to generate the decoder minterms in their complemented form. Furthermore, decoders include one or more *enable* inputs to control the circuit operation. A two-to-four-line decoder with an enable input constructed with NAND gates is shown in Fig. 4.19. The circuit operates with complemented outputs and a complement enable input. The decoder is enabled when E is equal to 0 (i.e., active-low enable). As indicated by the truth table, only one

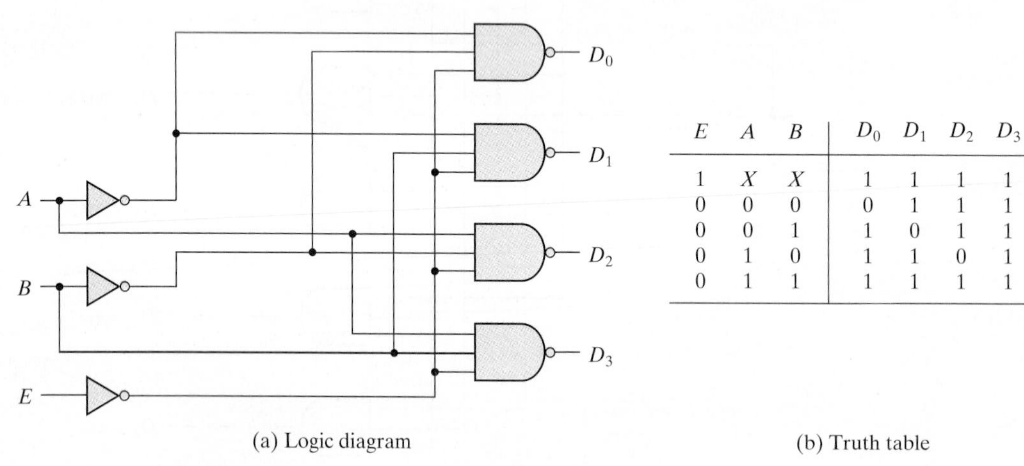

E	A	B	D_0	D_1	D_2	D_3
1	X	X	1	1	1	1
0	0	0	0	1	1	1
0	0	1	1	0	1	1
0	1	0	1	1	0	1
0	1	1	1	1	1	0

FIGURE 4.19
Two-to-four-line decoder with enable input

output can be equal to 0 at any given time; all other outputs are equal to 1. The output whose value is equal to 0 represents the minterm selected by inputs A and B. The circuit is disabled when E is equal to 1, regardless of the values of the other two inputs. When the circuit is disabled, none of the outputs are equal to 0 and none of the minterms are selected. In general, a decoder may operate with complemented or uncomplemented outputs. The enable input may be activated with a 0 or with a 1 signal. Some decoders have two or more enable inputs that must satisfy a given logic condition in order to enable the circuit.

A decoder with enable input can function as a *demultiplexer*—a circuit that receives information from a single line and directs it to one of 2^n possible output lines. The selection of a specific output is controlled by the bit combination of n selection lines. The decoder of Fig. 4.19 can function as a one-to-four-line demultiplexer when E is taken as a data input line and A and B are taken as the selection inputs. The single input variable E has a path to all four outputs, but the input information is directed to only one of the output lines, as specified by the binary combination of the two selection lines A and B. This feature can be verified from the truth table of the circuit. For example, if the selection lines $AB = 10$, output D_2 will be the same as the input value E, while all other outputs are maintained at 1. Because decoder and demultiplexer operations are obtained from the same circuit, a decoder with an enable input is referred to as a *decoder–demultiplexer*.

Decoders with enable inputs can be connected together to form a larger decoder circuit. Figure 4.20 shows two 3-to-8-line decoders with enable inputs connected to form a 4-to-16-line decoder. When $w = 0$, the top decoder is enabled and the other is disabled. The bottom decoder outputs are all 0's, and the top eight outputs generate minterms 0000 to 0111. When $w = 1$, the enable conditions are reversed: The bottom decoder outputs generate minterms 1000 to 1111, while the outputs of the top decoder are all 0's. This example demonstrates the usefulness of enable inputs in decoders and other

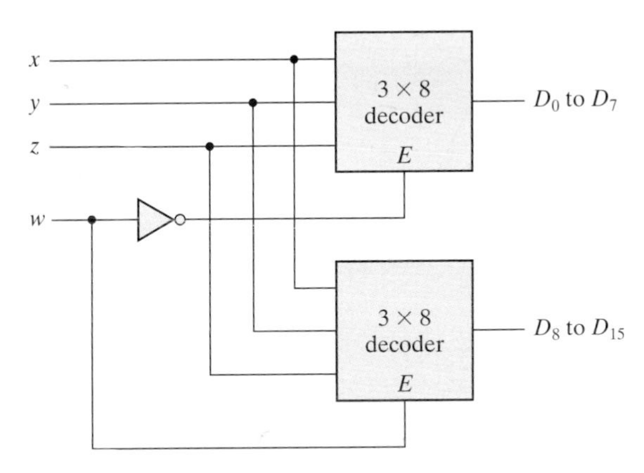

FIGURE 4.20
4×16 decoder constructed with two 3×8 decoders

combinational logic components. In general, enable inputs are a convenient feature for interconnecting two or more standard components for the purpose of combining them into a similar function with more inputs and outputs.

Combinational Logic Implementation

A decoder provides the 2^n minterms of n input variables. Each asserted output of the decoder is associated with a unique pattern of input bits. Since any Boolean function can be expressed in sum-of-minterms form, a decoder that generates the minterms of the function, together with an external OR gate that forms their logical sum, provides a hardware implementation of the function. In this way, any combinational circuit with n inputs and m outputs can be implemented with an n-to-2^n-line decoder and m OR gates.

The procedure for implementing a combinational circuit by means of a decoder and OR gates requires that the Boolean function for the circuit be expressed as a sum of minterms. A decoder is then chosen that generates all the minterms of the input variables. The inputs to each OR gate are selected from the decoder outputs according to the list of minterms of each function. This procedure will be illustrated by an example that implements a full-adder circuit.

From the truth table of the full adder (see Table 4.4), we obtain the functions for the combinational circuit in sum-of-minterms form:

$$S(x, y, z) = \Sigma(1, 2, 4, 7)$$

$$C(x, y, z) = \Sigma(3, 5, 6, 7)$$

Since there are three inputs and a total of eight minterms, we need a three-to-eight-line decoder. The implementation is shown in Fig. 4.21. The decoder generates the eight minterms for x, y, and z. The OR gate for output S forms the logical sum of minterms 1, 2, 4, and 7. The OR gate for output C forms the logical sum of minterms 3, 5, 6, and 7.

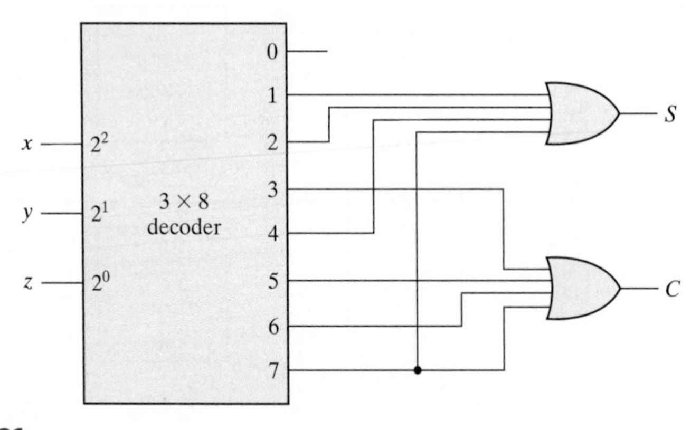

FIGURE 4.21
Implementation of a full adder with a decoder

A function with a long list of minterms requires an OR gate with a large number of inputs. A function having a list of k minterms can be expressed in its complemented form F' with $2^n - k$ minterms. If the number of minterms in the function is greater than $2^n/2$, then F' can be expressed with fewer minterms. In such a case, it is advantageous to use a NOR gate to sum the minterms of F'. The output of the NOR gate complements this sum and generates the normal output F. If NAND gates are used for the decoder, as in Fig. 4.19, then the external gates must be NAND gates instead of OR gates. This is because a two-level NAND gate circuit implements a sum-of-minterms function and is equivalent to a two-level AND–OR circuit.

4.10 ENCODERS

An encoder is a digital circuit that performs the inverse operation of a decoder. An encoder has 2^n (or fewer) input lines and n output lines. The output lines, as an aggregate, generate the binary code corresponding to the input value. An example of an encoder is the octal-to-binary encoder whose truth table is given in Table 4.7. It has eight inputs (one for each of the octal digits) and three outputs that generate the corresponding binary number. It is assumed that only one input has a value of 1 at any given time.

The encoder can be implemented with OR gates whose inputs are determined directly from the truth table. Output z is equal to 1 when the input octal digit is 1, 3, 5, or 7. Output y is 1 for octal digits 2, 3, 6, or 7, and output x is 1 for digits 4, 5, 6, or 7. These conditions can be expressed by the following Boolean output functions:

$$z = D_1 + D_3 + D_5 + D_7$$
$$y = D_2 + D_3 + D_6 + D_7$$
$$x = D_4 + D_5 + D_6 + D_7$$

The encoder can be implemented with three OR gates.

Table 4.7
Truth Table of an Octal-to-Binary Encoder

Inputs								Outputs		
D_0	D_1	D_2	D_3	D_4	D_5	D_6	D_7	x	y	z
1	0	0	0	0	0	0	0	0	0	0
0	1	0	0	0	0	0	0	0	0	1
0	0	1	0	0	0	0	0	0	1	0
0	0	0	1	0	0	0	0	0	1	1
0	0	0	0	1	0	0	0	1	0	0
0	0	0	0	0	1	0	0	1	0	1
0	0	0	0	0	0	1	0	1	1	0
0	0	0	0	0	0	0	1	1	1	1

The encoder defined in Table 4.7 has the limitation that only one input can be active at any given time. If two inputs are active simultaneously, the output produces an undefined combination. For example, if D_3 and D_6 are 1 simultaneously, the output of the encoder will be 111 because all three outputs are equal to 1. The output 111 does not represent either binary 3 or binary 6. To resolve this ambiguity, encoder circuits must establish an input priority to ensure that only one input is encoded. If we establish a higher priority for inputs with higher subscript numbers, and if both D_3 and D_6 are 1 at the same time, the output will be 110 because D_6 has higher priority than D_3.

Another ambiguity in the octal-to-binary encoder is that an output with all 0's is generated when all the inputs are 0; but this output is the same as when D_0 is equal to 1. The discrepancy can be resolved by providing one more output to indicate whether at least one input is equal to 1.

Priority Encoder

A priority encoder is an encoder circuit that includes the priority function. The operation of the priority encoder is such that if two or more inputs are equal to 1 at the same time, the input having the highest priority will take precedence. The truth table of a four-input priority encoder is given in Table 4.8. In addition to the two outputs x and y, the circuit has a third output designated by V; this is a *valid* bit indicator that is set to 1 when one or more inputs are equal to 1. If all inputs are 0, there is no valid input and V is equal to 0. The other two outputs are not inspected when V equals 0 and are specified as don't-care conditions. Note that whereas X's in output columns represent don't-care conditions, the X's in the input columns are useful for representing a truth table in condensed form. Instead of listing all 16 minterms of four variables, the truth table uses an X to represent either 1 or 0. For example, X100 represents the two minterms 0100 and 1100.

According to Table 4.8, the higher the subscript number, the higher the priority of the input. Input D_3 has the highest priority, so, regardless of the values of the other inputs, when this input is 1, the output for xy is 11 (binary 3). D_2 has the next priority level. The output is 10 if $D_2 = 1$, provided that $D_3 = 0$, regardless of the values of the other two lower priority inputs. The output for D_1 is generated only if higher priority inputs are 0, and so on down the priority levels.

Table 4.8
Truth Table of a Priority Encoder

Inputs				Outputs		
D_0	D_1	D_2	D_3	x	y	V
0	0	0	0	X	X	0
1	0	0	0	0	0	1
X	1	0	0	0	1	1
X	X	1	0	1	0	1
X	X	X	1	1	1	1

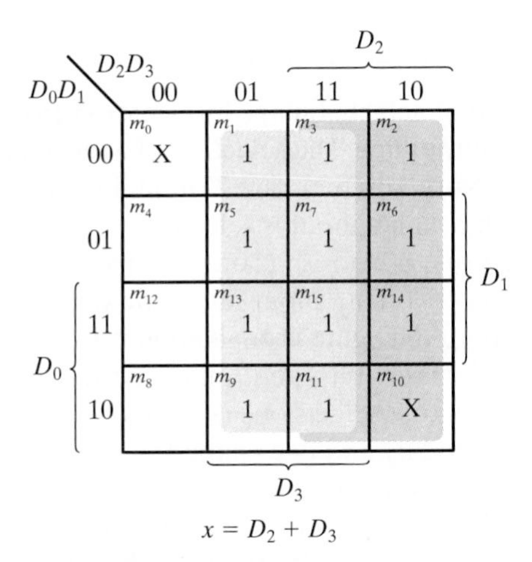

$x = D_2 + D_3$

$y = D_3 + D_1D'_2$

FIGURE 4.22
Maps for a priority encoder

The maps for simplifying outputs x and y are shown in Fig. 4.22. The minterms for the two functions are derived from Table 4.8. Although the table has only five rows, when each X in a row is replaced first by 0 and then by 1, we obtain all 16 possible input combinations. For example, the fourth row in the table, with inputs XX10, represents the four minterms 0010, 0110, 1010, and 1110. The simplified Boolean expressions for the priority encoder are obtained from the maps. The condition for output V is an OR function of all the input variables. The priority encoder is implemented in Fig. 4.23 according to the following Boolean functions:

$$x = D_2 + D_3$$
$$y = D_3 + D_1 D_2'$$
$$V = D_0 + D_1 + D_2 + D_3$$

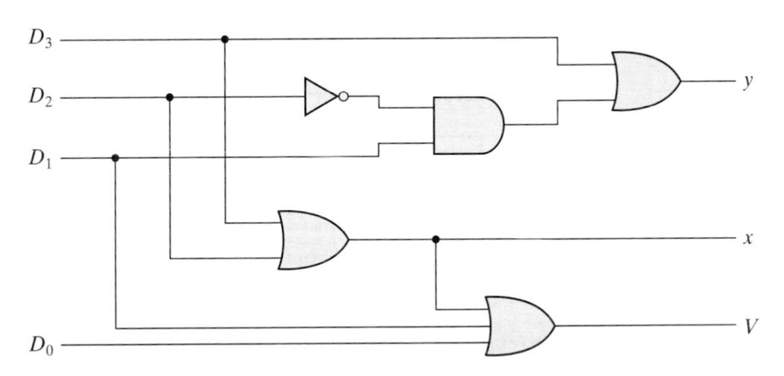

FIGURE 4.23
Four-input priority encoder

4.11 MULTIPLEXERS

A multiplexer is a combinational circuit that selects binary information from one of many input lines and directs it to a single output line. The selection of a particular input line is controlled by a set of selection lines. Normally, there are 2^n input lines and n selection lines whose bit combinations determine which input is selected.

A two-to-one-line multiplexer connects one of two 1-bit sources to a common destination, as shown in Fig. 4.24. The circuit has two data input lines, one output line, and one selection line S. When $S = 0$, the upper AND gate is enabled and I_0 has a path to the output. When $S = 1$, the lower AND gate is enabled and I_1 has a path to the output. The multiplexer acts like an electronic switch that selects one of two sources. The block diagram of a multiplexer is sometimes depicted by a wedge-shaped symbol, as shown in Fig. 4.24(b). It suggests visually how a selected one of multiple data sources is directed into a single destination. The multiplexer is often labeled "MUX" in block diagrams.

A four-to-one-line multiplexer is shown in Fig. 4.25. Each of the four inputs, I_0 through I_3, is applied to one input of an AND gate. Selection lines S_1 and S_0 are decoded to select a particular AND gate. The outputs of the AND gates are applied to a single OR gate that provides the one-line output. The function table lists the input that is passed to the output for each combination of the binary selection values. To demonstrate the operation of the circuit, consider the case when $S_1S_0 = 10$. The AND gate associated with input I_2 has two of its inputs equal to 1 and the third input connected to I_2. The other three AND gates have at least one input equal to 0, which makes their outputs equal to 0. The output of the OR gate is now equal to the value of I_2, providing a path from the selected input to the output. A multiplexer is also called a *data selector*, since it selects one of many inputs and steers the binary information to the output line.

The AND gates and inverters in the multiplexer resemble a decoder circuit, and indeed, they decode the selection input lines. In general, a 2^n-to-1-line multiplexer is constructed from an n-to-2^n decoder by adding 2^n input lines to it, one to each AND gate. The outputs of the AND gates are applied to a single OR gate. The size of a multiplexer is specified by

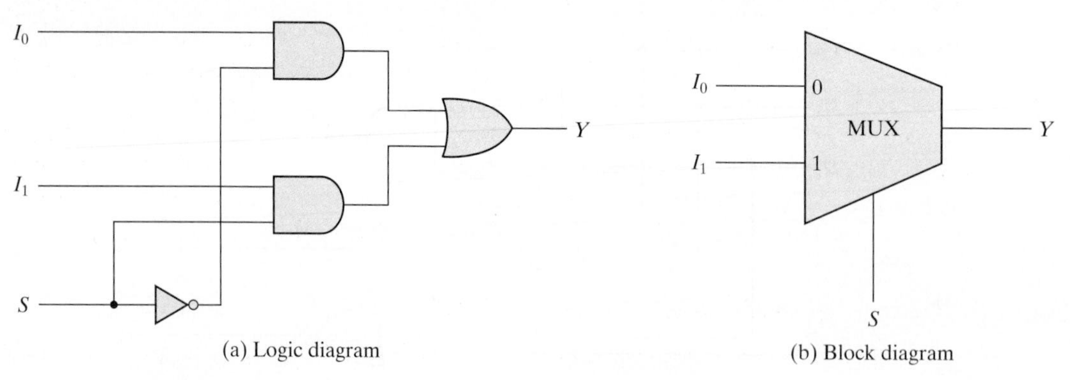

FIGURE 4.24
Two-to-one-line multiplexer

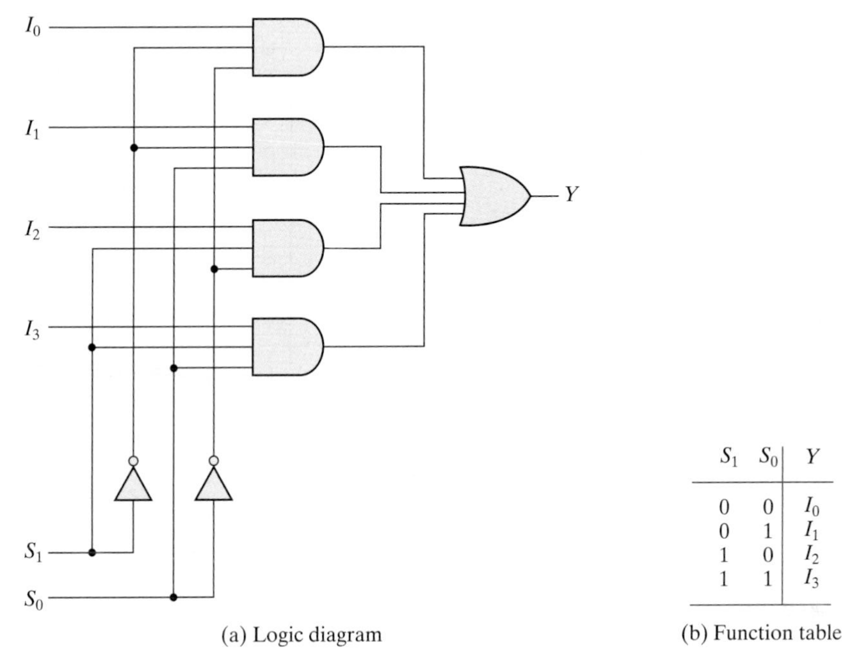

(a) Logic diagram

S_1	S_0	Y
0	0	I_0
0	1	I_1
1	0	I_2
1	1	I_3

(b) Function table

FIGURE 4.25
Four-to-one-line multiplexer

the number 2^n of its data input lines and the single output line. The n selection lines are implied from the 2^n data lines. As in decoders, multiplexers may have an enable input to control the operation of the unit. When the enable input is in the inactive state, the outputs are disabled, and when it is in the active state, the circuit functions as a normal multiplexer.

Multiplexer circuits can be combined with common selection inputs to provide multiple-bit selection logic. As an illustration, a quadruple 2-to-1-line multiplexer is shown in Fig. 4.26. The circuit has four multiplexers, each capable of selecting one of two input lines. Output Y_0 can be selected to come from either input A_0 or input B_0. Similarly, output Y_1 may have the value of A_1 or B_1, and so on. Input selection line S selects one of the lines in each of the four multiplexers. The enable input E must be active (i.e., asserted) for normal operation. Although the circuit contains four 2-to-1-line multiplexers, we are more likely to view it as a circuit that selects one of two 4-bit sets of data lines. As shown in the function table, the unit is enabled when $E = 0$. Then, if $S = 0$, the four A inputs have a path to the four outputs. If, by contrast, $S = 1$, the four B inputs are applied to the outputs. The outputs have all 0's when $E = 1$, regardless of the value of S.

Boolean Function Implementation

In Section 4.9, it was shown that a decoder can be used to implement Boolean functions by employing external OR gates. An examination of the logic diagram of a multiplexer reveals that it is essentially a decoder that includes the OR gate within the unit. The

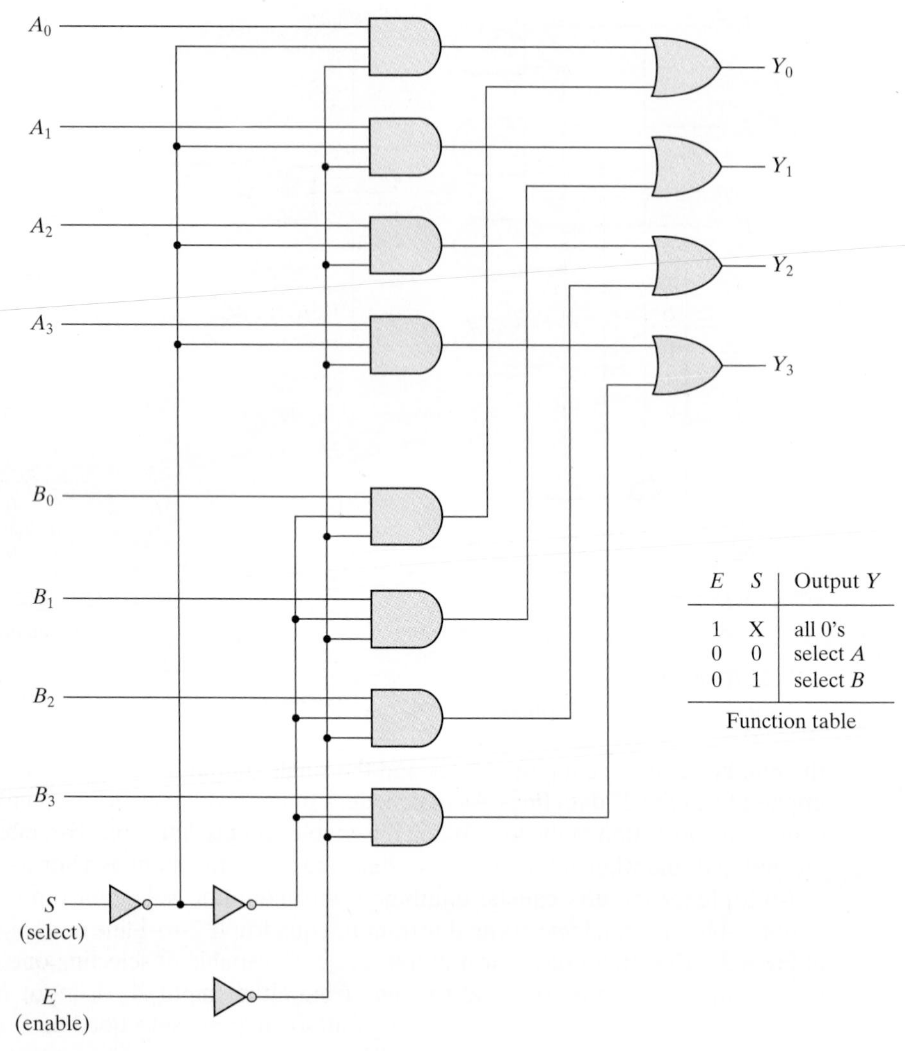

E	S	Output Y
1	X	all 0's
0	0	select A
0	1	select B

Function table

FIGURE 4.26
Quadruple two-to-one-line multiplexer

minterms of a function are generated in a multiplexer by the circuit associated with the selection inputs. The individual minterms can be selected by the data inputs, thereby providing a method of implementing a Boolean function of n variables with a multiplexer that has n selection inputs and 2^n data inputs, one for each minterm.

We will now show a more efficient method for implementing a Boolean function of n variables with a multiplexer that has $n - 1$ selection inputs. The first $n - 1$ variables of the function are connected to the selection inputs of the multiplexer. The remaining single variable of the function is used for the data inputs. If the single variable is denoted by z, each data input of the multiplexer will be z, z', 1, or 0. To demonstrate this procedure, consider the Boolean function

$$F(x, y, z) = \Sigma(1, 2, 6, 7)$$

This function of three variables can be implemented with a four-to-one-line multiplexer as shown in Fig. 4.27. The two variables x and y are applied to the selection lines in that order; x is connected to the S_1 input and y to the S_0 input. The values for the data input lines are determined from the truth table of the function. When $xy = 00$, output F is equal to z because $F = 0$ when $z = 0$ and $F = 1$ when $z = 1$. This requires that variable z be applied to data input 0. The operation of the multiplexer is such that when $xy = 00$, data input 0 has a path to the output, and that makes F equal to z. In a similar fashion, we can determine the required input to data lines 1, 2, and 3 from the value of F when $xy = 01$, 10, and 11, respectively. This particular example shows all four possibilities that can be obtained for the data inputs.

The general procedure for implementing any Boolean function of n variables with a multiplexer with $n - 1$ selection inputs and 2^{n-1} data inputs follows from the previous example. To begin with, Boolean function is listed in a truth table. Then first $n - 1$ variables in the table are applied to the selection inputs of the multiplexer. For each combination of the selection variables, we evaluate the output as a function of the last variable. This function can be 0, 1, the variable, or the complement of the variable. These values are then applied to the data inputs in the proper order.

As a second example, consider the implementation of the Boolean function

$$F(A, B, C, D) = \Sigma(1, 3, 4, 11, 12, 13, 14, 15)$$

This function is implemented with a multiplexer with three selection inputs as shown in Fig. 4.28. Note that the first variable A must be connected to selection input S_2 so that A, B, and C correspond to selection inputs S_2, S_1, and S_0, respectively. The values for the

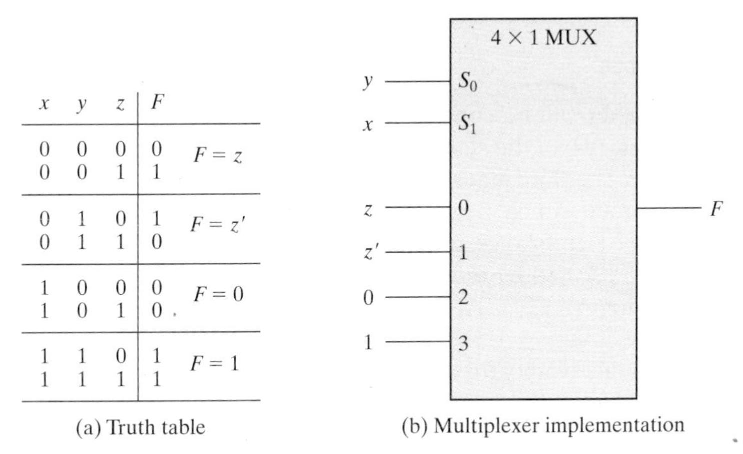

x	y	z	F	
0	0	0	0	$F = z$
0	0	1	1	
0	1	0	1	$F = z'$
0	1	1	0	
1	0	0	0	$F = 0$
1	0	1	0	
1	1	0	1	$F = 1$
1	1	1	1	

(a) Truth table

(b) Multiplexer implementation

FIGURE 4.27
Implementing a Boolean function with a multiplexer

A	B	C	D	F	
0	0	0	0	0	$F = D$
0	0	0	1	1	
0	0	1	0	0	$F = D$
0	0	1	1	1	
0	1	0	0	1	$F = D'$
0	1	0	1	0	
0	1	1	0	0	$F = 0$
0	1	1	1	0	
1	0	0	0	0	$F = 0$
1	0	0	1	0	
1	0	1	0	0	$F = D$
1	0	1	1	1	
1	1	0	0	1	$F = 1$
1	1	0	1	1	
1	1	1	0	1	$F = 1$
1	1	1	1	1	

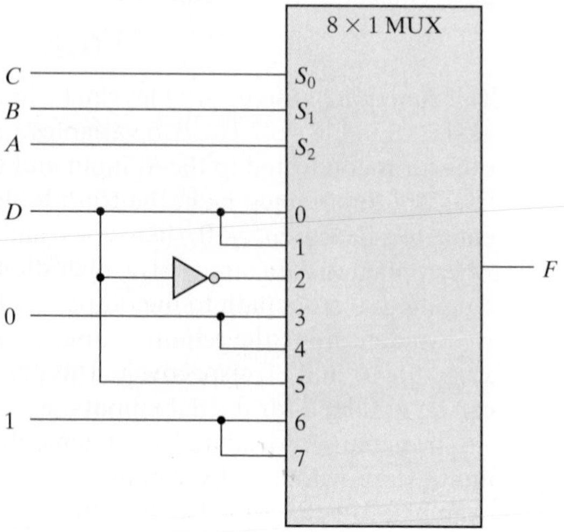

FIGURE 4.28
Implementing a four-input function with a multiplexer

data inputs are determined from the truth table listed in the figure. The corresponding data line number is determined from the binary combination of ABC. For example, the table shows that when $ABC = 101$, $F = D$, so the input variable D is applied to data input 5. The binary constants 0 and 1 correspond to two fixed signal values. When integrated circuits are used, logic 0 corresponds to signal ground and logic 1 is equivalent to the power signal, depending on the technology (e.g., 3 V).

Three-State Gates

A multiplexer can be constructed with three-state gates—digital circuits that exhibit three states. Two of the states are signals equivalent to logic 1 and logic 0 as in a conventional gate. The third state is a *high-impedance* state in which (1) the logic behaves like an open circuit, which means that the output appears to be disconnected, (2) the circuit has no logic significance, and (3) the circuit connected to the output of the three-state gate is not affected by the inputs to the gate. Three-state gates may perform any conventional logic, such as AND or NAND. However, the one most commonly used is the buffer gate.

The graphic symbol for a three-state buffer gate is shown in Fig. 4.29. It is distinguished from a normal buffer by an input control line entering the bottom of the symbol. The buffer has a normal input, an output, and a control input that determines the state of the output. When the control input is equal to 1, the output is enabled and the gate behaves like a conventional buffer, with the output equal to the normal input. When the control

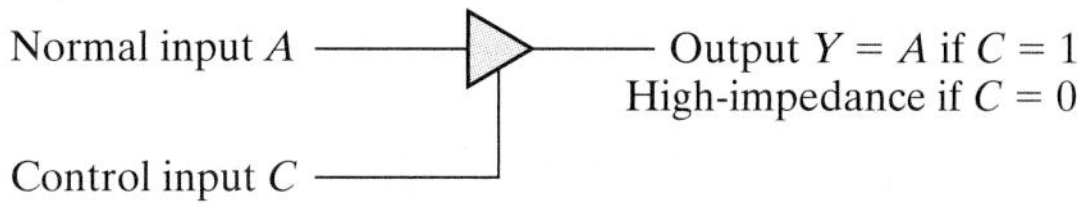

Normal input A

Output $Y = A$ if $C = 1$
High-impedance if $C = 0$

Control input C

FIGURE 4.29
Graphic symbol for a three-state buffer

input is 0, the output is disabled and the gate goes to a high-impedance state, regardless of the value in the normal input. The high-impedance state of a three-state gate provides a special feature not available in other gates. Because of this feature, a large number of three-state gate outputs can be connected with wires to form a common line without endangering loading effects.

The construction of multiplexers with three-state buffers is demonstrated in Fig. 4.30. Fig. 4.30(a) shows the construction of a two-to-one-line multiplexer with 2 three-state buffers and an inverter. The two outputs are connected together to form a single output line. (Note that this type of connection cannot be made with gates that do not have three-state outputs.) When the select input is 0, the upper buffer is enabled by its control input and the lower buffer is disabled. Output Y is then equal to input A. When the select input is 1, the lower buffer is enabled and Y is equal to B.

The construction of a four-to-one-line multiplexer is shown in Fig. 4.30(b). The outputs of 4 three-state buffers are connected together to form a single output line. The control inputs to the buffers determine which one of the four normal inputs I_0 through

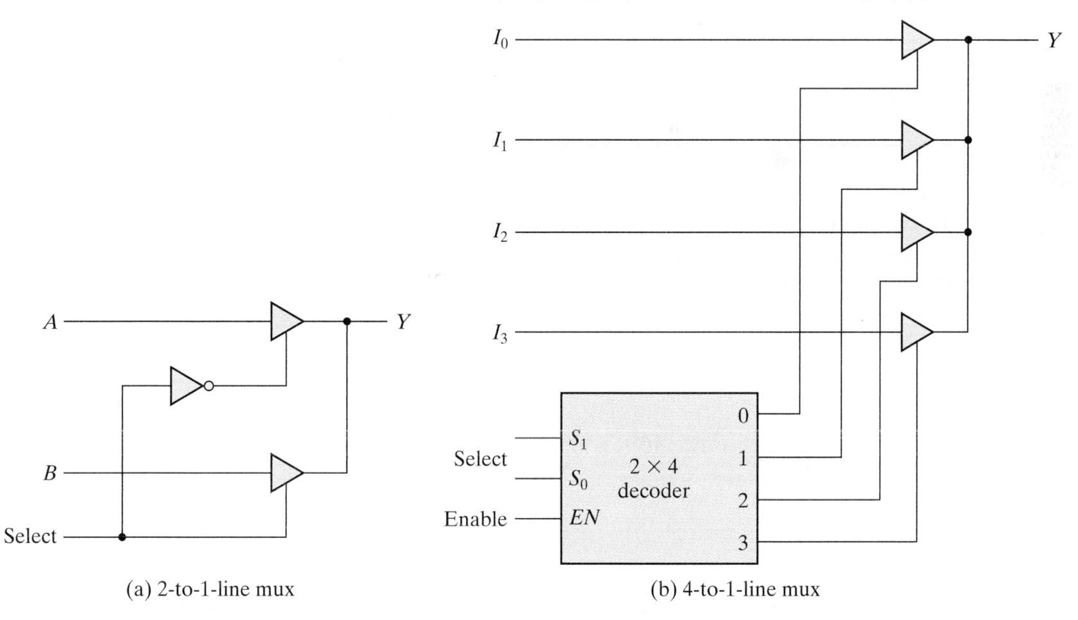

(a) 2-to-1-line mux

(b) 4-to-1-line mux

FIGURE 4.30
Multiplexers with three-state gates

I_3 will be connected to the output line. No more than one buffer may be in the active state at any given time. The connected buffers must be controlled so that only 1 three-state buffer has access to the output while all other buffers are maintained in a high-impedance state. One way to ensure that no more than one control input is active at any given time is to use a decoder, as shown in the diagram. When the enable input of the decoder is 0, all of its four outputs are 0 and the bus line is in a high-impedance state because all four buffers are disabled. When the enable input is active, one of the three-state buffers will be active, depending on the binary value in the select inputs of the decoder. Careful investigation reveals that this circuit is another way of constructing a four-to-one-line multiplexer.

4.12 HDL MODELS OF COMBINATIONAL CIRCUITS

The Verilog HDL was introduced in Section 3.10. In the current section, we introduce additional features of Verilog, present more elaborate examples, and compare alternative descriptions of combinational circuits in Verilog. Sequential circuits are presented in Chapter 5. As mentioned previously, the module is the basic building block for modeling hardware with the Verilog HDL. The logic of a module can be described in any one (or a combination) of the following modeling styles:

- Gate-level modeling using instantiations of predefined and user-defined primitive gates.
- Dataflow modeling using continuous assignment statements with the keyword **assign**.
- Behavioral modeling using procedural assignment statements with the keyword **always**.

Gate-level (structural) modeling describes a circuit by specifying its gates and how they are connected with each other. Dataflow modeling is used mostly for describing the Boolean equations of combinational logic. We'll also consider here behavioral modeling that is used to describe combinational and sequential circuits at a higher level of abstraction. Combinational logic can be designed with truth tables, Boolean equations, and schematics; Verilog has a construct corresponding to each of these "classical" approaches to design: user-defined primitives, continuous assignments, and primitives, as shown in Fig. 4.31. There is one other modeling style, called switch-level modeling. It is sometimes used in the simulation of MOS transistor circuit models, but not in logic synthesis. We will not consider switch-level modeling.

Gate-Level Modeling

Gate-level modeling was introduced in Section 3.10 with a simple example. In this type of representation, a circuit is specified by its logic gates and their interconnections. Gate-level modeling provides a textual description of a schematic diagram. The Verilog HDL

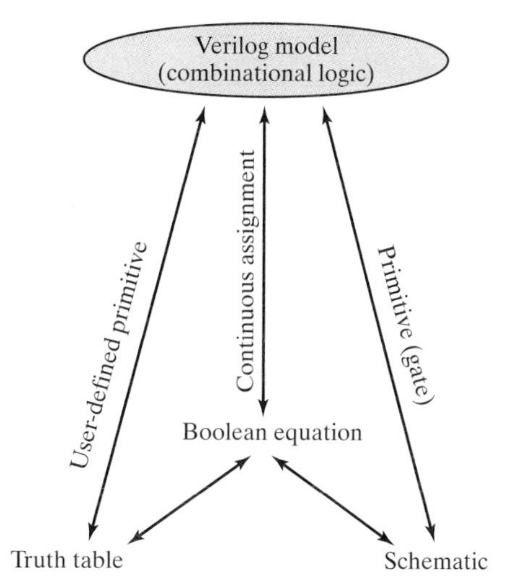

FIGURE 4.31
Relationship of Verilog constructs to truth tables, Boolean equations, and schematics

includes 12 basic gates as predefined primitives. Four of these primitive gates are of the three-state type. The other eight are the same as the ones listed in Section 2.8. They are all declared with the lowercase keywords **and**, **nand**, **or**, **nor**, **xor**, **xnor**, **not**, and **buf**. Primitives such as **and** are *n*-input primitives. They can have any number of scalar inputs (e.g., a three-input **and** primitive). The **buf** and **not** primitives are *n*-output primitives. A single input can drive multiple output lines distinguished by their identifiers.

The Verilog language includes a functional description of each type of gate, too. The logic of each gate is based on a four-valued system. When the gates are simulated, the simulator assigns one value to the output of each gate at any instant. In addition to the two logic values of 0 and 1, there are two other values: *unknown* and *high impedance*. An unknown value is denoted by **x** and a high impedance by **z**. An unknown value is assigned during simulation when the logic value of a signal is ambiguous—for instance, if it cannot be determined whether its value is 0 or 1 (e.g., a flip-flop without a reset condition). A high-impedance condition occurs at the output of three-state gates that are not enabled or if a wire is inadvertently left unconnected. The four-valued logic truth tables for the **and**, **or**, **xor**, and **not** primitives are shown in Table 4.9. The truth table for the other four gates is the same, except that the outputs are complemented. Note that for the **and** gate, the output is 1 only when both inputs are 1 and the output is 0 if any input is 0. Otherwise, if one input is **x** or **z**, the output is **x**. The output of the **or** gate is 0 if both inputs are 0, is 1 if any input is 1, and is **x** otherwise.

When a primitive gate is listed in a module, we say that it is *instantiated* in the module. In general, component instantiations are statements that reference lower level components in the design, essentially creating unique copies (or *instances*) of those components in the higher level module. Thus, a module that uses a gate in its description is said to

Table 4.9
Truth Table for Predefined Primitive Gates

and	0	1	x	z
0	0	0	0	0
1	0	1	x	x
x	0	x	x	x
z	0	x	x	x

or	0	1	x	z
0	0	1	x	x
1	1	1	1	1
x	x	1	x	x
z	x	1	x	x

xor	0	1	x	z
0	0	1	x	x
1	1	0	x	x
x	x	x	x	x
z	x	x	x	x

not	input	output
	0	1
	1	0
	x	x
	z	x

instantiate the gate. Think of instantiation as the HDL counterpart of placing and connecting parts on a circuit board.

We now present two examples of gate-level modeling. Both examples use identifiers having multiple bit widths, called *vectors*. The syntax specifying a vector includes within square brackets two numbers separated with a colon. The following Verilog statements specify two vectors:

```
output [0: 3] D;
wire   [7: 0] SUM;
```

The first statement declares an output vector *D* with four bits, 0 through 3. The second declares a wire vector *SUM* with eight bits numbered 7 through 0. (*Note*: The first (leftmost) number (array index) listed is always the most significant bit of the vector.) The individual bits are specified within square brackets, so *D[2]* specifies bit 2 of *D*. It is also possible to address parts (contiguous bits) of vectors. For example, *SUM[2: 0]* specifies the three least significant bits of vector *SUM*.

HDL Example 4.1 shows the gate-level description of a two-to-four-line decoder. (See Fig. 4.19.) This decoder has two data inputs *A* and *B* and an enable input *E*. The four outputs are specified with the vector *D*. The **wire** declaration is for internal connections. Three **not** gates produce the complement of the inputs, and four **nand** gates provide the outputs for *D*. Remember that *the output is always listed first in the port list of a primitive*, followed by the inputs. This example describes the decoder of Fig. 4.19 and follows the procedures established in Section 3.10. Note that the keywords **not** and **nand** are written only once and do not have to be repeated for each gate, but commas must be inserted at the end of each of the gates in the series, except for the last statement, which must be terminated with a semicolon.

HDL Example 4.1 (Two-to-Four-Line Decoder)

```
// Gate-level description of two-to-four-line decoder
// Refer to Fig. 4.19 with symbol E replaced by enable, for clarity.

module decoder_2x4_gates (D, A, B, enable);
 output      [0: 3]      D;
 input                   A, B;
 input                   enable;
 wire                    A_not,B_not, enable_not;

not
  G1 (A_not, A),
  G2 (B_not, B),
  G3 (enable_not, enable);
nand
  G4 (D[0], A_not, B_not, enable_not),
  G5 (D[1], A_not, B, enable_not),
  G6 (D[2], A, B_not, enable_not),
  G7 (D[3], A, B, enable_not);
endmodule
```

Two or more modules can be combined to build a hierarchical description of a design. There are two basic types of design methodologies: top down and bottom up. In a *top-down* design, the top-level block is defined and then the subblocks necessary to build the top-level block are identified. In a *bottom-up* design, the building blocks are first identified and then combined to build the top-level block. Take, for example, the binary adder of Fig. 4.9. It can be considered as a top-block component built with four full-adder blocks, while each full adder is built with two half-adder blocks. In a top-down design, the four-bit adder is defined first, and then the two adders are described. In a bottom-up design, the half adder is defined, then each full adder is constructed, and then the four-bit adder is built from the full adders.

A bottom-up hierarchical description of a four-bit adder is shown in HDL Example 4.2. The half adder is defined by instantiating primitive gates. The next module describes the full adder by instantiating and connecting two half adders. The third module describes the four-bit adder by instantiating and connecting four full adders. Note that the first character of an identifier cannot be a number, but can be an underscore, so the module name *_4bitadder* is valid. An alternative name that is meaningful, but does not require a leading underscore, is *adder_4_bit*. The instantiation is done by using the name of the module that is instantiated together with a new (or the same) set of port names. For example, the half adder *HA1* inside the full adder module is instantiated with ports *S1*, *C1*, *x*, and *y*. This produces a half adder with outputs *S1* and *C1* and inputs *x* and *y*.

HDL Example 4.2 (Ripple-Carry Adder)

```
// Gate-level description of four-bit ripple carry adder
// Description of half adder (Fig. 4.5b)

// module half_adder (S, C, x, y);                      // Verilog 1995 syntax
// output  S, C;
// input    x, y;

module half_adder (output S, C, input x, y);           // Verilog 2001, 2005 syntax
// Instantiate primitive gates
  xor (S, x, y);
  and (C, x, y);
endmodule

// Description of full adder (Fig. 4.8)                 // Verilog 1995 syntax
// module full_adder (S, C, x, y, z);
// output            S, C;
// input             x, y, z;

module full_adder (output S, C, input x, y, z);         // Verilog 2001, 2005 syntax
  wire S1, C1, C2;

// Instantiate half adders
  half_adder HA1 (S1, C1, x, y);
  half_adder HA2 (S, C2, S1, z);
  or G1 (C, C2, C1);
endmodule

// Description of four-bit adder (Fig. 4.9)       // Verilog 1995 syntax
// module ripple_carry_4_bit_adder (Sum, C4, A, B, C0);
// output [3: 0]    Sum;
// output           C4;
// input  [3: 0]    A, B;
// input            C0;
// Alternative Verilog 2001, 2005 syntax:

module ripple_carry_4_bit_adder ( output [3: 0] Sum, output C4,
  input [3: 0] A, B, input C0);
  wire           C1, C2, C3;        // Intermediate carries
// Instantiate chain of full adders
full_adder        FA0 (Sum[0], C1, A[0], B[0], C0),
                  FA1 (Sum[1], C2, A[1], B[1], C1),
                  FA2 (Sum[2], C3, A[2], B[2], C2),
                  FA3 (Sum[3], C4, A[3], B[3], C3);
endmodule
```

HDL Example 4.2 illustrates Verilog 2001, 2005 syntax, which eliminates extra typing of identifiers declaring the mode (e.g., **output**), type (**reg**), and declaration of a vector range (e.g., [3: 0]) of a port. The first version of the standard (1995) uses separate statements for these declarations.

Note that modules can be instantiated (nested) within other modules, but module declarations cannot be nested; that is, a module definition (declaration) cannot be placed within another module declaration. In other words, a module definition cannot be inserted into the text between the **module** and **endmodule** keywords of another module. The only way one module definition can be incorporated into another module is by instantiating it. Instantiating modules within other modules creates a hierarchical decomposition of a design. A description of a module is said to be a *structural* description if it is composed of instantiations of other modules. Note also that *instance names* must be specified when defined modules are instantiated (such as *FA0* for the first full adder in the third module), but using a name is optional when instantiating primitive gates. Module *ripple_carry_4_bit_adder* is composed of instantiated and interconnected full adders, each of which is itself composed of half adders and some *glue logic*. The top level, or parent module, of the design hierarchy is the module *ripple_carry_4_bit_adder*. Four copies of *full_adder* are its child modules, etc. *C0* is an input of the cell forming the least significant bit of the chain, and *C4* is the output of the cell forming the most significant bit.

Three-State Gates

As mentioned in Section 4.11, a three-state gate has a control input that can place the gate into a high-impedance state. The high-impedance state is symbolized by **z** in Verilog. There are four types of three-state gates, as shown in Fig. 4.32. The **bufif1** gate behaves like a normal buffer if $control = 1$. The output goes to a high-impedance state **z** when $control = 0$. The **bufif0** gate behaves in a similar fashion, except that the high-impedance state occurs when $control = 1$. The two **notif** gates operate in a similar manner, except that the output is the complement of the input when the gate is not in a high-impedance state. The gates are instantiated with the statement

$$gate\,name\ (output, input, control);$$

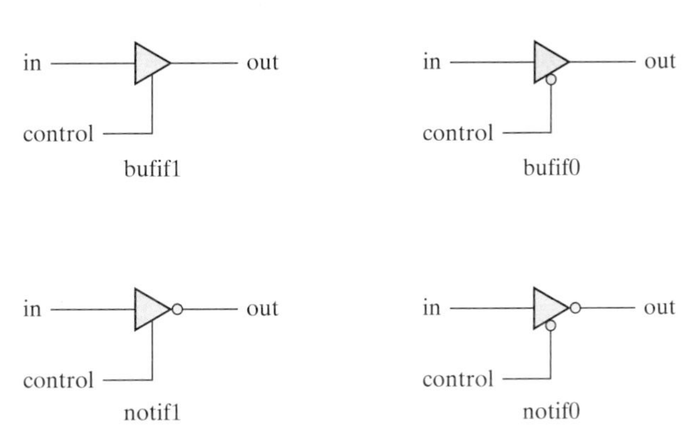

FIGURE 4.32
Three-state gates

The gate name can be that of any 1 of the 4 three-state gates. In simulation, the output can result in 0, 1, **x**, or **z.** Two examples of gate instantiation are

```
bufif1  (OUT, A, control);
notif0  (Y, B, enable);
```

In the first example, input A is transferred to OUT when $control = 1$. OUT goes to **z** when $control = 0$. In the second example, output $Y =$ **z** when $enable = 1$ and output $Y = B'$ when $enable = 0$.

The outputs of three-state gates can be connected together to form a common output line. To identify such a connection, Verilog HDL uses the keyword **tri** (for tristate) to indicate that the output has multiple drivers. As an example, consider the two-to-one-line multiplexer with three-state gates shown in Fig. 4.33.

The HDL description must use a **tri** data type for the output:

```
// Mux with three-state output

module mux_tri (m_out, A, B, select);
 output m_out;
 input  A, B, select;
 tri    m_out;

 bufif1 (m_out, A, select);
 bufif0 (m_out, B, select);
endmodule
```

The 2 three-state buffers have the same output. In order to show that they have a common connection, it is necessary to declare *m_out* with the keyword **tri.**

Keywords **wire** and **tri** are examples of a set of data types called *nets*, which represent connections between hardware elements. In simulation, their value is determined by a continuous assignment statement or by the device whose output they represent. The word *net* is not a keyword, but represents a class of data types, such as **wire, wor, wand, tri, supply1,** and **supply0.** The **wire** declaration is used most frequently. In fact, if an identifier is used, but not declared, the language specifies that it will be interpreted (by default) as a **wire**. The net **wor** models the hardware implementation of the wired-OR configuration (emitter-coupled logic). The **wand** models the wired-AND configuration (open-collector technology; see Fig. 3.26). The nets **supply1** and **supply0** represent power supply and ground, respectively. They are used to hardwire an input of a device to either 1 or 0.

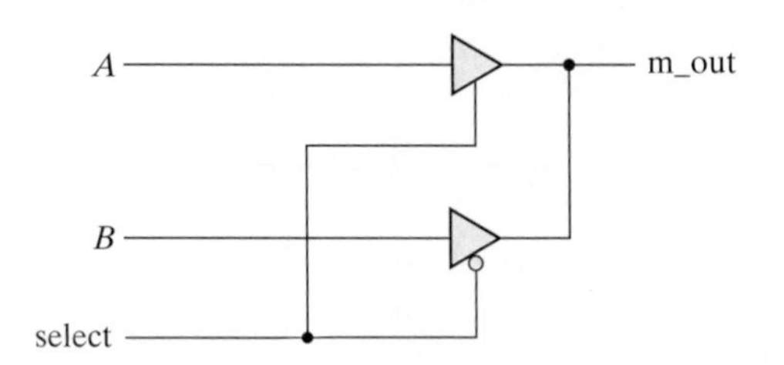

FIGURE 4.33
Two-to-one-line multiplexer with three-state buffers

Dataflow Modeling

Dataflow modeling of combinational logic uses a number of operators that act on binary operands to produce a binary result. Verilog HDL provides about 30 different operators. Table 4.10 lists some of these operators, their symbols, and the operation that they perform. (A complete list of operators supported by Verilog 2001, 2005 can be found in Table 8.1 in Section 8.2.) It is necessary to distinguish between arithmetic and logic operations, so different symbols are used for each. The plus symbol (+) indicates the arithmetic operation of addition; the bitwise logic AND operation (conjunction) uses the symbol &. There are special symbols for bitwise logical OR (disjunction), NOT, and XOR. The equality symbol uses two equals signs (without spaces between them) to distinguish it from the equals sign used with the **assign** statement. The bitwise operators operate bit by bit on a pair of vector operands to produce a vector result. The concatenation operator provides a mechanism for appending multiple operands. For example, two operands with two bits each can be concatenated to form an operand with four bits. The conditional operator acts like a multiplexer and is explained later, in conjunction with HDL Example 4.6.

It should be noted that a bitwise operator (e.g., ~) and its corresponding logical operator (e.g., !) may produce different results, depending on their operand. If the operands are scalar the results will be identical; if the operands are vectors the result will not necessarily match. For example, ~(1010) is (0101), and !(1010) is 0. A binary value is considered to be logically true if it is not 0. In general, use the bitwise operators to describe arithmetic operations and the logical operators to describe logical operations.

Dataflow modeling uses continuous assignments and the keyword **assign.** A continuous assignment is a statement that assigns a value to a net. The data type family *net* is used in Verilog HDL to represent a physical connection between circuit elements. A net

Table 4.10
Some Verilog HDL Operators

Symbol	Operation	Symbol	Operation
+	binary addition		
−	binary subtraction		
&	bitwise AND	&&	logical AND
|	bitwise OR	||	logical OR
^	bitwise XOR		
~	bitwise NOT	!	logical NOT
= =	equality		
>	greater than		
<	less than		
{}	concatenation		
?:	conditional		

is declared explicitly by a net keyword (e.g., **wire**) or by declaring an identifier to be an input port. The logic value associated with a net is determined by what the net is connected to. If the net is connected to an output of a gate, the net is said to be *driven* by the gate, and the logic value of the net is determined by the logic values of the inputs to the gate and the truth table of the gate. If the identifier of a net is the left-hand side of a continuous assignment statement or a procedural assignment statement, the value assigned to the net is specified by a Boolean expression that uses operands and operators. As an example, assuming that the variables were declared, a two-to-one-line multiplexer with scalar data inputs A and B, select input S, and output Y is described with the continuous assignment

```
assign Y = (A && S) || (B && S)
```

The relationship between Y, A, B, and S is declared by the keyword **assign,** followed by the target output Y and an equals sign. Following the equals sign is a Boolean expression. In hardware terms, this assignment would be equivalent to connecting the output of the OR gate to wire Y.

The next two examples show the dataflow models of the two previous gate-level examples. The dataflow description of a two-to-four-line decoder with active-low output enable and inverted output is shown in HDL Example 4.3. The circuit is defined with four continuous assignment statements using Boolean expressions, one for each output. The dataflow description of the four-bit adder is shown in HDL Example 4.4. The addition logic is described by a single statement using the operators of addition and concatenation. The plus symbol (+) specifies the binary addition of the four bits of A with the four bits of B and the one bit of *C_in*. The target output is the *concatenation* of the output carry *C_out* and the four bits of *Sum*. Concatenation of operands is expressed within braces and a comma separating the operands. Thus, *{C_out, Sum}* represents the five-bit result of the addition operation.

HDL Example 4.3 (Dataflow: Two-to-Four Line Decoder)

```
// Dataflow description of two-to-four-line decoder

// See Fig. 4.19. Note: The figure uses symbol E, but the
// Verilog model uses enable to clearly indicate functionality.

module decoder_2x4_df (                          // Verilog 2001, 2005 syntax
  output    [0: 3]          D,
  input                     A, B,
                            enable
);
  assign    D[0] = !((!A) && (!B) && (!enable)),
            D[1] = !(*!A) && B && (!enable)),
            D[2] = !(A && B && (!enable)
            D[3] = !(A && B && (!enable))
endmodule
```

HDL Example 4.4 (Dataflow: Four-Bit Adder)

```
// Dataflow description of four-bit adder
// Verilog 2001, 2005 module port syntax
module binary_adder (
  output [3: 0]         Sum,
  output                C_out,
  input [3: 0]          A, B,
  input                 C_in
);

  assign {C_out, Sum} = A + B + C_in;
endmodule
```

Dataflow HDL models describe combinational circuits by their *function* rather than by their gate structure. To show how dataflow descriptions facilitate digital design, consider the 4-bit magnitude comparator described in HDL Example 4.5. The module specifies two 4-bit inputs A and B and three outputs. One output (A_lt_B) is logic 1 if A is less than B, a second output (A_gt_B) is logic 1 if A is greater than B, and a third output (A_eq_B) is logic 1 if A is equal to B. Note that equality (identity) is symbolized with two equals signs (= =) to distinguish the operation from that of the assignment operator (=). A Verilog HDL synthesis compiler can accept this module description as input, execute synthesis algorithms, and provide an output netlist and a schematic of a circuit equivalent to the one in Fig. 4.17, all without manual intervention! The designer need not draw the schematic.

HDL Example 4.5 (Dataflow: Four-Bit Comparator)

```
// Dataflow description of a four-bit comparator //V2001, 2005 syntax
module mag_compare
( output              A_lt_B, A_eq_B, A_gt_B,
  input [3: 0]        A, B
);
  assign A_lt_B = (A < B);
  assign A_gt_B = (A > B);
  assign A_eq_B = (A = = B);
endmodule
```

The next example uses the conditional operator (? :). This operator takes three operands:

condition ? true-expression : false-expression;

The condition is evaluated. If the result is logic 1, the true expression is evaluated and used to assign a value to the left-hand side of an assignment statement. If the result is logic 0, the false expression is evaluated. The two conditions together are equivalent to an if–else condition. HDL Example 4.6 describes a two-to-one-line multiplexer using the conditional operator. The continuous assignment

```
assign OUT = select ? A : B;
```

specifies the condition that $OUT = A$ if $select = 1$, else $OUT = B$ if $select = 0$.

HDL Example 4.6 (Dataflow: Two-to-One Multiplexer)

```
// Dataflow description of two-to-one-line multiplexer

module mux_2x1_df(m_out, A, B, select);
  output      m_out;
  input       A, B;
  input       select;
  assign m_out = (select)? A : B;
endmodule
```

Behavioral Modeling

Behavioral modeling represents digital circuits at a functional and algorithmic level. It is used mostly to describe sequential circuits, but can also be used to describe combinational circuits. Here, we give two simple combinational circuit examples to introduce the subject. Behavioral modeling is presented in more detail in Section 5.6, after the study of sequential circuits.

Behavioral descriptions use the keyword **always**, followed by an optional event control expression and a list of procedural assignment statements. The event control expression specifies when the statements will execute. The target output of a procedural assignment statement must be of the **reg** data type. Contrary to the **wire** data type, whereby the target output of an assignment may be continuously updated, a **reg** data type retains its value until a new value is assigned.

HDL Example 4.7 shows the behavioral description of a two-to-one-line multiplexer. (Compare it with HDL Example 4.6.) Since variable *m_out* is a target output, it must be declared as **reg** data (in addition to the **output** declaration). The procedural assignment statements inside the **always** block are executed every time there is a change in any of the variables listed after the @ symbol. (Note that there is no semicolon (;) at the end of the **always** statement.) In this case, these variables are the input variables A, B, and *select.* The statements execute if A, B, or *select* changes value. Note that the keyword **or**, instead of the bitwise logical OR operator "|", is used between variables. The conditional statement **if–else** provides a decision based upon the value of the *select* input. The **if** statement can be written without the equality symbol:

```
if (select) OUT = A;
```

The statement implies that *select* is checked for logic 1.

HDL Example 4.7 (Behavioral: Two-to-One Line Multiplexer)

```
// Behavioral description of two-to-one-line multiplexer

module mux_2x1_beh (m_out, A, B, select);
  output      m_out;
  input       A, B, select;
  reg         m_out;

  always      @(A or B or select)
   if (select == 1) m_out = A;
   else m_out 5 B;
endmodule
```

HDL Example 4.8 describes the function of a four-to-one-line multiplexer. The *select* input is defined as a two-bit vector, and output *y* is declared to have type **reg**. The **always** statement, in this example, has a sequential block enclosed between the keywords **case** and **endcase**. The block is executed whenever any of the inputs listed after the **@** symbol changes in value. The **case** statement is a multiway conditional branch construct. Whenever *in_0, in_1, in_2, in_3* or *select* change, the case expression (*select*) is evaluated and its value compared, from top to bottom, with the values in the list of statements that follow, the so-called **case** items. The statement associated with the first **case** item that matches the **case** expression is executed. In the absence of a match, no statement is executed. Since *select* is a two-bit number, it can be equal to 00, 01, 10, or 11. The **case** items have an implied priority because the list is evaluated from top to bottom.

The list is called a *sensitivity list* (Verilog 2001, 2005) and is equivalent to the *event control expression* (Verilog 1995) formed by "ORing" the signals. Combinational logic is reactive—when an input changes an output may change.

HDL Example 4.8 (Behavioral: Four-to-One Line Multiplexer)

```
// Behavioral description of four-to-one line multiplexer

// Verilog 2001, 2005 port syntax

module mux_4x1_beh
( output reg m_out,
  input       in_0, in_1, in_2, in_3,
  input [1: 0] select
);
always @ (in_0, in_1, in_2, in_3, select)      // Verilog 2001, 2005 syntax
  case (select)
   2'b00:         m_out = in_0;
   2'b01:         m_out = in_1;
   2'b10:         m_out = in_2;
   2'b11:         m_out = in_3;
  endcase
endmodule
```

Binary numbers in Verilog are specified and interpreted with the letter **b** preceded by a prime. The size of the number is written first and then its value. Thus, 2′*b*01 specifies a two-bit binary number whose value is 01. Numbers are stored as a bit pattern in memory, but they can be referenced in decimal, octal, or hexadecimal formats with the letters ***d′ o′*** and ***h′*** respectively. For example, 4′HA = 4′d10 = 4′b1010 and have the same internal representation in a simulator. If the base of the number is not specified, its interpretation defaults to decimal. If the size of the number is not specified, the system assumes that the size of the number is at least 32 bits; if a host simulator has a larger word length—say, 64 bits—the language will use that value to store unsized numbers. The integer data type (keyword **integer**) is stored in a 32-bit representation. The underscore (_) may be inserted in a number to improve readability of the code (e.g., 16′b0101_1110_0101_0011). It has no other effect.

The **case** construct has two important variations: **casex** and **casez**. The first will treat as don't-cares any bits of the **case** expression or the **case** item that have logic value **x** or **z**. The **casez** construct treats as don't-cares only the logic value **z**, for the purpose of detecting a match between the **case** expression and a **case** item.

The list of case items need not be complete. If the list of **case** items does not include all possible bit patterns of the **case** expression, no match can be detected. Unlisted **case** items, i.e., bit patterns that are not explicitly decoded can be treated by using the **default** keyword as the last item in the list of **case** items. The associated statement will execute when no other match is found. This feature is useful, for example, when there are more possible state codes in a sequential machine than are actually used. Having a **default** case item lets the designer map all of the unused states to a desired next state without having to elaborate each individual state, rather than allowing the synthesis tool to arbitrarily assign the next state.

The examples of behavioral descriptions of combinational circuits shown here are simple ones. Behavioral modeling and procedural assignment statements require knowledge of sequential circuits and are covered in more detail in Section 5.6.

Writing a Simple Test Bench

A test bench is an HDL program used for describing and applying a stimulus to an HDL model of a circuit in order to test it and observe its response during simulation. Test benches can be quite complex and lengthy and may take longer to develop than the design that is tested. The results of a test are only as good as the test bench that is used to test a circuit. Care must be taken to write stimuli that will test a circuit thoroughly, exercising all of the operating features that are specified. However, the test benches considered here are relatively simple, since the circuits we want to test implement only combinational logic. The examples are presented to demonstrate some basic features of HDL stimulus modules. Chapter 8 considers test benches in greater depth.

In addition to employing the **always** statement, test benches use the **initial** statement to provide a stimulus to the circuit being tested. We use the term "**always** statement" loosely. Actually, **always** is a Verilog language construct specifying *how* the associated statement is to execute (subject to the event control expression). The **always** statement executes repeatedly in a loop. The **initial** statement executes only once, starting from simulation time 0, and may continue with any operations that are delayed by a given number of time units, as specified by the symbol #. For example, consider the **initial** block

```
initial
  begin
        A = 0; B = 0;
    #10 A = 1;
    #20 A = 0; B = 1;
  end
```

The block is enclosed between the keywords **begin** and **end.** At time 0, A and B are set to 0. Ten time units later, A is changed to 1. Twenty time units after that (at $t = 30$), A is changed to 0 and B to 1. Inputs specified by a three-bit truth table can be generated with the **initial** block:

```
initial
  begin
    D = 3'b000;
    repeat (7)
    #10 D = D + 3'b001;
  end
```

When the simulator runs, the three-bit vector D is initialized to 000 at time = 0. The keyword **repeat** specifies a looping statement: D is incremented by 1 seven times, once every 10 time units. The result is a sequence of binary numbers from 000 to 111.

A stimulus module has the following form:

```
module test_module_name;
  // Declare local reg and wire identifiers.
  // Instantiate the design module under test.
  // Specify a stopwatch, using $finish to terminate the simulation.
  // Generate stimulus, using initial and always statements.
  // Display the output response (text or graphics (or both)).
endmodule
```

A test module is written like any other module, but it typically has no inputs or outputs. The signals that are applied as inputs to the design module for simulation are declared in the stimulus module as local **reg** data type. The outputs of the design module that are displayed for testing are declared in the stimulus module as local **wire** data type. The module under test is then instantiated, using the local identifiers in its port list. Figure 4.34 clarifies this relationship. The stimulus module generates inputs for the design module by declaring local identifiers *t_A* and *t_B* as **reg** type and checks the output of the design unit with the **wire** identifier *t_C*. The local identifiers are then used to instantiate the design module being tested. The simulator associates the (actual) local identifiers within the test bench, *t_A*, *t_B*, and *t_C*, with the formal identifiers of the

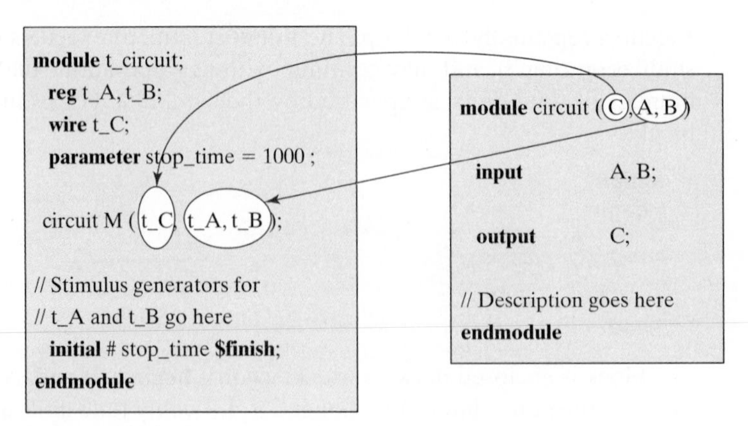

FIGURE 4.34
Interaction between stimulus and design modules

module (*A, B, C*). The association shown here is based on *position* in the port list, which is adequate for the examples that we will consider. The reader should note, however, that Verilog provides a more flexible *name association* mechanism for connecting ports in larger circuits.

The response to the stimulus generated by the **initial** and **always** blocks will appear in text format as standard output and as waveforms (timing diagrams) in simulators having graphical output capability. Numerical outputs are displayed by using Verilog *system tasks*. These are built-in system functions that are recognized by keywords that begin with the symbol **$.** Some of the system tasks that are useful for display are

$display—display a one-time value of variables or strings with an end-of-line return,

$write**—same as **$display, but without going to next line,

$monitor—display variables whenever a value changes during a simulation run,

$time—display the simulation time,

$finish—terminate the simulation.

The syntax for **$display, $write,** and **$monitor** is of the form

Task-name (*format specification, argumentlist*);

The format specification uses the symbol % to specify the radix of the numbers that are displayed and may have a string enclosed in quotes ("). The base may be binary, decimal, hexadecimal, or octal, identified with the symbols %b, %d, %h, and %o, respectively (%B, %D, %H, and %O are valid too). For example, the statement

```
$display ("%d %b %b", C, A, B);
```

specifies the display of *C* in decimal and of *A* and *B* in binary. Note that there are no commas in the format specification, that the format specification and argument list are separated by a comma, and that the argument list has commas between the variables. An example that specifies a string enclosed in quotes may look like the statement

```
$display ("time = %0d A = %b", $time, A, B);
```

and will produce the display

```
time = 3 A = 10 B = 1
```

where (*time* =), (*A* =), and (*B* =) are part of the string to be displayed. The format specifiers %0d, %b, and %b specify the base for **$time,** *A*, and *B*, respectively. In displaying time values, it is better to use the format %0d instead of %d. This provides a display of the significant digits without the leading spaces that %d will include. (%d will display about 10 leading spaces because time is calculated as a 32-bit number.)

An example of a stimulus module is shown in HDL Example 4.9. The circuit to be tested is the two-to-one-line multiplexer described in Example 4.6. The module *t_mux_2x1_df* has no ports. The inputs for the mux are declared with a **reg** keyword and the outputs with a **wire** keyword. The mux is instantiated with the local variables. The **initial** block specifies a sequence of binary values to be applied during the simulation. The output response is checked with the **$monitor** system task. Every time a variable in its argument changes value, the simulator displays the inputs, output, and time. The result of the simulation is listed under the simulation log in the example. It shows that *m_out* = *A* when *select* = 1 and *m_out* = *B* when *select* = 0 verifying the operation of the multiplexer.

HDL Example 4.9 (Test Bench)

```
// Test bench with stimulus for mux_2x1_df

module t_mux_2x1_df;
 wire          t_mux_out;
 reg           t_A, t_B;
 reg           t_select;
 parameter stop_time = 50;

mux_2x1_df M1 (t_mux_out, t_A, t_B, t_select);     // Instantiation of circuit to be tested

initial # stop_time $finish;

 initial begin                                       // Stimulus generator
        t_select = 1; t_A = 0; t_B = 1;
  #10 t_A = 1; t_B = 0;
  #10 t_select = 0;
  #10 t_A = 0; t_B = 1;
 end

 initial begin                                       // Response monitor
  // $display ("  time   Select  A  B  m_out");
  // $monitor ($time,, " %b   %b   %b   %b ", t_select, t_A, t_B, t_m_out);
```

```
    $monitor ("time = ", $time,, "select = %b A = %b B = %b OUT = %b",
    t_select, t_A, t_B, t_mux_out);
  end
endmodule

// Dataflow description of two-to-one-line multiplexer

// from Example 4.6
module mux_2x1_df (m_out, A, B, select);
  output      m_out;
  input       A, B;
  input       select;

  assign m_out = (select)? A : B;
endmodule

Simulation log:
select = 1 A = 0 B = 1 OUT = 0 time = 0
select = 1 A = 1 B = 0 OUT = 1 time = 10
select = 0 A = 1 B = 0 OUT = 0 time = 20
select = 0 A = 0 B = 1 OUT = 1 time = 30
```

Logic simulation is a fast and accurate method of verifying that a model of a combinational circuit is correct. There are two types of verification: functional and timing. In *functional* verification, we study the circuit logical operation independently of timing considerations. This can be done by deriving the truth table of the combinational circuit. In *timing* verification, we study the circuit's operation by including the effect of delays through the gates. This can be done by observing the waveforms at the outputs of the gates when they respond to a given input. An example of a circuit with gate delays was presented in Section 3.10 in HDL Example 3.3. We next show an HDL example that produces the truth table of a combinational circuit. A **$monitor** system task displays the output caused by the given stimulus. A commented alternative statement having a **$display** task would create a header that could be used with a **$monitor** statement to eliminate the repetition of names on each line of output.

The analysis of combinational circuits was covered in Section 4.3. A multilevel circuit of a full adder was analyzed, and its truth table was derived by inspection. The gate-level description of this circuit is shown in HDL Example 4.10. The circuit has three inputs, two outputs, and nine gates. The description of the circuit follows the interconnections between the gates according to the schematic diagram of Fig. 4.2. The stimulus for the circuit is listed in the second module. The inputs for simulating the circuit are specified with a three-bit **reg** vector D. *D[2]* is equivalent to input A, *D[1]* to input B, and *D[0]* to input C. The outputs of the circuit F_1 and F_2 are declared as **wire.** The complement of *F2* is named *F2_b* to illustrate a common industry practice for designating the complement of a signal (instead of appending *_not*). This procedure follows the steps outlined in Fig. 4.34. The **repeat** loop provides the seven binary numbers after 000 for the truth table. The result of the simulation generates the output truth table displayed with the example. The truth table listed shows that the circuit is a full adder.

HDL Example 4.10 (Gate-Level Circuit)

```
// Gate-level description of circuit of Fig. 4.2

module Circuit_of_Fig_4_2 (A, B, C, F1, F2);
  input A, B, C;
  output F1, F2;
  wire  T1, T2, T3, F2_b, E1, E2, E3;
  or  g1 (T1, A, B, C);
  and g2 (T2, A, B, C);
  and g3 (E1, A, B);
  and g4 (E2, A, C);
  and g5 (E3, B, C);
  or  g6 (F2, E1, E2, E3);
  not g7 (F2_b, F2);
  and g8 (T3, T1, F2_b);
  or  g9 (F1, T2, T3);
endmodule

// Stimulus to analyze the circuit

module test_circuit;
  reg [2: 0] D;
  wire F1, F2;
  Circuit_of_Fig_4_2 (D[2], D[1], D[0], F1, F2);
  initial
   begin
    D = 3'b000;
    repeat (7) #10 D = D 1 1'b1;
   end
  initial
  $monitor ("ABC = %b F1 = %b F2 =%b", D, F1, F2);

endmodule

Simulation log: ABC = 000 F1 = 0 F2 =0
ABC = 001 F1 = 1 F2 =0 ABC = 010 F1 = 1 F2 =0
ABC = 011 F1 = 0 F2 =1 ABC = 100 F1 = 1 F2 =0
ABC = 101 F1 = 0 F2 =1 ABC = 110 F1 = 0 F2 =1
ABC = 111 F1 = 1 F2 =1
```

PROBLEMS

(Answers to problems marked with * appear at the end of the text. Where appropriate, a logic design and its related HDL modeling problem are cross-referenced.)

4.1 Consider the combinational circuit shown in Fig. P4.1. (HDL—see Problem 4.49.)

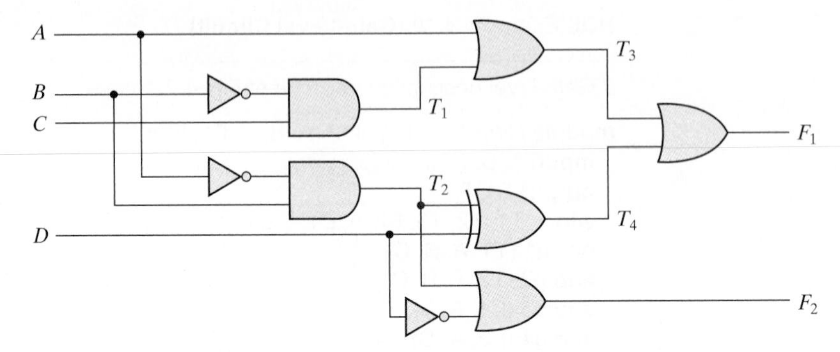

FIGURE P4.1

(a)* Derive the Boolean expressions for T_1 through T_4. Evaluate the outputs F_1 and F_2 as a function of the four inputs.

(b) List the truth table with 16 binary combinations of the four input variables. Then list the binary values for T_1 through T_4 and outputs F_1 and F_2 in the table.

(c) Plot the output Boolean functions obtained in part (b) on maps and show that the simplified Boolean expressions are equivalent to the ones obtained in part (a).

4.2* Obtain the simplified Boolean expressions for output F and G in terms of the input variables in the circuit of Fig. P4.2.

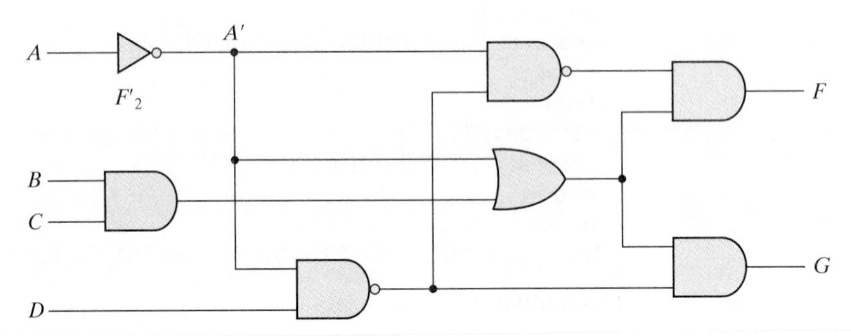

FIGURE P4.2

4.3 For the circuit shown in Fig. 4.26 (Section 4.11),

(a) Write the Boolean functions for the four outputs in terms of the input variables.

(b)* If the circuit is described in a truth table, how many rows and columns would there be in the table?

4.4 Design a combinational circuit with three inputs and one output.

(a)* The output is 1 when the binary value of the inputs is less than 3 and greater than 6. The output is 0 otherwise.

(b) The output is 1 when the binary value of the inputs is an odd number.

4.5 Design a combinational circuit with three inputs, x, y, and z, and three outputs, A, B, and C. When the binary input is 3, 4, 5, 6, or 7, the binary output is one less than the input. When the binary input is 0, 1, or 2, the binary output is two greater than the input.

4.6 A majority circuit is a combinational circuit whose output is equal to 1 if the input variables have more 1's than 0's. The output is 0 otherwise.

(a)* Design a 3-input majority circuit by finding the circuit's truth table, Boolean equation, and a logic diagram.

(b) Write and verify a Verilog gate-level model of the circuit.

4.7 Design a combinational circuit that converts a four-bit Gray code (Table 1.6) to a bit four-binary number.

(a)* Implement the circuit with exclusive-OR gates.

(b) Using a case statement, write and verify a Verilog model of the circuit.

4.8 Design a code converter that converts a decimal digit from

(a)* The 8, 4, –2, –1 code to BCD (see Table 1.5). (HDL—see Problem 4.50.)

(b) The 8, 4, –2, –1 code to Gray code.

4.9 An ABCD-to-seven-segment decoder is a combinational circuit that converts a decimal digit in BCD to an appropriate code for the selection of segments in an indicator used to display the decimal digit in a familiar form. The seven outputs of the decoder *(a, b, c, d, e, f, g)* select the corresponding segments in the display, as shown in Fig. P4.9(a). The numeric display chosen to represent the decimal digit is shown in Fig. P4.9(b). Using a truth table and Karnaugh maps, design the BCD-to-seven-segment decoder using a minimum number of gates. The six invalid combinations should result in a blank display. (HDL—see Problem 4.51.)

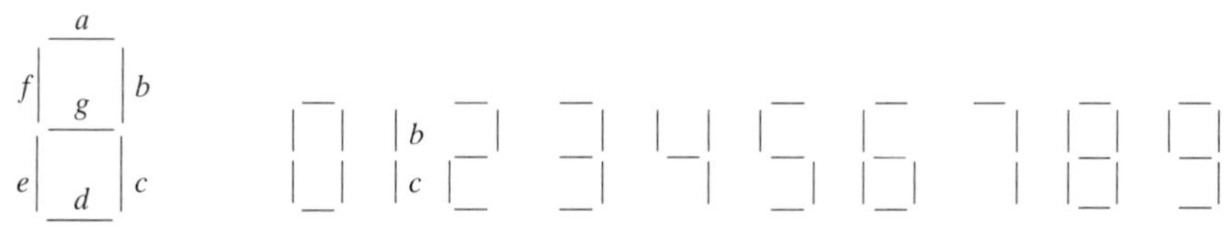

(a) Segment designation

(b) Numerical designation for display

FIGURE P4.9

4.10* Design a three-bit combinational circuit 2's complementer. (The output generates the 2's complement of the input binary number.) Show that the circuit can be constructed only with exclusive-OR and OR gates. Can you predict what the output functions are for a four-bit 2's complementer?

4.11 Using four half-adders (HDL—see Problem 4.52),

(a) Design a full-subtractor circuit incrementer. (A circuit that adds one to a four-bit binary number.)

(b)* Design a four-bit combinational decrementer (a circuit that subtracts 1 from a four-bit binary number).

4.12 Design a half-subtractor circuit with inputs x and y and outputs *Diff* and B_{out}. The circuit subtracts the bits $y - x$ and places the difference in D and the borrow in B_{out}.

(a) Design a full-subtractor circuit with three inputs x, y, B_{in} and two outputs *Diff* and B_{out}. The circuit subtracts $y - x - B_{in}$, where B_{in} is the input borrow, B_{out} is the output borrow, and *Diff* is the difference.

4.13* The adder–subtractor circuit of Fig. 4.13 has the following values for mode input M and data inputs A and B.

	M	A	B
(a)	0	0111	0110
(b)	0	1000	1001
(c)	1	1100	1000
(d)	1	0101	1010
(e)	1	0000	0001

In each case, determine the values of the four *SUM* outputs, the carry C, and overflow V. (HDL—see Problems 4.37 and 4.40.)

4.14* Assume that the exclusive-OR gate has a propagation delay of 10 ns and that the AND or OR gates have a propagation delay of 5 ns. What is the total propagation delay time in the four-bit adder of Fig. 4.12?

4.15 Derive the two-level Boolean expression for the output carry C_4 and sum S_0 through S_4 shown in the lookahead carry generator of Fig. 4.12.

4.16 Define the carry propagate and carry generate as

$$P_i = A_i + B_i$$
$$G_i = A_iB_i$$

respectively. Show that the output carry and output sum of a full adder becomes

$$C_{i+1} = (C_i'G_i' + P_i')'$$
$$S_i = (P_iG_i') \oplus C_i$$

The logic diagram of the first stage of a four-bit parallel adder as implemented in IC type 74283 is shown in Fig. P4.16. Identify the P_i' and G_i' terminals and show that the circuit implements a full-adder circuit.

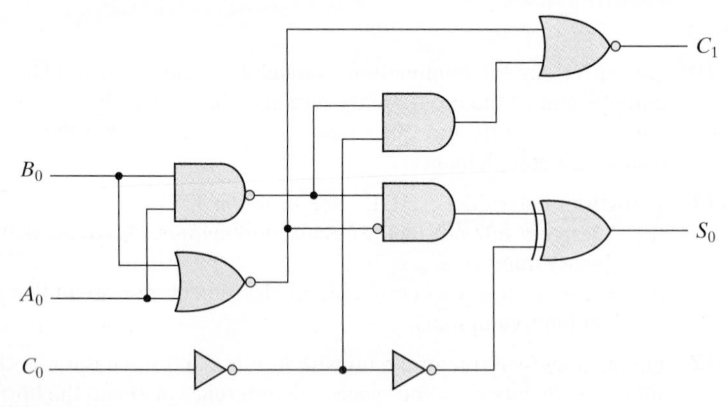

FIGURE P4.16
First stage of a parallel adder

4.17 Show that the output carry in a full adder circuit can be expressed in the AND-OR-INVERT form

$$C_{i+1} = G_i + P_iC_i = (G_i'P_i' + G_i'C_i')'$$

IC type 74182 is a lookahead carry generator circuit that generates the carries with AND-OR-INVERT gates (see Section 3.8). The circuit assumes that the input terminals have the complements of the G's, the P's, and of C_1. Derive the Boolean functions for the lookahead carries C_2, C_3, and C_4 in this IC. (*Hint:* Use the equation-substitution method to derive the carries in terms of C_i')

4.18 Design a combinational circuit that generates the 9's complement of a
(a)* BCD digit. (HDL—see Problem 4.54(a).)
(b) Gray-code digit. (HDL—see Problem 4.54(b).)

4.19 Construct a BCD adder–subtractor circuit. Use the BCD adder of Fig. 4.14 and the 9's complementer of problem 4.18. Use block diagrams for the components. (HDL—see Problem 4.55.)

4.20 For a binary multiplier that multiplies two unsigned four-bit numbers,
(a) Using AND gates and binary adders (see Fig. 4.16), design the circuit.
(b) Write and verify a Verilog dataflow model of the circuit.

4.21 Design a combinational circuit that compares two 4-bit numbers to check if they are unequal. The circuit output is equal to 1 if the two numbers are unequal and 0 otherwise.

4.22* Design an excess-3-to-binary decoder using the unused combinations of the code as don't-care conditions. (HDL—see Problem 4.42.)

4.23 Draw the logic diagram of a 2-to-4-line decoder using (a) NOR gates only and (b) NAND gates only. Include an enable input. (HDL—see Problems 4.36, 4.45.)

4.24 Design a 2, 4, 2, 1-to-decimal decoder using the unused combinations of the 2, 4, 2, 1 code as don't-care conditions.

4.25 Construct a 5-to-32-line decoder with four 3-to-8-line decoders with enable and a 2-to-4-line decoder. Use block diagrams for the components. (HDL—see Problem 4.63.)

4.26 Construct a 4-to-16-line decoder with five 2-to-4-line decoders with enable. (HDL—see Problem 4.64.)

4.27 A combinational circuit is specified by the following three Boolean functions:

$$F_1(A, B, C) = \Sigma(2, 4, 6)$$
$$F_2(A, B, C) = \Sigma(3, 5, 7)$$
$$F_3(A, B, C) = \Sigma(0, 2, 3, 4, 7)$$

Implement the circuit with a decoder constructed with NAND gates (similar to Fig. 4.19) and NAND or AND gates connected to the decoder outputs. Use a block diagram for the decoder. Minimize the number of inputs in the external gates.

4.28 Using a decoder constructed with NAND gates (similar to Fig. 4.19) and external gates, design the combinational circuit defined by the following three Boolean functions:

(a) $F_1 = xy + xz' + yz'$
$F_2 = xz + xy + yz$
$F_3 = y'z + x'y'z' + xy$

(b) $F_1 = z' + xy$
$F_2 = yz + x'y + y'z'$
$F_3 = (x' + y)z + xy'z$

4.29* Design an eight-input priority encoder with inputs D_0 through D_7, with input D_7 having the highest priority and input D_0 the lowest priority.

4.30 Specify the truth table of an octal-to-binary priority encoder. Provide an output V to indicate that at least one of the inputs is present. The input with the highest subscript number has the highest priority. What will be the value of the four outputs if inputs D_2 and D_6 are 1 at the same time? (HDL—see Problem 4.65.)

4.31 Construct a 16×1 multiplexer with two 8×1 and one 2×1 multiplexers. Use block diagrams. (HDL—see Problem 4.67.)

4.32 Implement the following Boolean function with a multiplexer (HDL—see Problem 4.46):

(a) $F(A, B, C, D) = \Sigma(0, 2, 5, 8, 10, 14)$
(b) $F(A, B, C, D) = \Pi(2, 6, 11)$

4.33 Implement a full subtractor with two 4×1 multiplexers.

4.34 An 8×1 multiplexer has inputs A, B, and C connected to the selection inputs S_0, S_1, and S_2, respectively. The data inputs I_0 through I_7 are as follows:

(a)* $I_1 = I_2 = I_7 = 1; I_3 = I_5 = 0; I_0 = I_4 = D'$; and $I_6 = D$.
(b) $I_1 = I_2 = 1; I_3 = I_7 = 0; I_4 = I_5 = D'$; and $I_0 = I_6 = D$.

Determine the Boolean function that the multiplexer implements.

4.35 Implement the following Boolean function with a 4×1 multiplexer and external gates.

(a)* $F_1(A, B, C, D) = \Sigma(1, 3, 4, 11, 12, 13, 14, 15)$
(b) $F_2(A, B, C, D) = \Sigma(1, 2, 5, 7, 8, 10, 11, 13, 15)$

Connect inputs A and B to the selection lines. The input requirements for the four data lines will be a function of variables C and D. These values are obtained by expressing F as a function of C and D for each of the four cases when $AB = 00, 01, 10$, and 11. These functions may have to be implemented with external gates. (HDL—see Problem 4.47.)

4.36 Write the HDL gate-level description of the priority encoder circuit shown in Fig. 4.23. (HDL—see Problem 4.45.)

4.37 Write the HDL gate-level hierarchical description of a four-bit adder–subtractor for unsigned binary numbers. The circuit is similar to Fig. 4.13 but without output V. You can instantiate the four-bit full adder described in HDL Example 4.2. (HDL—see Problems 4.13 and 4.40.)

4.38 Write the HDL dataflow description of a quadruple 2-to-1-line multiplexer with enable (see Fig. 4.26).

4.39* Write an HDL behavioral description of a four-bit comparator with a six-bit output $Y[5{:}0]$. Bit 5 of Y is for "equals," bit 4 for "not equal to," bit 3 for "greater than," bit 2 for "less than," bit 1 for "greater than or equal," and bit 0 for "less than or equal to."

4.40 Using the conditional operator (?:), write an HDL dataflow description of a four-bit adder–subtractor of unsigned numbers. (See Problems 4.13 and 4.37.)

4.41 Repeat problem 4.40 using an always statement.

4.42 (a) Write an HDL gate-level description of the BCD-to-excess-3 converter circuit shown in Fig. 4.4 (see Problem 4.22).
(b) Write a dataflow description of the BCD-to-excess-3 converter using the Boolean expressions listed in Fig. 4.3.
(c)* Write an HDL behavioral description of a BCD-to-excess-3 converter.
(d) Write a test bench to simulate and test the BCD-to-excess-3 converter circuit in order to verify the truth table. Check all three circuits.

4.43 Explain the function of the circuit specified by the following HDL description:

```
module Prob4_43 (A, B, S, E, Q);
 input  [1:0] A, B;
 input        S, E;
 output [1:0] Q;
 assign Q = E ? (S ? A : B) : 'bz;
endmodule
```

4.44 Using a case statement, write an HDL behavioral description of a eight-bit arithmetic-logic unit (ALU). The circuit has a three-bit select bus (Sel), sixteen-bit input datapaths (A[15:0] and B[15:0]), an eight-bit output datapath (y[15:0]), and performs the arithmetic and logic operations listed below.

Sel	Operation	Description
000	y = 8′b0	
001	y = A & B	Bitwise AND
010	y = A | B	Bitwise OR
011	y = A ^ B	Bitwise exclusive OR
100	y = ~A	Bitwise complement
101	y = A − B	Subtract
110	y = A + B	Add (Assume A and B are unsigned)
111	y = 8′hFF	

4.45 Write an HDL behavioral description of a four-input priority encoder. Use a four-bit vector for the *D* inputs and an **always** block with if–else statements. Assume that input *D*[3] has the highest priority (see Problem 4.36).

4.46 Write a Verilog dataflow description of the logic circuit described by the Boolean function in Problem 4.32.

4.47 Write a Verilog dataflow description of the logic circuit described by the Boolean function in Problem 4.35.

4.48 Develop and modify the eight-bit ALU specified in Problem 4.44 so that it has three-state output controlled by an enable input, *En*. Write a test bench and simulate the circuit.

4.49 For the circuit shown in Fig. P4.1,
(a) Write and verify a gate-level HDL model of the circuit.
(b) Compare your results with those obtained for Problem 4.1.

4.50 Using a case statement, develop and simulate a behavioral model of
(a)* The 8, 4, –2, –1 to BCD code converter described in Problem 4.8(a).
(b) The 8, 4, –2, –1 to Gray code converter described in Problem 4.8(b).

4.51 Develop and simulate a behavioral model of the ABCD-to-seven-segment decoder described in Problem 4.9.

4.52 Using a continuous assignment, develop and simulate a dataflow model of
(a) The four-bit incrementer described in Problem 4.11(a).
(b) The four-bit decrementer described in Problem 4.11(b).

4.53 Develop and simulate a structural model of the decimal adder shown in Fig. 4.14.

4.54 Develop and simulate a behavioral model of a circuit that generates the 9's complement of
(a) a BCD digit (see Problem 4.18(a)).
(b) a Gray-code digit (see Problem 4.18(b).)

4.55 Construct a hierarchical model of the BCD adder–subtractor described in Problem 4.19. The BCD adder and the 9's complementer are to be described as behavioral models in separate modules, and they are to be instantiated in a top-level module.

4.56* Write a continuous assignment statement that compares two 4-bit numbers to check if their bit patterns match. The variable to which the assignment is made is to equal 1 if the numbers match and 0 otherwise.

4.57* Develop and verify a behavioral model of the four-bit priority encoder described in Problem 4.29.

4.58 Write a Verilog model of a circuit whose 32-bit output is formed by shifting its 32-bit input three positions to the right and filling the vacant positions with the bit that was in the MSN before the shift occurred (shift arithmetic right). Write a Verilog model of a circuit whose 32-bit output is formed by shifting its 32-bit input three positions to the left and filling the vacant positions with 0 (shift logical left).

4.59 Write a Verilog model of a BCD-to-decimal decoder using the unused combinations of the BCD code as don't-care conditions (see Problem 4.24).

4.60 Using the port syntax of the IEEE 1364-2001 standard, write and verify a gate-level model of the four-bit even parity checker shown in Fig. 3.34.

4.61 Using continuous assignment statements and the port syntax of the IEEE 1364-2001 standard, write and verify a gate-level model of the four-bit even parity checker shown in Fig. 3.34.

4.62 Write and verify a gate-level hierarchical model of the circuit described in Problem 4.25.

4.63 Write and verify a gate-level hierarchical model of the circuit described in Problem 4.26.

4.64 Write and verify a Verilog model of the octal-to-binary circuit described in Problem 4.30.

4.65 Write a hierarchical gate-level model of the multiplexer described in Problem 4.31.

REFERENCES

1. Bhasker, J. 1997. *A Verilog HDL Primer*. Allentown, PA: Star Galaxy Press.
2. Bhasker, J. 1998. *Verilog HDL Synthesis*. Allentown, PA: Star Galaxy Press.
3. Ciletti, M. D. 1999. *Modeling, Synthesis, and Rapid Prototyping with Verilog HDL*. Upper Saddle River, NJ: Prentice Hall.
4. Dietmeyer, D. L. 1988. *Logic Design of Digital Systems*, 3rd ed. Boston: Allyn Bacon.

5. Gajski, D. D. 1997. *Principles of Digital Design*. Upper Saddle River, NJ: Prentice Hall.
6. Hayes, J. P. 1993. *Introduction to Digital Logic Design*. Reading, MA: Addison-Wesley.
7. Katz, R. H. 2005. *Contemporary Logic Design*. Upper Saddle River, NJ: Pearson Prentice Hall.
8. Mano, M. M. and C. R. Kime. 2007. *Logic and Computer Design Fundamentals*, 4th ed. Upper Saddle River, NJ: Prentice Hall.
9. Nelson, V. P., H. T. Nagle, J. D. Irwin, and B. D. Carroll. 1995. *Digital Logic Circuit Analysis and Design*. Englewood Cliffs, NJ: Prentice Hall.
10. Palnitkar, S. 1996. *Verilog HDL: A Guide to Digital Design and Synthesis*. Mountain View, CA: SunSoft Press (a Prentice Hall title).
11. Roth, C. H. 2009. *Fundamentals of Logic Design*, 6th ed. St. Paul, MN: West.
12. Thomas, D. E. and P. R. Moorby. 2002. *The Verilog Hardware Description Language*, 5th ed. Boston: Kluwer Academic Publishers.
13. Wakerly, J. F. 2005. *Digital Design: Principles and Practices*, 4th ed. Upper Saddle River, NJ: Prentice Hall.

WEB SEARCH TOPICS

Boolean equation
Combinational logic
Truth table
Exclusive–OR
Comparator
Multiplexer
Decoder
Priority encoder
Three-state inverter
Three-state buffer

Chapter 5

Synchronous Sequential Logic

5.1 INTRODUCTION

Hand-held devices, cell phones, navigation receivers, personal computers, digital cameras, personal media players, and virtually all electronic consumer products have the ability to send, receive, store, retrieve, and process information represented in a binary format. The technology enabling and supporting these devices is critically dependent on electronic components that can store information, i.e., have memory. This chapter examines the operation and control of these devices and their use in circuits and enables you to better understand what is happening in these devices when you interact with them. The digital circuits considered thus far have been combinational—their output depends only and immediately on their inputs—they have no memory, i.e., dependence on past values of their inputs. Sequential circuits, however, act as storage elements and have memory. They can store, retain, and then retrieve information when needed at a later time. Our treatment will distinguish sequential logic from combinational logic.

5.2 SEQUENTIAL CIRCUITS

A block diagram of a sequential circuit is shown in Fig. 5.1. It consists of a combinational circuit to which storage elements are connected to form a feedback path. The storage elements are devices capable of storing binary information. The binary information stored in these elements at any given time defines the *state* of the sequential circuit at that time. The sequential circuit receives binary information from external inputs that, together with the present state of the storage elements, determine the binary value of the outputs. These external inputs also determine the condition for changing the state

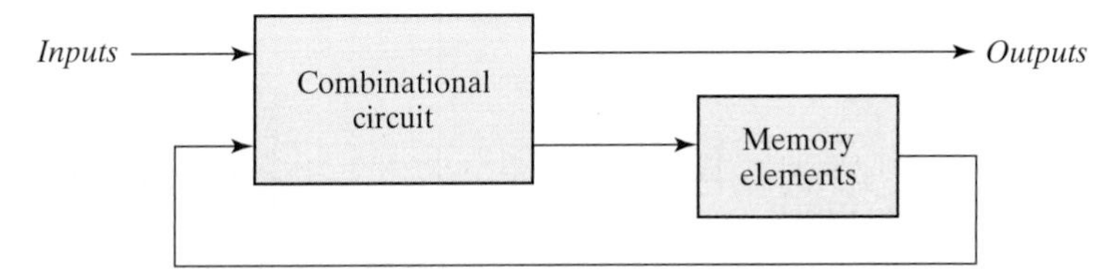

FIGURE 5.1
Block diagram of sequential circuit

in the storage elements. The block diagram demonstrates that the outputs in a sequential circuit are a function not only of the inputs, but also of the present state of the storage elements. The next state of the storage elements is also a function of external inputs and the present state. Thus, **a sequential circuit is specified by a time sequence of inputs, outputs, and internal states**. In contrast, the outputs of combinational logic depend only on the present values of the inputs.

There are two main types of sequential circuits, and their classification is a function of the timing of their signals. A *synchronous* sequential circuit is a system whose behavior can be defined from the knowledge of its signals at discrete instants of time. The behavior of an *asynchronous* sequential circuit depends upon the input signals at any instant of time *and* the order in which the inputs change. The storage elements commonly used in asynchronous sequential circuits are time-delay devices. The storage capability of a time-delay device varies with the time it takes for the signal to propagate through the device. In practice, the internal propagation delay of logic gates is of sufficient duration to produce the needed delay, so that actual delay units may not be necessary. In gate-type asynchronous systems, the storage elements consist of logic gates whose propagation delay provides the required storage. Thus, an asynchronous sequential circuit may be regarded as a combinational circuit with feedback. Because of the feedback among logic gates, an asynchronous sequential circuit may become unstable at times. The instability problem imposes many difficulties on the designer. These circuits will not be covered in this text.

A synchronous sequential circuit employs signals that affect the storage elements at only discrete instants of time. Synchronization is achieved by a timing device called a *clock generator*, which provides a clock signal having the form of a periodic train of *clock pulses*. The clock signal is commonly denoted by the identifiers *clock* and *clk*. The clock pulses are distributed throughout the system in such a way that storage elements are affected only with the arrival of each pulse. In practice, the clock pulses determine *when* computational activity will occur within the circuit, and other signals (external inputs and otherwise) determine *what* changes will take place affecting the storage elements and the outputs. For example, a circuit that is to add and store two binary numbers would compute their sum from the values of the numbers and store the sum at the occurrence of a clock pulse. Synchronous sequential circuits that use clock pulses to control storage elements are called *clocked sequential circuits* and are the type most frequently encountered in practice. They are called *synchronous circuits* because the activity within the circuit and the resulting updating of stored values is synchronized to the occurrence of

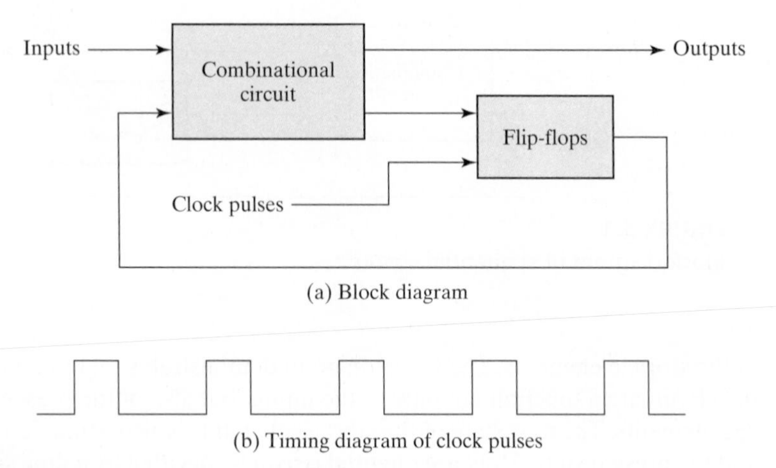

(a) Block diagram

(b) Timing diagram of clock pulses

FIGURE 5.2
Synchronous clocked sequential circuit

clock pulses. The design of synchronous circuits is feasible because they seldom manifest instability problems and their timing is easily broken down into independent discrete steps, each of which can be considered separately.

The storage elements (memory) used in clocked sequential circuits are called *flip-flops.* A flip-flop is a binary storage device capable of storing one bit of information. In a stable state, the output of a flip-flop is either 0 or 1. A sequential circuit may use many flip-flops to store as many bits as necessary. The block diagram of a synchronous clocked sequential circuit is shown in Fig. 5.2. The *outputs* are formed by a combinational logic function of the inputs to the circuit or the values stored in the flip-flops (or both). The value that is stored in a flip-flop when the clock pulse occurs is also determined by the inputs to the circuit or the values presently stored in the flip-flop (or both). The new value is stored (i.e., the flip-flop is updated) when a pulse of the clock signal occurs. Prior to the occurrence of the clock pulse, the combinational logic forming the next value of the flip-flop must have reached a stable value. Consequently, the speed at which the combinational logic circuits operate is critical. If the clock (synchronizing) pulses arrive at a regular interval, as shown in the timing diagram in Fig. 5.2, the combinational logic must respond to a change in the state of the flip-flop in time to be updated before the next pulse arrives. Propagation delays play an important role in determining the minimum interval between clock pulses that will allow the circuit to operate correctly. A change in state of the flip-flops is initiated only by a clock pulse transition—for example, when the value of the clock signals changes from 0 to 1. When a clock pulse is not active, the feedback loop between the value stored in the flip-flop and the value formed at the input to the flip-flop is effectively broken because the flip-flop outputs cannot change even if the outputs of the combinational circuit driving their inputs change in value. Thus, the transition from one state to the next occurs only at predetermined intervals dictated by the clock pulses.

5.3 STORAGE ELEMENTS: LATCHES

A storage element in a digital circuit can maintain a binary state indefinitely (as long as power is delivered to the circuit), until directed by an input signal to switch states. The major differences among various types of storage elements are in the number of inputs they possess and in the manner in which the inputs affect the binary state. *Storage elements that operate with signal levels (rather than signal transitions) are referred to as latches; those controlled by a clock transition are flip-flops*. Latches are said to be level sensitive devices; flip-flops are edge-sensitive devices. The two types of storage elements are related because latches are the basic circuits from which all flip-flops are constructed. Although latches are useful for storing binary information and for the design of asynchronous sequential circuits, they are not practical for use as storage elements in synchronous sequential circuits. Because they are the building blocks of flip-flops, however, we will consider the fundamental storage mechanism used in latches before considering flip-flops in the next section.

SR Latch

The *SR* latch is a circuit with two cross-coupled NOR gates or two cross-coupled NAND gates, and two inputs labeled *S* for set and *R* for reset. The *SR* latch constructed with two cross-coupled NOR gates is shown in Fig. 5.3. The latch has two useful states. When output $Q = 1$ and $Q' = 0$, the latch is said to be in the *set state*. When $Q = 0$ and $Q' = 1$, it is in the *reset state*. Outputs Q and Q' are normally the complement of each other. However, when both inputs are equal to 1 at the same time, a condition in which both outputs are equal to 0 (rather than be mutually complementary) occurs. If both inputs are then switched to 0 simultaneously, the device will enter an unpredictable or undefined state or a metastable state. Consequently, in practical applications, setting both inputs to 1 is forbidden.

Under normal conditions, both inputs of the latch remain at 0 unless the state has to be changed. The application of a momentary 1 to the *S* input causes the latch to go to the set state. The *S* input must go back to 0 before any other changes take place, in order to avoid the occurrence of an undefined next state that results from the forbidden input condition. As shown in the function table of Fig. 5.3(b), two input conditions cause the circuit to be in

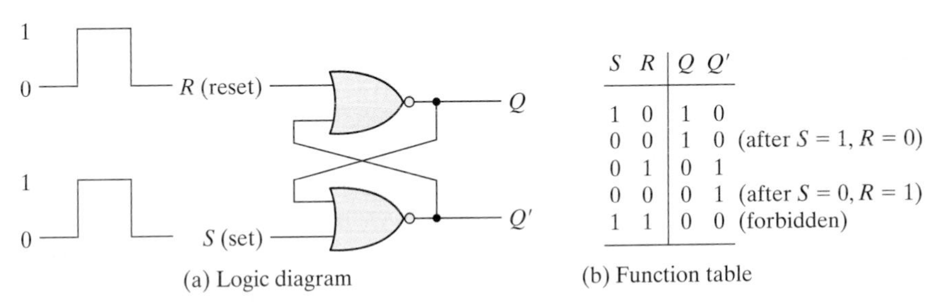

(a) Logic diagram

S	R	Q	Q'	
1	0	1	0	
0	0	1	0	(after $S = 1, R = 0$)
0	1	0	1	
0	0	0	1	(after $S = 0, R = 1$)
1	1	0	0	(forbidden)

(b) Function table

FIGURE 5.3
SR latch with NOR gates

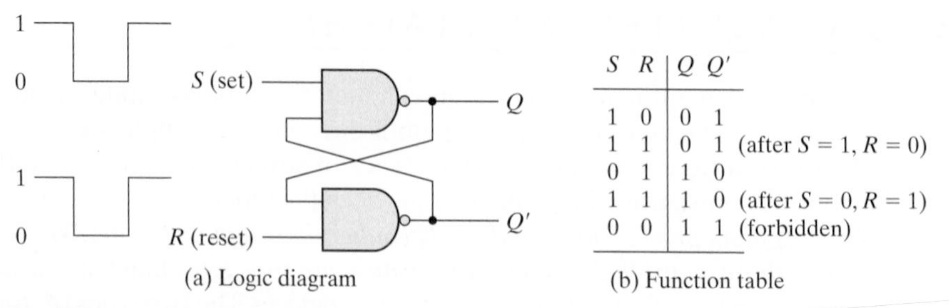

S	R	Q	Q'	
1	0	0	1	
1	1	0	1	(after $S = 1, R = 0$)
0	1	1	0	
1	1	1	0	(after $S = 0, R = 1$)
0	0	1	1	(forbidden)

(b) Function table

FIGURE 5.4
***SR* latch with NAND gates**

the set state. The first condition ($S = 1, R = 0$) is the action that must be taken by input S to bring the circuit to the set state. Removing the active input from S leaves the circuit in the same state. After both inputs return to 0, it is then possible to shift to the reset state by momentary applying a 1 to the R input. The 1 can then be removed from R, whereupon the circuit remains in the reset state. Thus, when both inputs S and R are equal to 0, the latch can be in either the set or the reset state, depending on which input was most recently a 1.

If a 1 is applied to both the S and R inputs of the latch, both outputs go to 0. This action produces an undefined next state, because the state that results from the input transitions depends on the order in which they return to 0. It also violates the requirement that outputs be the complement of each other. In normal operation, this condition is avoided by making sure that 1's are not applied to both inputs simultaneously.

The *SR* latch with two cross-coupled NAND gates is shown in Fig. 5.4. It operates with both inputs normally at 1, unless the state of the latch has to be changed. The application of 0 to the S input causes output Q to go to 1, putting the latch in the set state. When the S input goes back to 1, the circuit remains in the set state. After both inputs go back to 1, we are allowed to change the state of the latch by placing a 0 in the R input. This action causes the circuit to go to the reset state and stay there even after both inputs return to 1. The condition that is forbidden for the NAND latch is both inputs being equal to 0 at the same time, an input combination that should be avoided.

In comparing the NAND with the NOR latch, note that the input signals for the NAND require the complement of those values used for the NOR latch. Because the NAND latch requires a 0 signal to change its state, it is sometimes referred to as an $S'R'$ latch. The primes (or, sometimes, bars over the letters) designate the fact that the inputs must be in their complement form to activate the circuit.

The operation of the basic *SR* latch can be modified by providing an additional input signal that determines (controls) *when* the state of the latch can be changed by determining whether S and R (or S' and R') can affect the circuit. An *SR* latch with a control input is shown in Fig. 5.5. It consists of the basic *SR* latch and two additional NAND gates. The control input *En* acts as an *enable* signal for the other two inputs. **The outputs of the NAND gates stay at the logic-1 level as long as the enable signal remains at 0.** This is the quiescent condition for the *SR* latch. When the enable input goes to 1, information from the S or R input is allowed to affect the latch. The set state is reached with $S = 1$, $R = 0$, and $En = 1$

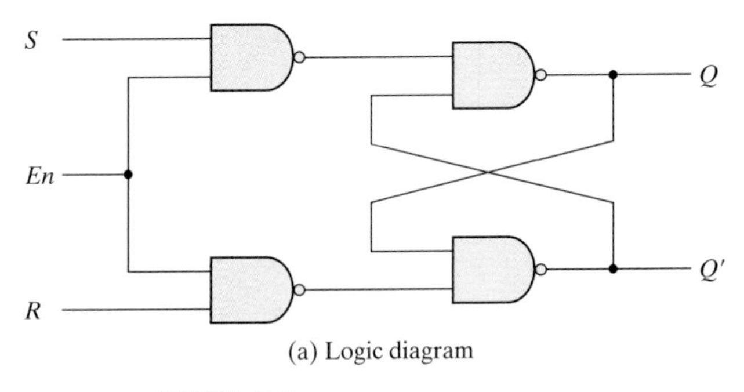

(a) Logic diagram

En	S	R	Next state of Q
0	X	X	No change
1	0	0	No change
1	0	1	$Q = 0$; reset state
1	1	0	$Q = 1$; set state
1	1	1	Indeterminate

(b) Function table

FIGURE 5.5
SR **latch with control input**

(active-high enabled). To change to the reset state, the inputs must be $S = 0$, $R = 1$, and $En = 1$. In either case, when En returns to 0, the circuit remains in its current state. The control input disables the circuit by applying 0 to En, so that the state of the output does not change regardless of the values of S and R. Moreover, when $En = 1$ and both the S and R inputs are equal to 0, the state of the circuit does not change. These conditions are listed in the function table accompanying the diagram.

An indeterminate condition occurs when all three inputs are equal to 1. This condition places 0's on both inputs of the basic *SR* latch, which puts it in the undefined state. When the enable input goes back to 0, one cannot conclusively determine the next state, because it depends on whether the *S* or *R* input goes to 0 first. This indeterminate condition makes this circuit difficult to manage, and it is seldom used in practice. Nevertheless, the *SR* latch is an important circuit because other useful latches and flip-flops are constructed from it.

D Latch (Transparent Latch)

One way to eliminate the undesirable condition of the indeterminate state in the *SR* latch is to ensure that inputs *S* and *R* are never equal to 1 at the same time. This is done in the *D* latch, shown in Fig. 5.6. This latch has only two inputs: *D* (data) and

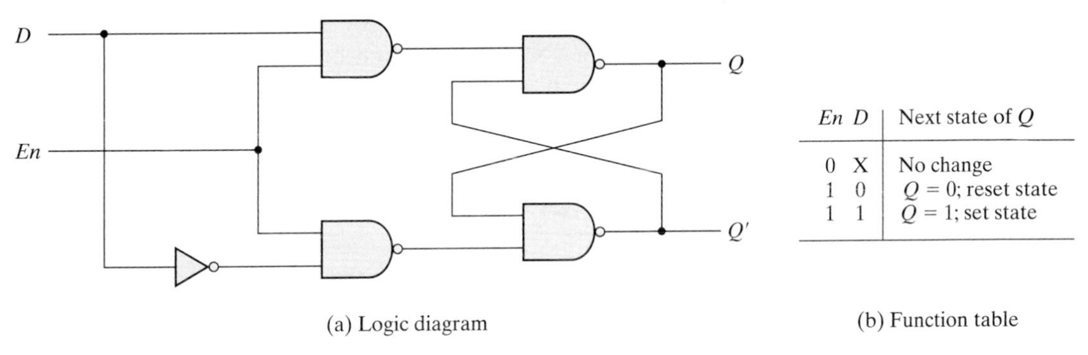

En	D	Next state of Q
0	X	No change
1	0	$Q = 0$; reset state
1	1	$Q = 1$; set state

(a) Logic diagram

(b) Function table

FIGURE 5.6
D **latch**

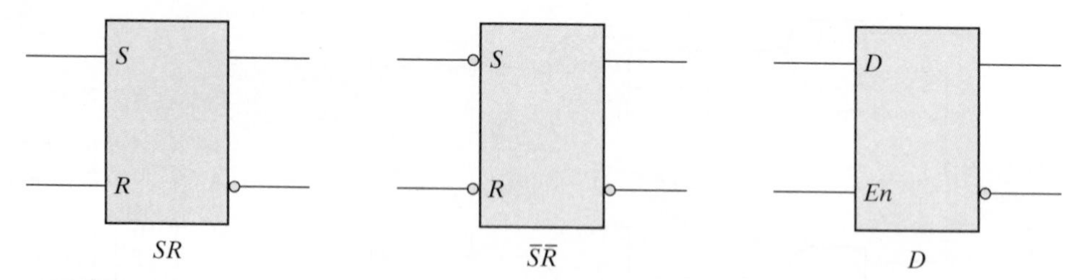

FIGURE 5.7
Graphic symbols for latches

En (enable). The *D* input goes directly to the *S* input, and its complement is applied to the *R* input. As long as the enable input is at 0, the cross-coupled *SR* latch has both inputs at the 1 level and the circuit cannot change state regardless of the value of *D*. The *D* input is sampled when $En = 1$. If $D = 1$, the *Q* output goes to 1, placing the circuit in the set state. If $D = 0$, output *Q* goes to 0, placing the circuit in the reset state.

The *D* latch receives that designation from its ability to hold *data* in its internal storage. It is suited for use as a temporary storage for binary information between a unit and its environment. The binary information present at the data input of the *D* latch is transferred to the *Q* output when the enable input is asserted. The output follows changes in the data input as long as the enable input is asserted. This situation provides a path from input *D* to the output, and for this reason, the circuit is often called a *transparent* latch. When the enable input signal is de-asserted, the binary information that was present at the data input at the time the transition occurred is retained (i.e., stored) at the *Q* output until the enable input is asserted again. Note that an inverter could be placed at the enable input. Then, depending on the physical circuit, the external enabling signal will be a value of 0 (active low) or 1 (active high).

The graphic symbols for the various latches are shown in Fig. 5.7. A latch is designated by a rectangular block with inputs on the left and outputs on the right. One output designates the normal output, and the other (with the bubble designation) designates the complement output. The graphic symbol for the *SR* latch has inputs *S* and *R* indicated inside the block. In the case of a NAND gate latch, bubbles are added to the inputs to indicate that setting and resetting occur with a logic-0 signal. The graphic symbol for the *D* latch has inputs *D* and *En* indicated inside the block.

5.4 STORAGE ELEMENTS: FLIP-FLOPS

The state of a latch or flip-flop is switched by a change in the control input. This momentary change is called a *trigger*, and the transition it causes is said to trigger the flip-flop. The *D* latch with pulses in its control input is essentially a flip-flop that is triggered every time the pulse goes to the logic-1 level. As long as the pulse input remains at this level, any changes in the data input will change the output and the state of the latch.

As seen from the block diagram of Fig. 5.2, a sequential circuit has a feedback path from the outputs of the flip-flops to the input of the combinational circuit. Consequently, the inputs of the flip-flops are derived in part from the outputs of the same and other flip-flops. When latches are used for the storage elements, a serious difficulty arises. The state transitions of the latches start as soon as the clock pulse changes to the logic-1 level. The new state of a latch appears at the output while the pulse is still active. This output is connected to the inputs of the latches through the combinational circuit. If the inputs applied to the latches change while the clock pulse is still at the logic-1 level, the latches will respond to new values and a new output state may occur. The result is an unpredictable situation, since the state of the latches may keep changing for as long as the clock pulse stays at the active level. Because of this unreliable operation, the output of a latch cannot be applied directly or through combinational logic to the input of the same or another latch when all the latches are triggered by a common clock source.

Flip-flop circuits are constructed in such a way as to make them operate properly when they are part of a sequential circuit that employs a common clock. The problem with the latch is that it responds to a change in the *level* of a clock pulse. As shown in Fig. 5.8(a), a positive level response in the enable input allows changes in the output when the D input changes while the clock pulse stays at logic 1. The key to the proper operation of a flip-flop is to trigger it only during a signal *transition*. This can be accomplished by eliminating the feedback path that is inherent in the operation of the sequential circuit using latches. A clock pulse goes through two transitions: from 0 to 1 and the return from 1 to 0. As shown in Fig. 5.8, the positive transition is defined as the positive edge and the negative transition as the negative edge. There are two ways that a latch can be modified to form a flip-flop. One way is to employ two latches in a special configuration that isolates the output of the flip-flop and prevents it from being affected while the input to the flip-flop is changing. Another way is to produce a flip-flop that

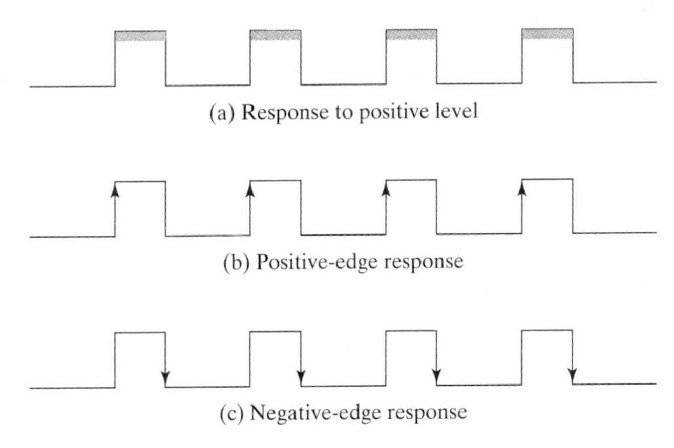

FIGURE 5.8
Clock response in latch and flip-flop

triggers only during a signal transition (from 0 to 1 or from 1 to 0) of the synchronizing signal (clock) and is disabled during the rest of the clock pulse. We will now proceed to show the implementation of both types of flip-flops.

Edge-Triggered *D* Flip-Flop

The construction of a *D* flip-flop with two *D* latches and an inverter is shown in Fig. 5.9. The first latch is called the master and the second the slave. The circuit samples the *D* input and changes its output *Q* only at the negative edge of the synchronizing or controlling clock (designated as *Clk*). When the clock is 0, the output of the inverter is 1. The slave latch is enabled, and its output *Q* is equal to the master output *Y*. The master latch is disabled because $Clk = 0$. When the input pulse changes to the logic-1 level, the data from the external *D* input are transferred to the master. The slave, however, is disabled as long as the clock remains at the 1 level, because its *enable* input is equal to 0. Any change in the input changes the master output at *Y*, but cannot affect the slave output. When the clock pulse returns to 0, the master is disabled and is isolated from the *D* input. At the same time, the slave is enabled and the value of *Y* is transferred to the output of the flip-flop at *Q*. Thus, *a change in the output of the flip-flop can be triggered only by and during the transition of the clock from 1* to 0.

The behavior of the master–slave flip-flop just described dictates that (1) the output may change only once, (2) a change in the output is triggered by the negative edge of the clock, and (3) the change may occur only during the clock's negative level. The value that is produced at the output of the flip-flop is the value that was *stored in the master stage immediately before the negative edge occurred*. It is also possible to design the circuit so that the flip-flop output changes on the positive edge of the clock. This happens in a flip-flop that has an additional inverter between the *Clk* terminal and the junction between the other inverter and input *En* of the master latch. Such a flip-flop is triggered with a negative pulse, so that the negative edge of the clock affects the master and the positive edge affects the slave and the output terminal.

Another construction of an edge-triggered *D* flip-flop uses three *SR* latches as shown in Fig. 5.10. Two latches respond to the external *D* (data) and *Clk* (clock) inputs. The third latch provides the outputs for the flip-flop. The *S* and *R* inputs of the output latch

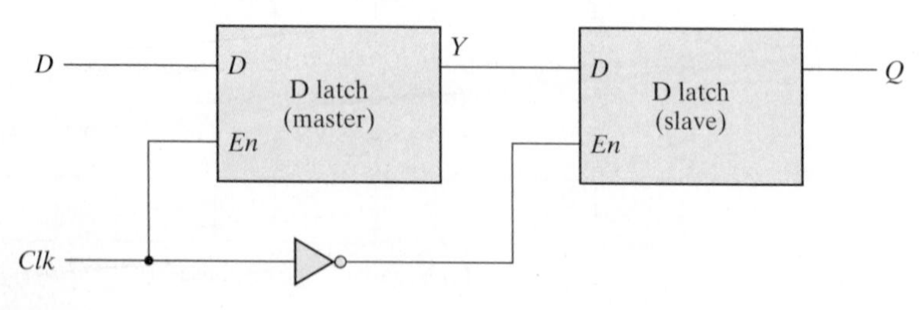

FIGURE 5.9
Master–slave *D* flip-flop

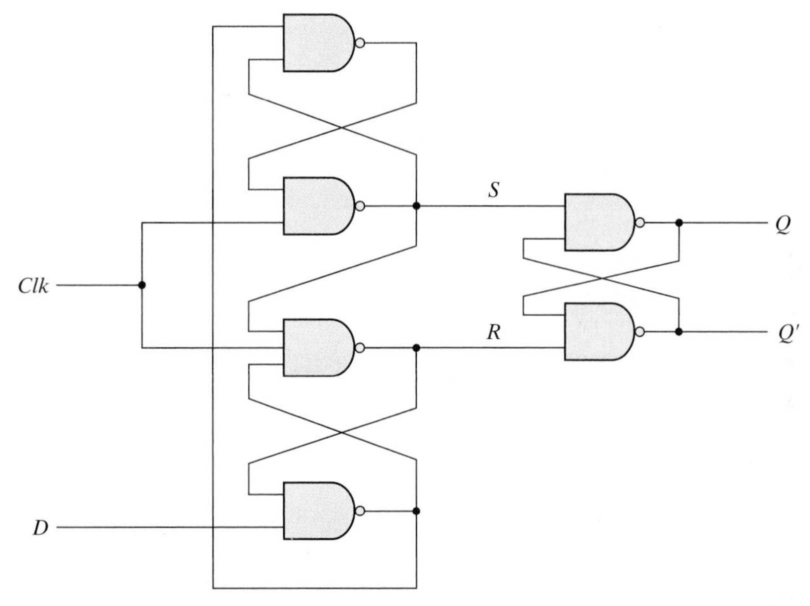

FIGURE 5.10
D-type positive-edge-triggered flip-flop

are maintained at the logic-1 level when $Clk = 0$. This causes the output to remain in its present state. Input D may be equal to 0 or 1. If $D = 0$ when Clk becomes 1, R changes to 0. This causes the flip-flop to go to the reset state, making $Q = 0$. If there is a change in the D input while $Clk = 1$, terminal R remains at 0 because Q is 0. Thus, the flip-flop is locked out and is unresponsive to further changes in the input. When the clock returns to 0, R goes to 1, placing the output latch in the quiescent condition without changing the output. Similarly, if $D = 1$ when Clk goes from 0 to 1, S changes to 0. This causes the circuit to go to the set state, making $Q = 1$. Any change in D while $Clk = 1$ does not affect the output.

In sum, when the input clock in the positive-edge-triggered flip-flop makes a positive transition, the value of D is transferred to Q. A negative transition of the clock (i.e., from 1 to 0) does not affect the output, nor is the output affected by changes in D when Clk is in the steady logic-1 level or the logic-0 level. Hence, this type of flip-flop responds to the transition from 0 to 1 and nothing else.

The timing of the response of a flip-flop to input data and to the clock must be taken into consideration when one is using edge-triggered flip-flops. There is a minimum time called the *setup time* during which the D input must be maintained at a constant value prior to the occurrence of the clock transition. Similarly, there is a minimum time called the *hold time* during which the D input must not change after the application of the positive transition of the clock. The propagation delay time of the flip-flop is defined as the interval between the trigger edge and the stabilization of the output to a new state. These and other parameters are specified in manufacturers' data books for specific logic families.

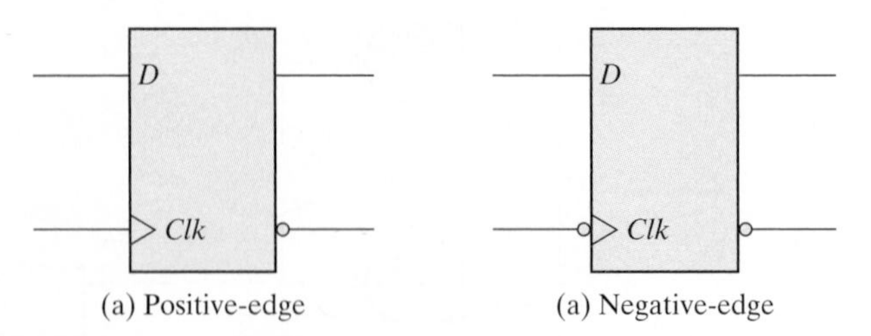

FIGURE 5.11
Graphic symbol for edge-triggered *D* flip-flop

The graphic symbol for the edge-triggered D flip-flop is shown in Fig. 5.11. It is similar to the symbol used for the D latch, except for the arrowhead-like symbol in front of the letter *Clk*, designating a *dynamic* input. The *dynamic indicator* (>) denotes the fact that the flip-flop responds to the edge transition of the clock. A bubble outside the block adjacent to the dynamic indicator designates a negative edge for triggering the circuit. The absence of a bubble designates a positive-edge response.

Other Flip-Flops

Very large-scale integration circuits contain several thousands of gates within one package. Circuits are constructed by interconnecting the various gates to provide a digital system. Each flip-flop is constructed from an interconnection of gates. The most economical and efficient flip-flop constructed in this manner is the edge-triggered D flip-flop, because it requires the smallest number of gates. Other types of flip-flops can be constructed by using the D flip-flop and external logic. Two flip-flops less widely used in the design of digital systems are the JK and T flip-flops.

There are three operations that can be performed with a flip-flop: Set it to 1, reset it to 0, or complement its output. With only a single input, the D flip-flop can set or reset the output, depending on the value of the D input immediately before the clock transition. Synchronized by a clock signal, the JK flip-flop has two inputs and performs all three operations. The circuit diagram of a JK flip-flop constructed with a D flip-flop and gates is shown in Fig. 5.12(a). The J input sets the flip-flop to 1, the K input resets it to 0, and when both inputs are enabled, the output is complemented. This can be verified by investigating the circuit applied to the D input:

$$D = JQ' + K'Q$$

When $J = 1$ and $K = 0$, $D = Q' + Q = 1$, so the next clock edge sets the output to 1. When $J = 0$ and $K = 1$, $D = 0$, so the next clock edge resets the output to 0. When both $J = K = 1$ and $D = Q'$, the next clock edge complements the output. When both $J = K = 0$ and $D = Q$, the clock edge leaves the output unchanged. The graphic symbol for the JK flip-flop is shown in Fig. 5.12(b). It is similar to the graphic symbol of the D flip-flop, except that now the inputs are marked J and K.

The T (toggle) flip-flop is a complementing flip-flop and can be obtained from a JK flip-flop when inputs J and K are tied together. This is shown in Fig. 5.13(a). When

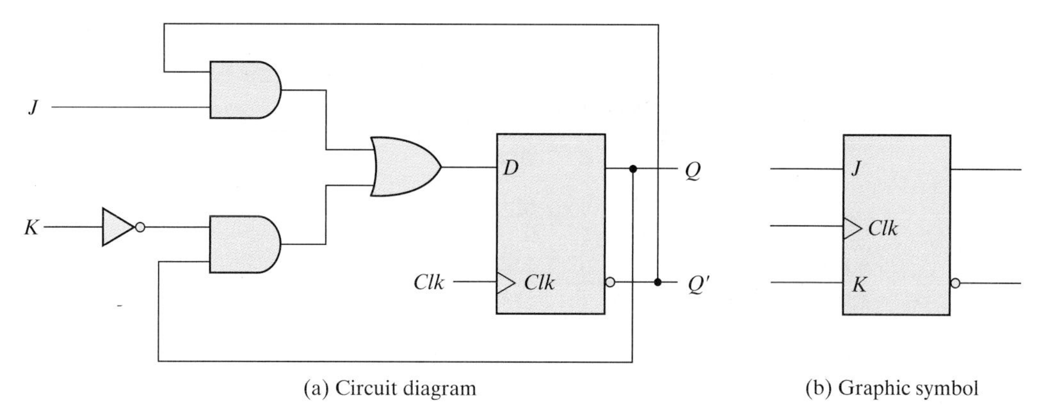

(a) Circuit diagram

(b) Graphic symbol

FIGURE 5.12
***JK* flip-flop**

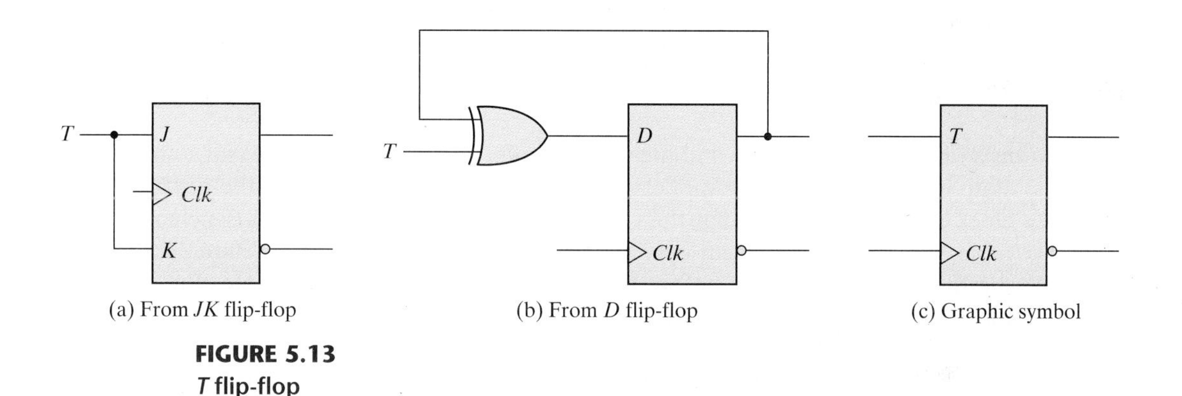

(a) From *JK* flip-flop

(b) From *D* flip-flop

(c) Graphic symbol

FIGURE 5.13
***T* flip-flop**

$T = 0$ ($J = K = 0$), a clock edge does not change the output. When $T = 1$ ($J = K = 1$), a clock edge complements the output. The complementing flip-flop is useful for designing binary counters.

The T flip-flop can be constructed with a D flip-flop and an exclusive-OR gate as shown in Fig. 5.13(b). The expression for the D input is

$$D = T \oplus Q = TQ' + T'Q$$

When $T = 0$, $D = Q$ and there is no change in the output. When $T = 1$, $D = Q'$ and the output complements. The graphic symbol for this flip-flop has a T symbol in the input.

Characteristic Tables

A characteristic table defines the logical properties of a flip-flop by describing its operation in tabular form. The characteristic tables of three types of flip-flops are presented in Table 5.1. They define the next state (i.e., the state that results from a clock transition)

Table 5.1
Flip-Flop Characteristic Tables

***JK* Flip-Flop**

J	K	$Q(t+1)$	
0	0	$Q(t)$	No change
0	1	0	Reset
1	0	1	Set
1	1	$Q'(t)$	Complement

***D* Flip-Flop**

D	$Q(t+1)$	
0	0	Reset
1	1	Set

***T* Flip-Flop**

T	$Q(t+1)$	
0	$Q(t)$	No change
1	$Q'(t)$	Complement

as a function of the inputs and the present state. $Q(t)$ refers to the present state (i.e., the state present prior to the application of a clock edge). $Q(t + 1)$ is the next state one clock period later. Note that the clock edge input is not included in the characteristic table, but is implied to occur between times t and $t + 1$. Thus, $Q(t)$ denotes the state of the flip-flop immediately before the clock edge, and $Q(t + 1)$ denotes the state that results from the clock transition.

The characteristic table for the *JK* flip-flop shows that the next state is equal to the present state when inputs J and K are both equal to 0. This condition can be expressed as $Q(t + 1) = Q(t)$, indicating that the clock produces no change of state. When $K = 1$ and $J = 0$, the clock resets the flip-flop and $Q(t + 1) = 0$. With $J = 1$ and $K = 0$, the flip-flop sets and $Q(t + 1) = 1$. When both J and K are equal to 1, the next state changes to the complement of the present state, a transition that can be expressed as $Q(t + 1) = Q'(t)$.

The next state of a D flip-flop is dependent only on the D input and is independent of the present state. This can be expressed as $Q(t + 1) = D$. It means that the next-state value is equal to the value of D. Note that the D flip-flop does not have a "no-change" condition. Such a condition can be accomplished either by disabling the clock or by operating the clock by having the output of the flip-flop connected into the D input. Either method effectively circulates the output of the flip-flop when the state of the flip-flop must remain unchanged.

The characteristic table of the T flip-flop has only two conditions: When $T = 0$, the clock edge does not change the state; when $T = 1$, the clock edge complements the state of the flip-flop.

Characteristic Equations

The logical properties of a flip-flop, as described in the characteristic table, can be expressed algebraically with a characteristic equation. For the D flip-flop, we have the characteristic equation

$$Q(t + 1) = D$$

which states that the next state of the output will be equal to the value of input D in the present state. The characteristic equation for the JK flip-flop can be derived from the characteristic table or from the circuit of Fig. 5.12. We obtain

$$Q(t + 1) = JQ' + K'Q$$

where Q is the value of the flip-flop output prior to the application of a clock edge. The characteristic equation for the T flip-flop is obtained from the circuit of Fig. 5.13:

$$Q(t + 1) = T \oplus Q = TQ' + T'Q$$

Direct Inputs

Some flip-flops have asynchronous inputs that are used to force the flip-flop to a particular state independently of the clock. The input that sets the flip-flop to 1 is called *preset* or *direct set*. The input that clears the flip-flop to 0 is called *clear* or *direct reset*. When power is turned on in a digital system, the state of the flip-flops is unknown. The direct inputs are useful for bringing all flip-flops in the system to a known starting state prior to the clocked operation.

A positive-edge-triggered D flip-flop with active-low asynchronous reset is shown in Fig. 5.14. The circuit diagram is the same as the one in Fig. 5.10, except for the additional reset input connections to three NAND gates. When the reset input is 0, it forces output Q' to stay at 1, which, in turn, clears output Q to 0, thus resetting the flip-flop. Two other connections from the reset input ensure that the S input of the third SR latch stays at logic 1 while the reset input is at 0, regardless of the values of D and Clk.

The graphic symbol for the D flip-flop with a direct reset has an additional input marked with R. The bubble along the input indicates that the reset is active at the logic-0 level. Flip-flops with a direct set use the symbol S for the asynchronous set input.

The function table specifies the operation of the circuit. When $R = 0$, the output is reset to 0. This state is independent of the values of D or Clk. Normal clock operation can proceed only after the reset input goes to logic 1. The clock at Clk is shown with an upward arrow to indicate that the flip-flop triggers on the positive edge of the clock. The value in D is transferred to Q with every positive-edge clock signal, provided that $R = 1$.

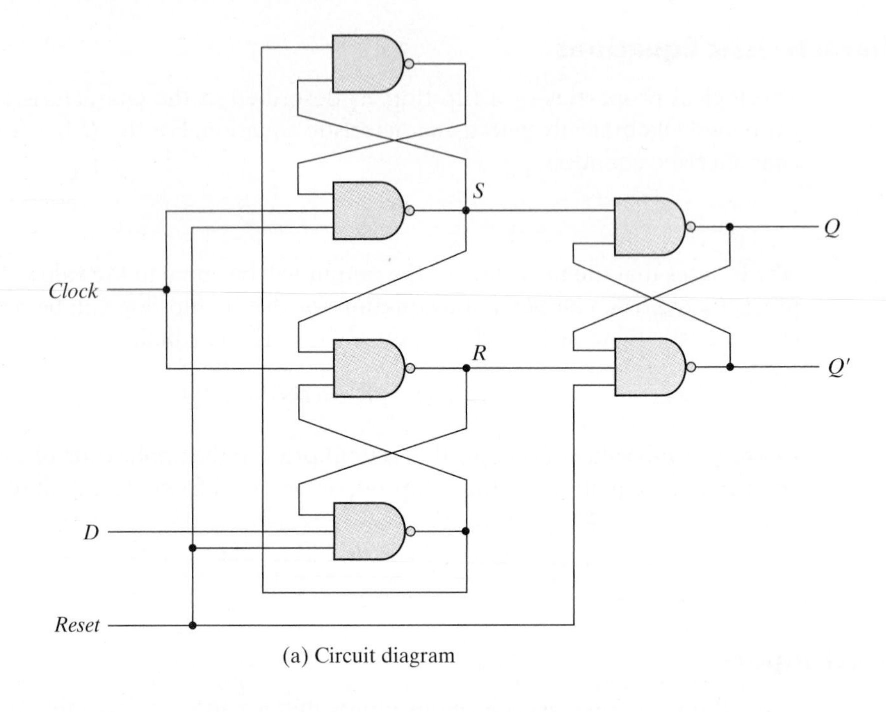

(a) Circuit diagram

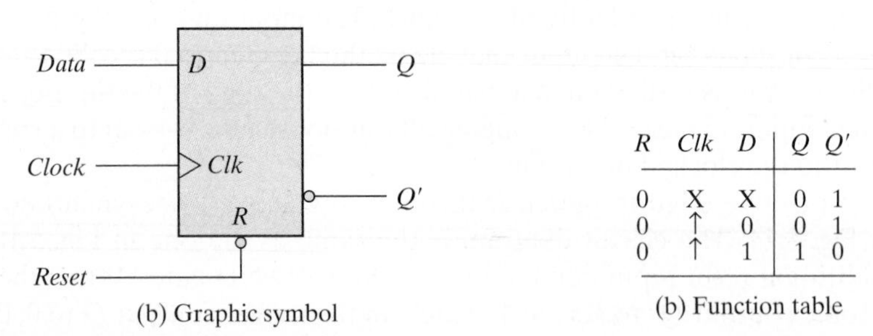

(b) Graphic symbol

R	Clk	D	Q	Q'
0	X	X	0	1
0	↑	0	0	1
0	↑	1	1	0

(b) Function table

FIGURE 5.14
D flip-flop with asynchronous reset

5.5 ANALYSIS OF CLOCKED SEQUENTIAL CIRCUITS

Analysis describes what a given circuit will do under certain operating conditions. The behavior of a clocked sequential circuit is determined from the inputs, the outputs, and the state of its flip-flops. The outputs and the next state are both a function of the inputs and the present state. The analysis of a sequential circuit consists of obtaining a table or a diagram for the time sequence of inputs, outputs, and internal states. It is also possible to write Boolean expressions that describe the behavior of the sequential circuit. These expressions must include the necessary time sequence, either directly or indirectly.

A logic diagram is recognized as a clocked sequential circuit if it includes flip-flops with clock inputs. The flip-flops may be of any type, and the logic diagram may or may not include combinational logic gates. In this section, we introduce an algebraic representation for specifying the next-state condition in terms of the present state and inputs. A state table and state diagram are then presented to describe the behavior of the sequential circuit. Another algebraic representation is introduced for specifying the logic diagram of sequential circuits. Examples are used to illustrate the various procedures.

State Equations

The behavior of a clocked sequential circuit can be described algebraically by means of state equations. A *state equation* (also called a *transition equation*) specifies the next state as a function of the present state and inputs. Consider the sequential circuit shown in Fig. 5.15. We will later show that it acts as a 0-detector by asserting its output when a

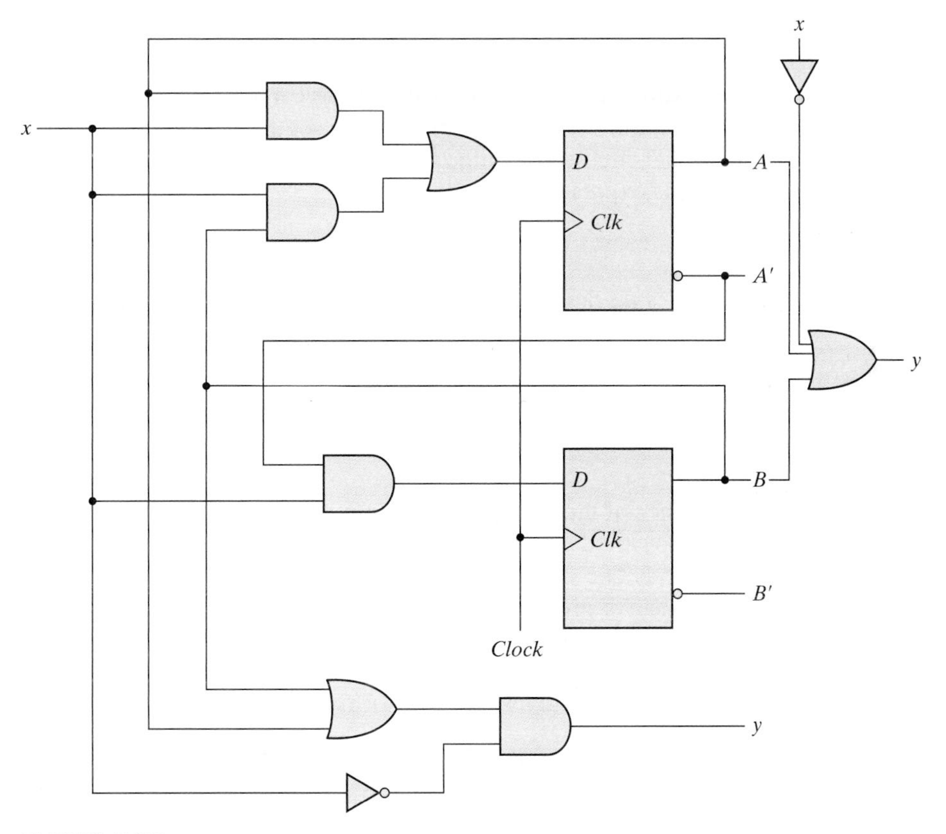

FIGURE 5.15
Example of sequential circuit

0 is detected in a stream of 1s. It consists of two D flip-flops A and B, an input x and an output y. Since the D input of a flip-flop determines the value of the next state (i.e., the state reached after the clock transition), it is possible to write a set of state equations for the circuit:

$$A(t + 1) = A(t)x(t) + B(t)x(t)$$
$$B(t + 1) = A'(t)x(t)$$

A state equation is an algebraic expression that specifies the condition for a flip-flop state transition. The left side of the equation, with $(t + 1)$, denotes the next state of the flip-flop one clock edge later. The right side of the equation is a Boolean expression that specifies the present state and input conditions that make the next state equal to 1. Since all the variables in the Boolean expressions are a function of the present state, we can omit the designation (t) after each variable for convenience and can express the state equations in the more compact form

$$A(t + 1) - Ax + Bx$$
$$B(t + 1) - A'x$$

The Boolean expressions for the state equations can be derived directly from the gates that form the combinational circuit part of the sequential circuit, since the D values of the combinational circuit determine the next state. Similarly, the present-state value of the output can be expressed algebraically as

$$y(t) = [A(t) + B(t)]x'(t)$$

By removing the symbol (t) for the present state, we obtain the output Boolean equation:

$$y = (A + B)x'$$

State Table

The time sequence of inputs, outputs, and flip-flop states can be enumerated in a *state table* (sometimes called a *transition table*). The state table for the circuit of Fig. 5.15 is shown in Table 5.2. The table consists of four sections labeled *present state*, *input*, *next state*, and *output*. The present-state section shows the states of flip-flops A and B at any given time t. The input section gives a value of x for each possible present state. The next-state section shows the states of the flip-flops one clock cycle later, at time $t + 1$. The output section gives the value of y at time t for each present state and input condition.

The derivation of a state table requires listing all possible binary combinations of present states and inputs. In this case, we have eight binary combinations from 000 to 111. The next-state values are then determined from the logic diagram or from the state equations. The next state of flip-flop A must satisfy the state equation

$$A(t + 1) = Ax + Bx$$

Table 5.2
State Table for the Circuit of Fig. 5.15

Present State		Input	Next State		Output
A	B	x	A	B	y
0	0	0	0	0	0
0	0	1	0	1	0
0	1	0	0	0	1
0	1	1	1	1	0
1	0	0	0	0	1
1	0	1	1	0	0
1	1	0	0	0	1
1	1	1	1	0	0

The next-state section in the state table under column A has three 1's where the present state of A and input x are both equal to 1 or the present state of B and input x are both equal to 1. Similarly, the next state of flip-flop B is derived from the state equation

$$B(t + 1) = A'x$$

and is equal to 1 when the present state of A is 0 and input x is equal to 1. The output column is derived from the output equation

$$y = Ax' + Bx'$$

The state table of a sequential circuit with D-type flip-flops is obtained by the same procedure outlined in the previous example. In general, a sequential circuit with m flip-flops and n inputs needs 2^{m+n} rows in the state table. The binary numbers from 0 through $2^{m+n} - 1$ are listed under the present-state and input columns. The next-state section has m columns, one for each flip-flop. The binary values for the next state are derived directly from the state equations. The output section has as many columns as there are output variables. Its binary value is derived from the circuit or from the Boolean function in the same manner as in a truth table.

It is sometimes convenient to express the state table in a slightly different form having only three sections: present state, next state, and output. The input conditions are enumerated under the next-state and output sections. The state table of Table 5.2 is repeated in Table 5.3 in this second form. For each present state, there are two possible next states and outputs, depending on the value of the input. One form may be preferable to the other, depending on the application.

State Diagram

The information available in a state table can be represented graphically in the form of a state diagram. In this type of diagram, a state is represented by a circle, and the (clock-triggered) transitions between states are indicated by directed lines connecting

Table 5.3
Second Form of the State Table

Present State		Next State				Output	
		x = 0		x = 1		x = 0	x = 1
A	B	A	B	A	B	y	y
0	0	0	0	0	1	0	0
0	1	0	0	1	1	1	0
1	0	0	0	1	0	1	0
1	1	0	0	1	0	1	0

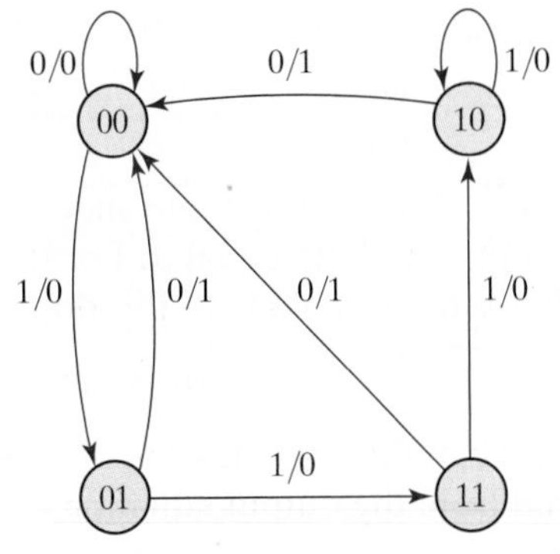

FIGURE 5.16
State diagram of the circuit of Fig. 5.15

the circles. The state diagram of the sequential circuit of Fig. 5.15 is shown in Fig. 5.16. The state diagram provides the same information as the state table and is obtained directly from Table 5.2 or Table 5.3. The binary number inside each circle identifies the state of the flip-flops. The directed lines are labeled with two binary numbers separated by a slash. The input value during the present state is labeled first, and the number after the slash gives the output during the present state with the given input. (It is important to remember that the bit value listed for the output along the directed line occurs during the present state and with the indicated input, and has nothing to do with the transition to the next state.) For example, the directed line from state 00 to 01 is labeled 1/0, meaning that when the sequential circuit is in the present state 00 and the input is 1, the output is 0. After the next clock cycle, the circuit goes to the next state, 01. If the input changes to 0, then the output becomes 1, but if the input remains at 1, the output stays at 0. This information is obtained from the state diagram along the two directed lines emanating from the circle with state 01. A directed line connecting a circle with itself indicates that no change of state occurs.

The steps presented in this example are summarized below:

Circuit diagram → Equations – State table → State diagram

This sequence of steps begins with a structural representation of the circuit and proceeds to an abstract representation of its behavior. An HDL model can be in the form of a gate-level description or in the form of a behavioral description. It is important to note that a gate-level approach requires that the designer understands how to select and connect gates and flip-flops to form a circuit having a particular behavior. That understanding comes with experience. On the other hand, an approach based on behavioral modeling does not require the designer to know how to invent a schematic—the designer needs only to know how to describe behavior using the constructs of the HDL, because the circuit is produced automatically by a synthesis tool. Therefore, one does not have to accumulate years of experience in order to become a productive designer of digital circuits; nor does one have to acquire an extensive background in electrical engineering.

There is no difference between a state table and a state diagram, except in the manner of representation. The state table is easier to derive from a given logic diagram and the state equation. The state diagram follows directly from the state table. *The state diagram gives a pictorial view of state transitions and is the form more suitable for human interpretation of the circuit's operation.* For example, the state diagram of Fig. 5.16 clearly shows that, starting from state 00, the output is 0 as long as the input stays at 1. The first 0 input after a string of 1's gives an output of 1 and transfers the circuit back to the initial state, 00. The machine represented by this state diagram acts to detect a zero in the bit stream of data. It corresponds to the behavior of the circuit in Fig. 5.15. Other circuits that detect a zero in a stream of data may have a simpler circuit diagram and state diagram.

Flip-Flop Input Equations

The logic diagram of a sequential circuit consists of flip-flops and gates. The interconnections among the gates form a combinational circuit and may be specified algebraically with Boolean expressions. The knowledge of the type of flip-flops and a list of the Boolean expressions of the combinational circuit provide the information needed to draw the logic diagram of the sequential circuit. The part of the combinational circuit that generates external outputs is described algebraically by a set of Boolean functions called *output equations*. The part of the circuit that generates the inputs to flip-flops is described algebraically by a set of Boolean functions called flip-flop *input equations* (or, sometimes, *excitation equations*). We will adopt the convention of using the flip-flop input symbol to denote the input equation variable and a subscript to designate the name of the flip-flop output. For example, the following input equation specifies an OR gate with inputs x and y connected to the D input of a flip-flop whose output is labeled with the symbol Q:

$$D_Q = x + y$$

The sequential circuit of Fig. 5.15 consists of two D flip-flops A and B, an input x, and an output y. The logic diagram of the circuit can be expressed algebraically with two flip-flop input equations and an output equation:

$$D_A = Ax + Bx$$
$$D_B = A'x$$
$$y = (A + B)x'$$

The three equations provide the necessary information for drawing the logic diagram of the sequential circuit. The symbol D_A specifies a D flip-flop labeled A. D_B specifies a second D flip-flop labeled B. The Boolean expressions associated with these two variables and the expression for output y specify the combinational circuit part of the sequential circuit.

The flip-flop input equations constitute a convenient algebraic form for specifying the logic diagram of a sequential circuit. They imply the type of flip-flop from the letter symbol, and they fully specify the combinational circuit that drives the flip-flops. Note that the expression for the input equation for a D flip-flop is identical to the expression for the corresponding state equation. This is because of the characteristic equation that equates the next state to the value of the D input: $Q(t + 1) = D_Q$.

Analysis with *D* Flip-Flops

We will summarize the procedure for analyzing a clocked sequential circuit with D flip-flops by means of a simple example. The circuit we want to analyze is described by the input equation

$$D_A = A \oplus x \oplus y$$

The D_A symbol implies a D flip-flop with output A. The x and y variables are the inputs to the circuit. No output equations are given, which implies that the output comes from the output of the flip-flop. The logic diagram is obtained from the input equation and is drawn in Fig. 5.17(a).

The state table has one column for the present state of flip-flop A, two columns for the two inputs, and one column for the next state of A. The binary numbers under Axy are listed from 000 through 111 as shown in Fig. 5.17(b). The next-state values are obtained from the state equation

$$A(t + 1) = A \oplus x \oplus y$$

The expression specifies an odd function and is equal to 1 when only one variable is 1 or when all three variables are 1. This is indicated in the column for the next state of A.

The circuit has one flip-flop and two states. The state diagram consists of two circles, one for each state as shown in Fig. 5.17(c). The present state and the output can be either 0 or 1, as indicated by the number inside the circles. A slash on the directed lines is not needed, because there is no output from a combinational circuit. The two inputs can have four possible combinations for each state. Two input combinations during each state transition are separated by a comma to simplify the notation.

Analysis with *JK* Flip-Flops

A state table consists of four sections: present state, inputs, next state, and outputs. The first two are obtained by listing all binary combinations. The output section is determined from the output equations. The next-state values are evaluated from the state equations. For a D-type flip-flop, the state equation is the same as the input equation. When a flip-flop other than the D type is used, such as JK or T, it is necessary to refer

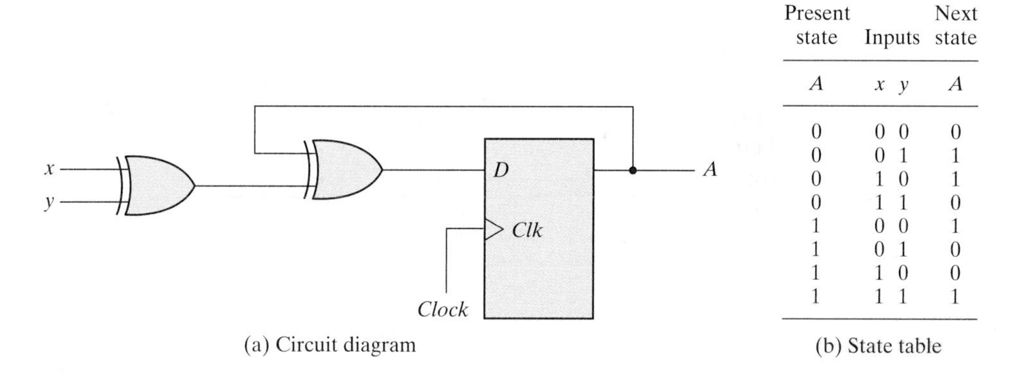

(a) Circuit diagram

Present state	Inputs	Next state
A	x y	A
0	0 0	0
0	0 1	1
0	1 0	1
0	1 1	0
1	0 0	1
1	0 1	0
1	1 0	0
1	1 1	1

(b) State table

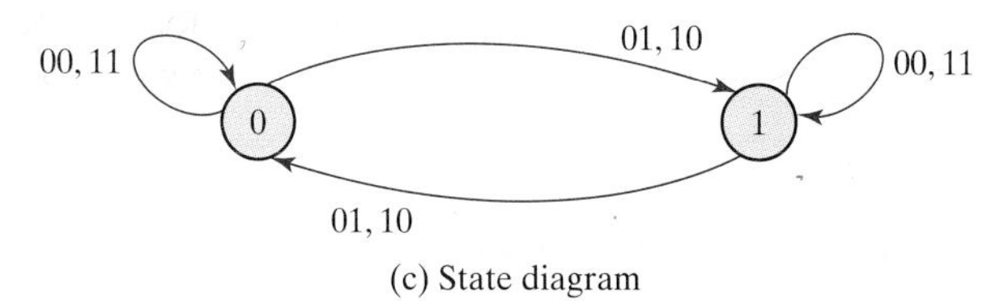

(c) State diagram

FIGURE 5.17
Sequential circuit with *D* flip-flop

to the corresponding characteristic table or characteristic equation to obtain the next-state values. We will illustrate the procedure first by using the characteristic table and again by using the characteristic equation.

The next-state values of a sequential circuit that uses *JK*- or *T*-type flip-flops can be derived as follows:

1. Determine the flip-flop input equations in terms of the present state and input variables.
2. List the binary values of each input equation.
3. Use the corresponding flip-flop characteristic table to determine the next-state values in the state table.

As an example, consider the sequential circuit with two *JK* flip-flops *A* and *B* and one input *x*, as shown in Fig. 5.18. The circuit has no outputs; therefore, the state table does not need an output column. (The outputs of the flip-flops may be considered as the outputs in this case.) The circuit can be specified by the flip-flop input equations

$$J_A = B \quad K_A = Bx'$$
$$J_B = x' \quad K_B = A'x + Ax' = A \oplus x$$

The state table of the sequential circuit is shown in Table 5.4. The present-state and input columns list the eight binary combinations. The binary values listed under the

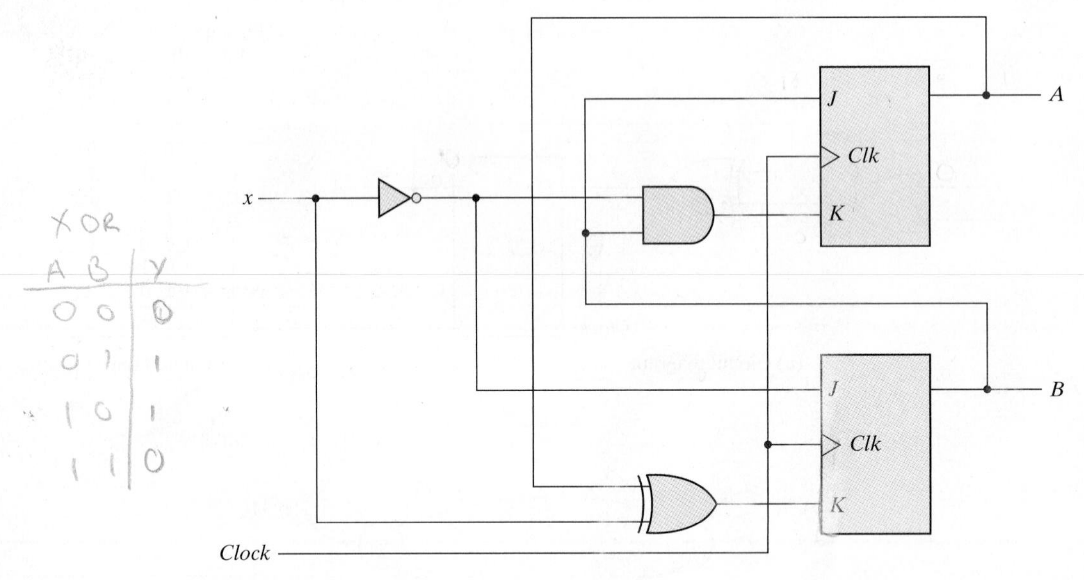

FIGURE 5.18
Sequential circuit with *JK* flip-flop

columns labeled *flip-flop inputs* are not part of the state table, but they are needed for the purpose of evaluating the next state as specified in step 2 of the procedure. These binary values are obtained directly from the four input equations in a manner similar to that for obtaining a truth table from a Boolean expression. The next state of each flip-flop is evaluated from the corresponding J and K inputs and the characteristic table of the JK flip-flop listed in Table 5.1. There are four cases to consider. When $J = 1$ and

Table 5.4
State Table for Sequential Circuit with JK Flip-Flops

Present State		Input	Next State		Flip-Flop Inputs			
A	**B**	**x**	**A**	**B**	J_A	K_A	J_B	K_B
0	0	0	0	1	0	0	1	0
0	0	1	0	0	0	0	0	1
0	1	0	1	1	1	1	1	0
0	1	1	1	0	1	0	0	1
1	0	0	1	1	0	0	1	1
1	0	1	1	0	0	0	0	0
1	1	0	0	0	1	1	1	1
1	1	1	1	1	1	0	0	0

$K = 0$, the next state is 1. When $J = 0$ and $K = 1$, the next state is 0. When $J = K = 0$, there is no change of state and the next-state value is the same as that of the present state. When $J = K = 1$, the next-state bit is the complement of the present-state bit. Examples of the last two cases occur in the table when the present state AB is 10 and input x is 0. JA and KA are both equal to 0 and the present state of A is 1. Therefore, the next state of A remains the same and is equal to 1. In the same row of the table, JB and KB are both equal to 1. Since the present state of B is 0, the next state of B is complemented and changes to 1.

The next-state values can also be obtained by evaluating the state equations from the characteristic equation. This is done by using the following procedure:

1. Determine the flip-flop input equations in terms of the present state and input variables.
2. Substitute the input equations into the flip-flop characteristic equation to obtain the state equations.
3. Use the corresponding state equations to determine the next-state values in the state table.

The input equations for the two JK flip-flops of Fig. 5.18 were listed a couple of paragraphs ago. The characteristic equations for the flip-flops are obtained by substituting A or B for the name of the flip-flop, instead of Q:

$$A(t + 1) = JA' + K'A$$
$$B(t + 1) = JB' + K'B$$

Substituting the values of J_A and K_A from the input equations, we obtain the state equation for A:

$$A(t + 1) = BA' + (Bx')'A = A'B + AB' + Ax$$

The state equation provides the bit values for the column headed "Next State" for A in the state table. Similarly, the state equation for flip-flop B can be derived from the characteristic equation by substituting the values of J_B and K_B:

$$B(t + 1) = x'B' + (A \oplus x)'B = B'x' + ABx + A'Bx'$$

The state equation provides the bit values for the column headed "Next State" for B in the state table. Note that the columns in Table 5.4 headed "Flip-Flop Inputs" are not needed when state equations are used.

The state diagram of the sequential circuit is shown in Fig. 5.19. Note that since the circuit has no outputs, the directed lines out of the circles are marked with one binary number only, to designate the value of input x.

Analysis with *T* Flip-Flops

The analysis of a sequential circuit with T flip-flops follows the same procedure outlined for JK flip-flops. The next-state values in the state table can be obtained by using either

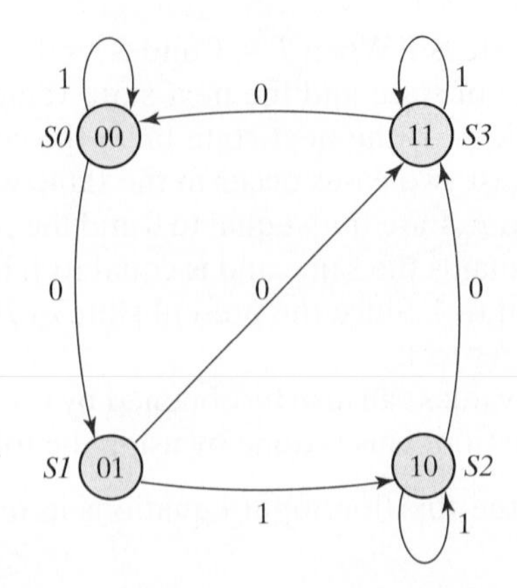

FIGURE 5.19
State diagram of the circuit of Fig. 5.18

the characteristic table listed in Table 5.1 or the characteristic equation

$$Q(t + 1) = T \oplus Q = T'Q + TQ'$$

Now consider the sequential circuit shown in Fig. 5.20. It has two flip-flops A and B, one input x, and one output y and can be described algebraically by two input equations and an output equation:

$$T_A = Bx$$
$$T_B = x$$
$$y = AB$$

The state table for the circuit is listed in Table 5.5. The values for y are obtained from the output equation. The values for the next state can be derived from the state equations by substituting T_A and T_B in the characteristic equations, yielding

$$A(t + 1) = (Bx)'A + (Bx)A' = AB' + Ax' + A'Bx$$
$$B(t + 1) = x \oplus B$$

The next-state values for A and B in the state table are obtained from the expressions of the two state equations.

The state diagram of the circuit is shown in Fig. 5.20(b). As long as input x is equal to 1, the circuit behaves as a binary counter with a sequence of states 00, 01, 10, 11, and back to 00. When $x = 0$, the circuit remains in the same state. Output y is equal to 1 when the present state is 11. Here, the output depends on the present state only and is independent of the input. The two values inside each circle and separated by a slash are for the present state and output.

Mealy and Moore Models of Finite State Machines

The most general model of a sequential circuit has inputs, outputs, and internal states. It is customary to distinguish between two models of sequential circuits: the Mealy model and the Moore model. Both are shown in Fig. 5.21. They differ only in the way the output

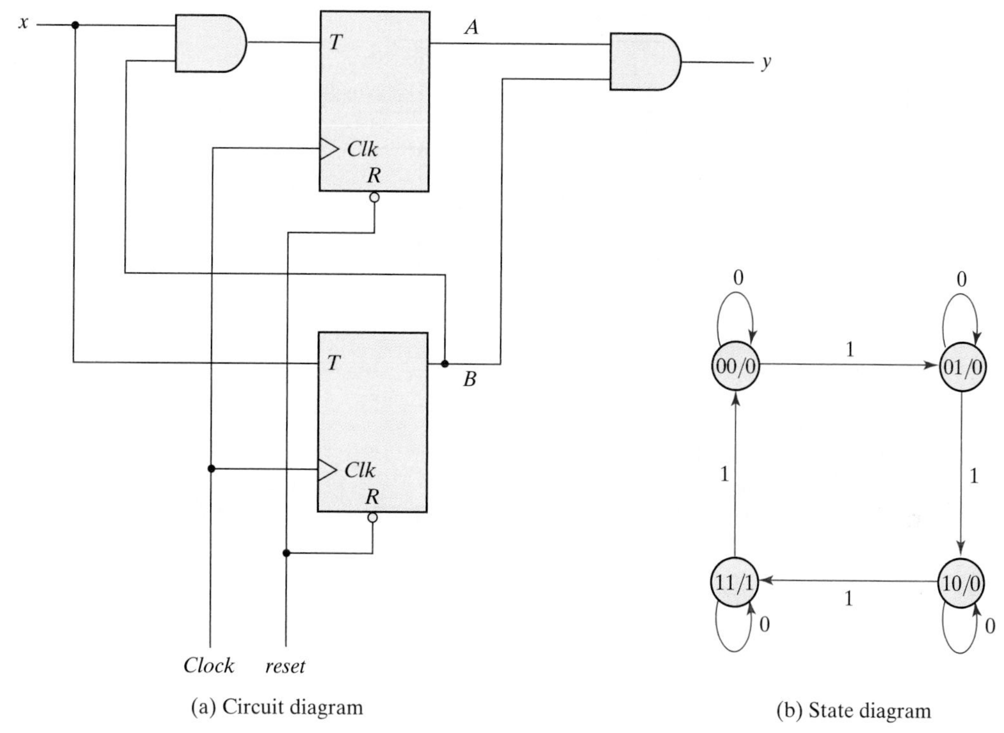

(a) Circuit diagram

(b) State diagram

FIGURE 5.20
Sequential circuit with *T* flip-flops (Binary Counter)

Table 5.5
State Table for Sequential Circuit with T Flip-Flops

Present State		Input	Next State		Output
A	***B***	***x***	***A***	***B***	***y***
0	0	0	0	0	0
0	0	1	0	1	0
0	1	0	0	1	0
0	1	1	1	0	0
1	0	0	1	0	0
1	0	1	1	1	0
1	1	0	1	1	1
1	1	1	0	0	1

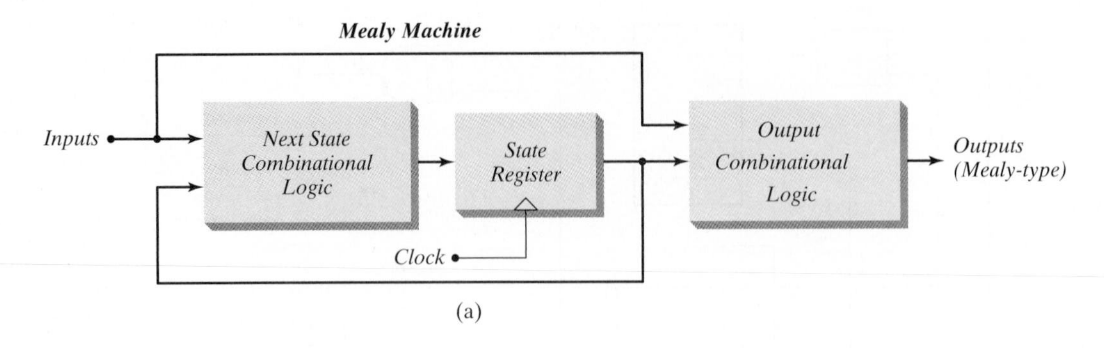

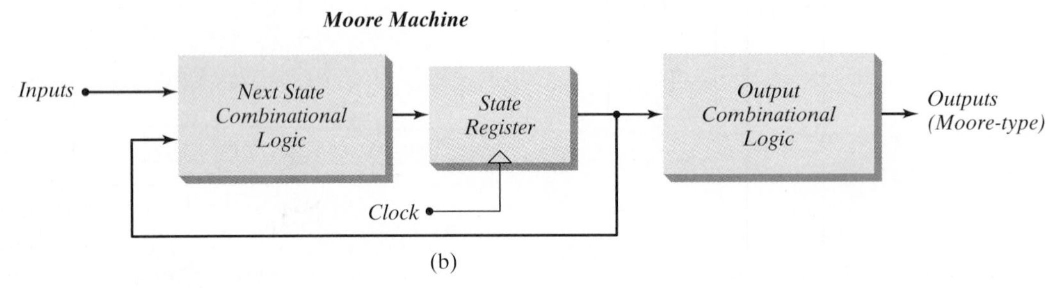

FIGURE 5.21
Block diagrams of Mealy and Moore state machines

is generated. In the Mealy model, the output is a function of both the present state and the input. In the Moore model, the output is a function of only the present state. A circuit may have both types of outputs. The two models of a sequential circuit are commonly referred to as a finite state machine, abbreviated FSM. The Mealy model of a sequential circuit is referred to as a Mealy FSM or Mealy machine. The Moore model is referred to as a Moore FSM or Moore machine.

The circuit presented previously in Fig. 5.15 is an example of a Mealy machine. Output y is a function of both input x and the present state of A and B. The corresponding state diagram in Fig. 5.16 shows both the input and output values, separated by a slash along the directed lines between the states.

An example of a Moore model is given in Fig. 5.18. Here, the output is a function of the present state only. The corresponding state diagram in Fig. 5.19 has only inputs marked along the directed lines. The outputs are the flip-flop states marked inside the circles. Another example of a Moore model is the sequential circuit of Fig. 5.20. The output depends only on flip-flop values, and that makes it a function of the present state only. The input value in the state diagram is labeled along the directed line, but the output value is indicated inside the circle together with the present state.

In a Moore model, the outputs of the sequential circuit are synchronized with the clock, because they depend only on flip-flop outputs that are synchronized with the clock. In a Mealy model, the outputs may change if the inputs change during the clock cycle. Moreover, the outputs may have momentary false values because of the delay encountered from the time that the inputs change and the time that the flip-flop outputs change. In order to synchronize a Mealy-type circuit, the inputs of the sequential circuit must be synchronized with the clock and the outputs must be sampled immediately before the clock edge. The inputs are changed at the inactive edge of the clock to ensure that the inputs to the flip-flops stabilize before the active edge of the clock occurs. Thus, **the output of the Mealy machine is the value that is present immediately before the active edge of the clock.**

5.6 SYNTHESIZABLE HDL MODELS OF SEQUENTIAL CIRCUITS

The Verilog HDL was introduced in Section 3.9. Combinational circuits were described in Section 4.12, and behavioral modeling with Verilog was introduced in that section as well. Behavioral models are abstract representations of the functionality of digital hardware. That is, they describe how a circuit behaves, but don't specify the internal details of the circuit. Historically, the abstraction has been described by truth tables, state tables, and state diagrams. An HDL describes the functionality differently, by language constructs that represent the operations of registers in a machine. This representation has "added value," i.e., it is important for you to know how to use, because it can be simulated to produce waveforms demonstrating the behavior of the machine.

Behavioral Modeling

There are two kinds of abstract behaviors in the Verilog HDL. Behavior declared by the keyword **initial** is called *single-pass behavior* and specifies a single statement or a block statement (i.e., a list of statements enclosed by either a **begin** . . . **end** or a **fork** . . . **join** keyword pair). A single-pass behavior expires after the associated statement executes. In practice, designers use single-pass behavior primarily to prescribe stimulus signals in a test bench—never to model the behavior of a circuit—because synthesis tools do not accept descriptions that use the **initial** statement. The **always** keyword declares a *cyclic behavior.* Both types of behaviors begin executing when the simulator launches at time $t = 0$. The **initial** behavior expires after its statement executes; the **always** behavior executes and reexecutes indefinitely, until the simulation is stopped. A module may contain an arbitrary number of **initial** or **always** behavioral statements. They execute concurrently with respect to each other, starting at time 0, and may interact through common variables. Here's a word description of how an **always** statement works for a simple model of a D flip-flop: Whenever the rising edge of the clock occurs, if the reset input is asserted, the output q gets 0; otherwise the output Q gets the value of the input D. The execution of statements triggered by the clock is repeated until the simulation ends. We'll see shortly how to write this description in Verilog.

An **initial** behavioral statement executes only once. It begins its execution at the start of simulation and expires after all of its statements have completed execution. As mentioned at the end of Section 4.12, the **initial** statement is useful for generating input signals to simulate a design. In simulating a sequential circuit, it is necessary to generate a clock source for triggering the flip-flops. The following are two possible ways to provide a free-running clock that operates for a specified number of cycles:

```
initial
 begin
  clock = 1'b0;
  repeat (30)
   #10 clock = ~clock;
 end
```

```
initial
 begin
  clock = 1'b0;
 end

initial 300 $finish;
always #10 clock = ~clock;
```

In the first version, the **initial** block contains two statements enclosed within the **begin** and **end** keywords. The first statement sets *clock* to 0 at time = 0. The second statement specifies a loop that reexecutes 30 times to wait 10 time units and then complement the value of *clock*. This produces 15 clock cycles, each with a cycle time of 20 time units. In the second version, the first **initial** behavior has a single statement that sets *clock* to 0 at time = 0, and it then expires (causes no further simulation activity). The second single-pass behavior declares a stopwatch for the simulation. The system task **finish** causes the simulation to terminate unconditionally after 300 time units have elapsed. Because this behavior has only one statement associated with it, there is no need to write the **begin** . . . **end** keyword pair. After 10 time units, the **always** statement repeatedly complements *clock,* providing a clock generator having a cycle time of 20 time units. The three behavioral statements in the second example can be written in any order.

Here is another way to describe a free-running clock:

```
initial begin clock = 0; forever #10 clock = ~clock; end
```

This version, with two statements in one block statement, initializes the clock and then executes an indefinite loop (**forever**) in which the clock is complemented after a delay of 10 time steps. Note that the single-pass behavior never finishes executing and so does not expire. Another behavior would have to terminate the simulation.

The activity associated with either type of behavioral statement can be controlled by a delay operator that waits for a certain time or by an event control operator that waits for certain conditions to become true or for specified events (changes in signals) to occur. Time delays specified with the # *delay control operator* are commonly used in single-pass behaviors. The delay control operator suspends execution of statements until a specified time has elapsed. We've already seen examples of its use to specify signals in a test bench. Another operator @ is called the *event control operator* and is used to *suspend* activity until an event occurs. An event can be an unconditional change in a signal value (e.g., @ A) or a specified transition of a signal value (e.g., @ (**posedge** clock)). The general form of this type of statement is

```
always @ (event control expression) begin
   // Procedural assignment statements that execute when the condition is met
end
```

The event control expression specifies the condition that must occur to launch execution of the procedural assignment statements. The variables in the left-hand side of the procedural statements must be of the **reg** data type and must be declared as such. The right-hand side can be any expression that produces a value using Verilog-defined operators.

The event control expression (also called the sensitivity list) specifies the events that must occur to initiate execution of the procedural statements associated with the **always** block. Statements within the block execute sequentially from top to bottom. After the last statement executes, the behavior waits for the event control expression to be satisfied. Then the statements are executed again. The sensitivity list can specify level-sensitive events, edge-sensitive events, or a combination of the two. In practice, designers do not make use of the third option, because this third form is not one that synthesis tools are able to translate into physical hardware. Level-sensitive events occur in combinational circuits and in latches. For example, the statement

```
always @ (A or B or C)
```

will initiate execution of the procedural statements in the associated **always** block if a change occurs in *A, B,* or *C*. In synchronous sequential circuits, changes in flip-flops occur only in response to a transition of a clock pulse. The transition may be either a positive edge or a negative edge of the clock, but not both. Verilog HDL takes care of these conditions by providing two keywords: **posedge** and **negedge**. For example, the expression

```
always @(posedge clock or negedge reset)      // Verilog 1995
```

will initiate execution of the associated procedural statements only if the clock goes through a positive transition or if *reset* goes through a negative transition. The 2001 and 2005 revisions to the Verilog language allow a comma-separated list for the event control expression (or sensitivity list):

```
always @(posedge clock, negedge reset)      // Verilog 2001, 2005
```

A procedural assignment is an assignment of a logic value to a variable within an **initial** or **always** statement. This is in contrast to a continuous assignment discussed in Section 4.12 with dataflow modeling. A continuous assignment has an implicit level-sensitive sensitivity list consisting of all of the variables on the right-hand side of its assignment statement. The updating of a continuous assignment is triggered whenever an event occurs in a variable included on the right-hand side of its expression. In contrast, a procedural assignment is made only when an assignment statement is executed and assigns value to it within a behavioral statement. For example, the clock signal in the preceding example was complemented only when the statement *clock* = ~*clock* executed; the statement did not execute until 10 time units after the simulation began. It is important to remember that a variable having type **reg** remains unchanged until a procedural assignment is made to give it a new value.

There are two kinds of procedural assignments: *blocking* and *nonblocking*. The two are distinguished by the symbols that they use. Blocking assignments use the symbol (=) as the assignment operator, and nonblocking assignments use (< =) as the operator.

Blocking assignment statements are executed sequentially in the order they are listed in a block of statements. Nonblocking assignments are executed concurrently by evaluating the set of expressions on the right-hand side of the list of statements; they do not make assignments to their left-hand sides until all of the expressions are evaluated. The two types of assignments may be better understood by means of an illustration. Consider these two procedural blocking assignments:

```
B = A
C = B + 1
```

The first statement transfers the value of A into B. The second statement increments the value of B and transfers the new value to C. At the completion of the assignments, C contains the value of $A + 1$.

Now consider the two statements as nonblocking assignments:

```
B <= A
C <= B + 1
```

When the statements are executed, the expressions on the right-hand side are evaluated and stored in a temporary location. The value of A is kept in one storage location and the value of $B + 1$ in another. After all the expressions in the block are evaluated and stored, the assignment to the targets on the left-hand side is made. In this case, C will contain the original value of B, plus 1. A general rule is to **use blocking assignments when sequential ordering is imperative and in cyclic behavior that is level sensitive** (i.e., in combinational logic). **Use nonblocking assignments when modeling concurrent execution** (e.g., edge-sensitive behavior such as synchronous, concurrent register transfers) **and when modeling latched behavior.** Nonblocking assignments are imperative in dealing with register transfer level design, as shown in Chapter 8. They model the concurrent operations of physical hardware synchronized by a common clock. Today's designers are expected to know what features of an HDL are useful in a practical way and how to avoid features that are not. Following these rules for using the assignment operators will prevent conditions that lead synthesis tools astray and create mismatches between the behavior of a model and the behavior of physical hardware that is produced by a synthesis tool.

HDL Models of Flip-Flops and Latches

HDL Examples 5.1 through 5.4 show Verilog descriptions of various flip-flops and a D latch. The D latch is said to be *transparent* because it responds to a change in data input with a change in the output as long as the enable input is asserted—viewing the output is the same as viewing the input. The description of a D latch is shown in HDL Example 5.1 It has two inputs, D and *enable,* and one output, Q. Since Q is assigned value in a behavior, its type must be declared to be **reg.** Hardware latches respond to input signal *levels,* so the two inputs are listed without edge qualifiers in the sensitivity list following the @ symbol in the **always** statement. In this model, there is only one blocking procedural assignment statement, and it specifies the transfer of input D to output Q if enable is true (logic 1).[1] Note that this statement is executed every time there is a change in D if *enable* is 1.

A D-type flip-flop is the simplest example of a sequential machine. HDL Example 5.2 describes two positive-edge D flip-flops in two modules. The first responds only to the clock; the second includes an asynchronous reset input. Output Q must be declared as a **reg** data type in addition to being listed as an output. This is because it is a target output of a procedural assignment statement. The keyword **posedge** ensures that the transfer of input D into Q is synchronized by the positive-edge transition of Clk. A change in D at any other time does not change Q.

HDL Example 5.1 (*D*-Latch)

```
// Description of D latch (See Fig. 5.6)
module D_latch (Q, D, enable);
 output Q;
 input  D, enable;
 reg    Q;
 always @ (enable or D)
  if (enable) Q <= D;                // Same as: if (enable == 1)
endmodule

// Alternative syntax (Verilog 2001, 2005)
module D_latch (output reg Q, input enable, D);
 always @ (enable, D)
  if (enable) Q <= D;                // No action if enable not asserted
endmodule
```

HDL Example 5.2 (*D*-Type Flip-Flop)

```
// D flip-flop without reset
module D_FF (Q, D, Clk);
 output Q;
 input  D, Clk;
 reg    Q;
 always @ (posedge Clk)
   Q <= D;
 endmodule

// D flip-flop with asynchronous reset (V2001, V2005)
module DFF (output reg Q, input D, Clk, rst);
 always @ (posedge Clk, negedge rst)
 if (!rst) Q <= 1'b0;          // Same as: if (rst == 0)
 else Q <= D;
endmodule
```

[1]The statement (single or block) associated with **if**(Boolean expression) executes if the Boolean expression is true.

The second module includes an asynchronous reset input in addition to the synchronous clock. A specific form of an **if** statement is used to describe such a flip-flop so that the model can be synthesized by a software tool. The event expression after the @ symbol in the **always** statement may have any number of edge events, either **posedge** or **negedge**. For modeling hardware, one of the events must be a clock event. The remaining events specify conditions under which asynchronous logic is to be executed. The designer knows which signal is the clock, but *clock* is not an identifier that software tools automatically recognize as the synchronizing signal of a circuit. The tool must be able to infer which signal is the clock, so *you need to write the description in a way that enables the tool to infer the clock correctly*. The rules are simple to follow: (1) Each **if** or **else if** statement in the procedural assignment statements is to correspond to an asynchronous event, (2) the last **else** statement corresponds to the clock event, and (3) the asynchronous events are tested first. There are two edge events in the second module of HDL Example 5.2. The **negedge** rst (reset) event is asynchronous, since it matches the **if** (!rst) statement. As long as *rst* is 0, Q is cleared to 0. If *Clk* has a positive transition, its effect is blocked. Only if $rst = 1$ can the **posedge** clock event synchronously transfer D into Q.

Hardware always has a reset signal. It is strongly recommended that all models of edge-sensitive behavior include a reset (or preset) input signal; otherwise, the initial state of the flip-flops of the sequential circuit cannot be determined. A sequential circuit cannot be tested with HDL simulation unless an initial state can be assigned with an input signal.

HDL Example 5.3 describes the construction of a T or JK flip-flop from a D flip-flop and gates. The circuit is described with the characteristic equations of the flip-flops:

$$Q(t + 1) = Q \oplus T \quad \text{for a } T \text{ flip-flop}$$
$$Q(t + 1) = JQ' + K'Q \quad \text{for a } JK \text{ flip-flop}$$

The first module, *TFF*, describes a T flip-flop by instantiating *DFF*. (Instantiation is explained in Section 4.12.) The declared **wire**, *DT*, is assigned the exclusive-OR of Q and T, as is required for building a T flip-flop with a D flip-flop. The instantiation with the value of *DT* replacing D in module *DFF* produces the required T flip-flop. The JK flip-flop is specified in a similar manner by using its characteristic equation to define a replacement for D in the instantiated *DFF*.

HDL Example 5.3 (Alternative Flip-Flop Models)

```
// T flip-flop from D flip-flop and gates
module TFF (Q, T, Clk, rst);
 output Q;
 input  T, Clk, rst;
 wire   DT;
 assign DT = Q ^ T ;                        // Continuous assignment
// Instantiate the D flip-flop
 DFF TF1 (Q, DT, Clk, rst);
endmodule
```

```
// JK flip-flop from D flip-flop and gates (V2001, 2005)
module JKFF (output reg Q, input J, K, Clk, rst);
  wire JK;
  assign JK = (J & ~Q) | (~K & Q);
// Instantiate D flip-flop
  DFF JK1 (Q, JK, Clk, rst);
endmodule

// D flip-flop (V2001, V2005)
module DFF (output reg Q, input D, Clk, rst);
  always @ (posedge Clk, negedge rst)
    if (!rst) Q <= 1'b0;
    else Q <= D;
endmodule
```

HDL Example 5.4 shows another way to describe a *JK* flip-flop. Here, we describe the flip-flop by using the characteristic table rather than the characteristic equation. The **case** multiway branch condition checks the two-bit number obtained by concatenating the bits of *J* and *K*. The **case** expression ({*J*, *K*}) is evaluated and compared with the values in the list of statements that follows. The first value that matches the true condition is executed. Since the concatenation of *J* and *K* produces a two-bit number, it can be equal to 00, 01, 10, or 11. The first bit gives the value of *J* and the second the value of *K*. The four possible conditions specify the value of the next state of *Q* after the application of a positive-edge clock.

HDL Example 5.4 (*JK* Flip-Flop)

```
// Functional description of JK flip-flop (V2001, 2005)
module JK_FF (input J, K, Clk, output reg Q, output Q_b);
  assign Q_b = ~ Q ;
  always @ (posedge Clk)
  case ({J,K})
    2'b00: Q <= Q;
    2'b01: Q <= 1'b0;
    2'b10: Q <= 1'b1;
    2'b11: Q <= !Q;
  endcase
endmodule
```

State diagram-Based HDL Models

An HDL model of the operation of a sequential circuit can be based on the format of the circuit's state diagram. A Mealy HDL model is presented in HDL Example 5.5 for the zero-detector machine described by the sequential circuit in Fig. 5.15 and its state diagram shown in Fig. 5.16. The input, output, clock, and reset are declared in the usual manner.

The state of the flip-flops is declared with identifiers *state* and *next_state*. These variables hold the values of the present state and the next value of the sequential circuit. The state's binary assignment is done with a **parameter** statement. (Verilog allows constants to be defined in a module by the keyword **parameter.**) The four states *S0* through *S3* are assigned binary 00 through 11. The notation $S2 = 2'b10$ is preferable to the alternative $S2 = 2$. The former uses only two bits to store the constant, whereas the latter results in a binary number with 32 (or 64) bits because an unsized number is interpreted and sized as an integer.

HDL Example 5.5 (Mealy Machine: Zero Detector)

```
// Mealy FSM zero detector (See Fig. 5.15 and Fig. 5.16)        Verilog 2001, 2005 syntax
module Mealy_Zero_Detector (
 output reg y_out,
 input  x_in, clock, reset
);
 reg [1: 0]          state, next_state;
 parameter           S0 = 2'b00, S1 = 2'b01, S2 = 2'b10, S3 = 2'b11;
 always @ (posedge clock, negedge reset)   Verilog 2001, 2005 syntax
  if (reset == 0) state <= S0;
  else state <= next_state;

 always @ (state, x_in)                      // Form the next state
  case (state)
    S0:      if (x_in) next_state = S1; else next_state = S0;
    S1:      if (x_in) next_state = S3; else next_state = S0;
    S2:      if (~x_in) next_state = S0; else next_state = S2;
    S3:      if (x_in) next_state = S2; else next_state = S0;
  endcase

 always @ (state, x_in)                      // Form the Mealy output
  case (state)
    S0:         y_out = 0;
    S1, S2, S3: y_out = ~x_in;
  endcase
endmodule

module t_Mealy_Zero_Detector;
 wire    t_y_out;
 reg     t_x_in, t_clock, t_reset;

Mealy_Zero_Detector M0 (t_y_out, t_x_in, t_clock, t_reset);
initial #200 $finish;
initial begin t_clock = 0; forever #5 t_clock = ~t_clock; end

initial fork
    t_reset = 0;
 #2 t_reset = 1;
 #87 t_reset = 0;
 #89 t_reset = 1;
```

```
#10 t_x_in = 1;
#30 t_x_in = 0;
#40 t_x_in = 1;
#50 t_x_in = 0;
#52 t_x_in = 1;
#54 t_x_in = 0;
#70 t_x_in = 1;
#80 t_x_in = 1;
#70 t_x_in = 0;
#90 t_x_in = 1;
#100 t_x_in = 0;
#120 t_x_in = 1;
#160 t_x_in = 0;
#170 t_x_in = 1;
join
endmodule
```

The circuit I HDL Example 5.5 detects a 0 following a sequence of 1s in a serial bit stream. Its Verilog model uses three **always** blocks that execute concurrently and interact through common variables. The first **always** statement resets the circuit to the initial state *S0* = 00 and specifies the synchronous clocked operation. The statement *state* <= *next_state* is synchronized to a positive-edge transition of the clock. This means that any change in the value of *next_state* in the second **always** block can affect the value of *state* only as a result of a **posedge** event of *clock*. The second **always** block determines the value of the next state transition as a function of the present state and input. The value assigned to *state* by the nonblocking assignment is the value of *next_state* immediately before the rising edge of *clock*. Notice how the multiway branch condition implements the state transitions specified by the annotated edges in the state diagram of Fig. 5.16. The third **always** block specifies the output as a function of the present state and the input. Although this block is listed as a separate behavior for clarity, it could be combined with the second block. Note that the value of output *y_out* may change if the value of input *x_in* changes while the circuit is in any given state.

So let's summarize how the model describes the behavior of the machine: At every rising edge of *clock*, if *reset* is not asserted, the state of the machine is updated by the first **always** block; when *state* is updated by the first **always** block, the change in *state* is detected by the sensitivity list mechanism of the second **always** block; then the second **always** block updates the value of *next_state* (it will be used by the first **always** block at the next tick of the clock); the third **always** block also detects the change in *state* and updates the value of the output. In addition, the second and third **always** blocks detect changes in *x_in* and update *next_state* and *y_out* accordingly. The test bench provided with *Mealy_Zero_Detector* provides some waveforms to stimulate the model, producing the results shown in Fig. 5.22. Notice how *t_y_out* responds to changes in both the state and the input, and has a glitch (a transient logic value). We display both to *state*[1:0] and *next_state*[1:0] to illustrate how changes in *t_x_in* influence the value of next_state and *t_y_out*. The Mealy glitch in *t_y_out* is due to the (intentional) dynamic behavior of *t_x_in*. The input, *t_x_in,* settles

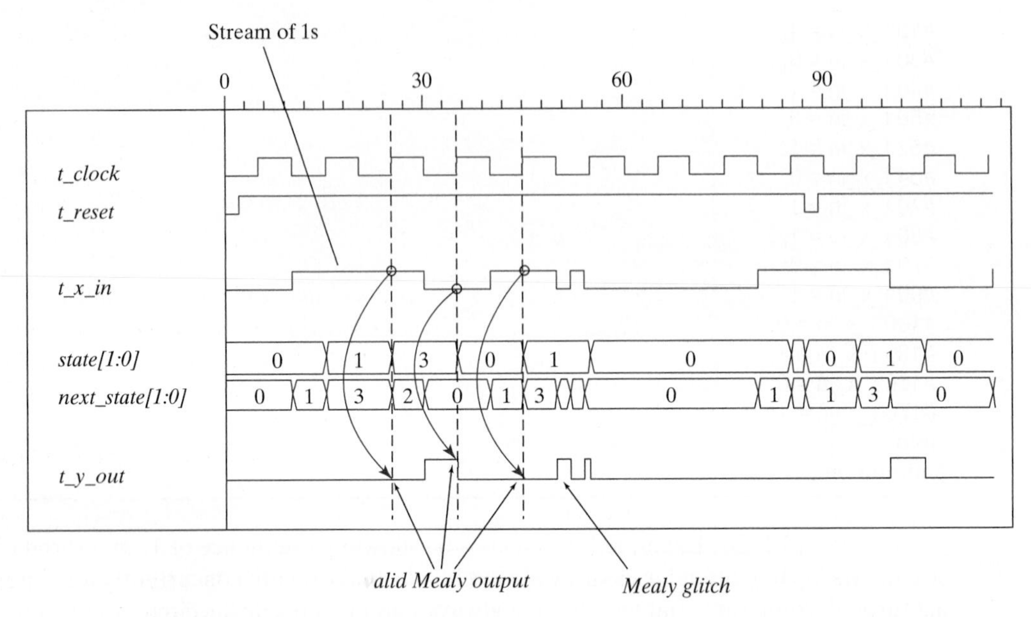

FIGURE 5.22
Simulation output of *Mealy_Zero_Detector*

to a value of 0 immediately before the clock, and at the clock, the state makes a transition from 0 to 1, which is consistent with Fig. 5.16. The output is 1 in state $S1$ immediately before the clock, and changes to 0 as the state enters $S0$.

The description of waveforms in the test bench uses the **fork** . . . **join** construct. Statements with the **fork** . . . **join** block execute in parallel, so the time delays are relative to a common reference of $t = 0$, the time at which the block begins execution.[2] It is usually more convenient to use the **fork** . . . **join** block instead of the **begin** . . . **end** block in describing waveforms. Notice that the waveform of reset is triggered "on the fly" to demonstrate that the machine recovers from an unexpected (asynchronous) reset condition during any state.

How does our Verilog model *Mealy_Zero_Detector* correspond to hardware? The first **always** block corresponds to a D flip-flop implementation of the state register in Fig. 5.21; the second **always** block is the combinational logic block describing the next state; the third **always** block describes the output combinational logic of the zero-detecting Mealy machine. The register operation of the state transition uses the nonblocking assignment operator (<=) because the (edge-sensitive) flip-flops of a sequential machine are updated concurrently by a common clock. The second and third **always** blocks describe combinational logic, which is level sensitive, so they use the blocking (=) assignment operator.

---

[2]A **fork** . . . **join** block completes execution when the last executing statement within it completes its execution.

Their sensitivity lists include both the state and the input because their logic must respond to a change in either or both of them.

Note: The modeling style illustrated by *Mealy_Zero_Detector* is commonly used by designers because it has a close relationship to the state diagram of the machine that is being described. Notice that the reset signal is associated with the **always** block that synchronizes the state transitions. In this example, it is modeled as an active-low reset. Because the reset condition is included in the description of the state transitions, there is no need to include it in the combinational logic that specifies the next state and the output, and the resulting description is less verbose, simpler, and more readable.

HDL Example 5.6 presents the Verilog behavioral model of the Moore FSM shown in Fig. 5.18 and having the state diagram given in Fig. 5.19. The model illustrates an alternative style in which the state transitions of the machine are described by a single clocked (i.e., edge-sensitive) cyclic behavior, i.e., by one **always** block. The present state of the circuit is identified by the variable state, and its transitions are triggered by the rising edge of the clock according to the conditions listed in the **case** statement. The combinational logic that determines the next state is included in the nonblocking assignment to state. In this example, the output of the circuits is independent of the input and is taken directly from the outputs of the flip-flops. The two-bit output *y_out* is specified with a continuous assignment statement and is equal to the value of the present state vector. Figure 5.23 shows some simulation results for *Moore_Model_Fig_5_19*. Here are some important observations: (1) the output depends on only the state, (2) reset "on-the-fly" forces the state of the machine back to S0 (00), and (3) the state transitions are consistent with Fig. 5.19.

HDL Example 5.6 (Moore Machine: Zero Detector)

```
// Moore model FSM (see Fig. 5.19)             Verilog 2001, 2005 syntax
module Moore_Model_Fig_5_19 (
 output [1: 0]          y_out,
 input                  x_in, clock, reset
);
 reg [1: 0]             state;
 parameter              S0 = 2'b00, S1 = 2'b01, S2 = 2'b10, S3 = 2'b11;

 always @ (posedge clock, negedge reset)
  if (reset == 0) state <= S0;                        // Initialize to state S0
  else case (state)
   S0:     if (~x_in) state <= S1; else state <= S0;
   S1:     if (x_in)  state <= S2; else state <= S3;
   S2:     if (~x_in) state <= S3; else state <= S2;
   S3:     if (~x_in) state <= S0; else state <= S3;
  endcase

 assign y_out = state;        // Output of flip-flops
endmodule
```

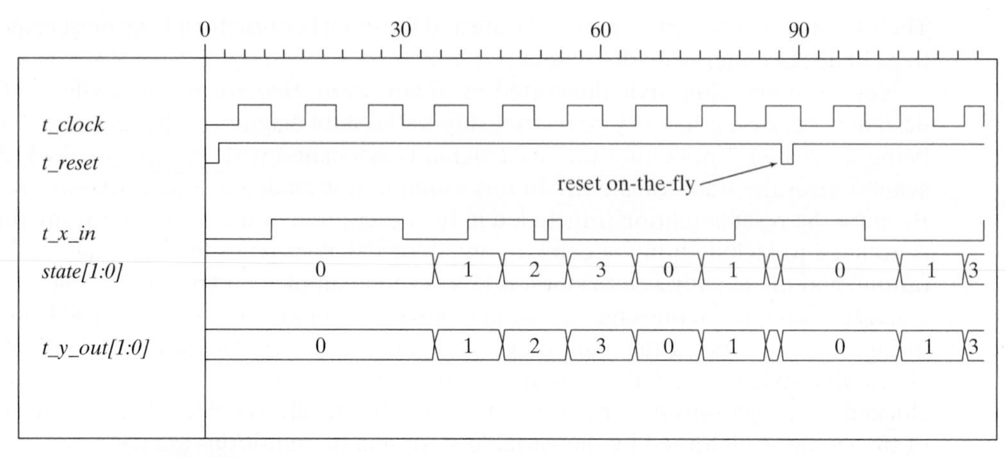

FIGURE 5.23
Simulation output of HDL Example 5.6

Structural Description of Clocked Sequential Circuits

Combinational logic circuits can be described in Verilog by a connection of gates (primitives and UDPs), by dataflow statements (continuous assignments), or by level-sensitive cyclic behaviors (**always** blocks). Sequential circuits are composed of combinational logic and flip-flops, and their HDL models use sequential UDPs and behavioral statements (edge-sensitive cyclic behaviors) to describe the operation of flip-flops. One way to describe a sequential circuit uses a combination of dataflow and behavioral statements. The flip-flops are described with an **always** statement. The combinational part can be described with **assign** statements and Boolean equations. The separate modules can be combined to form a structural model by instantiation within a **module**.

The structural description of a Moore-type zero detector sequential circuit is shown in HDL Example 5.7. We want to encourage the reader to consider alternative ways to model a circuit, so as a point of comparison, we first present *Moore_Model_Fig_5_20*, a Verilog behavioral description of a binary counter having the state diagram examined earlier shown in Fig. 5.20(b). This style of modeling follows directly from the state diagram. An alternative style, used in *Moore_Model_STR_Fig_5_20*, represents the structure shown in Fig. 5.20(a). This style uses two modules. The first describes the circuit of Fig. 5.20(a). The second describes the *T* flip-flop that will be used by the circuit. We also show two ways to model the *T* flip-flop. The first asserts that, at every clock tick, the value of the output of the flip-flop toggles if the toggle input is asserted. The second model describes the behavior of the toggle flip-flop in terms of its characteristic equation. The first style is attractive because it does not require the reader to remember the characteristic equation. Nonetheless, the models are interchangeable and will synthesize to the same hardware circuit. A test bench module provides a stimulus for verifying the functionality of the circuit. The sequential circuit is a two-bit binary counter controlled by input *x_in*. The output, *y_out*, is enabled when the count reaches binary 11. Flip-flops

A and *B* are included as outputs in order to check their operation. The flip-flop input equations and the output equation are evaluated with continuous assignment (**assign**) statements having the corresponding Boolean expressions. The instantiated *T* flip-flops use *TA* and *TB* as defined by the input equations.

The second module describes the *T* flip-flop. The *reset* input resets the flip-flop to 0 with an active-low signal. The operation of the flip-flop is specified by its characteristic equation, $Q(t + 1) = Q \oplus T$.

The test bench includes both models of the machine. The stimulus module provides common inputs to the circuits to simultaneously display their output responses. The first **initial** block provides eight clock cycles with a period of 10 ns. The second **initial** block specifies a toggling of input *x_in* that occurs at the negative edge transition of the clock. The result of the simulation is shown in Fig. 5.24. The pair (A, B) goes through the binary sequence 00, 01, 10, 11, and back to 00. The change in the count is triggered by a positive edge of the clock, provided that $x\_in = 1$. If $x\_in = 0$, the count does not change. *y_out* is equal to 1 when both *A* and *B* are equal to 1. This verifies the main functionality of the circuit, but not a recovery from an unexpected reset event.

HDL Example 5.7 (Binary Counter_Moore Model)

```
// State-diagram-based model (V2001, 2005)
module Moore_Model_Fig_5_20 (
 output y_out,
 input   x_in, clock, reset
);
 reg [1: 0]          state;
 parameter           S0 = 2'b00, S1 = 2'b01, S2 = 2'b10, S3 = 2'b11;

 always @ (posedge clock, negedge reset)
  if (reset == 0) state <= S0;              // Initialize to state S0
  else case (state)
   S0:     if (x_in)  state <= S1; else state <= S0;
   S1:     if (x_in)  state <= S2; else state <= S1;
   S2:     if (x_in)  state <= S3; else state <= S2;
   S3:     if (x_in)  state <= S0; else state <= S3;
  endcase

 assign y_out = (state == S3);           // Output of flip-flops
endmodule

// Structural model
module Moore_Model_STR_Fig_5_20 (
 output     y_out, A, B,
 input      x_in, clock, reset
);
 wire       TA, TB;

// Flip-flop input equations
 assign TA = x_in & B;
```

```
 assign TB = x_in;
// Output equation
 assign y_out = A & B;

// Instantiate Toggle flip-flops
 Toggle_flip_flop_3 M_A (A, TA, clock, reset);
 Toggle_flip_flop_3 M_B (B, TB, clock, reset);
endmodule

module Toggle_flip_flop (Q, T, CLK, RST_b);
 output   Q;
 input    T, CLK, RST_b;
 reg      Q;

 always @ (posedge CLK, negedge RST_b)
  if (RST_b == 0) Q <= 1'b0;
  else if (T) Q <= ~Q;
endmodule

// Alternative model using characteristic equation
// module Toggle_flip_flop (Q, T, CLK, RST_b);
// output   Q;
// input    T, CLK, RST_b;
// reg      Q;

// always @ (posedge CLK, negedge RST)
//   if (RST_b == 0) Q <= 1'b0;
//   else Q <= Q ^ T;
// endmodule

module t_Moore_Fig_5_20;
 wire     t_y_out_2, t_y_out_1;
 reg      t_x_in, t_clock, t_reset;

Moore_Model_Fig_5_20          M1(t_y_out_1, t_x_in, t_clock, t_reset);
Moore_Model_STR_Fig_5_20      M2 (t_y_out_2, A, B, t_x_in, t_clock, t_reset);

initial #200 $finish;
initial begin
   t_reset = 0;
   t_clock = 0;
   #5 t_reset = 1;
 repeat (16)
   #5 t_clock = ~t_clock;
end
initial begin
      t_x_in = 0;
 #15 t_x_in = 1;
 repeat (8)
  #10 t_x_in = ~t_x_in;
 end
endmodule
```

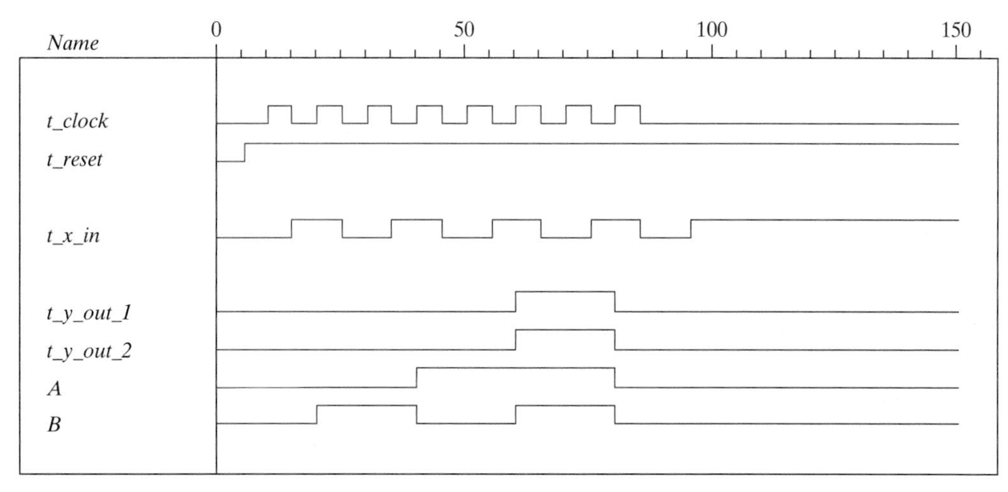

FIGURE 5.24
Simulation output of HDL Example 5.7

5.7 STATE REDUCTION AND ASSIGNMENT

The *analysis* of sequential circuits starts from a circuit diagram and culminates in a state table or diagram. The *design* (synthesis) of a sequential circuit starts from a set of specifications and culminates in a logic diagram. Design procedures are presented in Section 5.8. Two sequential circuits may exhibit the same input–output behavior, but have a different number of internal states in their state diagram. The current section discusses certain properties of sequential circuits that may simplify a design by reducing the number of gates and flip-flops it uses. In general, reducing the number of flip-flops reduces the cost of a circuit.

State Reduction

The reduction in the number of flip-flops in a sequential circuit is referred to as the *state-reduction* problem. State-reduction algorithms are concerned with procedures for reducing the number of states in a state table, while keeping the external input–output requirements unchanged. Since m flip-flops produce 2^m states, a reduction in the number of states may (or may not) result in a reduction in the number of flip-flops. An unpredictable effect in reducing the number of flip-flops is that sometimes the equivalent circuit (with fewer flip-flops) may require more combinational gates to realize its next state and output logic.

We will illustrate the state-reduction procedure with an example. We start with a sequential circuit whose specification is given in the state diagram of Fig. 5.25. In our example, only the input–output sequences are important; the internal states are used merely to provide the required sequences. For that reason, the states marked inside the circles are denoted

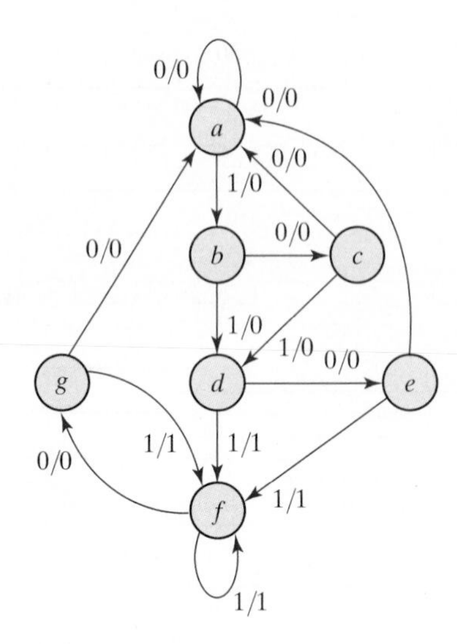

FIGURE 5.25
State diagram

by letter symbols instead of their binary values. This is in contrast to a binary counter, wherein the binary value sequence of the states themselves is taken as the outputs.

There are an infinite number of input sequences that may be applied to the circuit; each results in a unique output sequence. As an example, consider the input sequence 01010110100 starting from the initial state *a*. Each input of 0 or 1 produces an output of 0 or 1 and causes the circuit to go to the next state. From the state diagram, we obtain the output and state sequence for the given input sequence as follows: With the circuit in initial state *a*, an input of 0 produces an output of 0 and the circuit remains in state *a*. With present state *a* and an input of 1, the output is 0 and the next state is *b*. With present state *b* and an input of 0, the output is 0 and the next state is *c*. Continuing this process, we find the complete sequence to be as follows:

state	*a*	*a*	*b*	*c*	*d*	*e*	*f*	*f*	*g*	*f*	*g*	*a*
input	0	1	0	1	0	1	1	0	1	0	0	
output	0	0	0	0	0	1	1	0	1	0	0	

In each column, we have the present state, input value, and output value. The next state is written on top of the next column. It is important to realize that in this circuit the states themselves are of secondary importance, because we are interested only in output sequences caused by input sequences.

Now let us assume that we have found a sequential circuit whose state diagram has fewer than seven states, and suppose we wish to compare this circuit with the circuit whose state diagram is given by Fig. 5.25. If identical input sequences are applied to the two circuits and identical outputs occur for all input sequences, then the two circuits are said to be equivalent (as far as the input–output is concerned) and one may be replaced by the other. The problem of state reduction is to find ways of reducing the number of states in a sequential circuit without altering the input–output relationships.

We now proceed to reduce the number of states for this example. First, we need the state table; it is more convenient to apply procedures for state reduction with the use of a table rather than a diagram. The state table of the circuit is listed in Table 5.6 and is obtained directly from the state diagram.

The following algorithm for the state reduction of a completely specified state table is given here without proof: "Two states are said to be equivalent if, for each member of the set of inputs, they give exactly the same output and send the circuit either to the same state or to an equivalent state." When two states are equivalent, one of them can be removed without altering the input–output relationships.

Now apply this algorithm to Table 5.6. Going through the state table, we look for two present states that go to the same next state and have the same output for both input combinations. States *e* and *g* are two such states: They both go to states *a* and *f* and have outputs of 0 and 1 for $x = 0$ and $x = 1$, respectively. Therefore, states *g* and *e* are equivalent, and one of these states can be removed. The procedure of removing a state and replacing it by its equivalent is demonstrated in Table 5.7. The row with present state *g* is removed, and state *g* is replaced by state *e* each time it occurs in the columns headed "Next State."

Present state *f* now has next states *e* and *f* and outputs 0 and 1 for $x = 0$ and $x = 1$, respectively. The same next states and outputs appear in the row with present state *d*. Therefore, states *f* and *d* are equivalent, and state *f* can be removed and replaced by *d*. The final reduced table is shown in Table 5.8. The state diagram for the reduced table consists of only five states and is shown in Fig. 5.26. This state diagram satisfies the original input–output specifications and will produce the required output sequence for any given input sequence. The following list derived from the state diagram of Fig. 5.26 is for the input sequence used previously (note that the same output sequence results, although the state sequence is different):

state	***a***	***a***	***b***	***c***	***d***	***e***	***d***	***d***	***e***	***d***	***e***	***a***
input	0	1	0	1	0	1	1	0	1	0	0	
output	0	0	0	0	0	1	1	0	1	0	0	

Table 5.6
State Table

Present State	**Next State** $x = 0$	**Next State** $x = 1$	**Output** $x = 0$	**Output** $x = 1$
a	*a*	*b*	0	0
b	*c*	*d*	0	0
c	*a*	*d*	0	0
d	*e*	*f*	0	1
e	*a*	*f*	0	1
f	*g*	*f*	0	1
g	*a*	*f*	0	1

Table 5.7
Reducing the State Table

Present State	Next State $x = 0$	Next State $x = 1$	Output $x = 0$	Output $x = 1$
a	a	b	0	0
b	c	d	0	0
c	a	d	0	0
d	e	f	0	1
e	a	f	0	1
f	e	f	0	1

Table 5.8
Reduced State Table

Present State	Next State $x = 0$	Next State $x = 1$	Output $x = 0$	Output $x = 1$
a	a	b	0	0
b	c	d	0	0
c	a	d	0	0
d	e	d	0	1
e	a	d	0	1

In fact, this sequence is exactly the same as that obtained for Fig. 5.25 if we replace g by e and f by d.

Checking each pair of states for equivalency can be done systematically by means of a procedure that employs an implication table, which consists of squares, one for every suspected pair of possible equivalent states. By judicious use of the table, it is possible to determine all pairs of equivalent states in a state table.

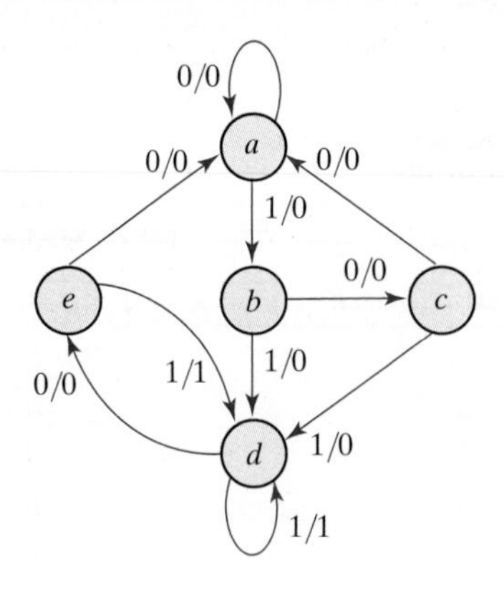

FIGURE 5.26
Reduced state diagram

The sequential circuit of this example was reduced from seven to five states. In general, reducing the number of states in a state table may result in a circuit with less equipment. However, the fact that a state table has been reduced to fewer states does not guarantee a saving in the number of flip-flops or the number of gates. In actual practice designers may skip this step because target devices are rich in resources.

State Assignment

In order to design a sequential circuit with physical components, it is necessary to assign unique coded binary values to the states. For a circuit with m states, the codes must contain n bits, where $2^n \geq m$. For example, with three bits, it is possible to assign codes to eight states, denoted by binary numbers 000 through 111. If the state table of Table 5.6 is used, we must assign binary values to seven states; the remaining state is unused. If the state table of Table 5.8 is used, only five states need binary assignment, and we are left with three unused states. Unused states are treated as don't-care conditions during the design. Since don't-care conditions usually help in obtaining a simpler circuit, it is more likely but not certain that the circuit with five states will require fewer combinational gates than the one with seven states.

The simplest way to code five states is to use the first five integers in binary counting order, as shown in the first assignment of Table 5.9. Another similar assignment is the Gray code shown in assignment 2. Here, only one bit in the code group changes when going from one number to the next. This code makes it easier for the Boolean functions to be placed in the map for simplification. Another possible assignment often used in the design of state machines to control data-path units is the one-hot assignment. This configuration uses as many bits as there are states in the circuit. At any given time, only one bit is equal to 1 while all others are kept at 0. This type of assignment uses one flip-flop per state, which is not an issue for register-rich field-programmable gate arrays. (See Chapter 7.) *One-hot encoding usually leads to simpler decoding logic for the next state and output.* One-hot machines can be faster than machines with sequential binary encoding, and the silicon area required by the extra flip-flops can be offset by the area

Table 5.9
Three Possible Binary State Assignments

State	Assignment 1, Binary	Assignment 2, Gray Code	Assignment 3, One-Hot
a	000	000	00001
b	001	001	00010
c	010	011	00100
d	011	010	01000
e	100	110	10000

Table 5.10
Reduced State Table with Binary Assignment 1

	Next State		Output	
Present State	$x = 0$	$x = 1$	$x = 0$	$x = 1$
000	000	001	0	0
001	010	011	0	0
010	000	011	0	0
011	100	011	0	1
100	000	011	0	1

saved by using simpler decoding logic. This trade-off is not guaranteed, so it must be evaluated for a given design.

Table 5.10 is the reduced state table with binary assignment 1 substituted for the letter symbols of the states. A different assignment will result in a state table with different binary values for the states. The binary form of the state table is used to derive the next-state and output-forming combinational logic part of the sequential circuit. The complexity of the combinational circuit depends on the binary state assignment chosen.

Sometimes, the name *transition table* is used for a state table with a binary assignment. This convention distinguishes it from a state table with symbolic names for the states. In this book, we use the same name for both types of state tables.

5.8 DESIGN PROCEDURE

Design procedures or methodologies specify hardware that will implement a desired behavior. The design effort for small circuits may be manual, but industry relies on automated synthesis tools for designing massive integrated circuits. The sequential building block used by synthesis tools is the *D* flip-flop. Together with additional logic, it can implement the behavior of *JK* and *T* flip-flops. In fact, designers generally do not concern themselves with the type of flip-flop; rather, their focus is on correctly describing the sequential functionality that is to be implemented by the synthesis tool. Here we will illustrate manual methods using *D, JK,* and *T* flip-flops.

The design of a clocked sequential circuit starts from a set of specifications and culminates in a logic diagram or a list of Boolean functions from which the logic diagram can be obtained. In contrast to a combinational circuit, which is fully specified by a truth table, a sequential circuit requires a state table for its specification. The first step in the design of sequential circuits is to obtain a state table or an equivalent representation, such as a state diagram.[3]

A synchronous sequential circuit is made up of flip-flops and combinational gates. The design of the circuit consists of choosing the flip-flops and then finding a combinational

[3]We will examine later another important representation of a machine's behavior—the algorithmic state machine (ASM) chart.

gate structure that, together with the flip-flops, produces a circuit which fulfills the stated specifications. The number of flip-flops is determined from the number of states needed in the circuit and the choice of state assignment codes. The combinational circuit is derived from the state table by evaluating the flip-flop input equations and output equations. In fact, once the type and number of flip-flops are determined, the design process involves a transformation from a sequential circuit problem into a combinational circuit problem. In this way, the techniques of combinational circuit design can be applied.

The procedure for designing synchronous sequential circuits can be summarized by a list of recommended steps:

1. From the word description and specifications of the desired operation, derive a state diagram for the circuit.
2. Reduce the number of states if necessary.
3. Assign binary values to the states.
4. Obtain the binary-coded state table.
5. Choose the type of flip-flops to be used.
6. Derive the simplified flip-flop input equations and output equations.
7. Draw the logic diagram.

The word specification of the circuit behavior usually assumes that the reader is familiar with digital logic terminology. It is necessary that the designer use intuition and experience to arrive at the correct interpretation of the circuit specifications, because word descriptions may be incomplete and inexact. Once such a specification has been set down and the state diagram obtained, it is possible to use known synthesis procedures to complete the design. Although there are formal procedures for state reduction and assignment (steps 2 and 3), they are seldom used by experienced designers. Steps 4 through 7 in the design can be implemented by exact algorithms and therefore can be automated. The part of the design that follows a well-defined procedure is referred to as *synthesis*. Designers using logic synthesis tools (software) can follow a simplified process that develops an HDL description directly from a state diagram, letting the synthesis tool determine the circuit elements and structure that implement the description.

The first step is a critical part of the process, because succeeding steps depend on it. We will give one simple example to demonstrate how a state diagram is obtained from a word specification.

Suppose we wish to design a circuit that detects a sequence of three or more consecutive 1's in a string of bits coming through an input line (i.e., the input is a *serial bit stream*). The state diagram for this type of circuit is shown in Fig. 5.27. It is derived by starting with state S_0, the reset state. If the input is 0, the circuit stays in S_0, but if the input is 1, it goes to state S_1 to indicate that a 1 was detected. If the next input is 1, the change is to state S_2 to indicate the arrival of two consecutive 1's, but if the input is 0, the state goes back to S_0. The third consecutive 1 sends the circuit to state S_3. If more 1's are detected, the circuit stays in S_3. Any 0 input sends the circuit back to S_0. In this way, the circuit stays in S_3 as long as there are three or more consecutive 1's received. This is a Moore model sequential circuit, since the output is 1 when the circuit is in state S_3 and is 0 otherwise.

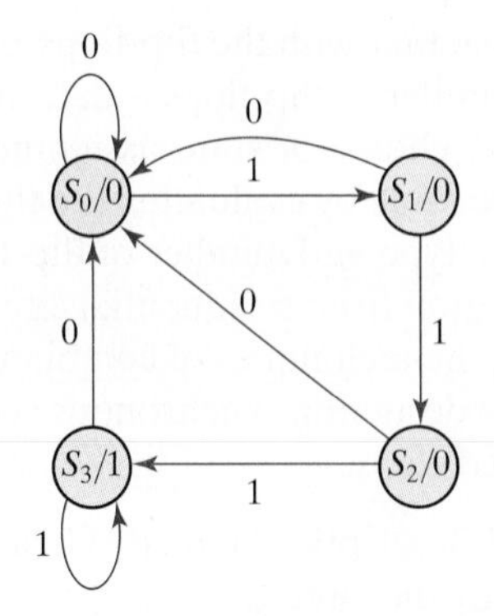

FIGURE 5.27
State diagram for sequence detector

Synthesis Using *D* Flip-Flops

Once the state diagram has been derived, the rest of the design follows a straightforward synthesis procedure. In fact, we can design the circuit by using an HDL description of the state diagram and the proper HDL synthesis tools to obtain a synthesized netlist. (The HDL description of the state diagram will be similar to HDL Example 5.6 in Section 5.6.) To design the circuit by hand, we need to assign binary codes to the states and list the state table. This is done in Table 5.11. The table is derived from the state diagram of Fig. 5.27 with a sequential binary assignment. We choose two D flip-flops to represent the four states, and we label their outputs A and B. There is one input x and one output y. The characteristic equation of the D flip-flop is $Q(t + 1) = D_Q$, which means that the next-state values in the state table specify the D input condition for the flip-flop. The flip-flop input equations

Table 5.11
State Table for Sequence Detector

Present State		Input	Next State		Output
A	*B*	*x*	*A*	*B*	*y*
0	0	0	0	0	0
0	0	1	0	1	0
0	1	0	0	0	0
0	1	1	1	0	0
1	0	0	0	0	0
1	0	1	1	1	0
1	1	0	0	0	1
1	1	1	1	1	1

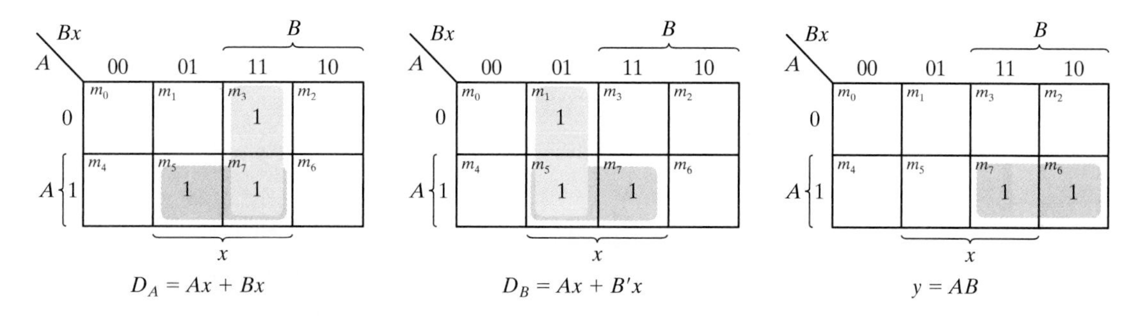

FIGURE 5.28
K-Maps for sequence detector

can be obtained directly from the next-state columns of A and B and expressed in sum-of-minterms form as

$$A(t + 1) = D_A(A, B, x) = \Sigma(3, 5, 7)$$
$$B(t + 1) = D_B(A, B, x) = \Sigma(1, 5, 7)$$
$$y(A, B, x) = \Sigma(6, 7)$$

where A and B are the present-state values of flip-flops A and B, x is the input, and D_A and D_B are the input equations. The minterms for output y are obtained from the output column in the state table.

The Boolean equations are simplified by means of the maps plotted in Fig. 5.28. The simplified equations are

$$D_A = Ax + Bx$$
$$D_B = Ax + B'x$$
$$y = AB$$

The advantage of designing with D flip-flops is that the Boolean equations describing the inputs to the flip-flops can be obtained directly from the state table. Software tools automatically infer and select the D-type flip-flop from a properly written HDL model. The schematic of the sequential circuit is drawn in Fig. 5.29.

Excitation Tables

The design of a sequential circuit with flip-flops other than the D type is complicated by the fact that the input equations for the circuit must be derived indirectly from the state table. When D-type flip-flops are employed, the input equations are obtained directly from the next state. This is not the case for the JK and T types of flip-flops. In order to determine the input equations for these flip-flops, it is necessary to derive a functional relationship between the state table and the input equations.

The flip-flop characteristic tables presented in Table 5.1 provide the value of the next state when the inputs and the present state are known. These tables are useful

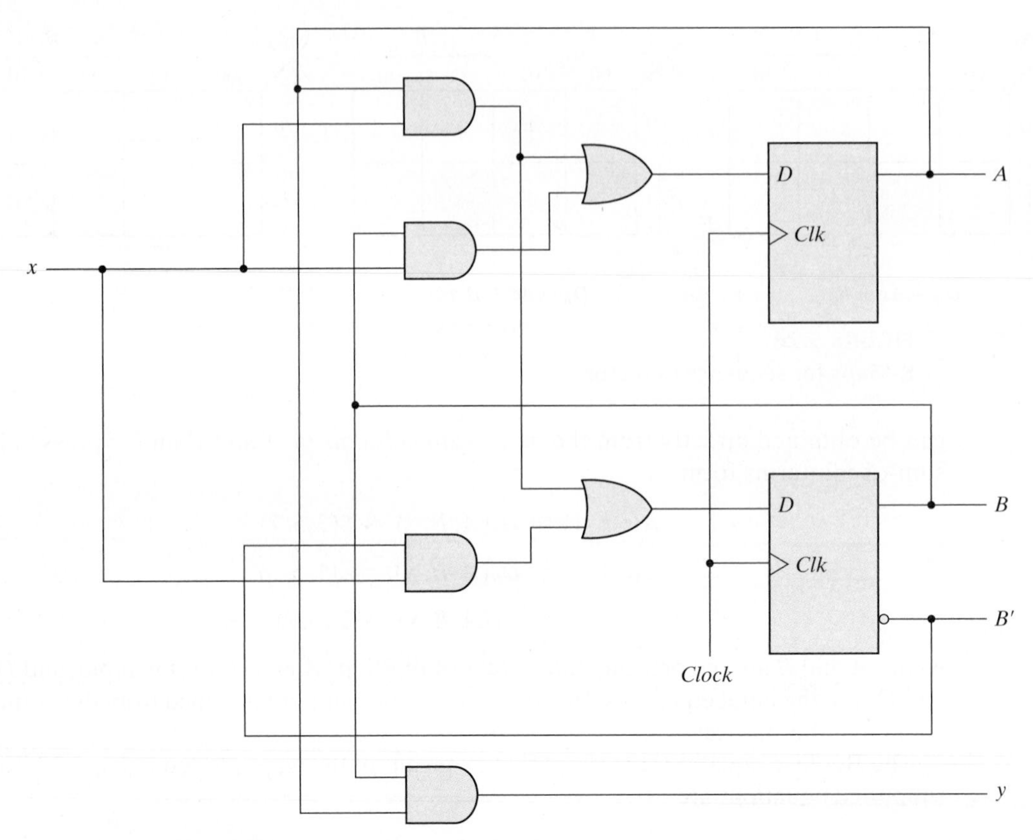

FIGURE 5.29
Logic diagram of a Moore-type sequence detector

for analyzing sequential circuits and for defining the operation of the flip-flops. During the design process, we usually know the transition from the present state to the next state and wish to find the flip-flop input conditions that will cause the required transition. For this reason, we need a table that lists the required inputs for a given change of state. Such a table is called an *excitation table*.

Table 5.12 shows the excitation tables for the two flip-flops (*JK* and *T*). Each table has a column for the present state $Q(t)$, a column for the next state $Q(t + 1)$, and a column for each input to show how the required transition is achieved. There are four possible transitions from the present state to the next state. The required input conditions for each of the four transitions are derived from the information available in the characteristic table. The symbol X in the tables represents a don't-care condition, which means that it does not matter whether the input is 1 or 0.

The excitation table for the *JK* flip-flop is shown in part (a). When both present state and next state are 0, the *J* input must remain at 0 and the *K* input can be either 0 or 1. Similarly, when both present state and next state are 1, the *K* input must remain at 0,

Table 5.12
Flip-Flop Excitation Tables

$Q(t)$	$Q(t = 1)$	J	K
0	0	0	X
0	1	1	X
1	0	X	1
1	1	X	0

(a) *JK* Flip-Flop

$Q(t)$	$Q(t = 1)$	T
0	0	0
0	1	1
1	0	1
1	1	0

(b) *T* Flip-Flop

while the J input can be 0 or 1. If the flip-flop is to have a transition from the 0-state to the 1-state, J must be equal to 1, since the J input sets the flip-flop. However, input K may be either 0 or 1. If $K = 0$, the $J = 1$ condition sets the flip-flop as required; if $K = 1$ and $J = 1$, the flip-flop is complemented and goes from the 0-state to the 1-state as required. Therefore, the K input is marked with a don't-care condition for the 0-to-1 transition. For a transition from the 1-state to the 0-state, we must have $K = 1$, since the K input clears the flip-flop. However, the J input may be either 0 or 1, since $J = 0$ has no effect and $J = 1$ together with $K = 1$ complements the flip-flop with a resultant transition from the 1-state to the 0-state.

The excitation table for the T flip-flop is shown in part (b). From the characteristic table, we find that when input $T = 1$, the state of the flip-flop is complemented, and when $T = 0$, the state of the flip-flop remains unchanged. Therefore, when the state of the flip-flop must remain the same, the requirement is that $T = 0$. When the state of the flip-flop has to be complemented, T must equal 1.

Synthesis Using *JK* Flip-Flops

The manual synthesis procedure for sequential circuits with JK flip-flops is the same as with D flip-flops, except that the input equations must be evaluated from the present-state to the next-state transition derived from the excitation table. To illustrate the procedure, we will synthesize the sequential circuit specified by Table 5.13. In addition to having columns for the present state, input, and next state, as in a conventional state table, the table shows the flip-flop input conditions from which the input equations are derived. The flip-flop inputs are derived from the state table in conjunction with the excitation table for the JK flip-flop. For example, in the first row of Table 5.13, we have a transition for flip-flop A from 0 in the present state to 0 in the next state. In Table 5.12, for the JK flip-flop, we find that a transition of states from present state 0 to next state 0 requires that input J be 0 and input K be a don't-care. So 0 and X are entered in the first row under J_A and K_A, respectively. Since the first row also shows a transition for flip-flop B from 0 in the present state to 0 in the next state, 0 and X are inserted into the first row under J_B and K_B, respectively. The second row of the table shows a transition for flip-flop B from 0 in the present state to 1 in the next state. From the excitation table, we find that a transition from 0 to 1 requires that J be 1 and K be a don't-care, so 1 and X are copied into

Table 5.13
State Table and JK Flip-Flop Inputs

Present State		Input	Next State		Flip-Flop Inputs			
A	**B**	**x**	**A**	**B**	J_A	K_A	J_B	K_B
0	0	0	0	0	0	X	0	X
0	0	1	0	1	0	X	1	X
0	1	0	1	0	1	X	X	1
0	1	1	0	1	0	X	X	0
1	0	0	1	0	X	0	0	X
1	0	1	1	1	X	0	1	X
1	1	0	1	1	X	0	X	0
1	1	1	0	0	X	1	X	1

the second row under J_B and K_B, respectively. The process is continued for each row in the table and for each flip-flop, with the input conditions from the excitation table copied into the proper row of the particular flip-flop being considered.

The flip-flop inputs in Table 5.13 specify the truth table for the input equations as a function of present state A, present state B, and input x. The input equations are simplified in the maps of Fig. 5.30. The next-state values are not used during the simplification,

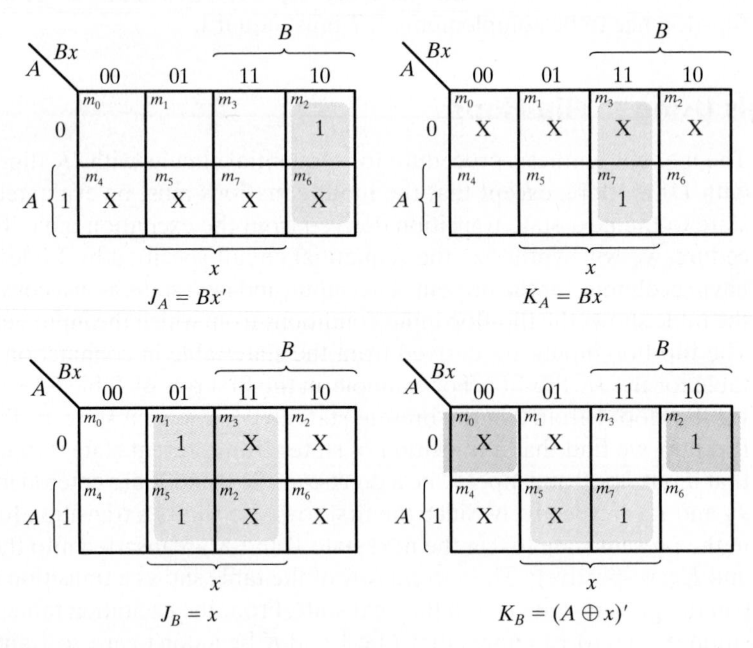

FIGURE 5.30
Maps for J and K input equations

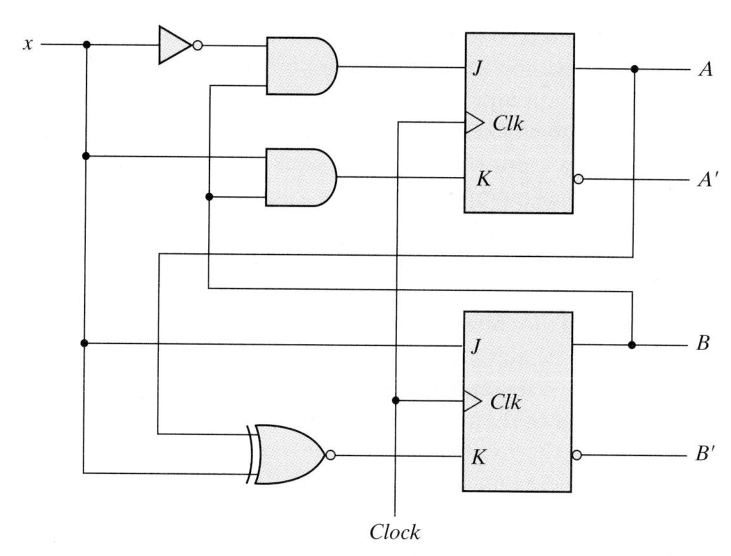

FIGURE 5.31
Logic diagram for sequential circuit with *JK* flip-flops

since the input equations are a function of the present state and the input only. Note the advantage of using *JK*-type flip-flops when sequential circuits are designed *manually*. The fact that there are so many don't-care entries indicates that the combinational circuit for the input equations is likely to be simpler, because don't-care minterms usually help in obtaining simpler expressions. If there are unused states in the state table, there will be additional don't-care conditions in the map. Nonetheless, *D*-type flip-flops are more amenable to an automated design flow.

The four input equations for the pair of *JK* flip-flops are listed under the maps of Fig. 5.30. The logic diagram (schematic) of the sequential circuit is drawn in Fig. 5.31.

Synthesis Using *T* Flip-Flops

The procedure for synthesizing circuits using *T* flip-flops will be demonstrated by designing a binary counter. An n-bit binary counter consists of n flip-flops that can count in binary from 0 to $2^n - 1$. The state diagram of a three-bit counter is shown in Fig. 5.32. As

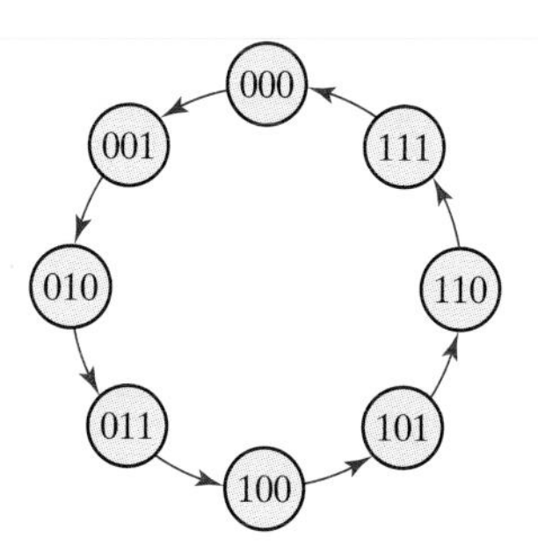

FIGURE 5.32
State diagram of three-bit binary counter

seen from the binary states indicated inside the circles, the flip-flop outputs repeat the binary count sequence with a return to 000 after 111. The directed lines between circles are not marked with input and output values as in other state diagrams. Remember that state transitions in clocked sequential circuits are initiated by a clock edge; the flip-flops remain in their present states if no clock is applied. For that reason, the clock does not appear explicitly as an input variable in a state diagram or state table. From this point of view, the state diagram of a counter does not have to show input and output values along the directed lines. The only input to the circuit is the clock, and the outputs are specified by the present state of the flip-flops. The next state of a counter depends entirely on its present state, and the state transition occurs every time the clock goes through a transition.

Table 5.14 is the state table for the three-bit binary counter. The three flip-flops are symbolized by A_2, A_1, and A_0. Binary counters are constructed most efficiently with T flip-flops because of their complement property. The flip-flop excitation for the T inputs is derived from the excitation table of the T flip-flop and by inspection of the state transition of the present state to the next state. As an illustration, consider the flip-flop input entries for row 001. The present state here is 001 and the next state is 010, which is the next count in the sequence. Comparing these two counts, we note that A_2 goes from 0 to 0, so T_{A2} is marked with 0 because flip-flop A_2 must not change when a clock occurs. Also, A_1 goes from 0 to 1, so T_{A1} is marked with a 1 because this flip-flop must be complemented in the next clock edge. Similarly, A_0 goes from 1 to 0, indicating that it must be complemented, so T_{A0} is marked with a 1. The last row, with present state 111, is compared with the first count 000, which is its next state. Going from all 1's to all 0's requires that all three flip-flops be complemented.

The flip-flop input equations are simplified in the maps of Fig. 5.33. Note that T_{A0} has 1's in all eight minterms because the least significant bit of the counter is complemented with each count. A Boolean function that includes all minterms defines a constant value of 1. The input equations listed under each map specify the combinational part of the counter. Including these functions with the three flip-flops, we obtain

Table 5.14
State Table for Three-Bit Counter

Present State			Next State			Flip-Flop Inputs		
A_2	A_1	A_0	A_2	A_1	A_0	T_{A2}	T_{A1}	T_{A0}
0	0	0	0	0	1	0	0	1
0	0	1	0	1	0	0	1	1
0	1	0	0	1	1	0	0	1
0	1	1	1	0	0	1	1	1
1	0	0	1	0	1	0	0	1
1	0	1	1	1	0	0	1	1
1	1	0	1	1	1	0	1	1
1	1	1	0	0	0	1	1	1

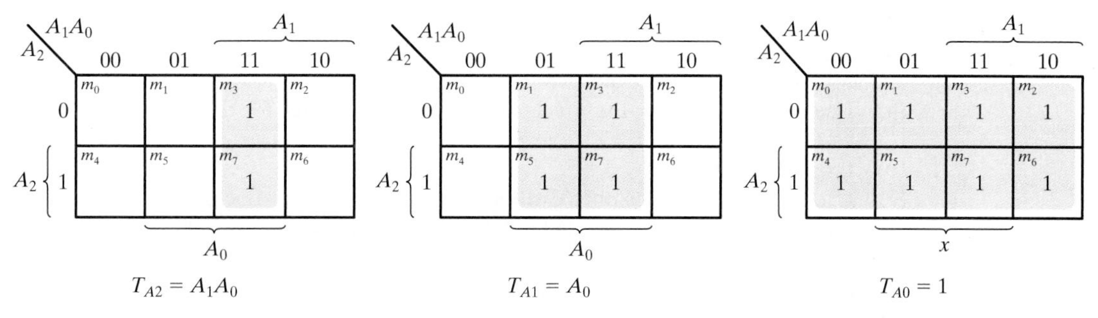

FIGURE 5.33
Maps for three-bit binary counter

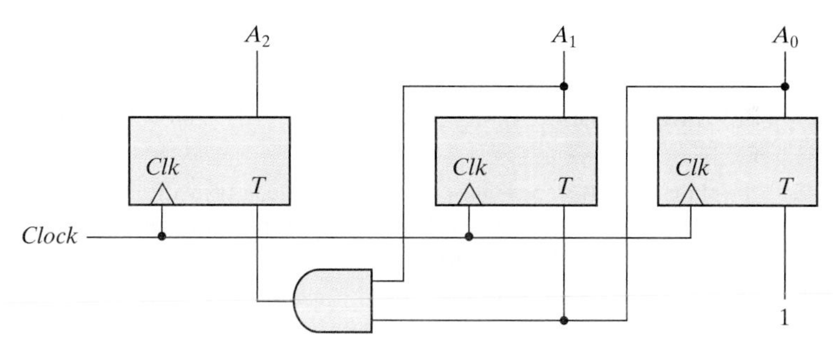

FIGURE 5.34
Logic diagram of three-bit binary counter

the logic diagram of the counter, as shown in Fig. 5.34. For simplicity, the reset signal is not shown, but be aware that every design should include a reset signal.

PROBLEMS

(Answers to problems marked with * appear at the end of the book. Where appropriate, a logic design and its related HDL modeling problem are cross-referenced.)

Note: For each problem that requires writing and verifying an HDL model, a test plan should be written to identify which functional features are to be tested during the simulation and how they will be tested. For example, a reset on the fly could be tested by asserting the reset signal while the simulated machine is in a state other than the reset state. The test plan is to guide development of a test bench that will implement the plan. Simulate the model, using the test bench, and verify that the behavior is correct. If synthesis tools and an ASIC cell library are available, the Verilog descriptions developed for Problems 5.34–5.42 can be assigned as synthesis exercises. The gate-level circuit produced by the synthesis tools should be simulated and compared to the simulation results for the pre-synthesis model. The same exercises can be assigned if an FPGA tool suite is available.

5.1 The D latch of Fig. 5.6 is constructed with four NAND gates and an inverter. Consider the following three other ways for obtaining a D latch. In each case, draw the logic diagram and verify the characteristic table for each.

(a) Use NAND gates for the SR latch part and AND gates for the other two. An inverter may be needed.

(b) Use only NOR gates.

(c) Use four NAND gates only (without an inverter). This can be done by connecting the output of the upper gate in Fig. 5.6 (the gate that goes to the SR latch) to the input of the lower gate (instead of the inverter output).

5.2 Construct a JK flip-flop using a D flip-flop, a two-to-one-line multiplexer, and an inverter. (HDL—see Problem 5.34.)

5.3 Show that the characteristic equation for the complement output of a T flip-flop is

$$Q'(t + 1) = T'Q' + TQ$$

5.4 A PN flip-flop has four operations: no change, clear to 0, set to 1, and complement, when inputs P and N are 00, 01, 10, and 11, respectively.

(a) Tabulate the characteristic table.

(b)* Derive the characteristic equation.

(c) Tabulate the excitation table.

(d) Show how the PN flip-flop can be converted to a D flip-flop.

5.5 Explain the differences among a truth table, a transition table, a characteristic table, and an excitation table. Also, explain the difference among a Boolean equation, a state equation, a characteristic equation, and a flip-flop input equation.

5.6 A sequential circuit with two D flip-flops A and B, two inputs, x and y; and one output z is specified by the following next-state and output equations (HDL—see Problem 5.35):

$$A(t + 1) = xy' + xB$$
$$B(t + 1) = xA + xB'$$
$$z = A$$

(a) Draw the logic diagram of the circuit.

(b) List the state table for the sequential circuit.

(c) Draw the corresponding state diagram.

5.7* A sequential circuit has one flip-flop Q, two inputs x and y, and one output $Diff$. It consists of a full-subtractor circuit ($x - y - B_{in}$) connected to a D flip-flop, as shown in Fig. P5.7. Derive the state table and state diagram of the sequential circuit.

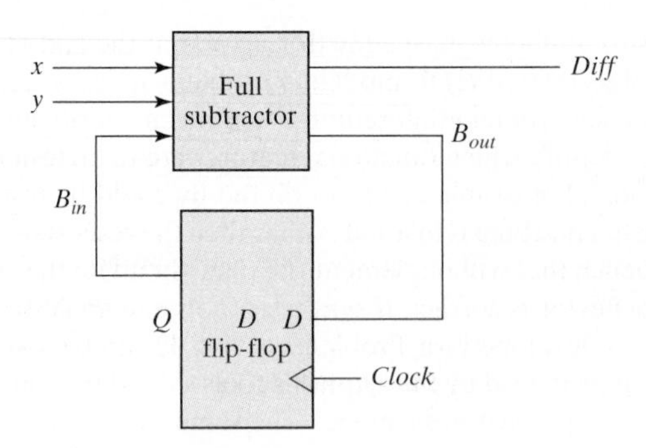

FIGURE P5.7

5.8* Derive the state table and the state diagram of the sequential circuit shown in Fig. P5.8. Explain the function that the circuit performs. (HDL—see Problem 5.36.)

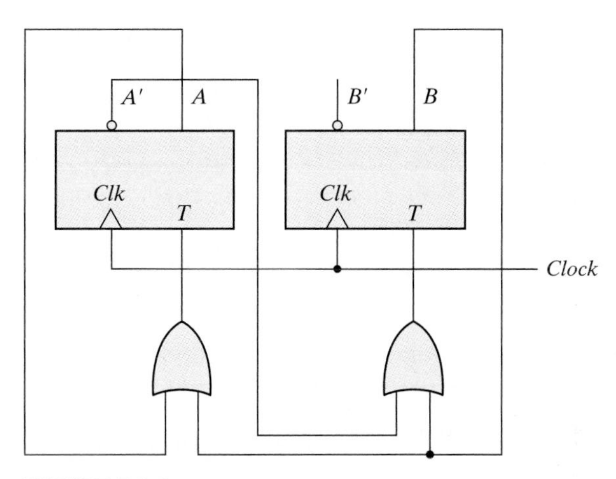

FIGURE P5.8

5.9 A sequential circuit has two *JK* flip-flops *A* and *B* and one input *x*. The circuit is described by the following flip-flop input equations:

$$J_A = x \quad K_A = B$$
$$J_B = x' \quad K_B = A'$$

(a) Derive the state equations $A(t + 1)$ and $B(t + 1)$ by substituting the input equations for the *J* and *K* variables.
(b) Draw the state diagram of the circuit.

5.10 A sequential circuit has two *JK* flip-flops *A* and *B*, two inputs *x* and *y*, and one output *z*. The flip-flop input equations and circuit output equation are

$$J_A = Bx + B'y' \qquad K_A = B'x + y$$
$$J_B = A'x \qquad K_B = A + xy'$$
$$z = (A + B)x'y'$$

(a) Draw the logic diagram of the circuit.
(b) Tabulate the state table.
(c) Derive the state equations for *A* and *B*.

5.11 For the circuit described by the state diagram of Fig. 5.16,
(a)* Determine the state transitions and output sequence that will be generated when an input sequence of 110111001 is applied to the circuit and it is initially in the state 00.
(b) Find all of the equivalent states in Fig. 5.16 and draw a simpler, but equivalent, state diagram.
(c) Using *JK* flip-flops, design the equivalent machine (including its logic diagram) described by the state diagram in (b).

5.12 For the following state table

	Next State		**Output**	
Present State	**x = 0**	**x = 1**	**x = 0**	**x = 1**
a	*f*	*b*	0	0
b	*d*	*c*	0	0
c	*f*	*e*	0	0
d	*g*	*a*	1	0
e	*d*	*c*	0	0
f	*f*	*b*	1	1
g	*g*	*h*	0	1
h	*g*	*a*	1	0

(a) Draw the corresponding state diagram.
(b)* Tabulate the reduced state table.
(c) Draw the state diagram corresponding to the reduced state table.

5.13 Starting from state *a*, and the input sequence 01010010111, determine the output sequence for
(a) The state table of the previous problem.
(b) The reduced state table from the previous problem. Show that the same output sequence is obtained for both.

5.14 Substitute the one-hot assignment 3 from Table 5.9 to the states in Table 5.8 and obtain the binary state table.

5.15 List a state table for the *T* flip-flop using *Q* as the present and next state and *T* as inputs. Design the sequential circuit specified by the state table using *D* flip-flop and show that it is equivalent to Fig. 5.13(b).

5.16 Design a sequential circuit with two *D* flip-flops *A* and *B*, and one input *x_in*.
(a)* When *x_in* = 0, the state of the circuit remains the same. When *x_in* = 1, the circuit goes through the state transitions from 00 to 01, to 11, to 10, back to 00, and repeats.
(b) When *x_in* = 0, the state of the circuit remains the same. When *x_in* =1, the circuit goes through the state transitions from 00 to 11, to 01, to 10, back to 00, and repeats. (HDL—see Problem 5.38.)

5.17 Design a one-input, one-output serial 2's complementer. The circuit accepts a string of bits from the input and generates the 2's complement at the output. The circuit can be reset asynchronously to start and end the operation. (HDL—see Problem 5.39.)

5.18* Design a sequential circuit with two *JK* flip-flops *A* and *B* and two inputs *E* and *F*. If *E* = 0, the circuit remains in the same state regardless of the value of *F*. When *E* = 1 and *F* = 1, the circuit goes through the state transitions from 00 to 01, to 10, to 11, back to 00, and repeats. When *E* = 1 and *F* = 0, the circuit goes through the state transitions from 00 to 11, to 10, to 01, back to 00, and repeats. (HDL—see Problem 5.40.)

5.19 A sequential circuit has three flip-flops *A, B, C*; one input *x_in*; and one output *y_out*. The state diagram is shown in Fig. P5.19. The circuit is to be designed by treating the unused states as don't-care conditions. Analyze the circuit obtained from the design to determine the effect of the unused states. (HDL—see Problem 5.41.)

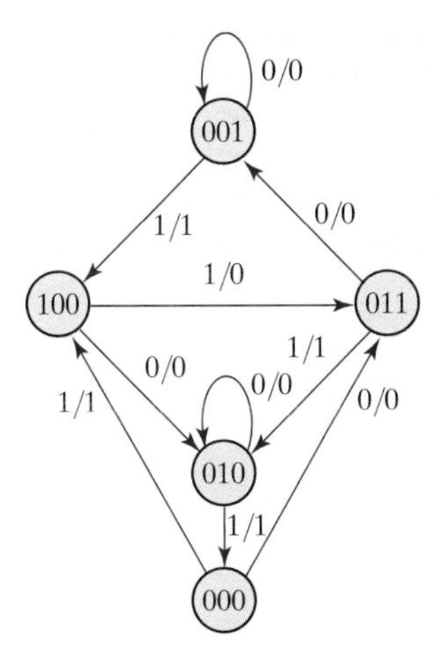

FIGURE P5.19

(a)* Use D flip-flops in the design.
(b) Use JK flip-flops in the design.

5.20 Design the sequential circuit specified by the state diagram of Fig. 5.19, using D flip-flops.

5.21 What is the main difference between an **initial** statement and an **always** statement in Verilog HDL?

5.22 Draw the waveform generated by the statements below:

(a)
```
initial begin
   w = 0; #10 w = 1; # 40 w = 0;  # 20 w = 1;  #15 w = 0;
  end
```

(b)
```
initial fork
   w = 0;  #10 w = 1;  # 40 w = 0;  # 20 w = 1;  #15 w = 0;
  join
```

5.23* Consider the following statements assuming that *RegA* contains the value of 50 initially.

(a)
```
RegA = 125;
RegB = RegA;
```

(b)
```
RegA <= 125;
RegB <= RegA;
```

What are the values of RegA and RegB after execution?

5.24 Write and verify an HDL behavioral description of a positive-edge-sensitive D flip-flop with asynchronous preset and clear.

5.25 A special positive-edge-triggered flip-flop circuit component has four inputs $D1$, $D2$, $D3$, and $D4$, and a two-bit control input that chooses between them. Write and verify an HDL behavioral description of this component.

5.26 Write and verify an HDL behavioral description of the JK flip-flop using an if-else statement based on the value of the present state.

(a)* Obtain the characteristic equation when $Q = 0$ or $Q = 1$.
(b) Specify how the J and K inputs affect the output of the flip-flop at each clock tick.

5.27 Rewrite and verify the description of HDL Example 5.5 by combining the state transitions and output into one **always** block.

5.28 Simulate the sequential circuit shown in Fig. 5.17.
(a) Write the HDL description of the state diagram (i.e., behavioral model).
(b) Write the HDL description of the logic (circuit) diagram (i.e., a structural model).
(c) Write an HDL stimulus with a sequence of inputs: 00, 01, 11, 10. Verify that the response is the same for both descriptions.

5.29 Write a behavioral description of the state machine described by the state diagram shown in Fig. P5.19. Write a test bench and verify the functionality of the description.

5.30 Draw the logic diagram for the sequential circuit described by the following HDL module:

```
module Seq_Ckt (input A, B, C, E output reg Q,input CLK,);
 reg E;

 always @ (posedge CLK)
 begin
  E <= A || B;
  Q <= E && C;
 end
endmodule
```

5.31 How should the description in problem 5.30 be written to have the same behavior when the assignments are made with = instead of with <= ?

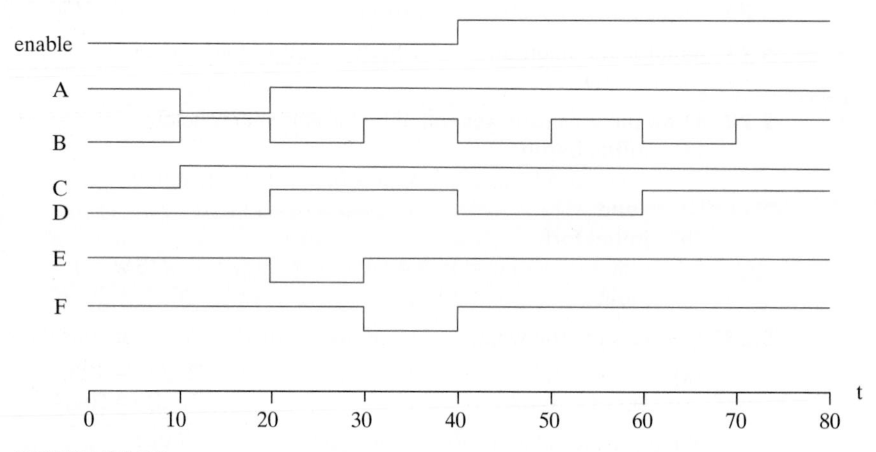

FIGURE P5.32
Waveforms for Problem 5.32

5.32 Using an **initial** statement with a **begin** . . . **end** block write a Verilog description of the waveforms shown in Fig. P5.32. Repeat using a **fork** . . . **join** block.

5.33 Explain why it is important that the stimulus signals in a test bench be synchronized to the inactive edge of the clock of the sequential circuit that is to be tested.

5.34 Write and verify an HDL structural description of the machine having the circuit diagram (schematic) shown in Fig. 5.5.

5.35 Write and verify an HDL model of the sequential circuit described in Problem 5.6.

5.36 Write and verify an HDL structural description of the machine having the circuit diagram (schematic) shown in Fig. P5.8.

5.37 Write and verify HDL behavioral descriptions of the state machines shown in Figs. 5.25 and 5.26. Write a test bench to compare the state sequences and input–output behaviors of the two machines.

5.38 Write and verify an HDL behavioral description of the machine described in Problem 5.16.

5.39 Write and verify a behavioral description of the machine specified in Problem 5.17.

5.40 Write and verify a behavioral description of the machine specified in Problem 5.18.

5.41 Write and verify a behavioral description of the machine specified in Problem 5.19. (*Hint*: See the discussion of the **default** case item preceding HDL Example 4.8 in Chapter 4.)

5.42 Write and verify an HDL structural description of the circuit shown in Fig. 5.29.

5.43 Write and verify an HDL behavioral description of the three-bit binary counter in Fig. 5.34.

5.44 Write and verify a Verilog model of a *D* flip-flop having asynchronous reset.

5.45 Write and verify an HDL behavioral description of the sequence detector described in Fig. 5.27.

5.46 A synchronous finite state machine has an input *x_in* and an output *y_out*. When *x_in* changes from 0 to 1, the output *y_out* is to assert for three cycles, regardless of the value of *x_in*, and then de-assert for two cycles before the machine will respond to another assertion of *x_in*. The machine is to have active-low synchronous reset.

(a) Draw the state diagram of the machine.

(b) Write and verify a Verilog model of the machine.

5.47 Write a Verilog model of a synchronous finite state machine whose output is the sequence 0, 2, 4, 6, 8 10, 12, 14, 0 . . . . The machine is controlled by a single input, *Run*, so that counting occurs while *Run* is asserted, suspends while *Run* is de-asserted, and resumes the count when *Run* is re-asserted. Clearly state any assumptions that you make.

5.48 Write a Verilog model of the Mealy FSM described by the state diagram in Fig. P5.48. Develop a test bench and demonstrate that the machine state transitions and output correspond to its state diagram.

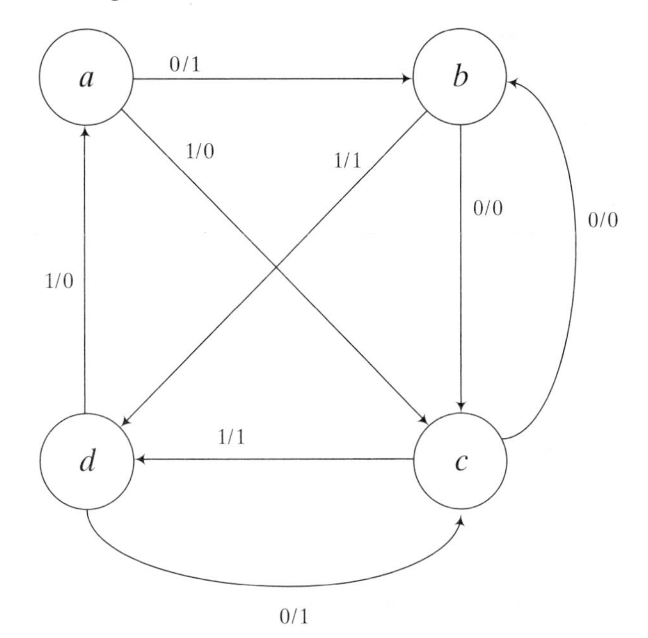

FIGURE P5.48

5.49 Write a Verilog model of the Moore FSM described by the state diagram in Fig. P5.49. Develop a test bench and demonstrate that the machine's state transitions and output correspond to its state diagram.

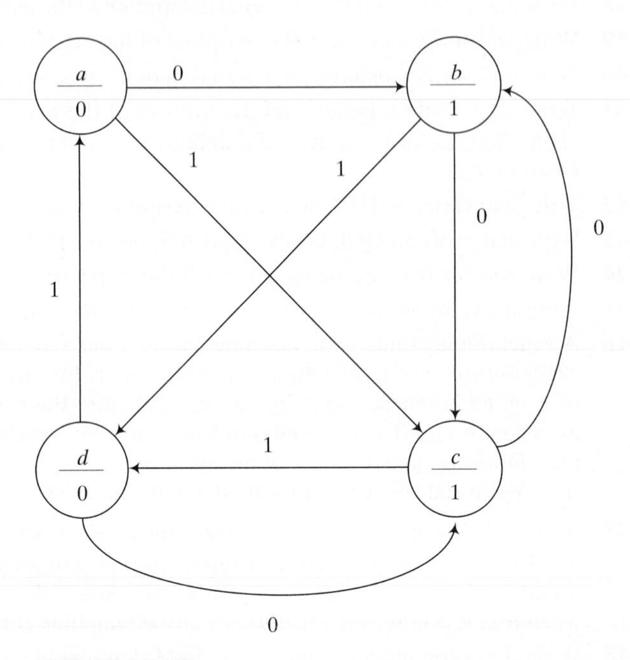

FIGURE P5.49

5.50 A synchronous Moore FSM has a single input, *x_in*, and a single output *y_out*. The machine is to monitor the input and remain in its initial state until a second sample of *x_in* is detected to be 1. Upon detecting the second assertion of *x_in* *y_out* is to asserted and remain asserted until a fourth assertion of *x_in* is detected. When the fourth assertion of *x_in* is detected the machine is to return to its initial state and resume monitoring of *x_in*.

(a) Draw the state diagram of the machine.

(b) Write and verify a Verilog model of the machine.

5.51 Draw the state diagram of the machine described by the Verilog model given below.

```
module Prob_5_51 (output reg y_out, input x_in, clk, reset);
 parameter s0 = 2'b00, s1 = 2'b01, s2 = 2'b10, s3 = 2'b11;
 reg [1:0] state, next_state;
 always @ (posedge clk, negedge reset) begin
  if (reset == 1'b0) state <= s0;
  else state <= next_state;
 always @(state, x_in) begin
  y_out = 0;
  next_state = s0;
  case (state)
  s0: if x_in = 1 begin y_out = 0; if (x_in) next_state = s1; else next_state = s0; end
  s1: if x_in = 1 begin y_out = 0; if (x_in) next_state = s2; else next_state = s1; end
```

```
   s2: if x_in = 1 begin y_out = 1; if (x_in) next_state = s3; else next_state = s2; end
   s3: if x_in = 1 begin y_out = 1; if (x_in) next_state = s0; else next_state = s3; end
   default: next_state = s0;
   endcase
  end
endmodule
```

5.52 Draw the state diagram of the machine described by the Verilog model given below.

```
module Prob_5_52 (output reg y_out, input x_in, clk, reset);
   parameter s0 = 2'b00, s1 = 2'b01, s2 = 2'b10, s3 = 2'b11;
  reg [1:0] state, next_state;
  always @ (posedge clk, negedge reset) begin
   if (reset == 1'b0) state <= s0;
   else state <= next_state;
  always @(state, x_in) begin
   y_out = 0;
   next_state = s0;
   case (state)
   s0: if x_in = 1 begin y_out = 0; if (x_in) next_state = s1; else next_state = s0; end
   s1: if x_in = 1 begin y_out = 0; if (x_in) next_state = s2; else next_state = s1; end
   s2: if x_in = 1 if (x_in) begin next_state = s3; y_out = 0;
                  else begin next_state = s2; y_out = 1; end
   s3: if x_in = 1 begin y_out = 1; if (x_in) next_state = s0; else next_state = s3; end
   default: next_state = s0;
   endcase
  end
endmodule
```

5.53 Draw a state diagram and write a Verilog model of a Mealy synchronous state machine having a single input, *x_in*, and a single output *y_out*, such that *y_out* is asserted if the total number of 1's received is a multiple of 3.

5.54 A synchronous Moore machine has two inputs, $x1$, and $x2$, and output *y_out*. If both inputs have the same value the output is asserted for one cycle; otherwise the output is 0. Develop a state diagram and a write a Verilog behavioral model of the machine. Demonstrate that the machine operates correctly.

5.55 Develop the state diagram for a Mealy state machine that detects a sequence of three or more consecutive 1's in a string of bits coming through an input line.

5.56 Using manual methods, obtain the logic diagram of a three-bit counter that counts in the sequence $0, 2, 4, 6, 0, \ldots$.

5.57 Write and verify a Verilog behavioral model of a three-bit counter that counts in the sequence $0, 2, 4, 6, 0, \ldots$.

5.58 Write and verify a Verilog behavioral model of the counter designed in Problem 5.55.

5.59 Write and verify a Verilog structural model of the counter described in Problem 5.56.

5.60 Write and verify a Verilog behavioral model of a four-bit counter that counts in the sequence $0, 1, \ldots, 9, 0, 1, 2, \ldots$.

REFERENCES

1. Bhasker, J. 1997. A *Verilog HDL Primer*. Allentown, PA: Star Galaxy Press.
2. Bhasker, J. 1998. *Verilog HDL Synthesis*. Allentown, PA: Star Galaxy Press.
3. Ciletti, M. D. 1999. *Modeling, Synthesis, and Rapid Prototyping with Verilog HDL*. Upper Saddle River, NJ: Prentice Hall.
4. Dietmeyer, D. L. 1988. *Logic Design of Digital Systems*, 3rd ed. Boston: Allyn Bacon.
5. Gajski, D. D. 1997. *Principles of Digital Design*. Upper Saddle River, NJ: Prentice Hall.
6. Hayes, J. P. 1993. *Introduction to Digital Logic Design*. Reading, MA: Addison-Wesley.
7. Katz, R. H. 2005. *Contemporary Logic Design*. Upper Saddle River, NJ: Prentice Hall.
8. Mano, M. M. and C. R. Kime. 2007. *Logic and Computer Design Fundamentals & Xilinx 6.3 Student Edition*, 4th ed. Upper Saddle River, NJ: Prentice Hall.
9. Nelson, V. P., H. T. Nagle, J. D. Irwin, and B. D. Carroll. 1995. *Digital Logic Circuit Analysis and Design*. Englewood Cliffs, NJ: Prentice Hall.
10. Palnitkar, S. 1996. *Verilog HDL: A Guide to Digital Design and Synthesis*. Mountain View, CA: SunSoft Press (a Prentice Hall title).
11. Roth, C. H. 2009. *Fundamentals of Logic Design,* 6th ed. St. Paul, MN: Brooks/Cole.
12. Thomas, D. E. and P. R. Moorby, 2002. *The Verilog Hardware Description Language*, 6th ed. Boston: Kluwer Academic Publishers.
13. Wakerly, J. F. 2006. *Digital Design: Principles and Practices*, 4th ed. Upper Saddle River, NJ: Prentice Hall.

WEB SEARCH TOPICS

Finite State Machine
Synchronous state machine
Asynchronous state machine
D-type flip-flop
Toggle flip-flop
J-K type flip-flop
Binary counter
State diagram
Mealy state machine
Moore state machine
One-hot/cold codes

Chapter 6

Registers and Counters

6.1 REGISTERS

A clocked sequential circuit consists of a group of flip-flops and combinational gates. The flip-flops are essential because, in their absence, the circuit reduces to a purely combinational circuit (provided that there is no feedback among the gates). A circuit with flip-flops is considered a sequential circuit even in the absence of combinational gates. Circuits that include flip-flops are usually classified by the function they perform rather than by the name of the sequential circuit. Two such circuits are registers and counters.

A *register* is a group of flip-flops, each one of which shares a common clock and is capable of storing one bit of information. An n-bit register consists of a group of n flip-flops capable of storing n bits of binary information. In addition to the flip-flops, a register may have combinational gates that perform certain data-processing tasks. In its broadest definition, a register consists of a group of flip-flops together with gates that affect their operation. The flip-flops hold the binary information, and the gates determine how the information is transferred into the register.

A *counter* is essentially a register that goes through a predetermined sequence of binary states. The gates in the counter are connected in such a way as to produce the prescribed sequence of states. Although counters are a special type of register, it is common to differentiate them by giving them a different name.

Various types of registers are available commercially. The simplest register is one that consists of only flip-flops, without any gates. Figure 6.1 shows such a register constructed with four D-type flip-flops to form a four-bit data storage register. The common clock input triggers all flip-flops on the positive edge of each pulse, and the binary data available at the four inputs are transferred into the register. The value of (I_3, I_2, I_1, I_0) immediately before the clock edge determines the value of (A_3, A_2, A_1, A_0) after the clock edge. The four

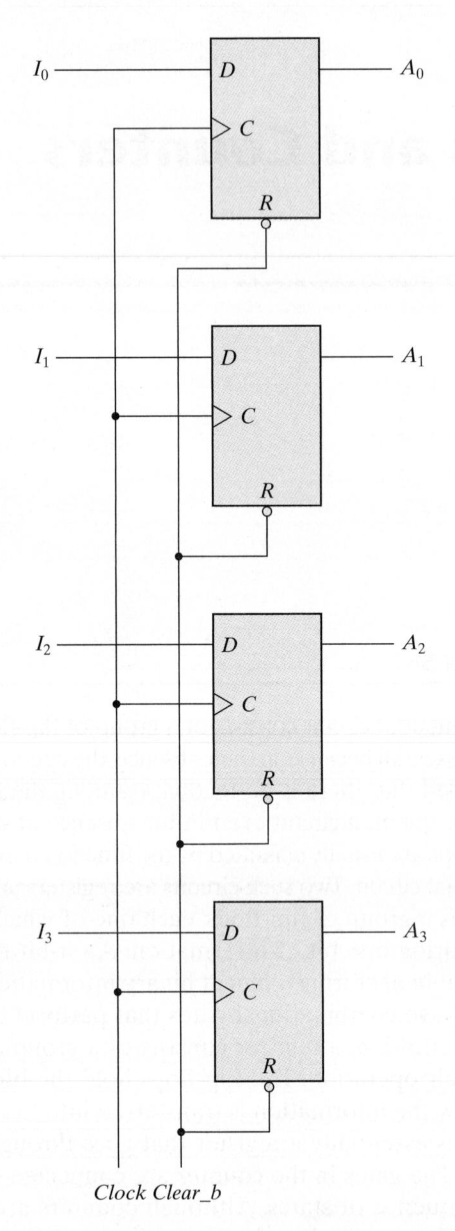

FIGURE 6.1
Four-bit register

outputs can be sampled at any time to obtain the binary information stored in the register. The input *Clear_b* goes to the active-low *R* (reset) input of all four flip-flops. When this input goes to 0, all flip-flops are reset asynchronously. The *Clear_b* input is useful for clearing the register to all 0's prior to its clocked operation. The *R* inputs must be maintained at logic 1 (i.e., de-asserted) during normal clocked operation. Note that, depending on the flip-flop, either *Clear*, *Clear_b*, *reset*, or *reset_b* can be used to indicate the transfer of the register to an all 0's state.

Register with Parallel Load

Registers with parallel load are a fundamental building block in digital systems. It is important that you have a thorough understanding of their behavior. Synchronous digital systems have a master clock generator that supplies a continuous train of clock pulses. The pulses are applied to all flip-flops and registers in the system. The master clock acts like a drum that supplies a constant beat to all parts of the system. A separate control signal must be used to decide which register operation will execute at each clock pulse. The transfer of new information into a register is referred to as *loading* or *updating* the register. If all the bits of the register are loaded simultaneously with a common clock pulse, we say that the loading is done *in parallel*. A clock edge applied to the C inputs of the register of Fig. 6.1 will load all four inputs in parallel. In this configuration, if the contents of the register must be left unchanged, the inputs must be held constant or the clock must be inhibited from the circuit. In the first case, the data bus driving the register would be unavailable for other traffic. In the second case, the clock can be inhibited from reaching the register by controlling the clock input signal with an enabling gate. However, inserting gates into the clock path is ill advised because it means that logic is performed with clock pulses. The insertion of logic gates produces uneven propagation delays between the master clock and the inputs of flip-flops. To fully synchronize the system, we must ensure that all clock pulses arrive at the same time anywhere in the system, so that all flip-flops trigger simultaneously. Performing logic with clock pulses inserts variable delays and may cause the system to go out of synchronism. For this reason, it is advisable to control the operation of the register with the D inputs, rather than controlling the clock in the C inputs of the flip-flops. This creates the effect of a gated clock, but without affecting the clock path of the circuit.

A four-bit data-storage register with a load control input that is directed through gates and into the D inputs of the flip-flops is shown in Fig. 6.2. The additional gates implement a two-channel mux whose output drives the input to the register with either the data bus or the output of the register. The load input to the register determines the action to be taken with each clock pulse. When the load input is 1, the data at the four external inputs are transferred into the register with the next positive edge of the clock. When the load input is 0, the outputs of the flip-flops are connected to their respective inputs. The feedback connection from output to input is necessary because a D flip-flop does not have a "no change" condition. With each clock edge, the D input determines the next state of the register. To leave the output unchanged, it is necessary to make the D input equal to the present value of the output (i.e., the output circulates to the input at each clock pulse). The clock pulses are applied to the C inputs without interruption. The load input determines whether the next pulse will accept new information or leave the information in the register intact. The transfer of information from the data inputs or the outputs of the register is done simultaneously with all four bits in response to a clock edge.

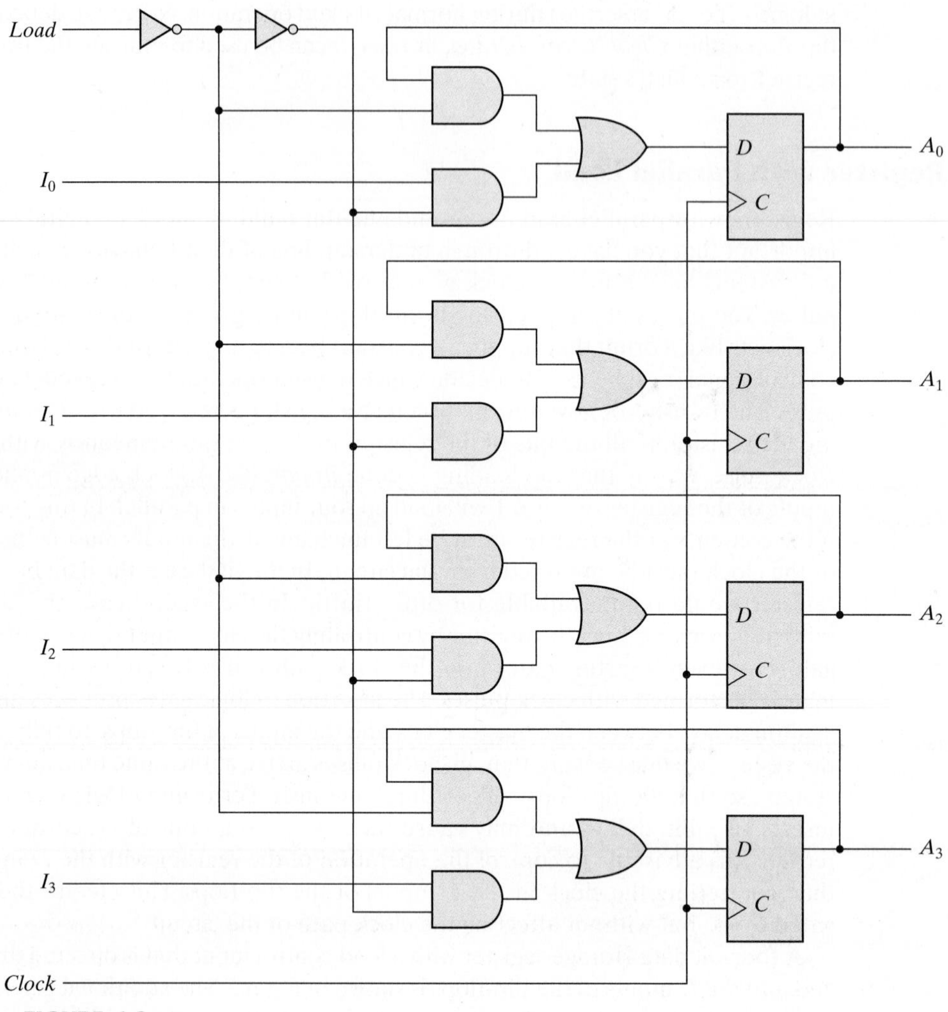

FIGURE 6.2
Four-bit register with parallel load

6.2 SHIFT REGISTERS

A register capable of shifting the binary information held in each cell to its neighboring cell, in a selected direction, is called a *shift register*. The logical configuration of a shift register consists of a chain of flip-flops in cascade, with the output of one flip-flop connected to the input of the next flip-flop. All flip-flops receive common clock pulses, which activate the shift of data from one stage to the next.

The simplest possible shift register is one that uses only flip-flops, as shown in Fig. 6.3. The output of a given flip-flop is connected to the D input of the flip-flop at its right. This shift register is unidirectional (left-to-right). Each clock pulse shifts the contents of the

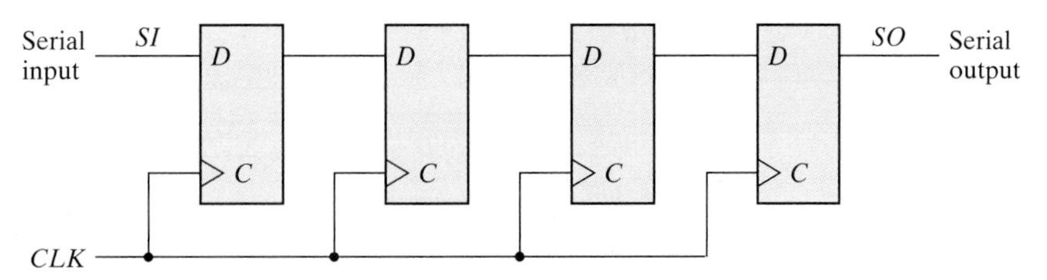

FIGURE 6.3
Four-bit shift register

register one bit position to the right. The configuration does not support a left shift. The *serial input* determines what goes into the leftmost flip-flop during the shift. The *serial output* is taken from the output of the rightmost flip-flop. Sometimes it is necessary to control the shift so that it occurs only with certain pulses, but not with others. As with the data register discussed in the previous section, the clock's signal can be suppressed by gating the clock signal to prevent the register from shifting. A preferred alternative in high-speed circuits is to suppress the clock *action*, rather than gate the clock signal, by leaving the clock path unchanged, but recirculating the output of each register cell back through a two-channel mux whose output is connected to the input of the cell. When the clock action is not suppressed, the other channel of the mux provides a datapath to the cell.

It will be shown later that the shift operation can be controlled through the *D* inputs of the flip-flops rather than through the clock input. If, however, the shift register of Fig. 6.3 is used, the shift can be controlled with an input by connecting the clock through an AND gate. This is not a preferred practice. Note that the simplified schematics do not show a reset signal, but such a signal is required in practical designs.

Serial Transfer

The datapath of a digital system is said to operate in serial mode when information is transferred and manipulated one bit at a time. Information is transferred one bit at a time by shifting the bits out of the source register and into the destination register. This type of transfer is in contrast to parallel transfer, whereby all the bits of the register are transferred at the same time.

The serial transfer of information from register *A* to register *B* is done with shift registers, as shown in the block diagram of Fig. 6.4(a). The serial output (*SO*) of register *A* is connected to the serial input (*SI*) of register *B*. To prevent the loss of information stored in the source register, the information in register *A* is made to circulate by connecting the serial output to its serial input. The initial content of register *B* is shifted out through its serial output and is lost unless it is transferred to a third shift register. The shift control input determines when and how many times the registers are shifted. For illustration here, this is done with an AND gate that allows clock pulses to pass into the *CLK* terminals only when the shift control is active. (This practice can be problematic because it may compromise the clock path of the circuit, as discussed earlier.)

Suppose the shift registers in Fig. 6.4 have four bits each. Then the control unit that supervises the transfer of data must be designed in such a way that it enables the shift

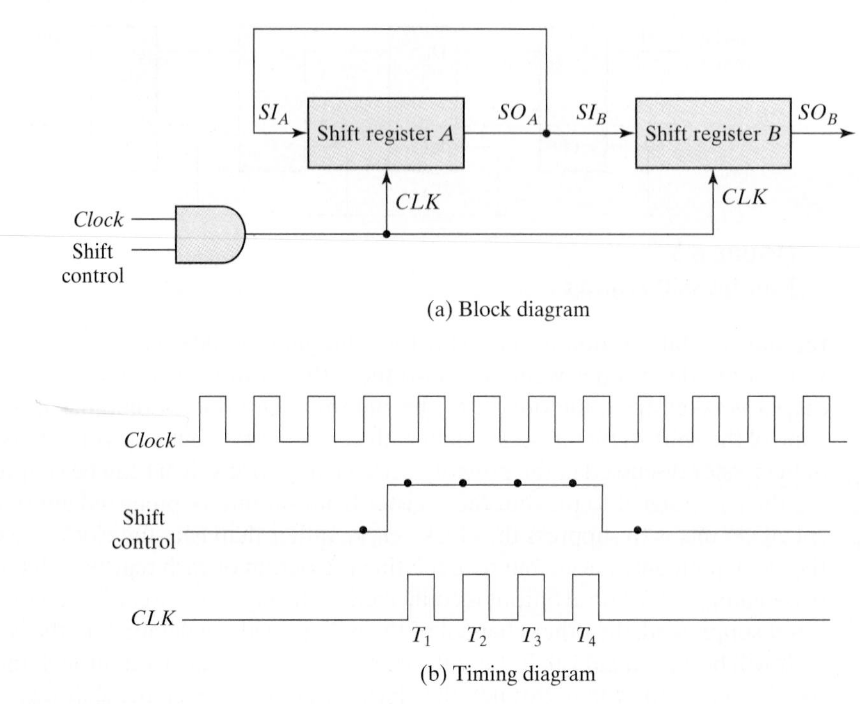

FIGURE 6.4
Serial transfer from register *A* to register *B*

registers, through the shift control signal, for a fixed time of four clock pulses in order to pass an entire word. This design is shown in the timing diagram of Fig. 6.4(b). The shift control signal is synchronized with the clock and changes value just after the negative edge of the clock. The next four clock pulses find the shift control signal in the active state, so the output of the AND gate connected to the CLK inputs produces four pulses: T_1, T_2, T_3, and T_4. Each rising edge of the pulse causes a shift in both registers. The fourth pulse changes the shift control to 0, and the shift registers are disabled.

Assume that the binary content of A before the shift is 1011 and that of B is 0010. The serial transfer from A to B occurs in four steps, as shown in Table 6.1. With the first pulse, T_1, the rightmost bit of A is shifted into the leftmost bit of B and is also circulated into the leftmost position of A. At the same time, all bits of A and B are shifted one position to the right. The previous serial output from B in the rightmost position is lost, and its value changes from 0 to 1. The next three pulses perform identical operations, shifting the bits of A into B, one at a time. After the fourth shift, the shift control goes to 0, and registers A and B both have the value 1011. Thus, the contents of A are copied into B, so that the contents of A remain unchanged i.e., the contents of A are restored to their original value.

The difference between the serial and the parallel mode of operation should be apparent from this example. In the parallel mode, information is available from all bits of a register and all bits can be transferred simultaneously during one clock pulse. In the serial

Table 6.1
Serial-Transfer Example

Timing Pulse	Shift Register A	Shift Register B
Initial value	1 0 1 1	0 0 1 0
After T_1	1 1 0 1	1 0 0 1
After T_2	1 1 1 0	1 1 0 0
After T_3	0 1 1 1	0 1 1 0
After T_4	1 0 1 1	1 0 1 1

mode, the registers have a single serial input and a single serial output. The information is transferred one bit at a time while the registers are shifted in the same direction.

Serial Addition

Operations in digital computers are usually done in parallel because that is a faster mode of operation. Serial operations are slower because a datapath operation takes several clock cycles, but serial operations have the advantage of requiring fewer hardware components. In VLSI circuits, they require less silicon area on a chip. To demonstrate the serial mode of operation, we present the design of a serial adder. The parallel counterpart was presented in Section 4.4.

The two binary numbers to be added serially are stored in two shift registers. Beginning with the least significant pair of bits, the circuit adds one pair at a time through a single full-adder (FA) circuit, as shown in Fig. 6.5. The carry out of the full adder is transferred to a D flip-flop, the output of which is then used as the carry input for the next pair of significant bits. The sum bit from the S output of the full adder could be transferred into a third shift register. By shifting the sum into A while the bits of A are shifted out, it is possible to use one register for storing both the augend and the sum bits. The serial input of register B can be used to transfer a new binary number while the addend bits are shifted out during the addition.

The operation of the serial adder is as follows: Initially, register A holds the augend, register B holds the addend, and the carry flip-flop is cleared to 0. The outputs (SO) of A and B provide a pair of significant bits for the full adder at x and y. Output Q of the flip-flop provides the input carry at z. The shift control enables both registers and the carry flip-flop, so at the next clock pulse, both registers are shifted once to the right, the sum bit from S enters the leftmost flip-flop of A, and the output carry is transferred into flip-flop Q. The shift control enables the registers for a number of clock pulses equal to the number of bits in the registers. For each succeeding clock pulse, a new sum bit is transferred to A, a new carry is transferred to Q, and both registers are shifted once to the right. This process continues until the shift control is disabled. Thus, the addition is accomplished by passing each pair of bits together with the previous carry through a single full-adder circuit and transferring the sum, one bit at a time, into register A.

Initially, register A and the carry flip-flop are cleared to 0, and then the first number is added from B. While B is shifted through the full adder, a second number is transferred

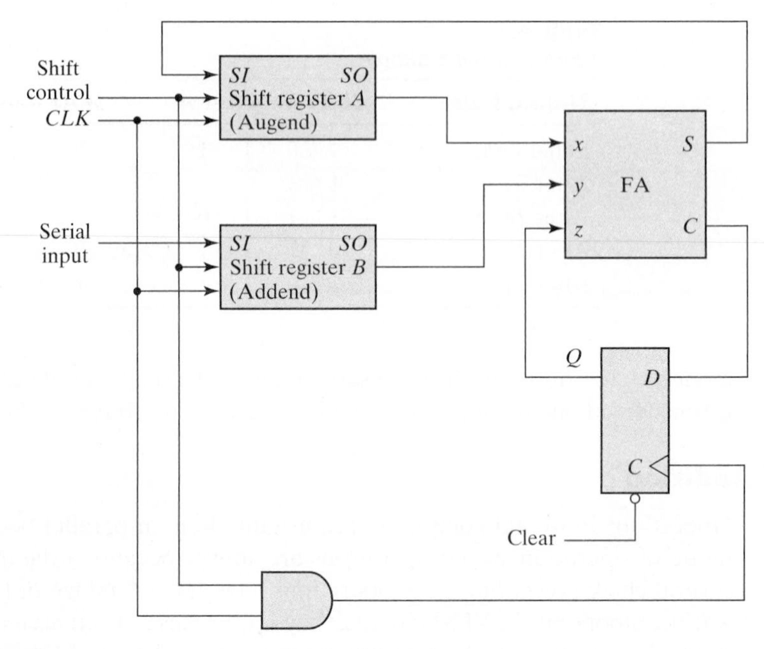

FIGURE 6.5
Serial adder

to it through its serial input. The second number is then added to the contents of register *A*, while a third number is transferred serially into register *B*. This can be repeated to perform the addition of two, three, or more four-bit numbers and accumulate their sum in register *A*.

Comparing the serial adder with the parallel adder described in Section 4.4, we note several differences. The parallel adder uses registers with a parallel load, whereas the serial adder uses shift registers. The number of full-adder circuits in the parallel adder is equal to the number of bits in the binary numbers, whereas the serial adder requires only one full-adder circuit and a carry flip-flop. Excluding the registers, the parallel adder is a combinational circuit, whereas the serial adder is a sequential circuit which consists of a full adder and a flip-flop that stores the output carry. This design is typical in serial operations because the result of a bit-time operation may depend not only on the present inputs, but also on previous inputs that must be stored in flip-flops.

To show that serial operations can be designed by means of sequential circuit procedure, we will redesign the serial adder with the use of a state table. First, we assume that two shift registers are available to store the binary numbers to be added serially. The serial outputs from the registers are designated by x and y. The sequential circuit to be designed will not include the shift registers, but they will be inserted later to show the complete circuit. The sequential circuit proper has the two inputs, x and y, that provide a pair of significant bits, an output S that generates the sum bit, and flip-flop Q for storing the carry. The state table that specifies the sequential circuit is listed in Table 6.2. The present state of Q is the present value of the carry. The present carry in

Table 6.2
State Table for Serial Adder

Present State	Inputs		Next State	Output	Flip-Flop Inputs	
Q	x	y	Q	S	J_Q	K_Q
0	0	0	0	0	0	X
0	0	1	0	1	0	X
0	1	0	0	1	0	X
0	1	1	1	0	1	X
1	0	0	0	1	X	1
1	0	1	1	0	X	0
1	1	0	1	0	X	0
1	1	1	1	1	X	0

Q is added together with inputs x and y to produce the sum bit in output S. The next state of Q is equal to the output carry. Note that the state table entries are identical to the entries in a full-adder truth table, except that the input carry is now the present state of Q and the output carry is now the next state of Q.

If a D flip-flop is used for Q, the circuit reduces to the one shown in Fig. 6.5. If a JK flip-flop is used for Q, it is necessary to determine the values of inputs J and K by referring to the excitation table (Table 5.12). This is done in the last two columns of Table 6.2. The two flip-flop input equations and the output equation can be simplified by means of maps to

$$J_Q = xy$$
$$K_Q = x'y' = (x + y)'$$
$$S = x \oplus y \oplus Q$$

The circuit diagram is shown in Fig. 6.6. The circuit consists of three gates and a JK flip-flop. The two shift registers are included in the diagram to show the complete serial adder. Note that output S is a function not only of x and y, but also of the present state of Q. The next state of Q is a function of the present state of Q and of the values of x and y that come out of the serial outputs of the shift registers.

Universal Shift Register

If the flip-flop outputs of a shift register are accessible, then information entered serially by shifting can be taken out in parallel from the outputs of the flip-flops. If a parallel load capability is added to a shift register, then data entered in parallel can be taken out in serial fashion by shifting the data stored in the register.

Some shift registers provide the necessary input and output terminals for parallel transfer. They may also have both shift-right and shift-left capabilities. The most general shift register has the following capabilities:

1. A *clear* control to clear the register to 0.
2. A *clock* input to synchronize the operations.

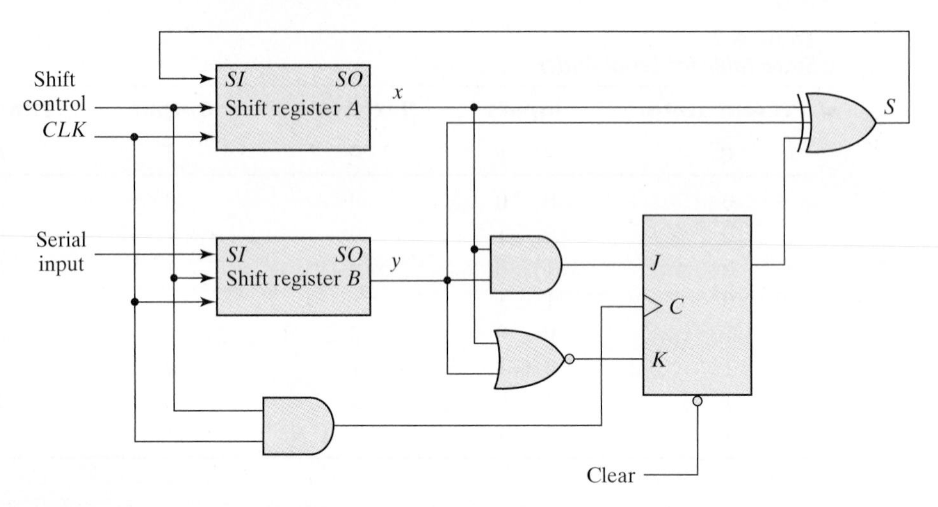

FIGURE 6.6
Second form of serial adder

3. A *shift-right* control to enable the shift-right operation and the *serial input* and *output* lines associated with the shift right.
4. A *shift-left* control to enable the shift-left operation and the *serial input* and *output* lines associated with the shift left.
5. A *parallel-load* control to enable a parallel transfer and the n input lines associated with the parallel transfer.
6. n parallel output lines.
7. A control state that leaves the information in the register unchanged in response to the clock. Other shift registers may have only some of the preceding functions, with at least one shift operation.

A register capable of shifting in one direction only is a *unidirectional* shift register. One that can shift in both directions is a *bidirectional* shift register. If the register has both shifts and parallel-load capabilities, it is referred to as a *universal shift register.*

The block diagram symbol and the circuit diagram of a four-bit universal shift register that has all the capabilities just listed are shown in Fig. 6.7. The circuit consists of four D flip-flops and four multiplexers. The four multiplexers have two common selection inputs s_1 and s_0. Input 0 in each multiplexer is selected when $s_1s_0 = 00$, input 1 is selected when $s_1s_0 = 01$, and similarly for the other two inputs. The selection inputs control the mode of operation of the register according to the function entries in Table 6.3. When $s_1s_0 = 00$, the present value of the register is applied to the D inputs of the flip-flops. This condition forms a path from the output of each flip-flop into the input of the same flip-flop, so that the output recirculates to the input in this mode of operation. The next clock edge transfers into each flip-flop the binary value it held previously, and no change of state occurs.

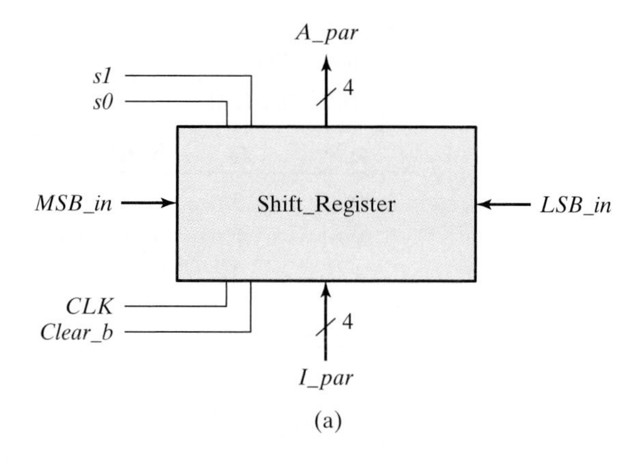

(a)

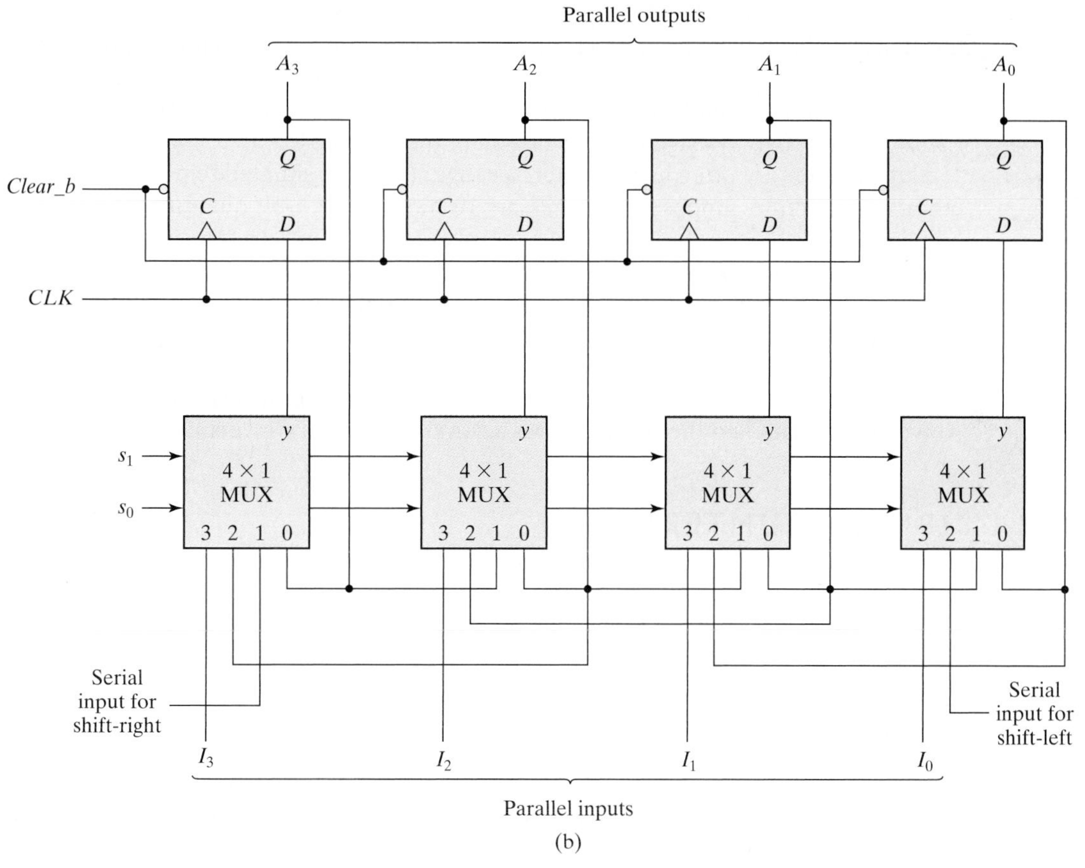

(b)

FIGURE 6.7
Four-bit universal shift register

Table 6.3
Function Table for the Register of Fig. 6.7

Mode Control		
s_1	s_0	**Register Operation**
0	0	No change
0	1	Shift right
1	0	Shift left
1	1	Parallel load

When $s_1s_0 = 01$, terminal 1 of the multiplexer inputs has a path to the D inputs of the flip-flops. This causes a shift-right operation, with the serial input transferred into flip-flop A_3. When $s_1s_0 = 10$, a shift-left operation results, with the other serial input going into flip-flop A_0. Finally, when $s_1s_0 = 11$, the binary information on the parallel input lines is transferred into the register simultaneously during the next clock edge. Note that data enters *MSB_in* for a shift-right operation and enters *LSB_in* for a shift-left operation. *Clear_b* is an active-low signal that clears all of the flip-flops.

Shift registers are often used to interface digital systems situated remotely from each other. For example, suppose it is necessary to transmit an n-bit quantity between two points. If the distance is far, it will be expensive to use n lines to transmit the n bits in parallel. It is more economical to use a single line and transmit the information serially, one bit at a time. The transmitter accepts the n-bit data in parallel into a shift register and then transmits the data serially along the common line. The receiver accepts the data serially into a shift register. When all n bits are received, they can be taken from the outputs of the register in parallel. Thus, the transmitter performs a parallel-to-serial conversion of data and the receiver does a serial-to-parallel conversion.

6.3 RIPPLE COUNTERS

A register that goes through a prescribed sequence of states upon the application of input pulses is called a *counter*. The input pulses may be clock pulses, or they may originate from some external source and may occur at a fixed interval of time or at random. The sequence of states may follow the binary number sequence or any other sequence of states. A counter that follows the binary number sequence is called a *binary counter*. An n-bit binary counter consists of n flip-flops and can count in binary from 0 through $2^n - 1$.

Counters are available in two categories: ripple counters and synchronous counters. In a ripple counter, a flip-flop output transition serves as a source for triggering other flip-flops. In other words, the C input of some or all flip-flops are triggered, not by the common clock pulses, but rather by the transition that occurs in other flip-flop outputs. In a synchronous counter, the C inputs of all flip-flops receive the common clock. Synchronous counters are presented in the next two sections. Here, we present the binary and BCD ripple counters and explain their operation.

Binary Ripple Counter

A binary ripple counter consists of a series connection of complementing flip-flops, with the output of each flip-flop connected to the C input of the next higher order flip-flop. The flip-flop holding the least significant bit receives the incoming count pulses. A complementing flip-flop can be obtained from a JK flip-flop with the J and K inputs tied together or from a T flip-flop. A third possibility is to use a D flip-flop with the complement output connected to the D input. In this way, the D input is always the complement of the present state, and the next clock pulse will cause the flip-flop to complement. The logic diagram of two 4-bit binary ripple counters is shown in Fig. 6.8. The counter is constructed with complementing flip-flops of the T type in part (a) and D type in part (b). The output of each flip-flop is connected to the C input of the next flip-flop in sequence. The flip-flop holding the least significant bit receives the incoming count pulses. The T inputs of all the flip-flops in (a) are connected to a permanent logic 1, making each flip-flop complement if the signal in its C input goes through a negative transition. The bubble in front of the dynamic indicator symbol next to C indicates that the flip-flops respond to the negative-edge transition of the input. The negative transition occurs when the output of the previous flip-flop to which C is connected goes from 1 to 0.

To understand the operation of the four-bit binary ripple counter, refer to the first nine binary numbers listed in Table 6.4. The count starts with binary 0 and increments by 1 with each count pulse input. After the count of 15, the counter goes back to 0 to repeat the count. The least significant bit, A_0, is complemented with each count pulse input. Every time that A_0 goes from 1 to 0, it complements A_1. Every time that A_1 goes from 1 to 0, it complements A_2. Every time that A_2 goes from 1 to 0, it complements A_3, and so on for any other higher order bits of a ripple counter. For example, consider the transition from count 0011 to 0100. A_0 is complemented with the count pulse. Since A_0 goes from 1 to 0, it triggers A_1 and complements it. As a result, A_1 goes from 1 to 0, which in turn complements A_2, changing it from 0 to 1. A_2 does not trigger A_3, because A_2 produces a positive transition and the flip-flop responds only to negative transitions. Thus, the count from 0011 to 0100 is achieved by changing the bits one at a time, so the

Table 6.4
Binary Count Sequence

A_3	A_2	A_1	A_0
0	0	0	0
0	0	0	1
0	0	1	0
0	0	1	1
0	1	0	0
0	1	0	1
0	1	1	0
0	1	1	1
1	0	0	0

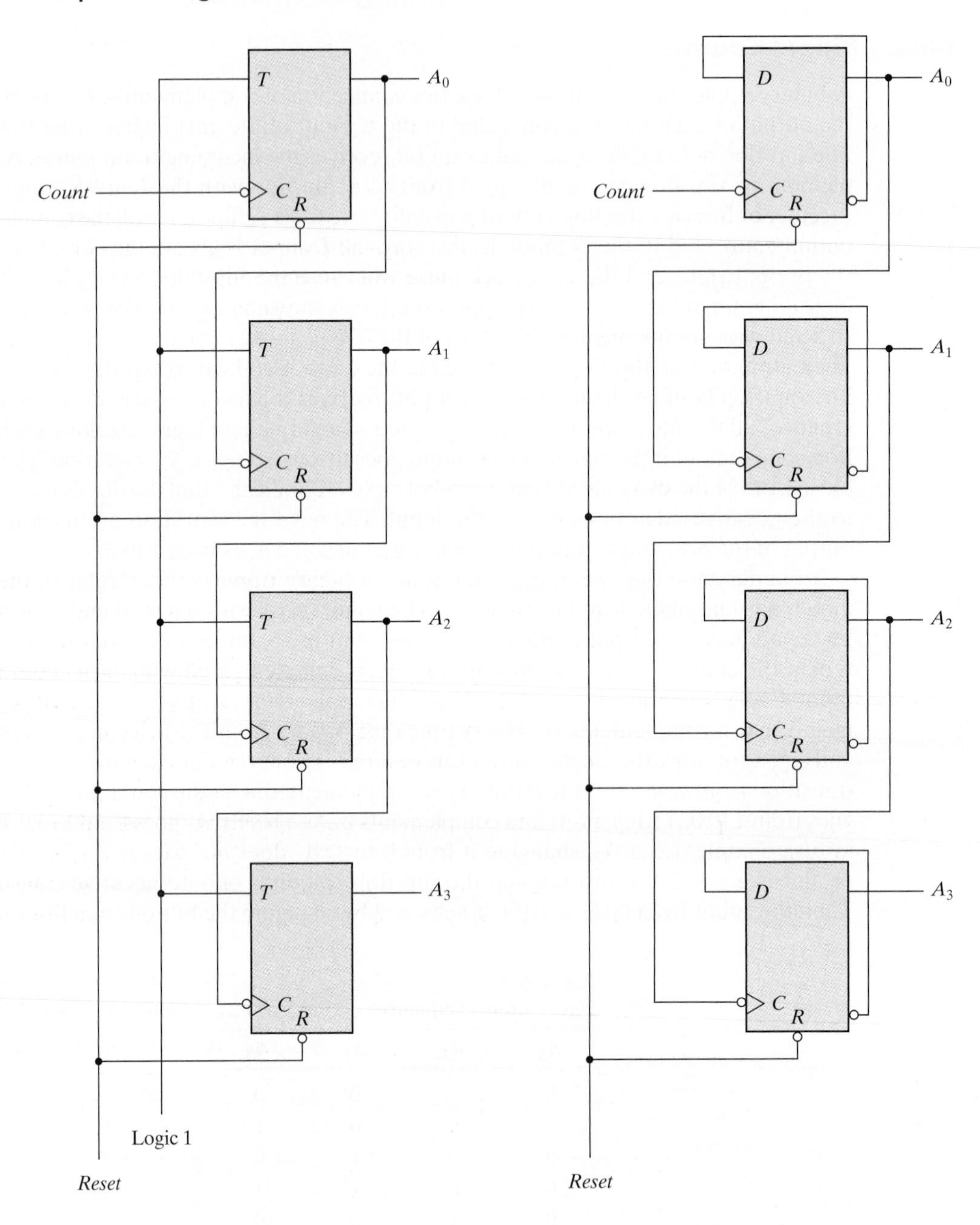

(a) With T flip-flops

(b) With D flip-flops

FIGURE 6.8
Four-bit binary ripple counter

count goes from 0011 to 0010, then to 0000, and finally to 0100. The flip-flops change one at a time in succession, and the signal propagates through the counter in a ripple fashion from one stage to the next.

A binary counter with a reverse count is called a *binary countdown counter*. In a countdown counter, the binary count is decremented by 1 with every input count pulse. The count of a four-bit countdown counter starts from binary 15 and continues to binary counts 14, 13, 12, . . . , 0 and then back to 15. A list of the count sequence of a binary countdown counter shows that the least significant bit is complemented with every count pulse. Any other bit in the sequence is complemented if its previous least significant bit goes from 0 to 1. Therefore, the diagram of a binary countdown counter looks the same as the binary ripple counter in Fig. 6.8, provided that all flip-flops trigger on the positive edge of the clock. (The bubble in the C inputs must be absent.) If negative-edge-triggered flip-flops are used, then the C input of each flip-flop must be connected to the complemented output of the previous flip-flop. Then, when the true output goes from 0 to 1, the complement will go from 1 to 0 and complement the next flip-flop as required.

BCD Ripple Counter

A decimal counter follows a sequence of 10 states and returns to 0 after the count of 9. Such a counter must have at least four flip-flops to represent each decimal digit, since a decimal digit is represented by a binary code with at least four bits. The sequence of states in a decimal counter is dictated by the binary code used to represent a decimal digit. If BCD is used, the sequence of states is as shown in the state diagram of Fig. 6.9. A decimal counter is similar to a binary counter, except that the state after 1001 (the code for decimal digit 9) is 0000 (the code for decimal digit 0).

The logic diagram of a BCD ripple counter using JK flip-flops is shown in Fig. 6.10. The four outputs are designated by the letter symbol Q, with a numeric subscript equal to the binary weight of the corresponding bit in the BCD code. Note that the output of Q_1 is applied to the C inputs of both Q_2 and Q_8 and the output of Q_2 is applied to the C input of Q_4. The J and K inputs are connected either to a permanent 1 signal or to outputs of other flip-flops.

A ripple counter is an asynchronous sequential circuit. Signals that affect the flip-flop transition depend on the way they change from 1 to 0. The operation of the counter can

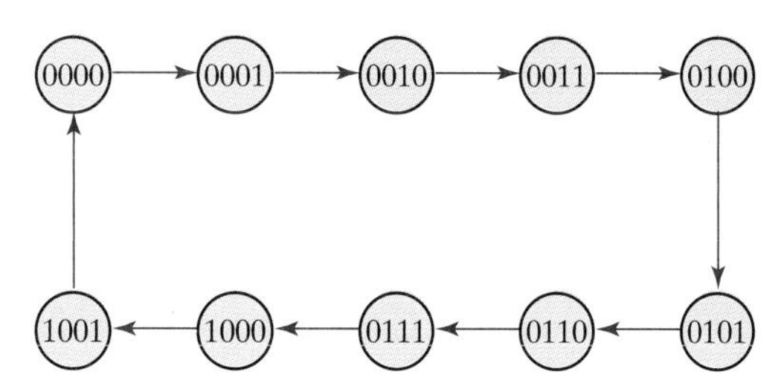

FIGURE 6.9
State diagram of a decimal BCD counter

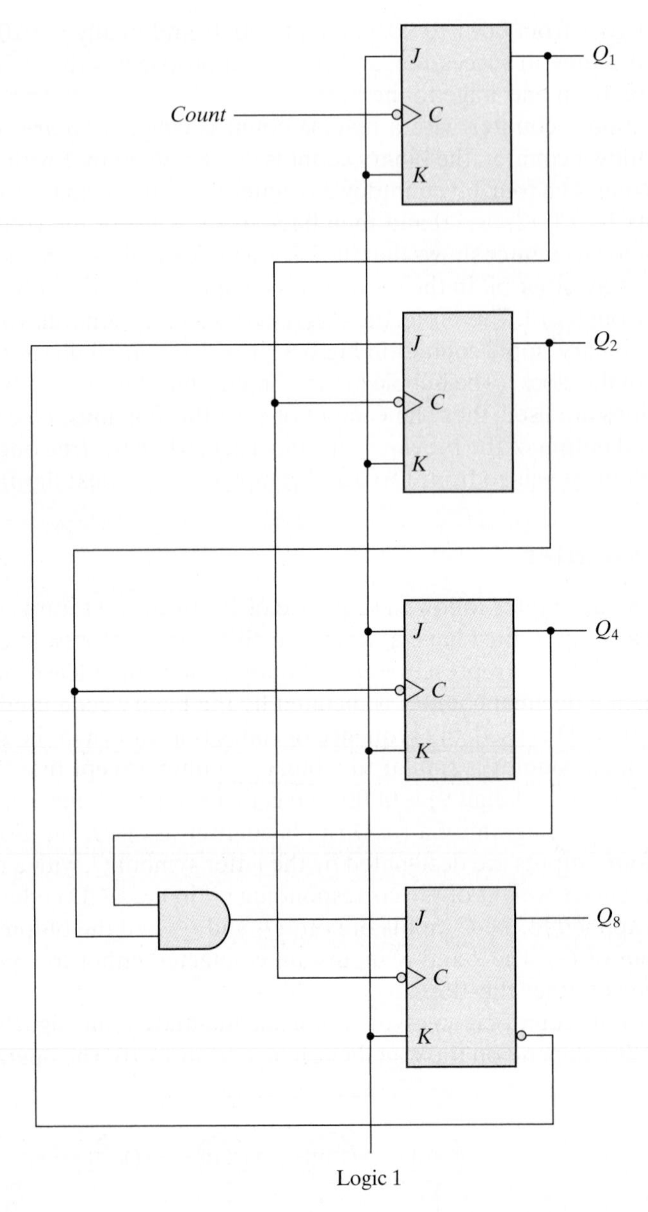

FIGURE 6.10
BCD ripple counter

be explained by a list of conditions for flip-flop transitions. These conditions are derived from the logic diagram and from knowledge of how a *JK* flip-flop operates. Remember that when the C input goes from 1 to 0, the flip-flop is set if $J = 1$, is cleared if $K = 1$, is complemented if $J = K = 1$, and is left unchanged if $J = K = 0$.

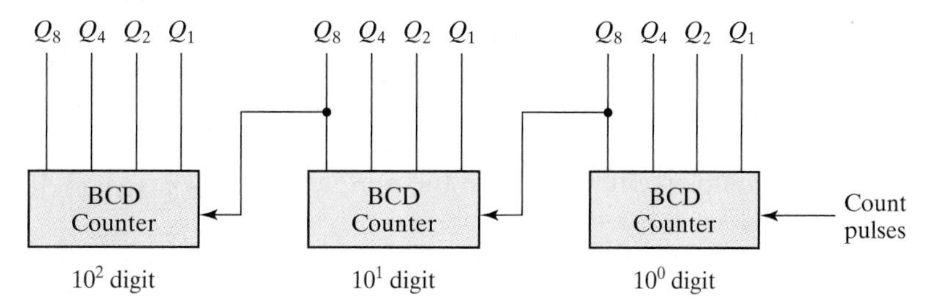

FIGURE 6.11
Block diagram of a three-decade decimal BCD counter

To verify that these conditions result in the sequence required by a BCD ripple counter, it is necessary to verify that the flip-flop transitions indeed follow a sequence of states as specified by the state diagram of Fig. 6.9. Q_1 changes state after each clock pulse. Q_2 complements every time Q_1 goes from 1 to 0, as long as $Q_8 = 0$. When Q_8 becomes 1, Q_2 remains at 0. Q_4 complements every time Q_2 goes from 1 to 0. Q_8 remains at 0 as long as Q_2 or Q_4 is 0. When both Q_2 and Q_4 become 1, Q_8 complements when Q_1 goes from 1 to 0. Q_8 is cleared on the next transition of Q_1.

The BCD counter of Fig. 6.10 is a *decade* counter, since it counts from 0 to 9. To count in decimal from 0 to 99, we need a two-decade counter. To count from 0 to 999, we need a three-decade counter. Multiple decade counters can be constructed by connecting BCD counters in cascade, one for each decade. A three-decade counter is shown in Fig. 6.11. The inputs to the second and third decades come from Q_8 of the previous decade. When Q_8 in one decade goes from 1 to 0, it triggers the count for the next higher order decade while its own decade goes from 9 to 0.

6.4 SYNCHRONOUS COUNTERS

Synchronous counters are different from ripple counters in that clock pulses are applied to the inputs of all flip-flops. A common clock triggers all flip-flops simultaneously, rather than one at a time in succession as in a ripple counter. The decision whether a flip-flop is to be complemented is determined from the values of the data inputs, such as T or J and K at the time of the clock edge. If $T = 0$ or $J = K = 0$, the flip-flop does not change state. If $T = 1$ or $J = K = 1$, the flip-flop complements.

The design procedure for synchronous counters was presented in Section 5.8, and the design of a three-bit binary counter was carried out in conjunction with Fig. 5.31. In this section, we present some typical synchronous counters and explain their operation.

Binary Counter

The design of a synchronous binary counter is so simple that there is no need to go through a sequential logic design process. In a synchronous binary counter, the flip-flop in the least significant position is complemented with every pulse. *A flip-flop in any other position is complemented when all the bits in the lower significant positions are equal to 1.* For example, if the present state of a four-bit counter is $A_3A_2A_1A_0 = 0011$, the next count is 0100. A_0 is always complemented. A_1 is complemented because the present state of $A_0 = 1$. A_2 is complemented because the present state of $A_1A_0 = 11$. However, A_3 is not complemented, because the present state of $A_2A_1A_0 = 011$, which does not give an all-1's condition.

Synchronous binary counters have a regular pattern and can be constructed with complementing flip-flops and gates. The regular pattern can be seen from the four-bit counter depicted in Fig. 6.12. The C inputs of all flip-flops are connected to a common clock. The counter is enabled by *Count_enable*. If the enable input is 0, all J and K inputs are equal to 0 and the clock does not change the state of the counter. The first stage, A_0, has its J and K equal to 1 if the counter is enabled. The other J and K inputs are equal to 1 if all previous least significant stages are equal to 1 and the count is enabled. The chain of AND gates generates the required logic for the J and K inputs in each stage. The counter can be extended to any number of stages, with each stage having an additional flip-flop and an AND gate that gives an output of 1 if all previous flip-flop outputs are 1.

Note that the flip-flops trigger on the positive edge of the clock. The polarity of the clock is not essential here, but it is with the ripple counter. The synchronous counter can be triggered with either the positive or the negative clock edge. The complementing flip-flops in a binary counter can be of either the JK type, the T type, or the D type with XOR gates. The equivalency of the three types is indicated in Fig. 5.13.

Up–Down Binary Counter

A synchronous countdown binary counter goes through the binary states in reverse order, from 1111 down to 0000 and back to 1111 to repeat the count. It is possible to design a countdown counter in the usual manner, but the result is predictable by inspection of the downward binary count. The bit in the least significant position is complemented with each pulse. *A bit in any other position is complemented if all lower significant bits are equal to 0.* For example, the next state after the present state of 0100 is 0011. The least significant bit is always complemented. The second significant bit is complemented because the first bit is 0. The third significant bit is complemented because the first two bits are equal to 0. But the fourth bit does not change, because not all lower significant bits are equal to 0.

A countdown binary counter can be constructed as shown in Fig. 6.12, except that the inputs to the AND gates must come from the complemented outputs, instead of the normal outputs, of the previous flip-flops. The two operations can be combined in one circuit to form a counter capable of counting either up or down. The circuit of an up–down binary counter using T flip-flops is shown in Fig. 6.13. It has an up control input and a down control input. When the up input is 1, the circuit counts up, since the T inputs receive their signals from the values of the previous normal outputs of the flip-flops. When the down input is 1 and the up input is 0, the circuit counts down, since the complemented outputs of the previous flip-flops are applied to the T inputs. When the up and down inputs are both 0, the circuit does not change state and remains

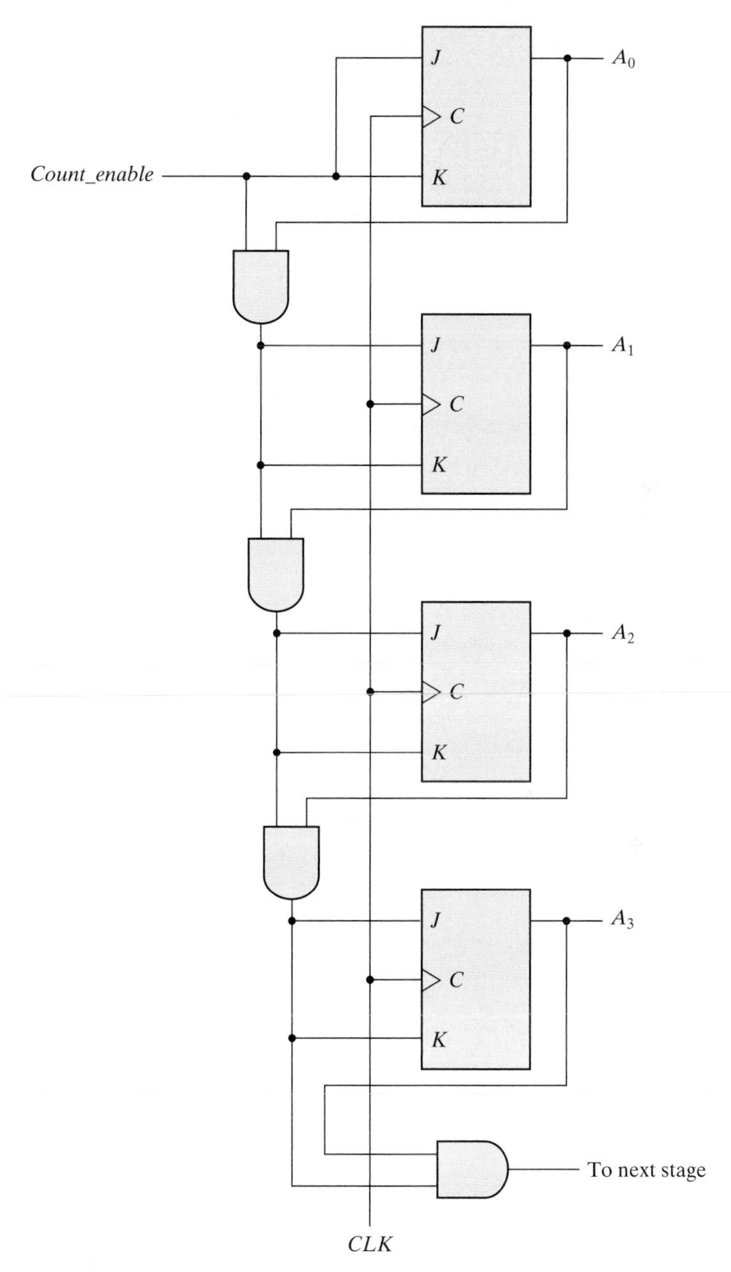

FIGURE 6.12
Four-bit synchronous binary counter

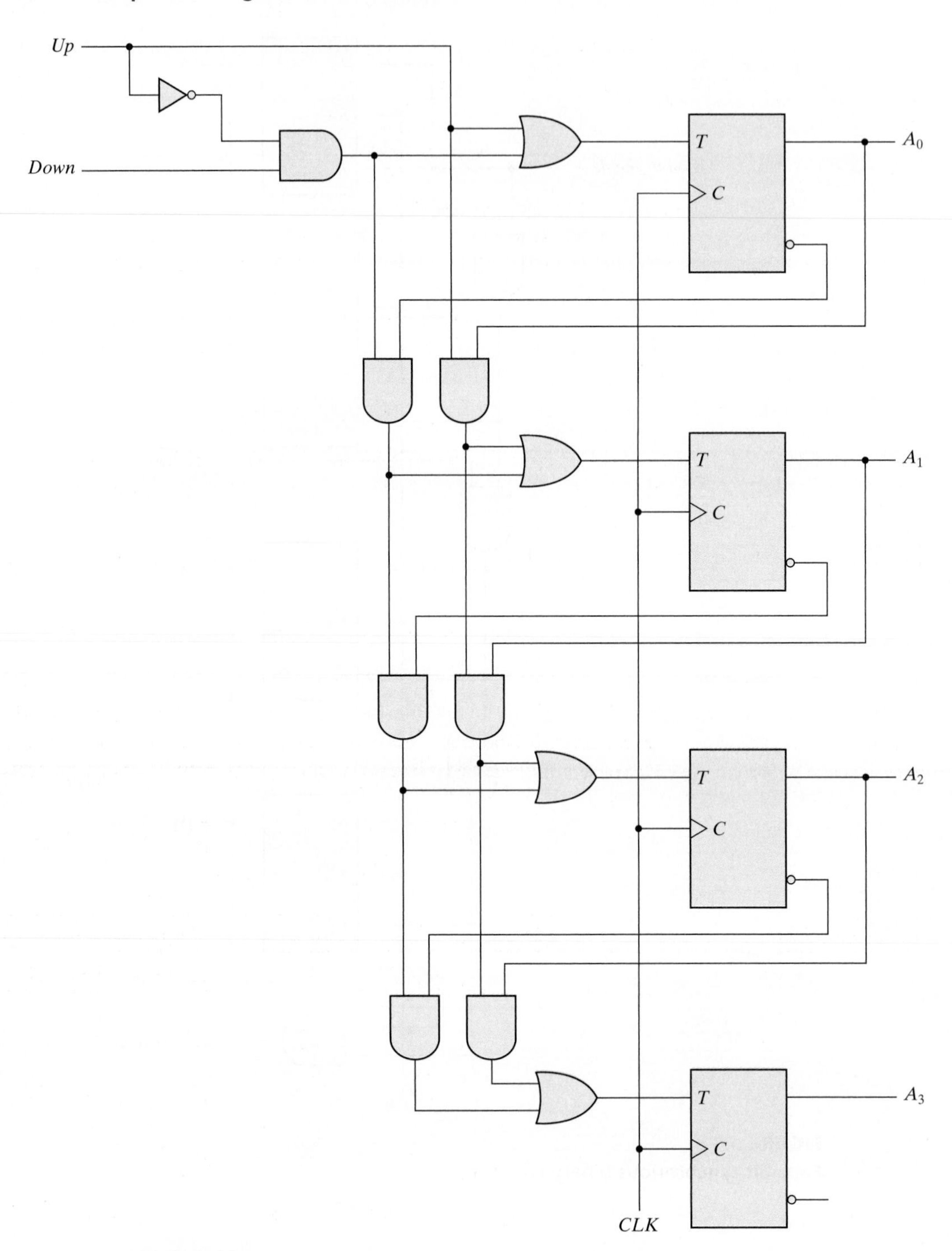

FIGURE 6.13
Four-bit up–down binary counter

in the same count. When the up and down inputs are both 1, the circuit counts up. This set of conditions ensures that only one operation is performed at any given time. Note that the up input has priority over the down input.

BCD Counter

A BCD counter counts in binary-coded decimal from 0000 to 1001 and back to 0000. Because of the return to 0 after a count of 9, a BCD counter does not have a regular pattern, unlike a straight binary count. To derive the circuit of a BCD synchronous counter, it is necessary to go through a sequential circuit design procedure.

The state table of a BCD counter is listed in Table 6.5. The input conditions for the T flip-flops are obtained from the present- and next-state conditions. Also shown in the table is an output y, which is equal to 1 when the present state is 1001. In this way, y can enable the count of the next-higher significant decade while the same pulse switches the present decade from 1001 to 0000.

The flip-flop input equations can be simplified by means of maps. The unused states for minterms 10 to 15 are taken as don't-care terms. The simplified functions are

$$T_{Q1} = 1$$
$$T_{Q2} = Q'_8 Q_1$$
$$T_{Q4} = Q_2 Q_1$$
$$T_{Q8} = Q_8 Q_1 + Q_4 Q_2 Q_1$$
$$y = Q_8 Q_1$$

The circuit can easily be drawn with four T flip-flops, five AND gates, and one OR gate. Synchronous BCD counters can be cascaded to form a counter for decimal numbers of any length. The cascading is done as in Fig. 6.11, except that output y must be connected to the count input of the next-higher significant decade.

Table 6.5
State Table for BCD Counter

Present State				Next State				Output	Flip-Flop Inputs			
Q_8	Q_4	Q_2	Q_1	Q_8	Q_4	Q_2	Q_1	y	TQ_8	TQ_4	TQ_2	TQ_1
0	0	0	0	0	0	0	1	0	0	0	0	1
0	0	0	1	0	0	1	0	0	0	0	1	1
0	0	1	0	0	0	1	1	0	0	0	0	1
0	0	1	1	0	1	0	0	0	0	1	1	1
0	1	0	0	0	1	0	1	0	0	0	0	1
0	1	0	1	0	1	1	0	0	0	0	1	1
0	1	1	0	0	1	1	1	0	0	0	0	1
0	1	1	1	1	0	0	0	0	1	1	1	1
1	0	0	0	1	0	0	1	0	0	0	0	1
1	0	0	1	0	0	0	0	1	1	0	0	1

Binary Counter with Parallel Load

Counters employed in digital systems quite often require a parallel-load capability for transferring an initial binary number into the counter prior to the count operation. Figure 6.14 shows the top-level block diagram symbol and the logic diagram of a four-bit register that has a parallel load capability and can operate as a counter. When equal to 1, the input load control disables the count operation and causes a transfer of data from the four data inputs into the four flip-flops. If both control inputs are 0, clock pulses do not change the state of the register.

The carry output becomes a 1 if all the flip-flops are equal to 1 while the count input is enabled. This is the condition for complementing the flip-flop that holds the next significant bit. The carry output is useful for expanding the counter to more than four bits. The speed of the counter is increased when the carry is generated directly from the outputs of all four flip-flops, because the delay to generate the carry bit is reduced. In going from state 1111 to 0000, only one gate delay occurs, whereas four gate delays occur in the AND gate chain shown in Fig. 6.12. Similarly, each flip-flop is associated with an AND gate that receives all previous flip-flop outputs directly instead of connecting the AND gates in a chain.

The operation of the counter is summarized in Table 6.6. The four control inputs—*Clear*, *CLK*, *Load*, and *Count*—determine the next state. The *Clear* input is asynchronous and, when equal to 0, causes the counter to be cleared regardless of the presence of clock pulses or other inputs. This relationship is indicated in the table by the X entries, which symbolize don't-care conditions for the other inputs. The *Clear* input must be in the 1 state for all other operations. With the *Load* and *Count* inputs both at 0, the outputs do not change, even when clock pulses are applied. A *Load* input of 1 causes a transfer from inputs $I_0 - I_3$ into the register during a positive edge of *CLK*. The input data are loaded into the register regardless of the value of the *Count* input, because the *Count* input is inhibited when the *Load* input is enabled. The *Load* input must be 0 for the *Count* input to control the operation of the counter.

A counter with a parallel load can be used to generate any desired count sequence. Figure 6.15 shows two ways in which a counter with a parallel load is used to generate the BCD count. In each case, the *Count* control is set to 1 to enable the count through the *CLK* input. Also, recall that the *Load* control inhibits the count and that the clear operation is independent of other control inputs.

The AND gate in Fig. 6.15(a) detects the occurrence of state 1001. The counter is initially cleared to 0, and then the *Clear* and *Count* inputs are set to 1, so the counter is active at all times. As long as the output of the AND gate is 0, each positive-edge clock

Table 6.6
Function Table for the Counter of Fig. 6.14

Clear	CLK	Load	Count	Function
0	X	X	X	Clear to 0
1	↑	1	X	Load inputs
1	↑	0	1	Count next binary state
1	↑	0	0	No change

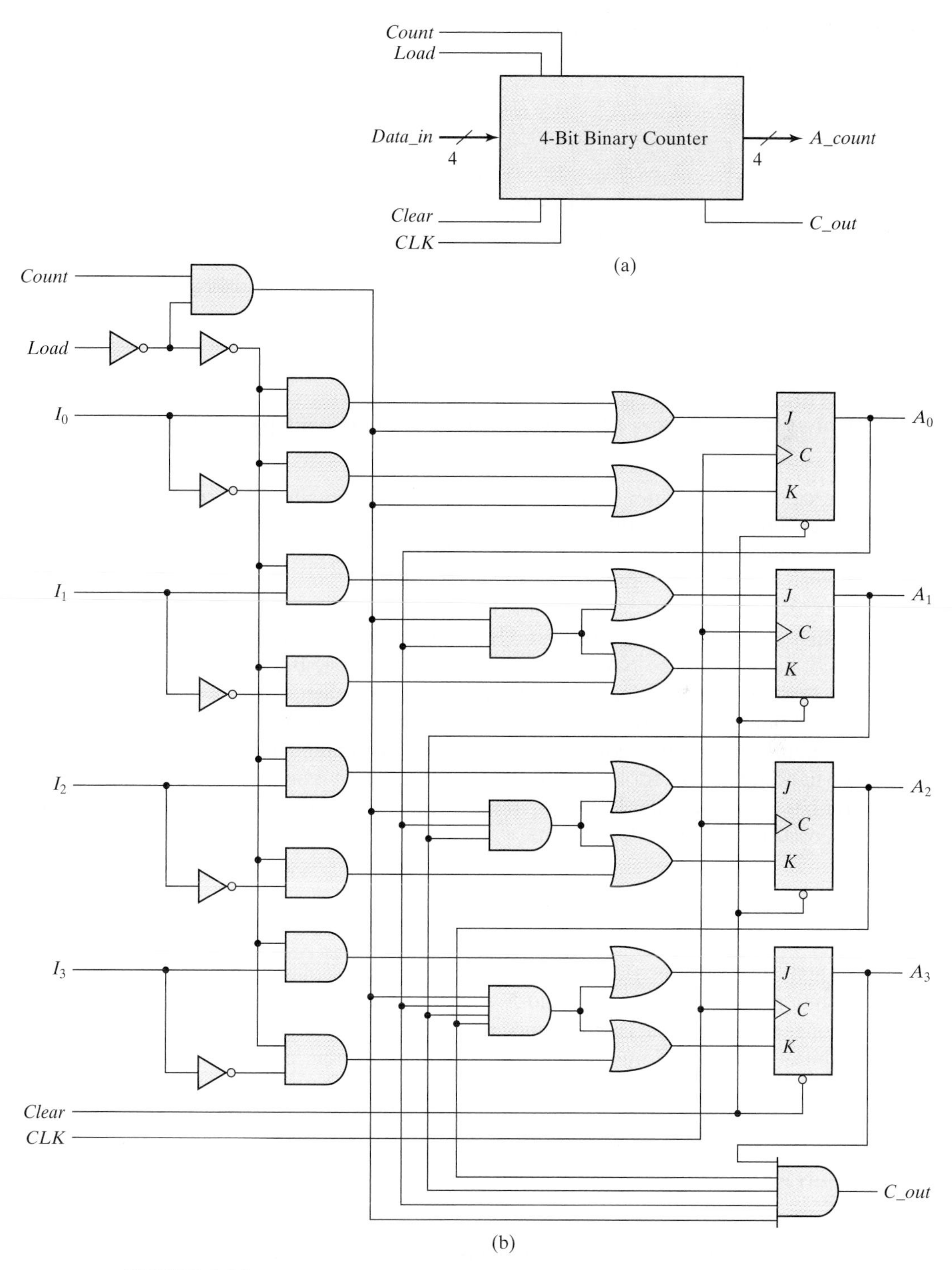

FIGURE 6.14
Four-bit binary counter with parallel load

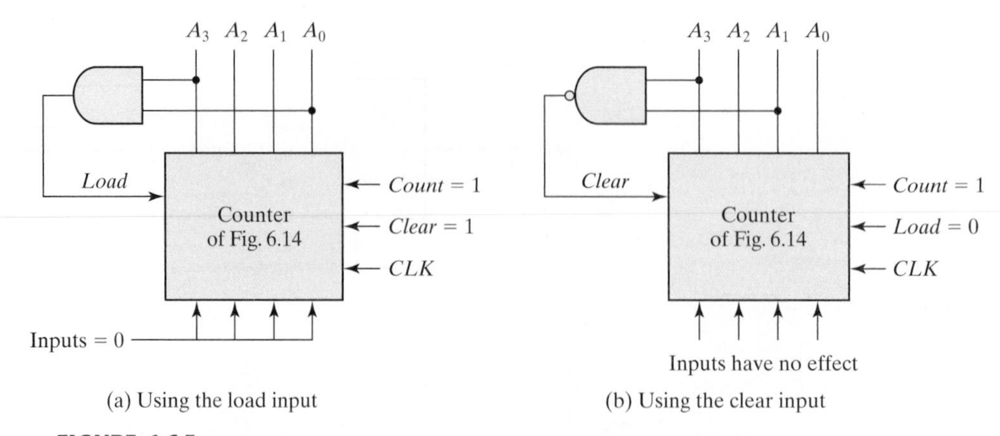

FIGURE 6.15
Two ways to achieve a BCD counter using a counter with parallel load

increments the counter by 1. When the output reaches the count of 1001, both A_0 and A_3 become 1, making the output of the AND gate equal to 1. This condition activates the *Load* input; therefore, on the next clock edge the register does not count, but is loaded from its four inputs. Since all four inputs are connected to logic 0, an all-0's value is loaded into the register following the count of 1001. Thus, the circuit goes through the count from 0000 through 1001 and back to 0000, as is required in a BCD counter.

In Fig. 6.15(b), the NAND gate detects the count of 1010, but as soon as this count occurs, the register is cleared. The count 1010 has no chance of staying on for any appreciable time, because the register goes immediately to 0. A momentary spike occurs in output A_0 as the count goes from 1010 to 1011 and immediately to 0000. The spike may be undesirable, and for that reason, this configuration is not recommended. If the counter has a synchronous clear input, it is possible to clear the counter with the clock after an occurrence of the 1001 count.

6.5 OTHER COUNTERS

Counters can be designed to generate any desired sequence of states. A divide-by-N counter (also known as a modulo-N counter) is a counter that goes through a repeated sequence of N states. The sequence may follow the binary count or may be any other arbitrary sequence. Counters are used to generate timing signals to control the sequence of operations in a digital system. Counters can also be constructed by means of shift registers. In this section, we present a few examples of nonbinary counters.

Counter with Unused States

A circuit with n flip-flops has 2^n binary states. There are occasions when a sequential circuit uses fewer than this maximum possible number of states. States that are not used in specifying the sequential circuit are not listed in the state table. In simplifying the input equations, the unused states may be treated as don't-care conditions or may be assigned specific next states. It is important to realize that once the circuit is designed and constructed, outside interference during its operation may cause the circuit to enter one of the unused states. In that case, it is necessary to ensure that the circuit eventually goes into one of the valid states so that it can resume normal operation. Otherwise, if the sequential circuit circulates among unused states, there will be no way to bring it back to its intended sequence of state transitions. If the unused states are treated as don't-care conditions, then once the circuit is designed, it must be investigated to determine the effect of the unused states. The next state from an unused state can be determined from the analysis of the circuit after it is designed.

As an illustration, consider the counter specified in Table 6.7. The count has a repeated sequence of six states, with flip-flops B and C repeating the binary count 00, 01, 10, and flip-flop A alternating between 0 and 1 every three counts. The count sequence of the counter is not straight binary, and two states, 011 and 111, are not included in the count. The choice of JK flip-flops results in the flip-flop input conditions listed in the table. Inputs K_B and K_C have only 1's and X's in their columns, so these inputs are always equal to 1. The other flip-flop input equations can be simplified by using minterms 3 and 7 as don't-care conditions. The simplified equations are

$$J_A = B \qquad K_A = B$$
$$J_B = C \qquad K_B = 1$$
$$J_C = B' \qquad K_C = 1$$

The logic diagram of the counter is shown in Fig. 6.16(a). Since there are two unused states, we analyze the circuit to determine their effect. If the circuit happens to be in state 011 because of an error signal, the circuit goes to state 100 after the application of a clock pulse. This action may be determined from an inspection of the logic diagram by noting that when $B = 1$, the next clock edge complements A and clears C to 0, and when $C = 1$, the next clock edge complements B. In a similar manner, we can evaluate the next state from present state 111 to be 000.

Table 6.7
State Table for Counter

Present State			Next State			Flip-Flop Inputs					
A	**B**	**C**	**A**	**B**	**C**	J_A	K_A	J_B	K_B	J_C	K_C
0	0	0	0	0	1	0	X	0	X	1	X
0	0	1	0	1	0	0	X	1	X	X	1
0	1	0	1	0	0	1	X	X	1	0	X
1	0	0	1	0	1	X	0	0	X	1	X
1	0	1	1	1	0	X	0	1	X	X	1
1	1	0	0	0	0	X	1	X	1	0	X

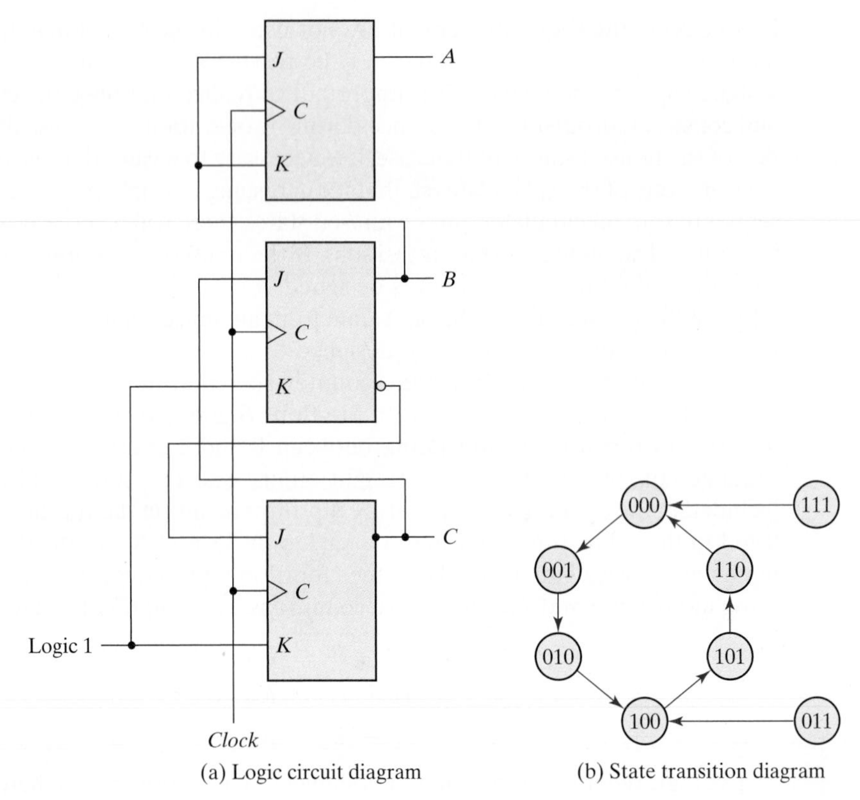

FIGURE 6.16
Counter with unused states

The state diagram including the effect of the unused states is shown in Fig. 6.16(b). If the circuit ever goes to one of the unused states because of outside interference, the next count pulse transfers it to one of the valid states and the circuit continues to count correctly. Thus, the counter is self-correcting. In a self-correcting counter, if the counter happens to be in one of the unused states, it eventually reaches the normal count sequence after one or more clock pulses. An alternative design could use additional logic to direct every unused state to a specific next state.

Ring Counter

Timing signals that control the sequence of operations in a digital system can be generated by a shift register or by a counter with a decoder. A *ring counter* is a circular shift register with only one flip-flop being set at any particular time; all others are cleared. The single bit is shifted from one flip-flop to the next to produce the sequence of timing signals. Figure 6.17(a) shows a four-bit shift register connected as a ring counter. The initial value of the register is 1000 and requires Preset/Clear flip-flops. The single bit is

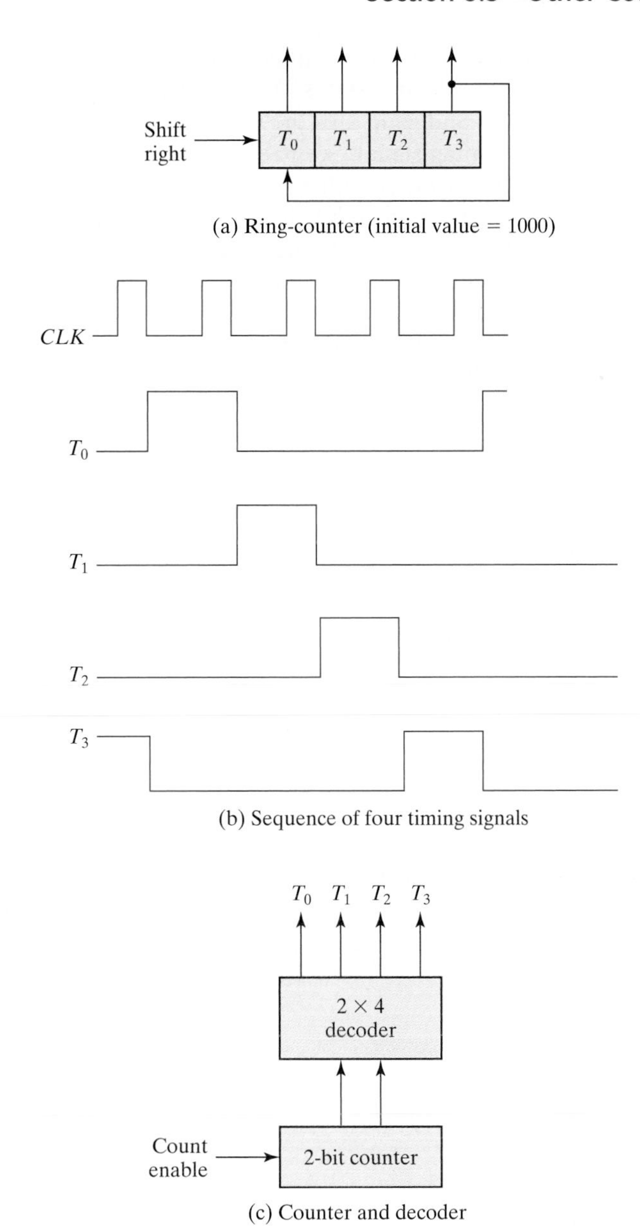

FIGURE 6.17
Generation of timing signals

shifted right with every clock pulse and circulates back from T_3 to T_0. Each flip-flop is in the 1 state once every four clock cycles and produces one of the four timing signals shown in Fig. 6.17(b). Each output becomes a 1 after the negative-edge transition of a clock pulse and remains 1 during the next clock cycle.

For an alternative design, the timing signals can be generated by a two-bit counter that goes through four distinct states. The decoder shown in Fig. 6.17(c) decodes the four states of the counter and generates the required sequence of timing signals.

To generate 2^n timing signals, we need either a shift register with 2^n flip-flops or an n-bit binary counter together with an n-to-2^n-line decoder. For example, 16 timing signals can be generated with a 16-bit shift register connected as a ring counter or with a 4-bit binary counter and a 4-to-16-line decoder. In the first case, we need 16 flip-flops. In the second, we need 4 flip-flops and 16 four-input AND gates for the decoder. It is also possible to generate the timing signals with a combination of a shift register and a decoder. That way, the number of flip-flops is less than that in a ring counter, and the decoder requires only two-input gates. This combination is called a *Johnson counter.*

Johnson Counter

A k-bit ring counter circulates a single bit among the flip-flops to provide k distinguishable states. The number of states can be doubled if the shift register is connected as a *switch-tail* ring counter. A switch-tail ring counter is a circular shift register with the complemented output of the last flip-flop connected to the input of the first flip-flop. Figure 6.18(a) shows such a shift register. The circular connection is made from the

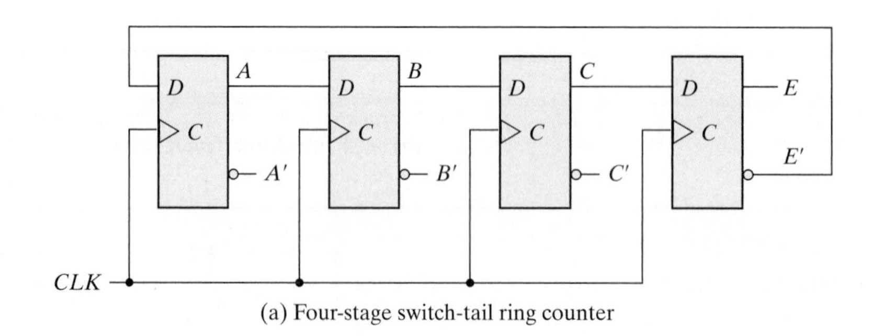

(a) Four-stage switch-tail ring counter

Sequence number	Flip-flop outputs A	B	C	E	AND gate required for output
1	0	0	0	0	$A'E'$
2	1	0	0	0	AB'
3	1	1	0	0	BC'
4	1	1	1	0	CE'
5	1	1	1	1	AE
6	0	1	1	1	$A'B$
7	0	0	1	1	$B'C$
8	0	0	0	1	$C'E$

(b) Count sequence and required decoding

FIGURE 6.18
Construction of a Johnson counter

complemented output of the rightmost flip-flop to the input of the leftmost flip-flop. The register shifts its contents once to the right with every clock pulse, and at the same time, the complemented value of the E flip-flop is transferred into the A flip-flop. Starting from a cleared state, the switch-tail ring counter goes through a sequence of eight states, as listed in Fig. 6.18(b). In general, a k-bit switch-tail ring counter will go through a sequence of $2k$ states. Starting from all 0's, each shift operation inserts 1's from the left until the register is filled with all 1's. In the next sequences, 0's are inserted from the left until the register is again filled with all 0's.

A Johnson counter is a k-bit switch-tail ring counter with $2k$ decoding gates to provide outputs for $2k$ timing signals. The decoding gates are not shown in Fig. 6.18, but are specified in the last column of the table. The eight AND gates listed in the table, when connected to the circuit, will complete the construction of the Johnson counter. Since each gate is enabled during one particular state sequence, the outputs of the gates generate eight timing signals in succession.

The decoding of a k-bit switch-tail ring counter to obtain $2k$ timing signals follows a regular pattern. The all-0's state is decoded by taking the complement of the two extreme flip-flop outputs. The all-1's state is decoded by taking the normal outputs of the two extreme flip-flops. All other states are decoded from an adjacent 1, 0 or 0, 1 pattern in the sequence. For example, sequence 7 has an adjacent 0, 1 pattern in flip-flops B and C. The decoded output is then obtained by taking the complement of B and the normal output of C, or $B'C$.

One disadvantage of the circuit in Fig. 6.18(a) is that if it finds itself in an unused state, it will persist in moving from one invalid state to another and never find its way to a valid state. The difficulty can be corrected by modifying the circuit to avoid this undesirable condition. One correcting procedure is to disconnect the output from flip-flop B that goes to the D input of flip-flop C and instead enable the input of flip-flop C by the function

$$D_C = (A + C)B$$

where D_C is the flip-flop input equation for the D input of flip-flop C.

Johnson counters can be constructed for any number of timing sequences. The number of flip-flops needed is one-half the number of timing signals. The number of decoding gates is equal to the number of timing signals, and only two-input gates are needed.

6.6 HDL FOR REGISTERS AND COUNTERS

Registers and counters can be described in Verilog at either the behavioral or the structural level. Behavioral modeling describes only the operations of the register, as prescribed by a function table, without a preconceived structure. A structural-level description shows the circuit in terms of a collection of components such as gates, flip-flops, and multiplexers. The various components are instantiated to form a hierarchical description of the design similar to a representation of a multilevel logic diagram. The examples in this section will illustrate both types of descriptions. Both are useful. When a machine is complex, a hierarchical description creates a physical partition of the machine into simpler and more easily described units.

Shift Register

The universal shift register presented in Section 6.2 is a bidirectional shift register with a parallel load. The four clocked operations that are performed with the register are specified in Table 6.6. The register also can be cleared asynchronously. Our chosen name for a behavioral description of the four-bit universal shift register shown in Fig. 6.7(a), the name *Shift_Register_4_beh*, signifies the behavioral model of the internal detail of the top-level block diagram symbol and distinguishes that model from a structural one. The behavioral model is presented in HDL Example 6.1, and the structural model is given in HDL Example 6.2. The top-level block diagram symbol in Fig. 6.7(a) indicates that the four-bit universal shift register has two selection inputs (*s1*, *s0*), two serial inputs (*shift_left*, *shift_right*), for controlling the shift register, two serial datapath inputs (MSB_in and LSB_in), a four-bit parallel input (*I_par*), and a four-bit parallel output (*A_par*). The elements of vector *I_par[3: 0]* correspond to the bits $I_3, \ldots, I_0$ in Fig. 6.7, and similarly for *A_par[3: 0]*. The **always** block describes the five operations that can be performed with the register. The *Clear* input clears the register asynchronously with an active-low signal. *Clear* must be high for the register to respond to the positive edge of the clock. The four clocked operations of the register are determined from the values of the two select inputs in the **case** statement. (*s1* and *s0* are concatenated into a two-bit vector and are used as the expression argument of the **case** statement.) The shifting operation is specified by the concatenation of the serial input and three bits of the register. For example, the statement

```
A_par <= {MSB_in, A_par [3: 1]}
```

specifies a concatenation of the serial data input for a right shift operation (*MSB_in*) with bits *A_par[3: 1] of the output data bus*. A reference to a contiguous range of bits within a vector is referred to as a *part select*. The four-bit result of the concatenation is transferred to register *A_par [3: 0]* when the clock pulse triggers the operation. This transfer produces a shift-right operation and updates the register with new information. The shift operation overwrites the contents of *A_par[0]* with the contents of *A_par[1]*. Note that only the functionality of the circuit has been described, irrespective of any particular hardware. A synthesis tool would create a netlist of ASIC cells to implement the shift register in the structure of Fig. 6.7(b).

HDL Example 6.1 (Universal Shift Register-Behavioral Model)

```
// Behavioral description of a 4-bit universal shift register
// Fig. 6.7 and Table 6.3
module Shift_Register_4_beh (                          // V2001, 2005
 output reg        [3: 0]    A_par,                    // Register output
 input             [3: 0]    I_par,                    // Parallel input
 input                       s1, s0,                   // Select inputs
                             MSB_in, LSB_in,           // Serial inputs
                             CLK, Clear_b              // Clock and Clear
);
```

```
always @ (posedge CLK, negedge Clear_b)   // V2001, 2005
  if (Clear_b == 0) A_par <= 4'b0000;
  else
   case ({s1, s0})
     2'b00: A_par <= A_par;                       // No change
     2'b01: A_par <= {MSB_in, A_par[3: 1]};       // Shift right
     2'b10: A_par <= {A_par[2: 0], LSB_in};       // Shift left
     2'b11: A_par <= I_par;                       // Parallel load of input
   endcase
endmodule
```

Variables of type **reg** retain their value until they are assigned a new value by an assignment statement. Consider the following alternative **case** statement for the shift register model:

```
case ({s1, s0})
  // 2'b00: A_par <= A_par;                       // No change
  2'b01: A_par <= {MSB_in, A_par [3: 1]};        // Shift right
  2'b10: A_par <= {A_par [2: 0], LSB_in};        // Shift left
  2'b11: A_par <= I_par;                         // Parallel load of input
endcase
```

Without the case item $2'b00$, the **case** statement would not find a match between $\{s1, s0\}$ and the case items, so register *A_par* would be left unchanged.

A structural model of the universal shift register can be described by referring to the logic diagram of Fig. 6.7(b). The diagram shows that the register has four multiplexers and four *D* flip-flops. A mux and flip-flop together are modeled as a stage of the shift register. The stage is a structural model, too, with an instantiation and interconnection of a module for a mux and another for a *D* flip-flop. For simplicity, the lowest-level modules of the structure are behavioral models of the multiplexer and flip-flop. Attention must be paid to the details of connecting the stages correctly. The structural description of the register is shown in HDL Example 6.2. The top-level module declares the inputs and outputs and then instantiates four copies of a stage of the register. The four instantiations specify the interconnections between the four stages and provide the detailed construction of the register as specified in the logic diagram. The behavioral description of the flip-flop uses a single edge-sensitive cyclic behavior (an **always** block). The assignment statements use the nonblocking assignment operator (<=) the model of the mux employs a single level-sensitive behavior, and the assignments use the blocking assignment operator (=).

HDL Example 6.2 (Universal Shift Register-Structural Model)

```
// Structural description of a 4-bit universal shift register (see Fig. 6.7)
module Shift_Register_4_str (                    // V2001, 2005
 output [3: 0] A_par,                            // Parallel output
 input [3: 0]  I_par,                            // Parallel input
```

```
 input        s1, s0,                          // Mode select
 input        MSB_in, LSB_in, CLK, Clear_b     // Serial inputs, clock, clear
);

// bus for mode control
 assign  [1:0]  select = {s1, s0};

// Instantiate the four stages
 stage ST0 (A_par[0], A_par[1], LSB_in, I_par[0], A_par[0], select, CLK, Clear_b);
 stage ST1 (A_par[1], A_par[2], A_par[0], I_par[1], A_par[1], select, CLK, Clear_b);
 stage ST2 (A_par[2], A_par[3], A_par[1], I_par[2], A_par[2], select, CLK, Clear_b);
 stage ST3 (A_par[3], MSB_in, A_par[2], I_par[3], A_par[3], select, CLK, Clear_b);
endmodule

// One stage of shift register
module stage (i0, i1, i2, i3, Q, select, CLK, Clr_b);
 input        i0,          // circulation bit selection
              i1,          // data from left neighbor or serial input for shift-right
              i2,          // data from right neighbor or serial input for shift-left
              i3;          // data from parallel input
 output       Q;
 input [1: 0] select;      // stage mode control bus
 input        CLK, Clr_b;  // Clock, Clear for flip-flops
 wire         mux_out;

// instantiate mux and flip-flop
 Mux_4_x_1  M0     (mux_out, i0, i1, i2, i3, select);
 D_flip_flop  M1   (Q, mux_out, CLK, Clr_b);
endmodule

// 4x1 multiplexer            // behavioral model
module Mux_4_x_1 (mux_out, i0, i1, i2, i3, select);
 output       mux_out;
 input        i0, i1, i2, i3;
 input [1: 0] select;
 reg          mux_out;
 always @ (select, i0, i1, i2, i3)
  case (select)
   2'b00:     mux_out = i0;
   2'b01:     mux_out = i1;
   2'b10:     mux_out = i2;
   2'b11:     mux_out = i3;
  endcase
endmodule
```

```
// Behavioral model of D flip-flop
module D_flip_flop (Q, D, CLK, Clr_b);
 output      Q;
 input       D, CLK, Clr;
 reg         Q;

 always @ (posedge CLK, negedge Clr_b)
  if (!Clr_b) Q <= 1'b0; else Q <= D;
endmodule
```

The above examples presented two descriptions of a universal shift register to illustrate the different styles for modeling a digital circuit. A simulation should verify that the models have the same functionality. In practice, a designer develops only the behavioral model, which is then synthesized. The function of the synthesized circuit can be compared with the behavioral description from which it was compiled. Eliminating the need for the designer to develop a structural model produces a huge improvement in the efficiency of the design process.

Synchronous Counter

HDL Example 6.3 presents *Binary_Counter_4_Par_Load,* a behavioral model of the synchronous counter with a parallel load from Fig. 6.14. *Count*, *Load*, *CLK*, and *Clear_b* are inputs that determine the operation of the counter according to the function specified in Table 6.6. The counter has four data inputs, four data outputs, and a carry output. The internal data lines (*I3*, *I2*, *I1*, *I0*) are bundled as *Data_in[3: 0]* in the behavioral model. Likewise, the register that holds the bits of the count (*A3*, *A2*, *A1*, *A0*) is *A_count[3: 0]*. It is good practice to have identifiers in the HDL model of a circuit correspond exactly to those in the documentation of the model. That is not always feasible, however, if the circuit-level identifiers are those found in a handbook, for they are often short and cryptic and do not exploit the text that is available with an HDL. The top-level block diagram symbol in Fig. 6.14(a) serves as an interface between the names used in a circuit diagram and the expressive names that can be used in the HDL model. The carry output *C_out* is generated by a combinational circuit and is specified with an **assign** statement. $C_out = 1$when the count reaches 15 and the counter is in the count state. Thus, $C_out = 1$if $Count = 1$, $Load = 0$, and $A = 1111$; otherwise $C\_out = 0$. The **always** block specifies the operation to be performed in the register, depending on the values of *Clear_b, Load,* and *Count*. A 0 (active-low signal) at *Clear_b* resets *A* to 0. Otherwise, if $Clear\_b = 1$, one out of three operations is triggered by the positive edge of the clock. The **if**, **else if**, and **else** statements establish a precedence among the control signals *Clear*, *Load*, and *Count* corresponding to the specification in Table 6.6. *Clear_b* overrides *Load* and *Count*; *Load* overrides *Count*. A synthesis tool will produce the circuit of Fig. 6.14(b) from the behavioral model.

HDL Example 6.3 (Synchronous Counter)

```
// Four-bit binary counter with parallel load (V2001, 2005)
// See Figure 6.14 and Table 6.6
module Binary_Counter_4_Par_Load (
  output reg [3: 0]          A_count,    // Data output
  output                     C_out,      // Output carry
  input [3: 0]               Data_in,    // Data input
  input                      Count,      // Active high to count
                             Load,       // Active high to load
                             CLK,        // Positive-edge sensitive
                             Clear_b     // Active low
);
assign C_out = Count && (~Load) && (A_count == 4'b1111);
always @ (posedge CLK, negedge Clear_b)
   if (~Clear_b)             A_count <= 4'b0000;
   else if (Load)            A_count <= Data_in;
   else if (Count)           A_count <= A_count + 1'b1;
   else                      A_count <= A_count;  // redundant statement
endmodule
```

Ripple Counter

The structural description of a ripple counter is shown in HDL Example 6.4. The first module instantiates four internally complementing flip-flops defined in the second module as *Comp_D_flip_flop (Q, CLK, Reset)*. The clock (input *CLK*) of the first flip-flop is connected to the external control signal *Count*. (*Count* replaces *CLK* in the port list of instance *F0*.) The clock input of the second flip-flop is connected to the output of the first. (*A0* replaces *CLK* in instance *F1*.) Similarly, the clock of each of the other flip-flops is connected to the output of the previous flip-flop. In this way, the flip-flops are chained together to create a ripple counter as shown in Fig. 6.8(b).

The second module describes a complementing flip-flop with delay. The circuit of a complementing flip-flop is constructed by connecting the complement output to the D input. A reset input is included with the flip-flop in order to be able to initialize the counter; otherwise the simulator would assign the unknown value (x) to the output of the flip-flop and produce useless results. The flip-flop is assigned a delay of two time units from the time that the clock is applied to the time that the flip-flop complements its output. The delay is specified by the statement $Q <= \#2 \sim Q$. Notice that the delay operator is placed to the right of the nonblocking assignment operator. This form of delay, called *intra-assignment delay*, has the effect of postponing the assignment of the complemented value of Q to Q. The effect of modeling the delay will be apparent in the simulation results. This style of modeling might be useful in simulation, but it is to be avoided when the model is to be synthesized. The results of synthesis depend on the ASIC cell library that is accessed by the tool, not on any propagation delays that might appear within the model that is to be synthesized.

HDL Example 6.4 (Ripple Counter)

```
// Ripple counter (See Fig. 6.8(b))
'timescale 1ns / 100 ps
module Ripple_Counter_4bit (A3, A2, A1, A0, Count, Reset);
  output A3, A2, A1, A0;
  input Count, Reset;
// Instantiate complementing flip-flop
  Comp_D_flip_flop F0 (A0, Count, Reset);
  Comp_D_flip_flop F1 (A1, A0, Reset);
  Comp_D_flip_flop F2 (A2, A1, Reset);
  Comp_D_flip_flop F3 (A3, A2, Reset);
endmodule
// Complementing flip-flop with delay
// Input to D flip-flop = Q'
module Comp_D_flip_flop (Q, CLK, Reset);
  output  Q;
  input   CLK, Reset;
  reg     Q;
  always @ (negedge CLK, posedge Reset)
  if (Reset) Q <= 1'b0;
  else Q <= #2 ~Q;             // intra-assignment delay
endmodule
// Stimulus for testing ripple counter
module t_Ripple_Counter_4bit;
  reg     Count;
  reg     Reset;
  wire    A0, A1, A2, A3;
// Instantiate ripple counter
  Ripple_Counter_4bit M0 (A3, A2, A1, A0, Count, Reset);
  always
  #5 Count = ~Count;
  initial
   begin
  Count = 1'b0;
  Reset = 1'b1;
  #4 Reset = 1'b0;
   end

  initial #170 $finish;

endmodule
```

The test bench module in HDL Example 6.4 provides a stimulus for simulating and verifying the functionality of the ripple counter. The **always** statement generates a free-running clock with a cycle of 10 time units. The flip-flops trigger on the negative edge of the clock, which occurs at $t = 10, 20, 30$, and every 10 time units thereafter. The waveforms obtained from this simulation are shown in Fig. 6.19. The control signal *Count* goes negative every 10 ns. *A0* is complemented with each negative edge of *Count*, but is delayed by 2 ns. Each flip-flop is complemented when its previous flip-flop goes from 1 to 0. After $t = 80$ ns, all four flip-flops complement because the counter goes from 0111 to 1000. Each output is delayed by 2 ns, and because of that, *A*3 goes from 0 to 1 at $t = 88$ ns and from 1 to 0 at 168 ns. Notice how the propagation delays accumulate to the last bit of the counter, resulting in very slow counter action. This limits the practical utility of the counter.

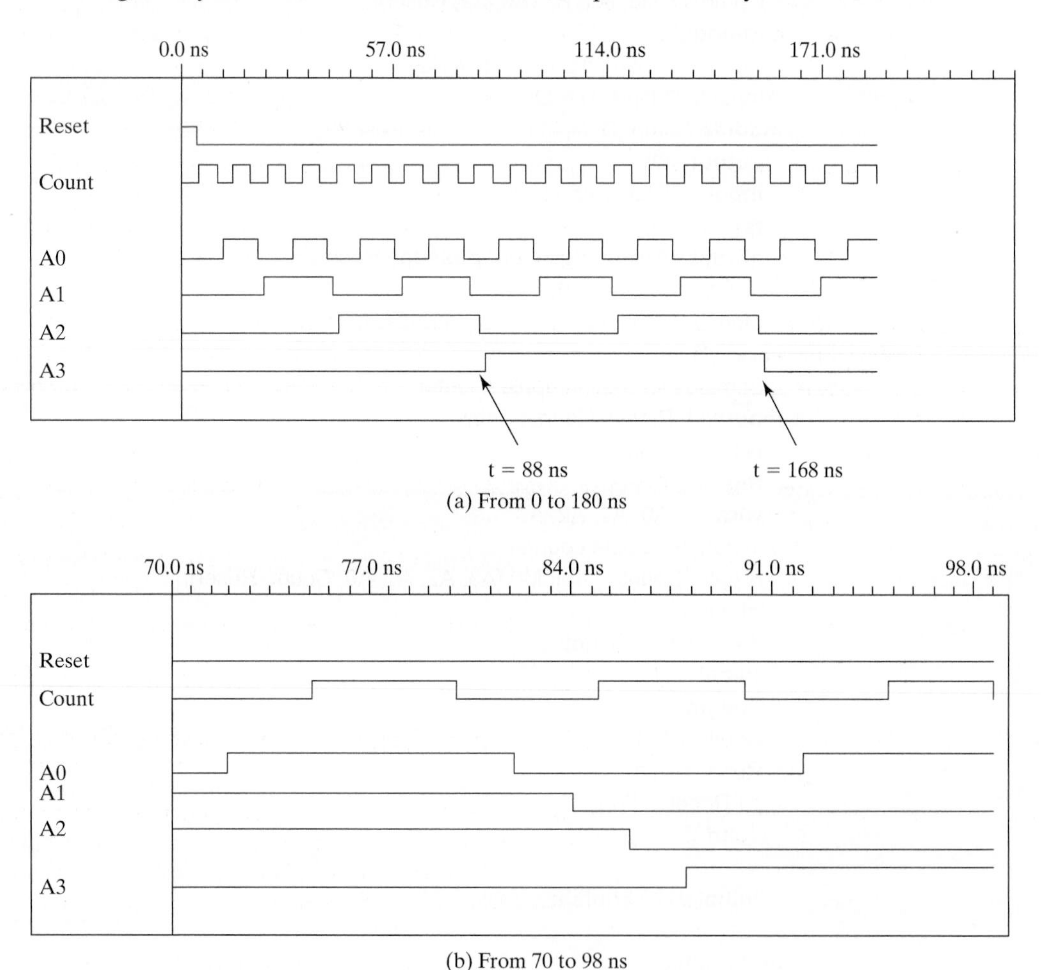

FIGURE 6.19
Simulation output of HDL Example 6.4

PROBLEMS

(Answers to problems marked with * appear at the end of the book. Where appropriate, a logic design and its related HDL modeling problem are cross-referenced.)
Note: For each problem that requires writing and verifying a Verilog description, a test plan is to be written to identify which functional features are to be tested during the simulation and how they will be tested. For example, a reset on the fly could be tested by asserting the reset signal while the simulated machine is in a state other than the reset state. The test plan is to guide the development of a test bench that will implement the plan. Simulate the model using the test bench and verify that the behavior is correct. If synthesis tools and an ASIC cell library or a field programmable gate array (FPGA) tool suite are available, the Verilog descriptions developed for Problems 6.34–6.51 can be assigned as synthesis exercises. The gate-level circuit produced by the synthesis tools should be simulated and compared to the simulation results for the pre-synthesis model.

In some of the HDL problems, there may be a need to deal with the issue of unused states (see the discussion of the **default case** item preceding HDL Example 4.8 in Chapter 4).

6.1 Include a 2-input NOR gate in the register of Fig. 6.1 and connect the gate output to the *C* inputs of all the flip-flops. One input of the NOR gate receives the clock pulses from the clock generator, and the other input of the NOR gate provides a parallel load control. Explain the operation of the modified register. Explain why this circuit might have operational problems.

6.2 Include a synchronous clear input to the register of Fig. 6.2. The modified register will have a parallel load capability and a synchronous clear capability. The register is cleared synchronously when the clock goes through a positive transition and the clear input is equal to 1. (HDL—see Problem 6.35(a), (b).)

6.3 What is the difference between serial and parallel transfer? Explain how to convert serial data to parallel and parallel data to serial. What type of register is needed?

6.4* The contents of a four-bit register is initially 1100. The register is shifted six times to the right with the serial input being 110010. What is the content of the register after each shift?

6.5 The four-bit universal shift register shown in Fig. 6.7 is enclosed within one IC component package. (HDL—see Problem 6.52.)

(a) Draw a block diagram of the IC showing all inputs and outputs. Include two pins for the power supply.

(b) Draw a block diagram using two of these ICs to produce an eight-bit universal shift register.

6.6 Design a four-bit shift register with parallel load using *D* flip-flops. There are two control inputs: *shift* and *load*. When *shift* = 1, the content of the register is shifted by one position. New data are transferred into the register when *load* = 1 and *shift* = 0. If both control inputs are equal to 0, the content of the register does not change. (HDL—see Problem 6.35(c), (d).)

6.7 Draw the logic diagram of a four-bit register with four *D* flip-flops and four 4×1 multiplexers with mode selection inputs s_1 and s_0. The register operates according to the following function table. (HDL—see Problem 6.35(e), (f).)

s_1	s_0	Register Operation
0	0	No change
1	0	Complement the four outputs
0	1	Clear register to 0 (synchronous with the clock)
1	1	Load parallel data

6.8* The serial adder of Fig. 6.6 uses two four-bit registers. Register A holds the binary number 0101 and register B holds 0111. The carry flip-flop is initially reset to 0. List the binary values in register A and the carry flip-flop after each shift. (HDL—see Problem 6.54).

6.9 Two ways for implementing a serial adder ($A + B$) is shown in Section 6.2. It is necessary to modify the circuits to convert them to serial subtractors ($A - B$).

(a) Using the circuit of Fig. 6.5, show the changes needed to perform $A + 2$'s complement of B. (HDL—see Problem 6.35(h).)

(b) *Using the circuit of Fig. 6.6, show the changes needed by modifying Table 6.2 from an adder to a subtractor circuit. (See Problem 4.12). (HDL—see Problem 6.35(i).)

6.10 Design a serial 2's complementer with a shift register and a flip-flop. The binary number is shifted out from one side and it's 2's complement shifted into the other side of the shift register. (HDL—see Problem 6.35(j).)

6.11 A 3-bit binary ripple counter uses flip-flops that trigger on the negative-edge of the clock. What will be the count if

(a) the normal outputs of the flip-flops are connected to the clock and

(b) the complement outputs of the flip-flops are connected to the clock?

6.12 Draw the logic diagram of a four-bit binary ripple countup counter using

(a) flip-flops that trigger on the positive-edge of the clock and

(b) flip-flops that trigger on the negative-edge of the clock.

6.13 Show that a BCD ripple counter can be constructed using a four-bit binary ripple counter with asynchronous clear and a NAND gate that detects the occurrence of count 1010. (HDL—see Problem 6.35(k).)

6.14 How many flip-flop will be complemented in a 10-bit binary ripple counter to reach the next count after the following counts?

(a) *1110000111

(b) 1011011111

(c) 0101111111

6.15* A flip-flops has a 5 ns delay from the time the clock edge occurs to the time the output is complemented. What is the maximum delay in a 8-bit binary ripple counter that uses these flip-flops? What is the maximum frequency at which the counter can operate reliably?

6.16* The BCD ripple counter shown in Fig. 6.10 has four flip-flops and 16 states, of which only 10 are used. Analyze the circuit and determine the next state for each of the other six unused states. What will happen if a noise signal sends the circuit to one of the unused states? (HDL—see Problem 6.54.)

6.17* Design a three-bit binary synchronous up counter with D flip-flops.

6.18 What operation is performed in the up–down counter of Fig. 6.13 when both the up and down inputs are enabled? Modify the circuit so that when both inputs are equal to 1, the counter does not change state. (HDL—see Problem 6.35(l).)

6.19 The flip-flop input equations for a BCD counter using T flip-flops are given in Section 6.4. Obtain the input equations for a BCD counter that uses (a) JK flip-flops and (b)* D flip-flops. Compare the three designs to determine which one is the most efficient.

6.20 Enclose the binary counter with parallel load of Fig. 6.14 in a block diagram showing, all inputs and outputs.

(a) Show the connections of two such blocks to produce a 8-bit counter with parallel load.

(b) Construct a binary counter that counts from 0 through binary 132.

6.21* The counter of Fig. 6.14 has two control inputs—*Load* (L) and *Count* (C)—and a data input, (I_i).

(a) Derive the flip-flop input equations for J and K of the first stage in terms of L, C, and I.

(b) The logic diagram of the first stage of an equivalent circuit is shown in Fig. P6.21. Verify that this circuit is equivalent to the one in (a).

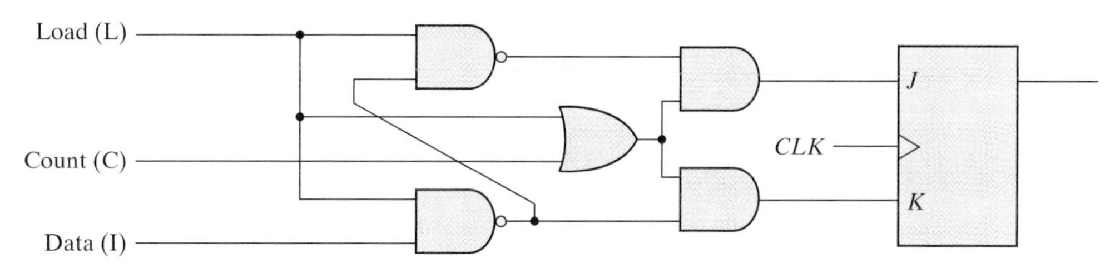

FIGURE P6.21

6.22 For the circuit of Fig. 6.14, give three alternatives for a mod-12 counter (i.e., the count evolves through a sequence of 12 distinct states).

(a) Using a NOR gate and the load input.

(b) Using the output carry.

(c) Using a NAND gate and the asynchronous clear input.

6.23 Design a timing circuit that provides an output signal that stays on for exactly twelve clock cycles. A start signal sends the output to the 1 state, and after twelve clock cycles the signal returns to the 0 state. (HDL—see Problem 6.45.)

6.24* Design a counter with T flip-flops that goes through the following binary repeated sequence: 0, 1, 3, 7, 6, 4. Show that when binary states 010 and 101 are considered as don't care conditions, the counter may not operate properly. Find a way to correct the design. (HDL—see Problem 6.55.)

6.25 It is necessary to generate six repeated timing signals T_0 through T_5 similar to the ones shown in Fig. 6.17(c). Design the circuit using (HDL—see Problem 6.46.):

(a) flip-flops only.

(b) a counter and a decoder.

6.26* A digital system has a clock generator that produces pulses at a frequency of 50 MHz. Design a circuit that provides a clock with a cycle time of 160 ns.

6.27 Using *JK* flip-flops,
- (a) Design a counter with the following repeated binary sequence: 0, 1, 2, 3, 4, 5, 6. (HDL—see Problem 6.50(a), 6.51.).
- (b) Draw the logic diagram of the counter.

6.28 Using *D* flip-flops,
- (a) *Design a counter with the following repeated binary sequence: 0, 1, 2, 4, 6. (HDL—see Problem 6.50(b).)
- (b) Draw the logic diagram of the counter.
- (c) Design a counter with the following repeated binary sequence: 0, 2, 4, 6, 8.
- (d) Draw the logic diagram of the counter.

6.29 List the eight unused states in the switch-tail ring counter of Fig. 6.18(a). Determine the next state for each of these states and show that, if the counter finds itself in an invalid state, it does not return to a valid state. Modify the circuit as recommended in the text and show that the counter produces the same sequence of states and that the circuit reaches a valid state from any one of the unused states.

6.30 Show that a Johnson counter with *n* flip-flops produces a sequence of $2n$ states. List the 12 states produced with six flip-flops and the Boolean terms of each of the 12 AND gate outputs.

6.31 Write and verify the HDL behavioral and structural descriptions of the four-bit register Fig. 6.1.

6.32
- (a) Write and verify an HDL behavioral description of a four-bit register with parallel load and asynchronous clear.
- (b) Write and verify the HDL structural description of the four-bit register with parallel load shown in Fig. 6.2. Use a 2×1 multiplexer for the flip-flop inputs. Include an asynchronous clear input.
- (c) Verify both descriptions, using a test bench.

6.33 The following stimulus program is used to simulate the binary counter with parallel load described in HDL Example 6.3. Draw waveforms showing the output of the counter and the carry output from $t = 0$ to $t = 155$ ns.

```
// Stimulus for testing the binary counter of Example 6.3
module testcounter;
 reg Count, Load, CLK, Clr;
 reg [3: 0] IN;
 wire C0;
 wire [3: 0] A;
 counter cnt (Count, Load, IN, CLK, Clr, A, CO);
 always
  #5 CLK = ~CLK;
 initial
  begin
   Clr = 0;
   CLK = 1;
   Load = 0; Count = 1;
```

```
    #5 Clr = 1;
    #40 Load = 1; IN = 4'b1001;
    #10 Load = 0;
    #70 Count = 0;
    #20 $finish;
  end
endmodule
```

6.34* Write and verify the HDL behavioral description of a four-bit shift register (see Fig. 6.3).

6.35 Write and verify
(a) A structural HDL model for the register described in Problem 6.2
(b) *A behavioral HDL model for the register described in Problem 6.2
(c) A structural HDL model for the register described in Problem 6.6
(d) A behavioral HDL model for the register described in Problem 6.6
(e) A structural HDL model for the register described in Problem 6.7
(f) A behavioral HDL model for the register described in Problem 6.7
(g) A behavioral HDL model of the binary counter described in Fig. 6.8(b)
(h) A behavioral HDL model of the serial subtractor described in Problem 6.9(a)
(i) A behavioral HDL model of the serial subtractor described in Problem 6.9(b)
(j) A behavioral HDL model of the serial 2's complementer described in Problem 6.10
(k) A behavioral HDL model of the BCD ripple counter described in Problem 6.13
(l) A behavioral HDL model of the up–down counter described in Problem 6.18.

6.36 Write and verify the HDL behavioral and structural descriptions of the four-bit up–down counter whose logic diagram is described by Fig. 6.13, Table 6.5, and Table 6.6.

6.37 Write and verify a behavioral description of the counter described in Problem 6.24.
(a) *Using an **if** … **else** statement
(b) Using a **case** statement
(c) A finite state machine.

6.38 Write and verify the HDL behavioral description of a four-bit up–down counter with parallel load using the following control inputs:
(a) *The counter has three control inputs for the three operations: *Up*, *Down*, and *Load*. The order of precedence is: *Load*, *Up*, and *Down*.
(b) The counter has two selection inputs to specify four operations: *Load*, *Up*, *Down*, and no change.

6.39 Write and verify HDL behavioral and structural descriptions of the counter of Fig. 6.16.

6.40 Write and verify the HDL description of an eight-bit ring-counter similar to the one shown in Fig. 6.17(a).

6.41 Write and verify the HDL description of a four-bit switch-tail ring (Johnson) counter (Fig. 6.18a).

6.42* The comment with the last clause of the if statement in *Binary_Counter_4_Par_Load* in HDL Example 6.3 notes that the statement is redundant. Explain why this statement can be removed without changing the behavior implemented by the description.

6.43 The scheme shown in Fig. 6.4 gates the clock to control the serial transfer of data from shift register A to shift register B. Using multiplexers at the input of each cell of the shift registers, develop a structural model of an alternative circuit that does not alter the clock path. The top level of the design hierarchy is to instantiate the shift registers. The module describing the shift registers is to have instantiations of flip-flops and muxes. Describe the mux and flip-flop modules with behavioral models. Be sure to consider the need to reset the machine. Develop a test bench to simulate the circuit and demonstrate the transfer of data.

6.44 Modify the design of the serial adder shown in Fig. 6.5 by removing the gated clock to the *D* flip-flop and supplying the clock signal to it directly. Augment the *D* flip-flop with a mux to recirculate the contents of the flip-flop when shifting is suspended and provide the carry out of the full adder when shifting is active. The shift registers are to incorporate this feature also, rather than use a gated clock. The top-level of the design is to instantiate modules using behavioral models for the shift registers, full adder, *D* flip-flop, and mux. Assume asynchronous reset. Develop a test bench to simulate the circuit and demonstrate the transfer of data.

6.45* Write and verify a behavioral description of a finite state machine to implement the counter described in Problem 6.24.

6.46 Problem 6.25 specifies an implementation of a circuit to generate timing signals using

(a) Only flip-flops.
(b) A counter and a decoder.

As an alternative, write a behavioral description (without consideration of the actual hardware) of a state machine whose output generates the timing signals T_0 through T_5.

6.47 Write a behavioral description of the circuit shown in Fig. P6.47 and verify that the circuit's output is asserted if successive samples of the input have an odd number of 1s.

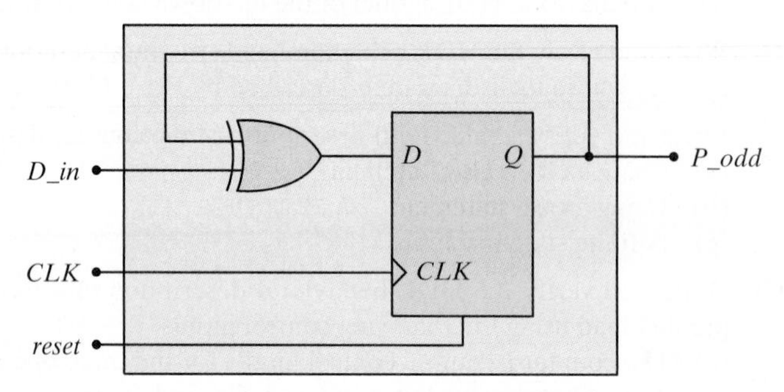

FIGURE P6.47
Circuit for Problem 6.47

6.48 Write and verify a behavioral description of the counter shown in Fig. P6.48(a); repeat for the counter in Fig. P6.48(b).

6.49 Write a test plan for verifying the functionality of the universal shift register described in HDL Example 6.1. Using the test plan, simulate the model given in HDL Example 6.1.

6.50 Write and verify a behavioral model of the counter described in

(a) Problem 6.27
(b) Problem 6.28

6.51 Without requiring a state machine, and using a shift register and additional logic, write and verify a model of an alternative to the sequence detector described in Fig. 5.27. Compare the implementations.

6.52 Write a Verilog structural model of the universal shift register in Fig. 6.7. Verify all modes of its operation.

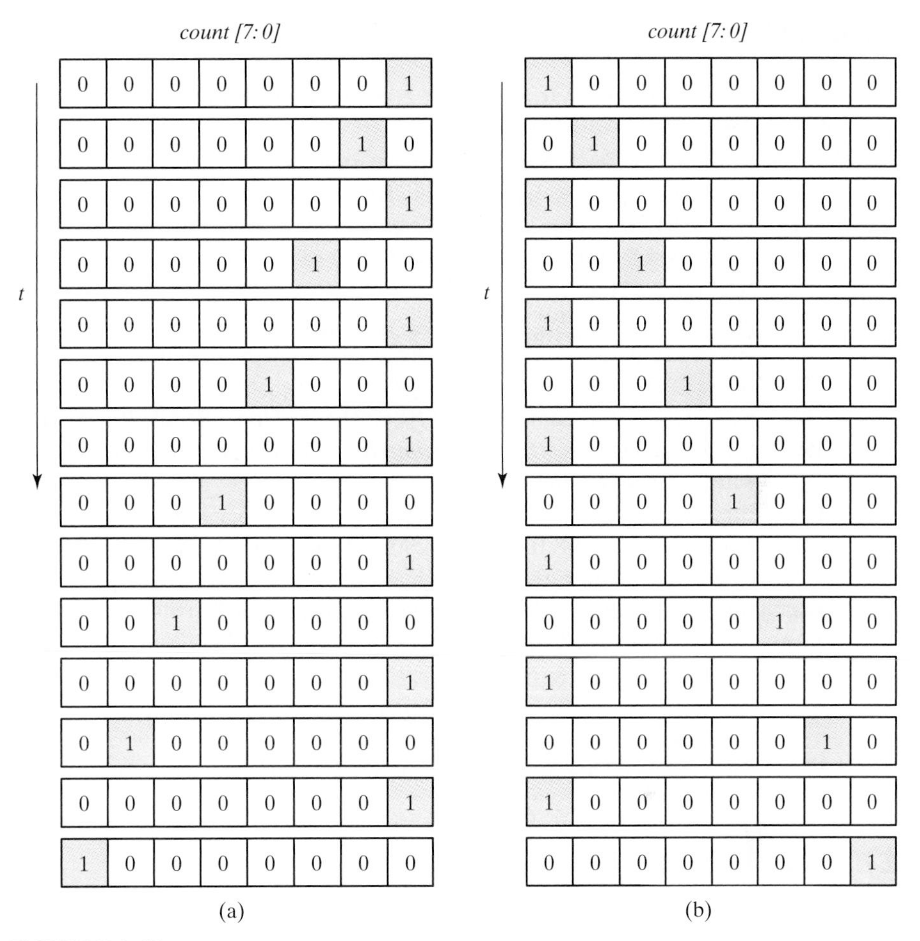

FIGURE P6.48
Circuit for Problem 6.48

6.53 Verify that the serial adder in Fig. 6.5 operates as an accumulator when words are shifted into the addend register repeatedly.

6.54 Write and verify a structural model of the serial adder in Fig. 6.6.

6.55 Write and verify a structural model of the BCD ripple counter in Fig. 6.10.

6.56 Write and verify a structural model of the synchronous binary counter in Fig. 6.12.

6.57 Write and verify a structural model of the up–down counter in Fig. 6.13.

6.58 Write and verify all modes of operation of
(a) A structural model of the binary counter in Fig. 6.14
(b) A behavioral model of the binary counter in Fig. 6.14.

6.59 Write and verify
(a) A structural model of the switch-tail ring counter in Fig. 6.18(a)
(b) A behavioral model of the switch-tail ringer counter in Fig. 6.18(a)

REFERENCES

1. Mano, M. M. and C. R. Kime. 2007. *Logic and Computer Design Fundamentals*, 4th ed. Upper Saddle River, NJ: Prentice Hall.
2. Nelson V. P., H. T. Nagle, J. D. Irwin, and B. D. Carroll. 1995. *Digital Logic Circuit Analysis and Design*. Upper Saddle River, NJ: Prentice Hall.
3. Hayes, J. P. 1993. *Introduction to Digital Logic Design*. Reading, MA: Addison-Wesley.
4. Wakerly, J. F. 2000. *Digital Design: Principles and Practices*, 3rd ed. Upper Saddle River, NJ: Prentice Hall.
5. Dietmeyer, D. L. 1988. *Logic Design of Digital Systems*, 3rd ed. Boston: Allyn Bacon.
6. Gajski, D. D. 1997. *Principles of Digital Design*. Upper Saddle River, NJ: Prentice Hall.
7. Roth, C. H. 2009. *Fundamentals of Logic Design*, 6th ed. St. Paul: West.
8. Katz, R. H. 1994. *Contemporary Logic Design*. Upper Saddle River, NJ: Prentice Hall.
9. Ciletti, M. D. 1999. *Modeling, Synthesis, and Rapid Prototyping with Verilog HDL*. Upper Saddle River, NJ: Prentice Hall.
10. Bhasker, J. 1997. *A Verilog HDL Primer*. Allentown, PA: Star Galaxy Press.
11. Thomas, D. E. and P. R. Moorby. 2002. *The VeriLog Hardware Description Language*, 5th ed. Boston: Kluwer Academic Publishers.
12. Bhasker, J. 1998. *Verilog HDL Synthesis*. Allentown, PA: Star Galaxy Press.
13. Palnitkar, S. 1996. *Verilog HDL: A Guide to Digital Design and Synthesis*. Mountain View, CA: SunSoft Press (A Prentice Hall Title).
14. Ciletti, M. D. 2010. *Advanced Digital Design with the Verilog HDL, 2e*. Upper Saddle River, NJ: Prentice Hall.
15. Ciletti, M. D. 2004. *Starter's Guide to Verilog 2001*. Upper Saddle River, NJ: Prentice Hall.

WEB SEARCH TOPICS

BCD counter
Johnson counter
Ring counter
Sequence detector
Synchronous counter
Switch-tail ring counter
Up–down counter

Chapter 7

Memory and Programmable Logic

7.1 INTRODUCTION

A memory unit is a device to which binary information is transferred for storage and from which information is retrieved when needed for processing. When data processing takes place, information from memory is transferred to selected registers in the processing unit. Intermediate and final results obtained in the processing unit are transferred back to be stored in memory. Binary information received from an input device is stored in memory, and information transferred to an output device is taken from memory. A memory unit is a collection of cells capable of storing a large quantity of binary information.

There are two types of memories that are used in digital systems: *random-access memory* (RAM) and *read-only memory* (ROM). RAM stores new information for later use. The process of storing new information into memory is referred to as a memory *write* operation. The process of transferring the stored information out of memory is referred to as a memory *read* operation. RAM can perform both write and read operations. ROM can perform only the read operation. This means that suitable binary information is already stored inside memory and can be retrieved or read at any time. However, that information cannot be altered by writing.

ROM is a *programmable logic device* (PLD). The binary information that is stored within such a device is specified in some fashion and then embedded within the hardware in a process is referred to as *programming* the device. The word "programming" here refers to a hardware procedure which specifies the bits that are inserted into the hardware configuration of the device.

ROM is one example of a PLD. Other such units are the programmable logic array (PLA), programmable array logic (PAL), and the field-programmable gate array (FPGA). A PLD is an integrated circuit with internal logic gates connected through electronic

FIGURE 7.1
Conventional and array logic diagrams for OR gate

paths that behave similarly to fuses. In the original state of the device, all the fuses are intact. Programming the device involves blowing those fuses along the paths that must be removed in order to obtain the particular configuration of the desired logic function. In this chapter, we introduce the configuration of PLDs and indicate procedures for their use in the design of digital systems. We also present CMOS FPGAs, which are configured by downloading a stream of bits into the device to configure transmission gates to establish the internal connectivity required by a specified logic function (combinational or sequential).

A typical PLD may have hundreds to millions of gates interconnected through hundreds to thousands of internal paths. In order to show the internal logic diagram of such a device in a concise form, it is necessary to employ a special gate symbology applicable to array logic. Figure 7.1 shows the conventional and array logic symbols for a multiple-input OR gate. Instead of having multiple input lines into the gate, we draw a single line entering the gate. The input lines are drawn perpendicular to this single line and are connected to the gate through internal fuses. In a similar fashion, we can draw the array logic for an AND gate. This type of graphical representation for the inputs of gates will be used throughout the chapter in array logic diagrams.

7.2 RANDOM-ACCESS MEMORY

A memory unit is a collection of storage cells, together with associated circuits needed to transfer information into and out of a device. The architecture of memory is such that information can be selectively retrieved from any of its internal locations. The time it takes to transfer information to or from any desired random location is always the same—hence the name *random-access memory,* abbreviated RAM. In contrast, the time required to retrieve information that is stored on magnetic tape depends on the location of the data.

A memory unit stores binary information in groups of bits called *words*. A word in memory is an entity of bits that move in and out of storage as a unit. A memory word is a group of 1's and 0's and may represent a number, an instruction, one or more alphanumeric characters, or any other binary-coded information. A group of 8 bits is called a *byte*. Most computer memories use words that are multiples of 8 bits in length. Thus, a 16-bit word contains two bytes, and a 32-bit word is made up of four bytes. The capacity of a memory unit is usually stated as the total number of bytes that the unit can store.

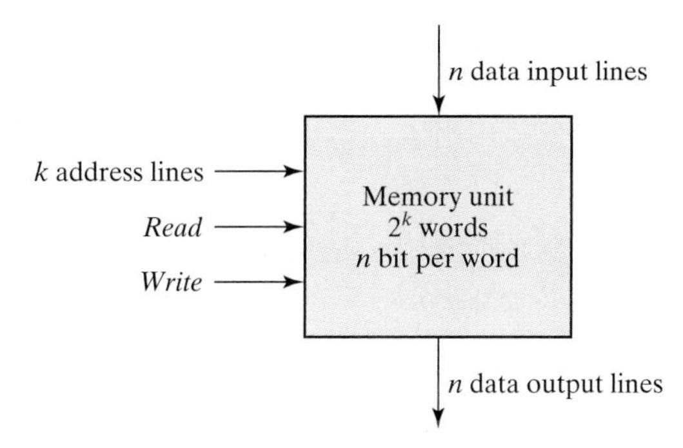

FIGURE 7.2
Block diagram of a memory unit

Communication between memory and its environment is achieved through data input and output lines, address selection lines, and control lines that specify the direction of transfer. A block diagram of a memory unit is shown in Fig. 7.2. The n data input lines provide the information to be stored in memory, and the n data output lines supply the information coming out of memory. The k address lines specify the particular word chosen among the many available. The two control inputs specify the direction of transfer desired: The *Write* input causes binary data to be transferred into the memory, and the *Read* input causes binary data to be transferred out of memory.

The memory unit is specified by the number of words it contains and the number of bits in each word. The address lines select one particular word. Each word in memory is assigned an identification number, called an *address,* starting from 0 up to $2^k - 1$, where k is the number of address lines. The selection of a specific word inside memory is done by applying the k-bit address to the address lines. An internal decoder accepts this address and opens the paths needed to select the word specified. Memories vary greatly in size and may range from 1,024 words, requiring an address of 10 bits, to 2^{32} words, requiring 32 address bits. It is customary to refer to the number of words (or bytes) in memory with one of the letters K (kilo), M (mega), and G (giga). K is equal to 2^{10}, M is equal to 2^{20}, and G is equal to 2^{30}. Thus, $64K = 2^{16}$, $2M = 2^{21}$, and $4G = 2^{32}$.

Consider, for example, a memory unit with a capacity of 1K words of 16 bits each. Since $1K = 1{,}024 = 2^{10}$ and 16 bits constitute two bytes, we can say that the memory can accommodate 2,048 = 2K bytes. Figure 7.3 shows possible contents of the first three and the last three words of this memory. Each word contains 16 bits that can be divided into two bytes. The words are recognized by their decimal address from 0 to 1,023. The equivalent binary address consists of 10 bits. The first address is specified with ten 0's; the last address is specified with ten 1's, because 1,023 in binary is equal to 1111111111. A word in memory is selected by its binary address. When a word is read or written, the memory operates on all 16 bits as a single unit.

The $1K \times 16$ memory of Fig. 7.3 has 10 bits in the address and 16 bits in each word. As another example, a $64K \times 10$ memory will have 16 bits in the address (since $64K = 2^{16}$) and each word will consist of 10 bits. The number of address bits needed in

Memory address		Memory content
Binary	Decimal	
0000000000	0	1011010101011101
0000000001	1	1010101110001001
0000000010	2	0000110101000110
	⋮	⋮
1111111101	1021	1001110100010100
1111111110	1022	0000110100011110
1111111111	1023	1101111000100101

FIGURE 7.3
Contents of a 1024 × 16 memory

a memory is dependent on the total number of words that can be stored in the memory and is independent of the number of bits in each word. The number of bits in the address is determined from the relationship $2^k \geq m$, where m is the total number of words and k is the number of address bits needed to satisfy the relationship.

Write and Read Operations

The two operations that RAM can perform are the write and read operations. As alluded to earlier, the write signal specifies a transfer-in operation and the read signal specifies a transfer-out operation. On accepting one of these control signals, the internal circuits inside the memory provide the desired operation.

The steps that must be taken for the purpose of transferring a new word to be stored into memory are as follows:

1. Apply the binary address of the desired word to the address lines.
2. Apply the data bits that must be stored in memory to the data input lines.
3. Activate the *write* input.

The memory unit will then take the bits from the input data lines and store them in the word specified by the address lines.

The steps that must be taken for the purpose of transferring a stored word out of memory are as follows:

1. Apply the binary address of the desired word to the address lines.
2. Activate the *read* input.

Table 7.1
Control Inputs to Memory Chip

Memory Enable	Read/Write	Memory Operation
0	X	None
1	0	Write to selected word
1	1	Read from selected word

The memory unit will then take the bits from the word that has been selected by the address and apply them to the output data lines. The contents of the selected word do not change after the read operation, i.e., the word operation is nondestructive.

Commercial memory components available in integrated-circuit chips sometimes provide the two control inputs for reading and writing in a somewhat different configuration. Instead of having separate read and write inputs to control the two operations, most integrated circuits provide two other control inputs: One input selects the unit and the other determines the operation. The memory operations that result from these control inputs are specified in Table 7.1.

The memory enable (sometimes called the chip select) is used to enable the particular memory chip in a multichip implementation of a large memory. When the memory enable is inactive, the memory chip is not selected and no operation is performed. When the memory enable input is active, the read/write input determines the operation to be performed.

Memory Description in HDL

Memory is modeled in the Verilog hardware description language (HDL) by an array of registers. It is declared with a **reg** keyword, using a two-dimensional array. The first number in the array specifies the number of bits in a word (the *word length*) and the second gives the number of words in memory (memory *depth*). For example, a memory of 1,024 words with 16 bits per word is declared as

```
reg[15: 0] memword [0: 1023];
```

This statement describes a two-dimensional array of 1,024 registers, each containing 16 bits. The second array range in the declaration of *memword* specifies the total number of words in memory and is equivalent to the address of the memory. For example, *memword[512]* refers to the 16-bit memory word at address 512.

The operation of a memory unit is illustrated in HDL Example 7.1. The memory has 64 words of four bits each. There are two control inputs: *Enable* and *ReadWrite*. The *DataIn* and *DataOut* lines have four bits each. The input *Address* must have six bits (since $2^6 = 64$). The memory is declared as a two-dimensional array of registers, with *Mem* used as an identifier that can be referenced with an index to access any of the 64 words. A memory operation requires that the *Enable* input be active. The *ReadWrite* input determines the type of operation. If *ReadWrite* is 1, the memory performs a read operation symbolized by the statement

DataOut ← Mem [Address];

Execution of this statement causes a transfer of four bits from the selected memory word specified by *Address* onto the *DataOut* lines. If *ReadWrite* is 0, the memory performs a write operation symbolized by the statement

Mem [Address] ← DataIn;

Execution of this statement causes a transfer from the four-bit *DataIn* lines into the memory word selected by *Address*. When *Enable* is equal to 0, the memory is disabled and the outputs are assumed to be in a high-impedance state, indicated by the symbol **z**. Thus, the memory has three-state outputs.

HDL Example 7.1

```
// Read and write operations of memory
// Memory size is 64 words of four bits each.

module memory (Enable, ReadWrite, Address, DataIn, DataOut);
 input  Enable, ReadWrite;
 input  [3: 0] DataIn;
 input  [5: 0] Address;
 output [3: 0] DataOut;
 reg [3: 0]      DataOut;
 reg [3: 0]      Mem [0: 63];                    // 64 x 4 memory
 always @ (Enable or ReadWrite)
  if (Enable)
   if (ReadWrite) DataOut = Mem [Address];        // Read
  else Mem [Address] = DataIn;                    // Write
 else DataOut = 4'bz;                             // High impedance state
endmodule
```

Timing Waveforms

The operation of the memory unit is controlled by an external device such as a central processing unit (CPU). The CPU is usually synchronized by its own clock. The memory, however, does not employ an internal clock. Instead, its read and write operations are specified by control inputs. The *access time* of memory is the time required to select a word and read it. The *cycle time* of memory is the time required to complete a write operation. The CPU must provide the memory control signals in such a way as to synchronize its internal clocked operations with the read and write operations of memory. This means that the access time and cycle time of the memory must be within a time equal to a fixed number of CPU clock cycles.

Suppose as an example that a CPU operates with a clock frequency of 50 MHz, giving a period of 20 ns for one clock cycle. Suppose also that the CPU communicates with a memory whose access time and cycle time do not exceed 50 ns. This means that the

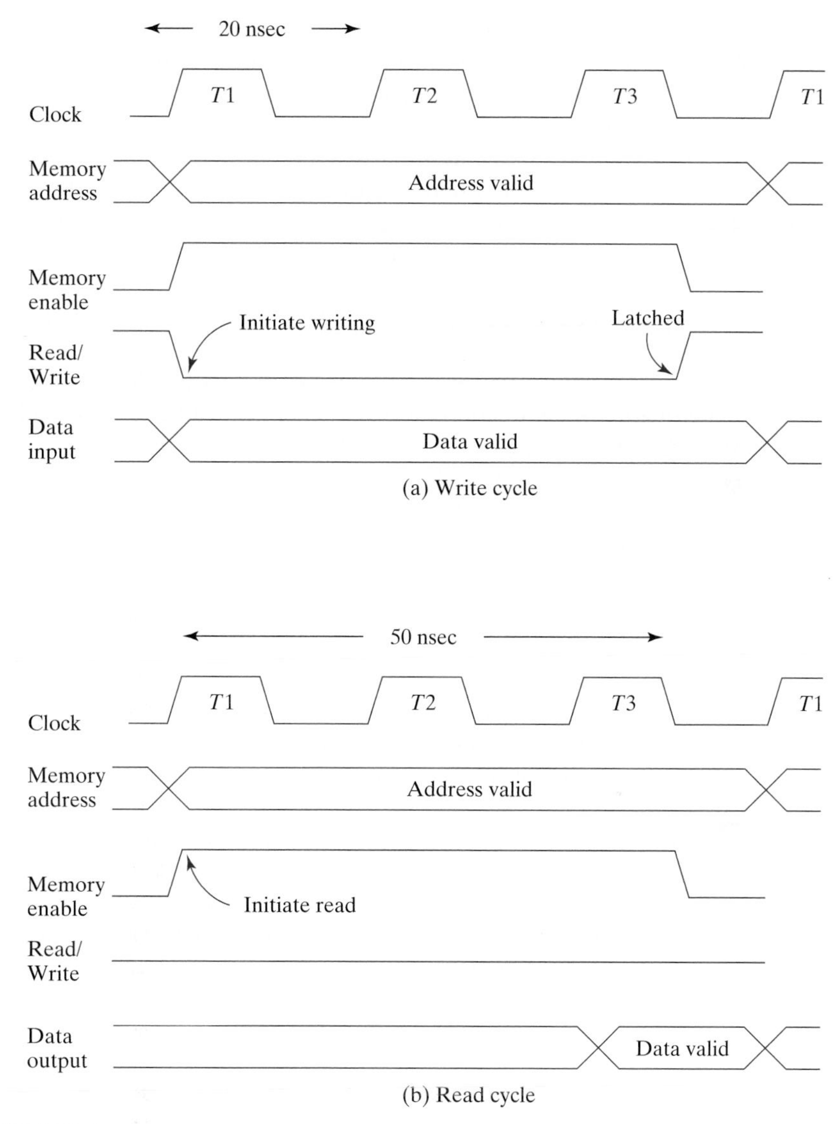

FIGURE 7.4
Memory cycle timing waveforms

write cycle terminates the storage of the selected word within a 50-ns interval and that the read cycle provides the output data of the selected word within 50 ns or less. (The two numbers are not always the same.) Since the period of the CPU cycle is 20 ns, it will be necessary to devote at least two-and-a-half, and possibly three, clock cycles for each memory request.

The memory timing shown in Fig. 7.4 is for a CPU with a 50-MHz clock and a memory with 50 ns maximum cycle time. The write cycle in part (a) shows three 20-ns cycles: *T*1,

$T2$, and $T3$. For a write operation, the CPU must provide the address and input data to the memory. This is done at the beginning of $T1$. (The two lines that cross each other in the address and data waveforms designate a possible change in value of the multiple lines.) The memory enable and the read/write signals must be activated after the signals in the address lines are stable in order to avoid destroying data in other memory words. The memory enable signal switches to the high level and the read/write signal switches to the low level to indicate a write operation. The two control signals must stay active for at least 50 ns. The address and data signals must remain stable for a short time after the control signals are deactivated. At the completion of the third clock cycle, the memory write operation is completed and the CPU can access the memory again with the next $T1$ cycle.

The read cycle shown in Fig. 7.4(b) has an address for the memory provided by the CPU. The memory-enable and read/write signals must be in their high level for a read operation. The memory places the data of the word selected by the address into the output data lines within a 50-ns interval (or less) from the time that the memory enable is activated. The CPU can transfer the data into one of its internal registers during the negative transition of $T3$. The next $T1$ cycle is available for another memory request.

Types of Memories

The mode of access of a memory system is determined by the type of components used. In a random-access memory, the word locations may be thought of as being separated in space, each word occupying one particular location. In a sequential-access memory, the information stored in some medium is not immediately accessible, but is available only at certain intervals of time. A magnetic disk or tape unit is of this type. Each memory location passes the read and write heads in turn, but information is read out only when the requested word has been reached. In a random-access memory, the access time is always the same regardless of the particular location of the word. In a sequential-access memory, the time it takes to access a word depends on the position of the word with respect to the position of the read head; therefore, the access time is variable.

Integrated circuit RAM units are available in two operating modes: *static* and *dynamic*. Static RAM (SRAM) consists essentially of internal latches that store the binary information. The stored information remains valid as long as power is applied to the unit. Dynamic RAM (DRAM) stores the binary information in the form of electric charges on capacitors provided inside the chip by MOS transistors. The stored charge on the capacitors tends to discharge with time, and the capacitors must be periodically recharged by *refreshing* the dynamic memory. Refreshing is done by cycling through the words every few milliseconds to restore the decaying charge. DRAM offers reduced power consumption and larger storage capacity in a single memory chip. SRAM is easier to use and has shorter read and write cycles.

Memory units that lose stored information when power is turned off are said to be *volatile*. CMOS integrated circuit RAMs, both static and dynamic, are of this category, since the binary cells need external power to maintain the stored information. In contrast, a nonvolatile memory, such as magnetic disk, retains its stored information after the removal of power. This type of memory is able to retain information because the data stored on magnetic components are represented by the direction of magnetization, which is retained after power is turned off. ROM is another nonvolatile memory. A nonvolatile memory enables digital computers to store programs that will be needed again after the computer is turned on. Programs and data that cannot be altered are stored in ROM, while other large programs are maintained on magnetic disks. The latter programs are transferred into the computer RAM as needed. Before the power is turned off, the binary information from the computer RAM is transferred to the disk so that the information will be retained.

7.3 MEMORY DECODING

In addition to requiring storage components in a memory unit, there is a need for decoding circuits to select the memory word specified by the input address. In this section, we present the internal construction of a RAM and demonstrate the operation of the decoder. To be able to include the entire memory in one diagram, the memory unit presented here has a small capacity of 16 bits, arranged in four words of 4 bits each. An example of a two-dimensional coincident decoding arrangement is presented to show a more efficient decoding scheme that is used in large memories. We then give an example of address multiplexing commonly used in DRAM integrated circuits.

Internal Construction

The internal construction of a RAM of m words and n bits per word consists of $m \times n$ binary storage cells and associated decoding circuits for selecting individual words. The binary storage cell is the basic building block of a memory unit. The equivalent logic of a binary cell that stores one bit of information is shown in Fig. 7.5. The storage part of the cell is modeled by an SR latch with associated gates to form a D latch. Actually, the

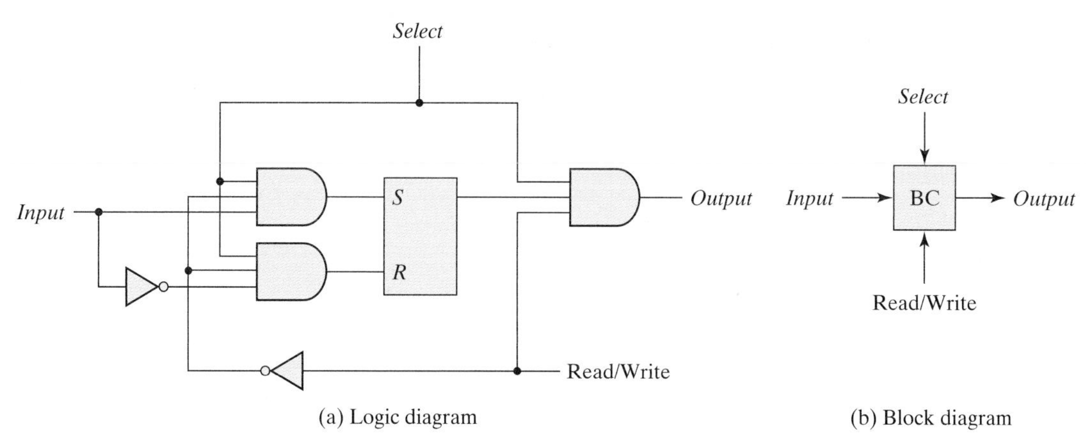

FIGURE 7.5
Memory cell

cell is an electronic circuit with four to six transistors. Nevertheless, it is possible and convenient to model it in terms of logic symbols. A binary storage cell must be very small in order to be able to pack as many cells as possible in the small area available in the integrated circuit chip. The binary cell stores one bit in its internal latch. The select input enables the cell for reading or writing, and the read/write input determines the operation of the cell when it is selected. A 1 in the read/write input provides the read operation by forming a path from the latch to the output terminal. A 0 in the read/write input provides the write operation by forming a path from the input terminal to the latch.

The logical construction of a small RAM is shown in Fig. 7.6. This RAM consists of four words of four bits each and has a total of 16 binary cells. The small blocks labeled BC represent the binary cell with its three inputs and one output, as specified in Fig. 7.5(b). A memory with four words needs two address lines. The two address inputs go through a 2×4 decoder to select one of the four words. The decoder is enabled with

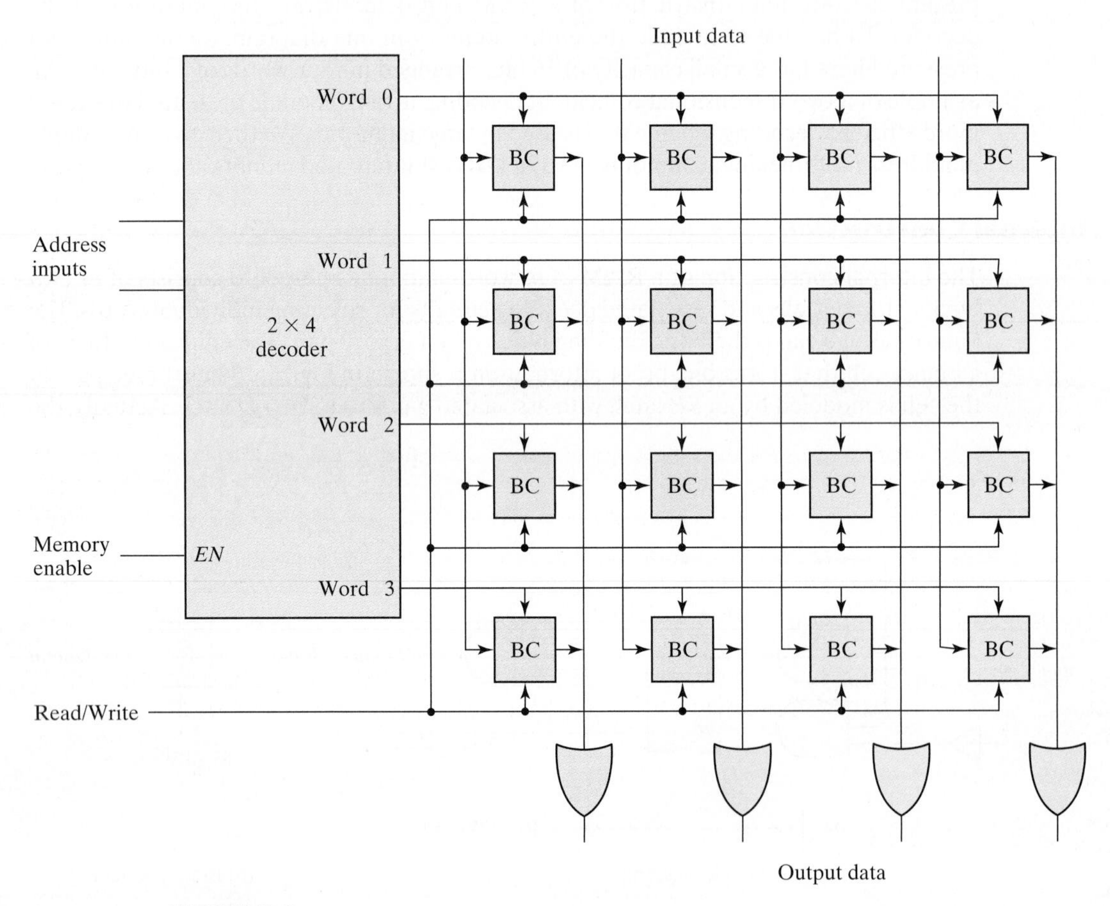

FIGURE 7.6
Diagram of a 4×4 RAM

the memory-enable input. When the memory enable is 0, all outputs of the decoder are 0 and none of the memory words are selected. With the memory select at 1, one of the four words is selected, dictated by the value in the two address lines. Once a word has been selected, the read/write input determines the operation. During the read operation, the four bits of the selected word go through OR gates to the output terminals. (Note that the OR gates are drawn according to the array logic established in Fig. 7.1.) During the write operation, the data available in the input lines are transferred into the four binary cells of the selected word. The binary cells that are not selected are disabled, and their previous binary values remain unchanged. When the memory select input that goes into the decoder is equal to 0, none of the words are selected and the contents of all cells remain unchanged regardless of the value of the read/write input.

Commercial RAMs may have a capacity of thousands of words, and each word may range from 1 to 64 bits. The logical construction of a large-capacity memory would be a direct extension of the configuration shown here. A memory with 2^k words of n bits per word requires k address lines that go into a $k \times 2^k$ decoder. Each one of the decoder outputs selects one word of n bits for reading or writing.

Coincident Decoding

A decoder with k inputs and 2^k outputs requires 2^k AND gates with k inputs per gate. The total number of gates and the number of inputs per gate can be reduced by employing two decoders in a two-dimensional selection scheme. The basic idea in two-dimensional decoding is to arrange the memory cells in an array that is close as possible to square. In this configuration, two $k/2$-input decoders are used instead of one k-input decoder. One decoder performs the row selection and the other the column selection in a two-dimensional matrix configuration.

The two-dimensional selection pattern is demonstrated in Fig. 7.7 for a 1K-word memory. Instead of using a single $10 \times 1{,}024$ decoder, we use two 5×32 decoders. With the single decoder, we would need 1,024 AND gates with 10 inputs in each. In the two-decoder case, we need 64 AND gates with 5 inputs in each. The five most significant bits of the address go to input X and the five least significant bits go to input Y. Each word within the memory array is selected by the coincidence of one X line and one Y line. Thus, each word in memory is selected by the coincidence between 1 of 32 rows and 1 of 32 columns, for a total of 1,024 words. Note that each intersection represents a word that may have any number of bits.

As an example, consider the word whose address is 404. The 10-bit binary equivalent of 404 is 01100 10100. This makes $X = 01100$ (binary 12) and $Y = 10100$ (binary 20). The n-bit word that is selected lies in the X decoder output number 12 and the Y decoder output number 20. All the bits of the word are selected for reading or writing.

Address Multiplexing

The SRAM memory cell modeled in Fig. 7.5 typically contains six transistors. In order to build memories with higher density, it is necessary to reduce the number of transistors in a cell. The DRAM cell contains a single MOS transistor and a capacitor. The charge stored

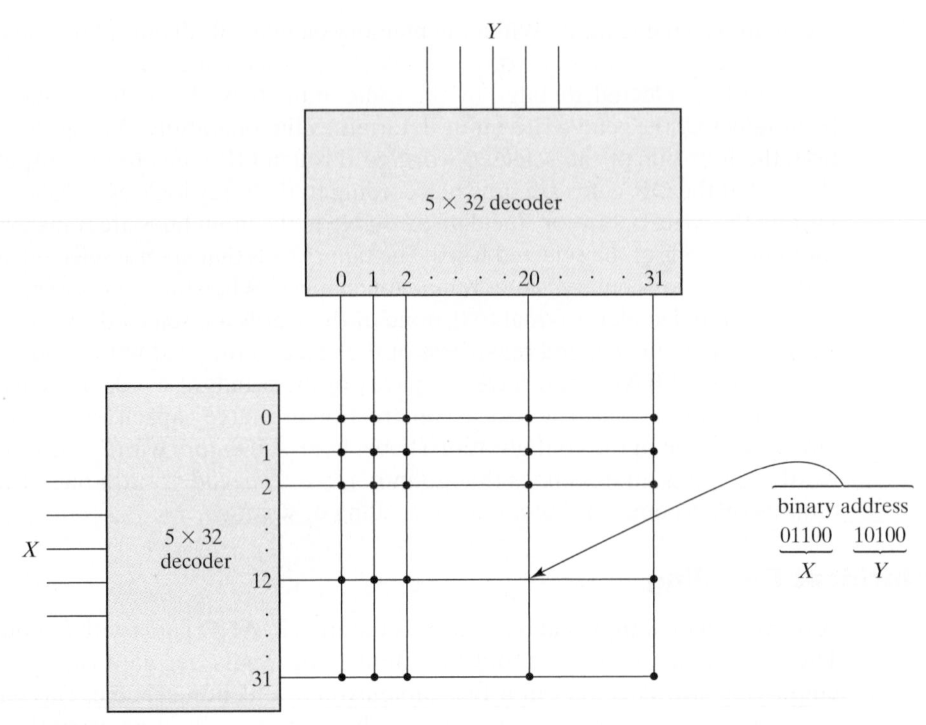

FIGURE 7.7
Two-dimensional decoding structure for a 1K-word memory

on the capacitor discharges with time, and the memory cells must be periodically recharged by refreshing the memory. Because of their simple cell structure, DRAMs typically have four times the density of SRAMs. This allows four times as much memory capacity to be placed on a given size of chip. The cost per bit of DRAM storage is three to four times less than that of SRAM storage. A further cost savings is realized because of the lower power requirement of DRAM cells. These advantages make DRAM the preferred technology for large memories in personal digital computers. DRAM chips are available in capacities from 64K to 256M bits. Most DRAMs have a 1-bit word size, so several chips have to be combined to produce a larger word size.

Because of their large capacity, the address decoding of DRAMs is arranged in a two-dimensional array, and larger memories often have multiple arrays. To reduce the number of pins in the IC package, designers utilize address multiplexing whereby one set of address input pins accommodates the address components. In a two-dimensional array, the address is applied in two parts at different times, with the row address first and the column address second. Since the same set of pins is used for both parts of the address, the size of the package is decreased significantly.

We will use a 64K-word memory to illustrate the address-multiplexing idea. A diagram of the decoding configuration is shown in Fig. 7.8. The memory consists of

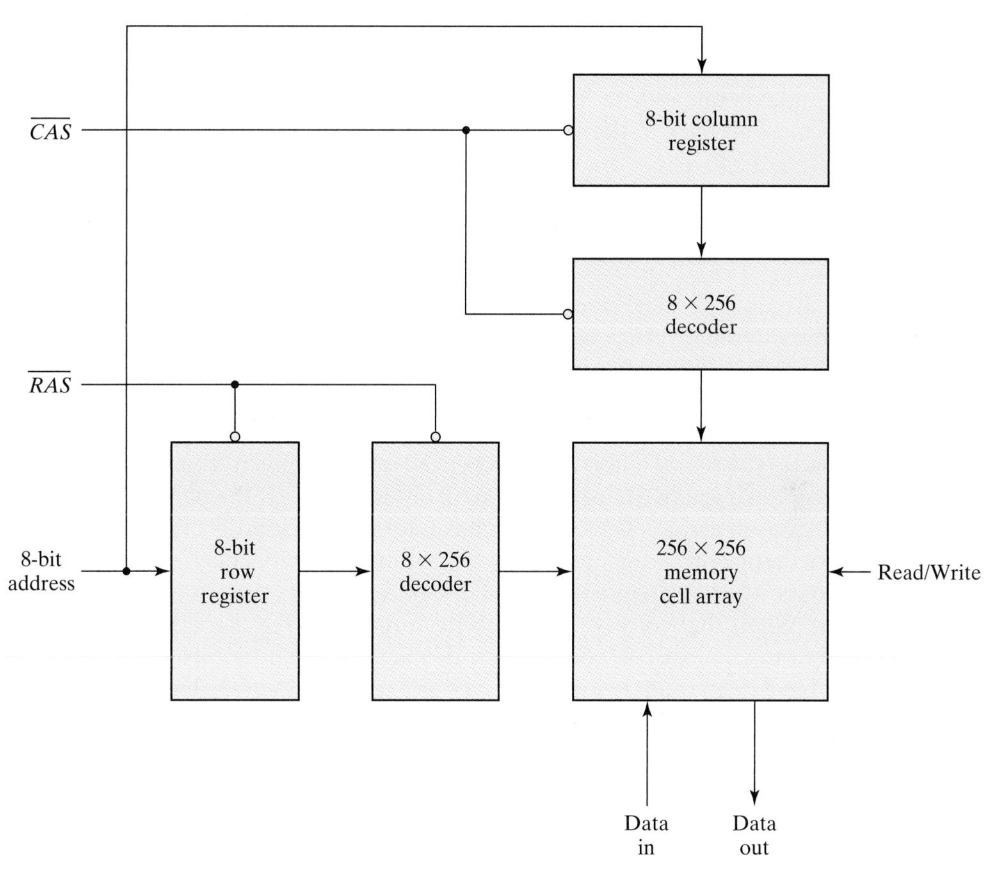

FIGURE 7.8
Address multiplexing for a 64K DRAM

a two-dimensional array of cells arranged into 256 rows by 256 columns, for a total of $2^8 \times 2^8 = 2^{16} = 64\text{K}$ words. There is a single data input line, a single data output line, and a read/write control, as well as an eight-bit address input and two address *strobes*, the latter included for enabling the row and column address into their respective registers. The row address strobe (RAS) enables the eight-bit row register, and the column address strobe (CAS) enables the eight-bit column register. The bar on top of the name of the strobe symbol indicates that the registers are enabled on the zero level of the signal.

The 16-bit address is applied to the DRAM in two steps using RAS and CAS. Initially, both strobes are in the 1 state. The 8-bit row address is applied to the address inputs and RAS is changed to 0. This loads the row address into the row address register. RAS also enables the row decoder so that it can decode the row address and select one row of the array. After a time equivalent to the settling time of the row selection, RAS goes back to the 1 level. The 8-bit column address is then applied to the address inputs, and CAS is driven to the 0 state. This transfers the column address into the column register and enables the column decoder. Now the two parts of the address are in their respective registers, the decoders have decoded them to select the one cell corresponding to the row and column address, and a read or write operation can be performed on that cell. CAS must go back to the 1 level before initiating another memory operation.

7.4 ERROR DETECTION AND CORRECTION

The dynamic physical interaction of the electrical signals affecting the data path of a memory unit may cause occasional errors in storing and retrieving the binary information. The reliability of a memory unit may be improved by employing error-detecting and error-correcting codes. The most common error detection scheme is the parity bit. (See Section 3.9.) A parity bit is generated and stored along with the data word in memory. The parity of the word is checked after reading it from memory. The data word is accepted if the parity of the bits read out is correct. If the parity checked results in an inversion, an error is detected, but it cannot be corrected.

An error-correcting code generates multiple parity check bits that are stored with the data word in memory. Each check bit is a parity over a group of bits in the data word. When the word is read back from memory, the associated parity bits are also read from memory and compared with a new set of check bits generated from the data that have been read. If the check bits are correct, no error has occurred. If the check bits do not match the stored parity, they generate a unique pattern, called a *syndrome*, that can be used to identify the bit that is in error. A single error occurs when a bit changes in value from 1 to 0 or from 0 to 1 during the write or read operation. If the specific bit in error is identified, then the error can be corrected by complementing the erroneous bit.

Hamming Code

One of the most common error-correcting codes used in RAMs was devised by R. W. Hamming. In the Hamming code, k parity bits are added to an n-bit data word, forming a new word of $n + k$ bits. The bit positions are numbered in sequence from 1 to $n + k$. Those positions numbered as a power of 2 are reserved for the parity bits. The remaining bits are the data bits. The code can be used with words of any length. Before giving the general characteristics of the code, we will illustrate its operation with a data word of eight bits.

Consider, for example, the 8-bit data word 11000100. We include 4 parity bits with the 8-bit word and arrange the 12 bits as follows:

Bit position:	1	2	3	4	5	6	7	8	9	10	11	12
	P_1	P_2	1	P_4	1	0	0	P_8	0	1	0	0

The 4 parity bits, P_1, P_2, P_4, and P_8, are in positions 1, 2, 4, and 8, respectively. The 8 bits of the data word are in the remaining positions. Each parity bit is calculated as follows:

$$P_1 = \text{XOR of bits } (3, 5, 7, 9, 11) = 1 \oplus 1 \oplus 0 \oplus 0 \oplus 0 = 0$$

$$P_2 = \text{XOR of bits } (3, 5, 7, 10, 11) = 1 \oplus 0 \oplus 0 \oplus 1 \oplus 0 = 0$$

$$P_4 = \text{XOR of bits } (5, 6, 7, 12) = 1 \oplus 0 \oplus 0 \oplus 0 = 1$$

$$P_8 = \text{XOR of bits } (9, 10, 11, 12) = 0 \oplus 1 \oplus 0 \oplus 0 = 1$$

Remember that the exclusive-OR operation performs the odd function: It is equal to 1 for an odd number of 1's in the variables and to 0 for an even number of 1's. Thus, each parity bit is set so that the total number of 1's in the checked positions, including the parity bit, is always even.

The 8-bit data word is stored in memory together with the 4 parity bits as a 12-bit composite word. Substituting the 4 P bits in their proper positions, we obtain the 12-bit composite word stored in memory:

	0	0	1	1	1	0	0	1	0	1	0	0
Bit position:	1	2	3	4	5	6	7	8	9	10	11	12

When the 12 bits are read from memory, they are checked again for errors. The parity is checked over the same combination of bits, including the parity bit. The 4 check bits are evaluated as follows:

$$C_1 = \text{XOR of bits } (1, 3, 5, 7, 9, 11)$$

$$C_2 = \text{XOR of bits } (2, 3, 6, 7, 10, 11)$$

$$C_4 = \text{XOR of bits } (4, 5, 6, 7, 12)$$

$$C_8 = \text{XOR of bits } (8, 9, 10, 11, 12)$$

A 0 check bit designates even parity over the checked bits and a 1 designates odd parity. Since the bits were stored with even parity, the result, $C = C_8C_4C_2C_1 = 0000$, indicates that no error has occurred. However, if $C \neq 0$, then the 4-bit binary number formed by the check bits gives the position of the erroneous bit. For example, consider the following three cases:

Bit position:	1	2	3	4	5	6	7	8	9	10	11	12	
	0	0	1	1	1	0	0	1	0	1	0	0	No error
	1	0	1	1	1	0	0	1	0	1	0	0	Error in bit 1
	0	0	1	1	0	0	0	1	0	1	0	0	Error in bit 5

In the first case, there is no error in the 12-bit word. In the second case, there is an error in bit position number 1 because it changed from 0 to 1. The third case shows an error in bit position 5, with a change from 1 to 0. Evaluating the XOR of the corresponding bits, we determine the 4 check bits to be as follows:

	C_8	C_4	C_2	C_1
For no error:	0	0	0	0
With error in bit 1:	0	0	0	1
With error in bit 5:	0	1	0	1

Thus, for no error, we have $C = 0000$; with an error in bit 1, we obtain $C = 0001$; and with an error in bit 5, we get $C = 0101$. When the binary number C is not equal to 0000, it gives the position of the bit in error. The error can be corrected by complementing the corresponding bit. Note that an error can occur in the data word or in one of the parity bits.

The Hamming code can be used for data words of any length. In general, the Hamming code consists of k check bits and n data bits, for a total of $n + k$ bits. The syndrome value C consists of k bits and has a range of 2^k values between 0 and $2^k - 1$. One of these values, usually zero, is used to indicate that no error was detected, leaving $2^k - 1$ values to indicate which of the $n + k$ bits was in error. Each of these $2^k - 1$ values can be used to uniquely describe a bit in error. Therefore, the range of k must be equal to or greater than $n + k$, giving the relationship

$$2^k - 1 \geq n + k$$

Solving for n in terms of k, we obtain

$$2^k - 1 - k \geq n$$

This relationship gives a formula for establishing the number of data bits that can be used in conjunction with k check bits. For example, when $k = 3$, the number of data bits that can be used is $n \leq (2^3 - 1 - 3) = 4$. For $k = 4$, we have $2^4 - 1 - 4 = 11$, giving $n \leq 11$. The data word may be less than 11 bits, but must have at least 5 bits; otherwise, only 3 check bits will be needed. This justifies the use of 4 check bits for the 8 data bits in the previous example. Ranges of n for various values of k are listed in Table 7.2.

The grouping of bits for parity generation and checking can be determined from a list of the binary numbers from 0 through $2^k - 1$. The least significant bit is a 1 in the binary numbers 1, 3, 5, 7, and so on. The second significant bit is a 1 in the binary numbers

Table 7.2
Range of Data Bits for k Check Bits

Number of Check Bits, k	Range of Data Bits, n
3	2–4
4	5–11
5	12–26
6	27–57
7	58–120

2, 3, 6, 7, and so on. Comparing these numbers with the bit positions used in generating and checking parity bits in the Hamming code, we note the relationship between the bit groupings in the code and the position of the 1-bits in the binary count sequence. Note that each group of bits starts with a number that is a power of 2: 1, 2, 4, 8, 16, etc. These numbers are also the position numbers for the parity bits.

Single-Error Correction, Double-Error Detection

The Hamming code can detect and correct only a single error. By adding another parity bit to the coded word, the Hamming code can be used to correct a single error and detect double errors. If we include this additional parity bit, then the previous 12-bit coded word becomes 001110010100P_{13}, where P_{13} is evaluated from the exclusive-OR of the other 12 bits. This produces the 13-bit word 0011100101001 (even parity). When the 13-bit word is read from memory, the check bits are evaluated, as is the parity P over the entire 13 bits. If $P = 0$, the parity is correct (even parity), but if $P = 1$, then the parity over the 13 bits is incorrect (odd parity). The following four cases can arise:

If $C = 0$ and $P = 0$, no error occurred.

If $C \neq 0$ and $P = 1$, a single error occurred that can be corrected.

If $C \neq 0$ and $P = 0$, a double error occurred that is detected, but that cannot be corrected.

If $C = 0$ and $P = 1$, an error occurred in the P_{13} bit.

This scheme may detect more than two errors, but is not guaranteed to detect all such errors.

Integrated circuits use a modified Hamming code to generate and check parity bits for single-error correction and double-error detection. The modified Hamming code uses a more efficient parity configuration that balances the number of bits used to calculate the XOR operation. A typical integrated circuit that uses an 8-bit data word and a 5-bit check word is IC type 74637. Other integrated circuits are available for data words of 16 and 32 bits. These circuits can be used in conjunction with a memory unit to correct a single error or detect double errors during write and read operations.

7.5 READ-ONLY MEMORY

A read-only memory (ROM) is essentially a memory device in which permanent binary information is stored. The binary information must be specified by the designer and is then embedded in the unit to form the required interconnection pattern. Once the pattern is established, it stays within the unit even when power is turned off and on again.

A block diagram of a ROM consisting of k inputs and n outputs is shown in Fig. 7.9. The inputs provide the address for memory, and the outputs give the data bits of the stored word that is selected by the address. The number of words in a ROM is determined from the fact that k address input lines are needed to specify 2^k words. Note that ROM does not have data inputs, because it does not have a write operation. Integrated

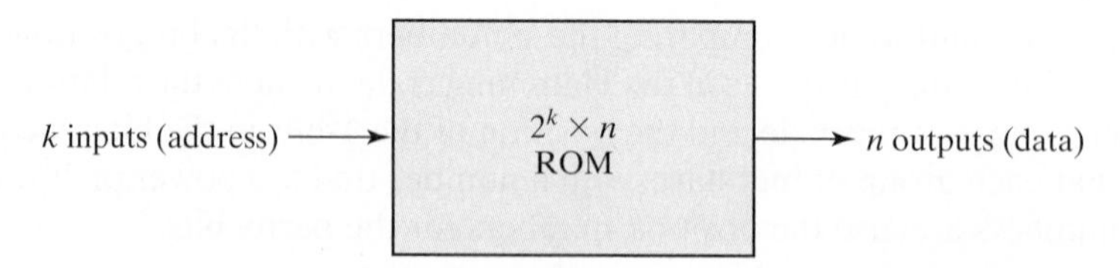

FIGURE 7.9
ROM block diagram

circuit ROM chips have one or more enable inputs and sometimes come with three-state outputs to facilitate the construction of large arrays of ROM.

Consider, for example, a 32 × 8 ROM. The unit consists of 32 words of 8 bits each. There are five input lines that form the binary numbers from 0 through 31 for the address. Figure 7.10 shows the internal logic construction of this ROM. The five inputs are decoded into 32 distinct outputs by means of a 5 × 32 decoder. Each output of the decoder represents a memory address. The 32 outputs of the decoder are connected to each of the eight OR gates. The diagram shows the array logic convention used in complex circuits. (See Fig. 6.1.) Each OR gate must be considered as having 32 inputs. Each output of the decoder is connected to one of the inputs of each OR gate. Since each OR gate has 32 input connections and there are 8 OR gates, the ROM contains $32 \times 8 = 256$ internal connections. In general, a $2^k \times n$ ROM will have an internal $k \times 2^k$ decoder and n OR gates. Each OR gate has 2^k inputs, which are connected to each of the outputs of the decoder.

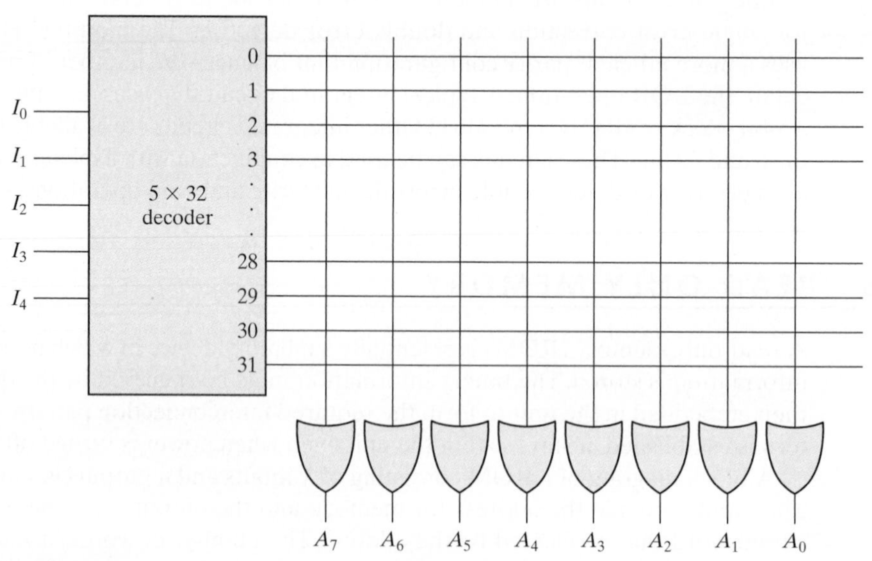

FIGURE 7.10
Internal logic of a 32 × 8 ROM

The 256 intersections in Fig. 7.10 are programmable. A programmable connection between two lines is logically equivalent to a switch that can be altered to be either closed (meaning that the two lines are connected) or open (meaning that the two lines are disconnected). The programmable intersection between two lines is sometimes called a *crosspoint*. Various physical devices are used to implement crosspoint switches. One of the simplest technologies employs a fuse that normally connects the two points, but is opened or "blown" by the application of a high-voltage pulse into the fuse.

The internal binary storage of a ROM is specified by a truth table that shows the word content in each address. For example, the content of a 32×8 ROM may be specified with a truth table similar to the one shown in Table 7.3. The truth table shows the five inputs under which are listed all 32 addresses. Each address stores a word of 8 bits, which is listed in the outputs columns. The table shows only the first four and the last four words in the ROM. The complete table must include the list of all 32 words.

The hardware procedure that programs the ROM blows fuse links in accordance with a given truth table. For example, programming the ROM according to the truth table given by Table 7.3 results in the configuration shown in Fig. 7.11. Every 0 listed in the truth table specifies the absence of a connection, and every 1 listed specifies a path that is obtained by a connection. For example, the table specifies the eight-bit word 10110010 for permanent storage at address 3. The four 0's in the word are programmed by blowing the fuse links between output 3 of the decoder and the inputs of the OR gates associated with outputs A_6, A_3, A_2, and A_0. The four 1's in the word are marked with a $\times$ to denote a temporary connection, in place of a dot used for a permanent connection in logic diagrams. When the input of the ROM is 00011, all the outputs of the decoder are 0 except for output 3, which is at logic 1. The signal equivalent to logic 1 at decoder output 3 propagates through the connections to the OR gate outputs of A_7, A_5, A_4, and A_1. The other four outputs remain at 0. The result is that the stored word 10110010 is applied to the eight data outputs.

Table 7.3
ROM Truth Table (Partial)

Inputs					Outputs							
I_4	I_3	I_2	I_1	I_0	A_7	A_6	A_5	A_4	A_3	A_2	A_1	A_0
0	0	0	0	0	1	0	1	1	0	1	1	0
0	0	0	0	1	0	0	0	1	1	1	0	1
0	0	0	1	0	1	1	0	0	0	1	0	1
0	0	0	1	1	1	0	1	1	0	0	1	0
		⋮						⋮				
1	1	1	0	0	0	0	0	0	1	0	0	1
1	1	1	0	1	1	1	1	0	0	0	1	0
1	1	1	1	0	0	1	0	0	1	0	1	0
1	1	1	1	1	0	0	1	1	0	0	1	1

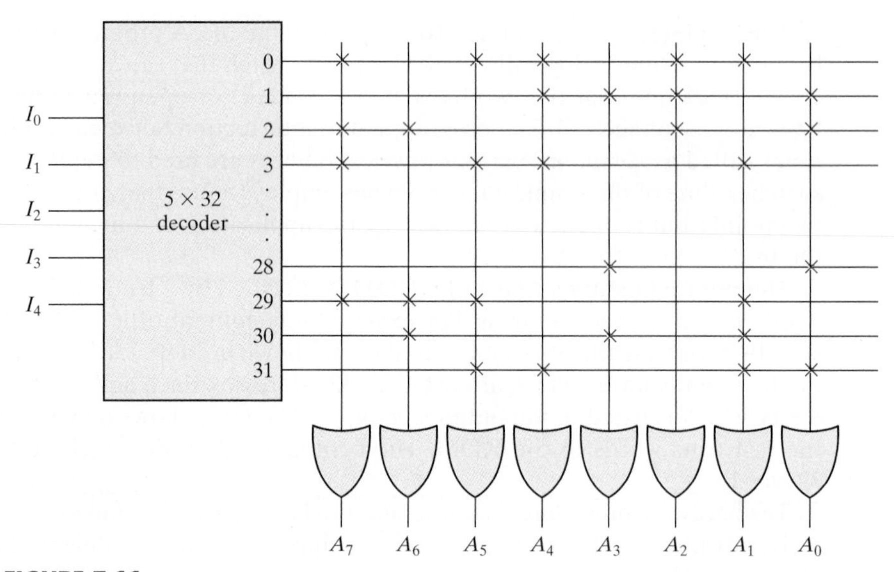

FIGURE 7.11
Programming the ROM according to Table 7.3

Combinational Circuit Implementation

In Section 4.9, it was shown that a decoder generates the 2^k minterms of the k input variables. By inserting OR gates to sum the minterms of Boolean functions, we were able to generate any desired combinational circuit. The ROM is essentially a device that includes both the decoder and the OR gates within a single device to form a minterm generator. By choosing connections for those minterms which are included in the function, the ROM outputs can be programmed to represent the Boolean functions of the output variables in a combinational circuit.

The internal operation of a ROM can be interpreted in two ways. The first interpretation is that of a memory unit that contains a fixed pattern of stored words. The second interpretation is that of a unit which implements a combinational circuit. From this point of view, each output terminal is considered separately as the output of a Boolean function expressed as a sum of minterms. For example, the ROM of Fig. 7.11 may be considered to be a combinational circuit with eight outputs, each a function of the five input variables. Output A_7 can be expressed in sum of minterms as

$$A_7(I_4, I_3, I_2, I_1, I_0) = \Sigma(0, 2, 3, \ldots, 29)$$

(The three dots represent minterms 4 through 27, which are not specified in the figure.) A connection marked with × in the figure produces a minterm for the sum. All other crosspoints are not connected and are not included in the sum.

In practice, when a combinational circuit is designed by means of a ROM, it is not necessary to design the logic or to show the internal gate connections inside the unit. All that the designer has to do is specify the particular ROM by its IC number and provide the applicable truth table. The truth table gives all the information for programming the ROM. No internal logic diagram is needed to accompany the truth table.

EXAMPLE 7.1

Design a combinational circuit using a ROM. The circuit accepts a three-bit number and outputs a binary number equal to the square of the input number.

The first step is to derive the truth table of the combinational circuit. In most cases, this is all that is needed. In other cases, we can use a partial truth table for the ROM by utilizing certain properties in the output variables. Table 7.4 is the truth table for the combinational circuit. Three inputs and six outputs are needed to accommodate all possible binary numbers. We note that output B_0 is always equal to input A_0, so there is no need to generate B_0 with a ROM, since it is equal to an input variable. Moreover, output B_1 is always 0, so this output is a known constant. We actually need to generate only four outputs with the ROM; the other two are readily obtained. The minimum size of ROM needed must have three inputs and four outputs. Three inputs specify eight words, so the ROM must be of size 8×4. The ROM implementation is shown in Fig. 7.12. The three inputs specify eight words of four bits each. The truth table in Fig. 7.12(b) specifies the information needed for programming the ROM. The block diagram of Fig. 7.12(a) shows the required connections of the combinational circuit.

Table 7.4
Truth Table for Circuit of Example 7.1

Inputs			Outputs						
A_2	A_1	A_0	B_5	B_4	B_3	B_2	B_1	B_0	Decimal
0	0	0	0	0	0	0	0	0	0
0	0	1	0	0	0	0	0	1	1
0	1	0	0	0	0	1	0	0	4
0	1	1	0	0	1	0	0	1	9
1	0	0	0	1	0	0	0	0	16
1	0	1	0	1	1	0	0	1	25
1	1	0	1	0	0	1	0	0	36
1	1	1	1	1	0	0	0	1	49

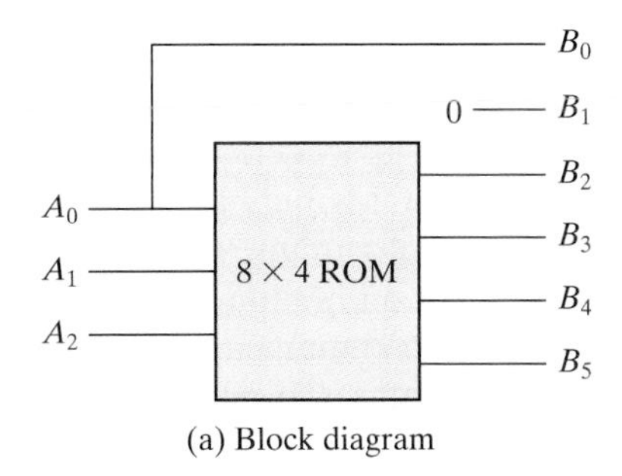

(a) Block diagram

A_2	A_1	A_0	B_5	B_4	B_3	B_2
0	0	0	0	0	0	0
0	0	1	0	0	0	0
0	1	0	0	0	0	1
0	1	1	0	0	1	0
1	0	0	0	1	0	0
1	0	1	0	1	1	0
1	1	0	1	0	0	1
1	1	1	1	1	0	0

(b) ROM truth table

FIGURE 7.12
ROM implementation of Example 7.1

■

Types of ROMs

The required paths in a ROM may be programmed in four different ways. The first is called *mask programming* and is done by the semiconductor company during the last fabrication process of the unit. The procedure for fabricating a ROM requires that the customer fill out the truth table he or she wishes the ROM to satisfy. The truth table may be submitted in a special form provided by the manufacturer or in a specified format on a computer output medium. The manufacturer makes the corresponding mask for the paths to produce the 1's and 0's according to the customer's truth table. This procedure is costly because the vendor charges the customer a special fee for custom masking the particular ROM. For this reason, mask programming is economical only if a large quantity of the same ROM configuration is to be ordered.

For small quantities, it is more economical to use a second type of ROM called *programmable read-only memory,* or PROM. When ordered, PROM units contain all the fuses intact, giving all 1's in the bits of the stored words. The fuses in the PROM are blown by the application of a high-voltage pulse to the device through a special pin. A blown fuse defines a binary 0 state and an intact fuse gives a binary 1 state. This procedure allows the user to program the PROM in the laboratory to achieve the desired relationship between input addresses and stored words. Special instruments called PROM programmers are available commercially to facilitate the procedure. In any case, all procedures for programming ROMs are hardware procedures, even though the word *programming* is used.

The hardware procedure for programming ROMs or PROMs is irreversible, and once programmed, the fixed pattern is permanent and cannot be altered. Once a bit pattern has been established, the unit must be discarded if the bit pattern is to be changed. A third type of ROM is the *erasable PROM,* or EPROM, which can be restructured to the initial state even though it has been programmed previously. When the EPROM is placed under a special ultraviolet light for a given length of time, the shortwave radiation discharges the internal floating gates that serve as the programmed connections. After erasure, the EPROM returns to its initial state and can be reprogrammed to a new set of values.

The fourth type of ROM is the electrically erasable PROM (EEPROM or E^2PROM). This device is like the EPROM, except that the previously programmed connections can be erased with an electrical signal instead of ultraviolet light. The advantage is that the device can be erased without removing it from its socket.

Flash memory devices are similar to EEPROMs, but have additional built-in circuitry to selectively program and erase the device in-circuit, without the need for a special programmer. They have widespread application in modern technology in cell phones, digital cameras, set-top boxes, digital TV, telecommunications, nonvolatile data storage, and microcontrollers. Their low consumption of power makes them an attractive storage medium for laptop and notebook computers. Flash memories incorporate additional circuitry, too, allowing simultaneous erasing of blocks of memory, for example, of size 16 to 64 K bytes. Like EEPROMs, flash memories are subject to fatigue, typically having about 10^5 block erase cycles.

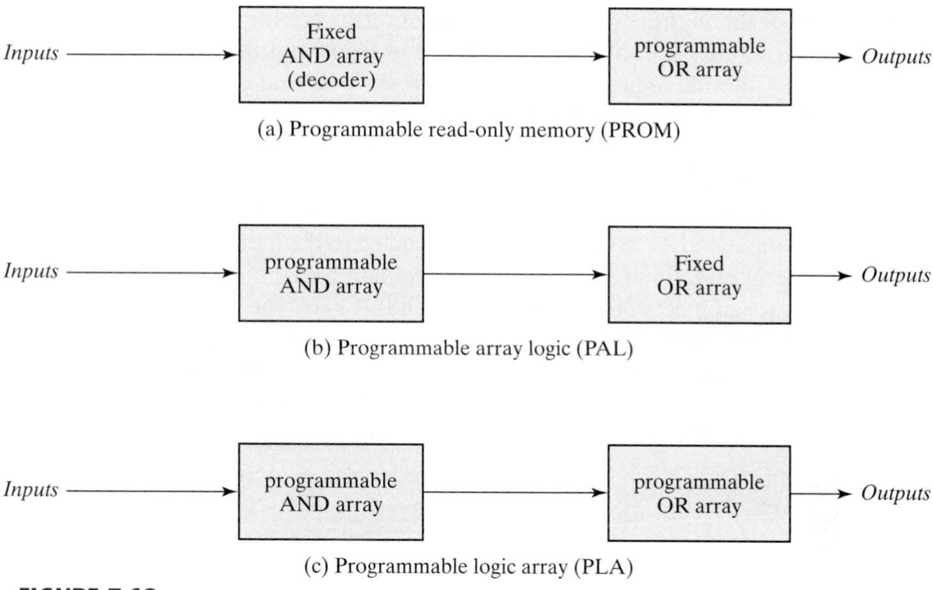

FIGURE 7.13
Basic configuration of three PLDs

Combinational PLDs

The PROM is a combinational programmable logic device (PLD)—an integrated circuit with programmable gates divided into an AND array and an OR array to provide an AND–OR sum-of-product implementation. There are three major types of combinational PLDs, differing in the placement of the programmable connections in the AND–OR array. Figure 7.13 shows the configuration of the three PLDs. The PROM has a fixed AND array constructed as a decoder and a programmable OR array. The programmable OR gates implement the Boolean functions in sum-of-minterms form. The PAL has a programmable AND array and a fixed OR array. The AND gates are programmed to provide the product terms for the Boolean functions, which are logically summed in each OR gate. The most flexible PLD is the PLA, in which both the AND and OR arrays can be programmed. The product terms in the AND array may be shared by any OR gate to provide the required sum-of-products implementation. The names PAL and PLA emerged from different vendors during the development of PLDs. The implementation of combinational circuits with PROM was demonstrated in this section. The design of combinational circuits with PLA and PAL is presented in the next two sections.

7.6 PROGRAMMABLE LOGIC ARRAY

The PLA is similar in concept to the PROM, except that the PLA does not provide full decoding of the variables and does not generate all the minterms. The decoder is replaced by an array of AND gates that can be programmed to generate any product term of the input variables. The product terms are then connected to OR gates to provide the sum of products for the required Boolean functions.

The internal logic of a PLA with three inputs and two outputs is shown in Fig. 7.14. Such a circuit is too small to be useful commercially, but is presented here to demonstrate the typical logic configuration of a PLA. The diagram uses the array logic graphic symbols for complex circuits. Each input goes through a buffer–inverter combination, shown in the diagram with a composite graphic symbol, that has both the true and complement outputs. Each input and its complement are connected to the inputs of each AND gate, as indicated by the intersections between the vertical and horizontal lines. The outputs of the AND gates are connected to the inputs of each OR gate. The output of the OR gate goes to an XOR gate, where the other input can be programmed to receive a signal equal to either logic 1 or logic 0. The output is inverted when the XOR input is connected to 1 (since $x \oplus 1 = x'$). The output does not change when the XOR input is connected to 0 (since $x \oplus 0 = x$). The particular Boolean functions implemented in the PLA of Fig. 7.14 are

$$F_1 = AB' + AC + A'BC'$$
$$F_2 = (AC + BC)'$$

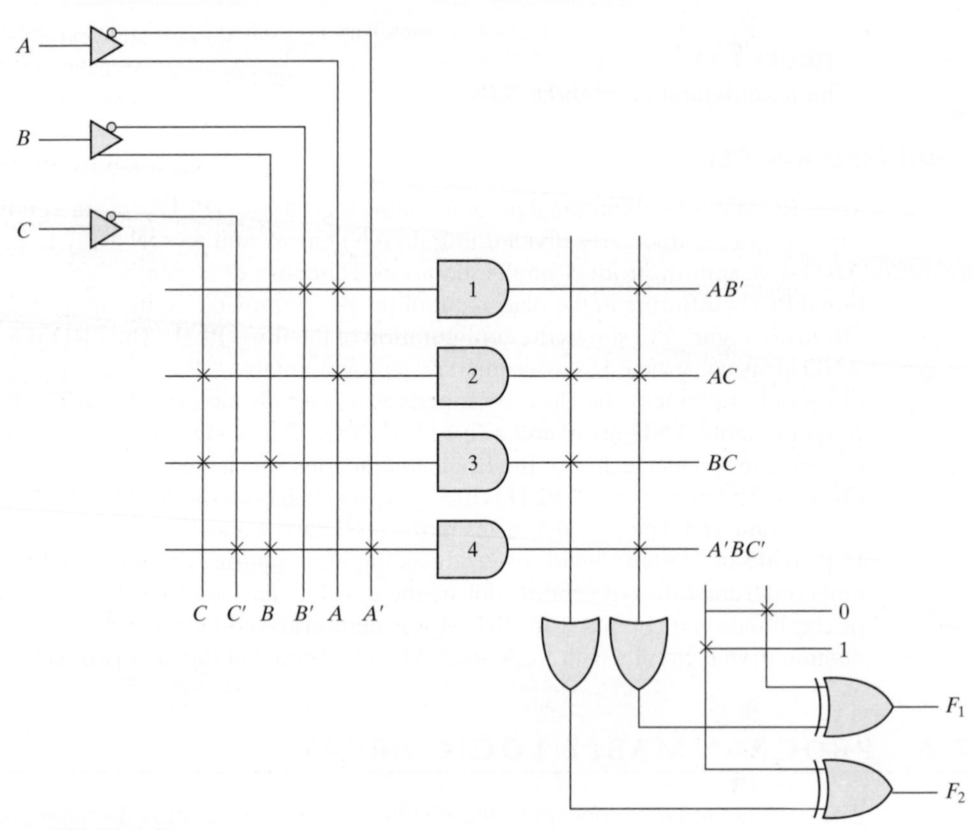

FIGURE 7.14
PLA with three inputs, four product terms, and two outputs

The product terms generated in each AND gate are listed along the output of the gate in the diagram. The product term is determined from the inputs whose crosspoints are connected and marked with a $\times$. The output of an OR gate gives the logical sum of the selected product terms. The output may be complemented or left in its true form, depending on the logic being realized.

The fuse map of a PLA can be specified in a tabular form. For example, the programming table that specifies the PLA of Fig. 7.14 is listed in Table 7.5. The PLA programming table consists of three sections. The first section lists the product terms numerically. The second section specifies the required paths between inputs and AND gates. The third section specifies the paths between the AND and OR gates. For each output variable, we may have a T (for true) or C (for complement) for programming the XOR gate. The product terms listed on the left are not part of the table; they are included for reference only. For each product term, the inputs are marked with 1, 0, or — (dash). If a variable in the product term appears in the form in which it is true, the corresponding input variable is marked with a 1. If it appears complemented, the corresponding input variable is marked with a 0. If the variable is absent from the product term, it is marked with a dash.

The paths between the inputs and the AND gates are specified under the column head "Inputs" in the programming table. A 1 in the input column specifies a connection from the input variable to the AND gate. A 0 in the input column specifies a connection from the complement of the variable to the input of the AND gate. A dash specifies a blown fuse in both the input variable and its complement. It is assumed that an open terminal in the input of an AND gate behaves like a 1.

The paths between the AND and OR gates are specified under the column head "Outputs." The output variables are marked with 1's for those product terms which are included in the function. Each product term that has a 1 in the output column requires a path from the output of the AND gate to the input of the OR gate. Those marked with a dash specify a blown fuse. It is assumed that an open terminal in the input of an OR gate behaves like a 0. Finally, a T (true) output dictates that the other input of the corresponding XOR gate be connected to 0, and a C (complement) specifies a connection to 1.

Table 7.5
PLA Programming Table

		Inputs			**Outputs** (T)	(C)
	Product Term	A	B	C	F_1	F_2
AB'	1	1	0	—	1	—
AC	2	1	—	1	1	1
BC	3	—	1	1	—	1
$A'BC'$	4	0	1	0	1	—

Note: See text for meanings of dashes.

The size of a PLA is specified by the number of inputs, the number of product terms, and the number of outputs. A typical integrated circuit PLA may have 16 inputs, 48 product terms, and eight outputs. For n inputs, k product terms, and m outputs, the internal logic of the PLA consists of n buffer–inverter gates, k AND gates, m OR gates, and m XOR gates. There are $2n \times k$ connections between the inputs and the AND array, $k \times m$ connections between the AND and OR arrays, and m connections associated with the XOR gates.

In designing a digital system with a PLA, there is no need to show the internal connections of the unit as was done in Fig. 7.14. All that is needed is a PLA programming table from which the PLA can be programmed to supply the required logic. As with a ROM, the PLA may be mask programmable or field programmable. With mask programming, the customer submits a PLA program table to the manufacturer. This table is used by the vendor to produce a custom-made PLA that has the required internal logic specified by the customer. A second type of PLA that is available is the field-programmable logic array, or FPLA, which can be programmed by the user by means of a commercial hardware programmer unit.

In implementing a combinational circuit with a PLA, careful investigation must be undertaken in order to reduce the number of distinct product terms, since a PLA has a finite number of AND gates. This can be done by simplifying each Boolean function to a minimum number of terms. The number of literals in a term is not important, since all the input variables are available anyway. Both the true value and the complement of each function should be simplified to see which one can be expressed with fewer product terms and which one provides product terms that are common to other functions.

EXAMPLE 7.2

Implement the following two Boolean functions with a PLA:

$$F_1(A, B, C) = \Sigma(0, 1, 2, 4)$$
$$F_2(A, B, C) = \Sigma(0, 5, 6, 7)$$

The two functions are simplified in the maps of Fig. 7.15. Both the true value and the complement of the functions are simplified into sum-of-products form. The combination that gives the minimum number of product terms is

$$F_1 = (AB + AC + BC)'$$

and

$$F_2 = AB + AC + A'B'C'$$

This combination gives four distinct product terms: $AB, AC, BC,$ and $A'B'C'$. The PLA programming table for the combination is shown in the figure. Note that output F_1 is the true output, even though a C is marked over it in the table. This is because F_1 is generated with an AND–OR circuit and is available at the output of the OR gate. The XOR gate complements the function to produce the true F_1 output.

PLA programming table

	Product term	Inputs			Outputs (C)	Outputs (T)
		A	B	C	F_1	F_2
AB	1	1	1	–	1	1
AC	2	1	–	1	1	1
BC	3	–	1	1	1	–
$A'B'C'$	4	0	0	0	–	1

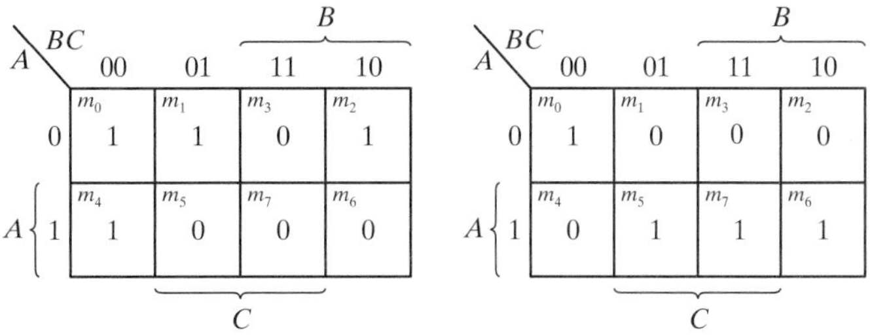

$A \backslash BC$	00	01	11	10
0	m_0 1	m_1 1	m_3 0	m_2 1
1	m_4 1	m_5 0	m_7 0	m_6 0

$A \backslash BC$	00	01	11	10
0	m_0 1	m_1 0	m_3 0	m_2 0
1	m_4 0	m_5 1	m_7 1	m_6 1

FIGURE 7.15
Solution to Example 7.1

■

The combinational circuit used in Example 7.2 is too simple for implementing with a PLA. It was presented merely for purposes of illustration. A typical PLA has a large number of inputs and product terms. The simplification of Boolean functions with so many variables should be carried out by means of computer-assisted simplification procedures. The computer-aided design (CAD) program simplifies each function and its complement to a minimum number of terms. The program then selects a minimum number of product terms that cover all functions in the form in which they are true or in their complemented form. The PLA programming table is then generated and the required fuse map obtained. The fuse map is applied to an FPLA programmer that goes through the hardware procedure of blowing the internal fuses in the integrated circuit.

7.7 PROGRAMMABLE ARRAY LOGIC

The PAL is a programmable logic device with a fixed OR array and a programmable AND array. Because only the AND gates are programmable, the PAL is easier to program than, but is not as flexible as, the PLA. Figure 7.16 shows the logic configuration of a typical PAL with four inputs and four outputs. Each input has a buffer–inverter gate, and each output is generated by a fixed OR gate. There are four sections in the unit, each composed of an AND–OR array that is *three wide*, the term used to indicate that there are three programmable AND gates in each section and one fixed OR gate. Each AND gate has 10 programmable input connections, shown in the diagram by 10 vertical lines intersecting each horizontal line. The horizontal line symbolizes the multiple-input configuration of the AND gate. One of the outputs is connected to a buffer–inverter gate and then fed back into two inputs of the AND gates.

Commercial PAL devices contain more gates than the one shown in Fig. 7.16. A typical PAL integrated circuit may have eight inputs, eight outputs, and eight sections, each consisting of an eight-wide AND–OR array. The output terminals are sometimes driven by three-state buffers or inverters.

In designing with a PAL, the Boolean functions must be simplified to fit into each section. Unlike the situation with a PLA, a product term cannot be shared among two or more OR gates. Therefore, each function can be simplified by itself, without regard

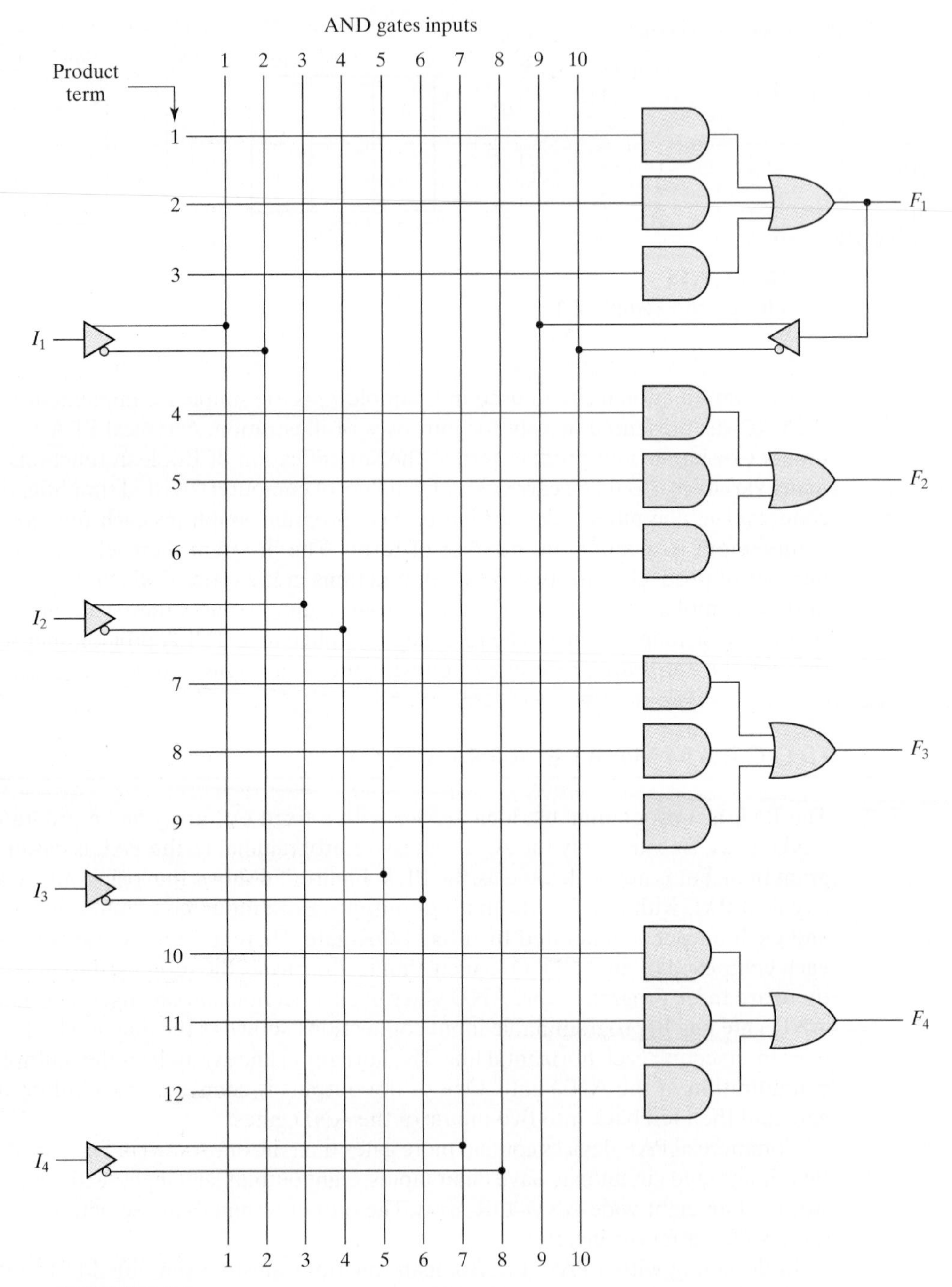

FIGURE 7.16
PAL with four inputs, four outputs, and a three-wide AND–OR structure

to common product terms. The number of product terms in each section is fixed, and if the number of terms in the function is too large, it may be necessary to use two sections to implement one Boolean function.

As an example of using a PAL in the design of a combinational circuit, consider the following Boolean functions, given in sum-of-minterms form:

$$w(A, B, C, D) = \Sigma(2, 12, 13)$$
$$x(A, B, C, D) = \Sigma(7, 8, 9, 10, 11, 12, 13, 14, 15)$$
$$y(A, B, C, D) = \Sigma(0, 2, 3, 4, 5, 6, 7, 8, 10, 11, 15)$$
$$z(A, B, C, D) = \Sigma(1, 2, 8, 12, 13)$$

Simplifying the four functions to a minimum number of terms results in the following Boolean functions:

$$w = ABC' + A'B'CD'$$
$$x = A + BCD$$
$$y = A'B + CD + B'D'$$
$$z = ABC' + A'B'CD' + AC'D' + A'B'C'D$$
$$= w + AC'D' + A'B'C'D$$

Note that the function for z has four product terms. The logical sum of two of these terms is equal to w. By using w, it is possible to reduce the number of terms for z from four to three.

The PAL programming table is similar to the one used for the PLA, except that only the inputs of the AND gates need to be programmed. Table 7.6 lists the PAL

Table 7.6
PAL Programming Table

	AND Inputs					
Product Term	**A**	**B**	**C**	**D**	**w**	**Outputs**
1	1	1	0	—	—	$w = ABC' + A'B'CD'$
2	0	0	1	0	—	
3	—	—	—	—	—	
4	1	—	—	—	—	$x = A + BCD$
5	—	1	1	1	—	
6	—	—	—	—	—	
7	0	1	—	—	—	$y = A'B + CD + B'D'$
8	—	—	1	1	—	
9	—	0	—	0	—	
10	—	—	—	—	1	$z = w + AC'D' + A'B'C'D$
11	1	—	0	0	—	
12	0	0	0	1	—	

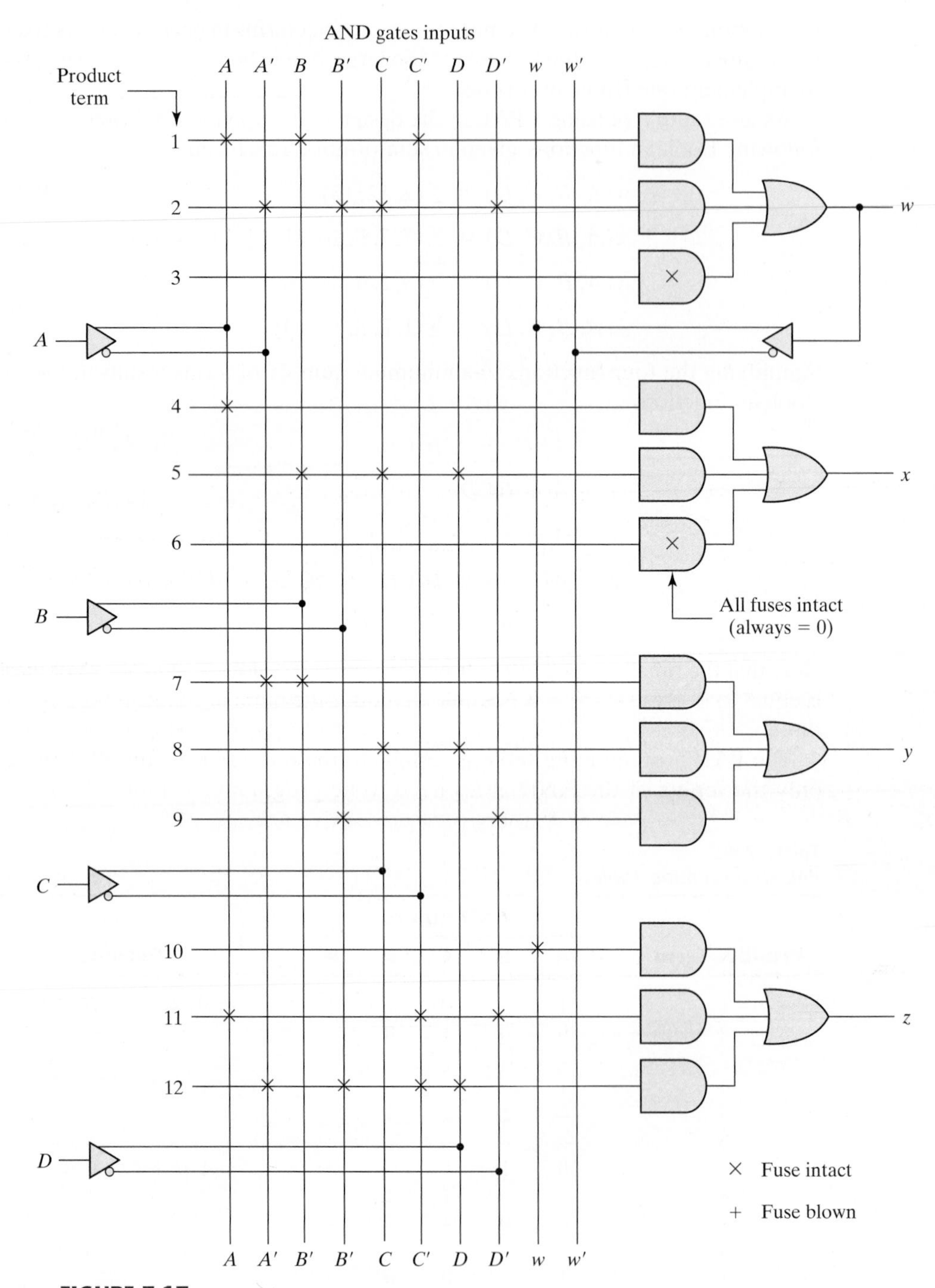

FIGURE 7.17
Fuse map for PAL as specified in Table 7.6

programming table for the four Boolean functions. The table is divided into four sections with three product terms in each, to conform with the PAL of Fig. 7.16. The first two sections need only two product terms to implement the Boolean function. The last section, for output z, needs four product terms. Using the output from w, we can reduce the function to three terms.

The fuse map for the PAL as specified in the programming table is shown in Fig. 7.17. For each 1 or 0 in the table, we mark the corresponding intersection in the diagram with the symbol for an intact fuse. For each dash, we mark the diagram with blown fuses in both the true and complement inputs. If the AND gate is not used, we leave all its input fuses intact. Since the corresponding input receives both the true value and the complement of each input variable, we have $AA' = 0$ and the output of the AND gate is always 0.

As with all PLDs, the design with PALs is facilitated by using CAD techniques. The blowing of internal fuses is a hardware procedure done with the help of special electronic instruments.

7.8 SEQUENTIAL PROGRAMMABLE DEVICES

Digital systems are designed with flip-flops and gates. Since the combinational PLD consists of only gates, it is necessary to include external flip-flops when they are used in the design. Sequential programmable devices include both gates and flip-flops. In this way, the device can be programmed to perform a variety of sequential-circuit functions. There are several types of sequential programmable devices available commercially, and each device has vendor-specific variants within each type. The internal logic of these devices is too complex to be shown here. Therefore, we will describe three major types without going into their detailed construction:

1. Sequential (or simple) programmable logic device (SPLD)
2. Complex programmable logic device (CPLD)
3. Field-programmable gate array (FPGA)

The sequential PLD is sometimes referred to as a simple PLD to differentiate it from the complex PLD. The SPLD includes flip-flops, in addition to the AND–OR array, within the integrated circuit chip. The result is a sequential circuit as shown in Fig. 7.18. A PAL or PLA is modified by including a number of flip-flops connected to form a register. The circuit outputs can be taken from the OR gates or from the outputs of the

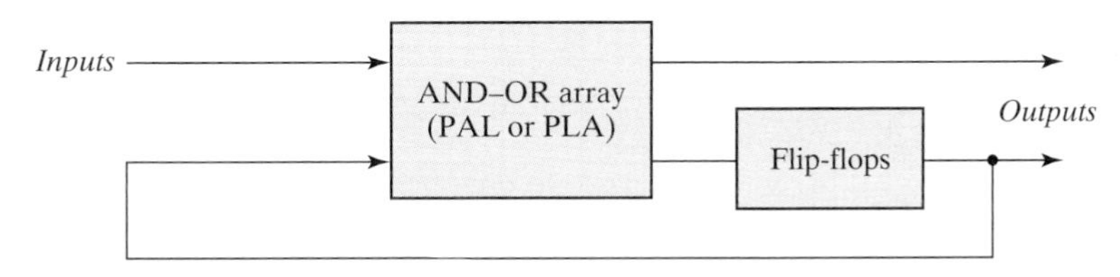

FIGURE 7.18
Sequential programmable logic device

flip-flops. Additional programmable connections are available to include the flip-flop outputs in the product terms formed with the AND array. The flip-flops may be of the *D* or the *JK* type.

The first programmable device developed to support sequential circuit implementation is the field-programmable logic sequencer (FPLS). A typical FPLS is organized around a PLA with several outputs driving flip-flops. The flip-flops are flexible in that they can be programmed to operate as either the *JK* or the *D* type. The FPLS did not succeed commercially, because it has too many programmable connections. The configuration mostly used in an SPLD is the combinational PAL together with *D* flip-flops. A PAL that includes flip-flops is referred to as a *registered* PAL, to signify that the device contains flip-flops in addition to the AND–OR array. Each section of an SPLD is called a *macrocell,* which is a circuit that contains a sum-of-products combinational logic function and an optional flip-flop. We will assume an AND–OR sum-of-products function, but in practice, it can be any one of the two-level implementations presented in Section 3.7.

Figure 7.19 shows the logic of a basic macrocell. The AND–OR array is the same as in the combinational PAL shown in Fig. 7.16. The output is driven by an edge-triggered *D* flip-flop connected to a common clock input and changes state on a clock edge. The output of the flip-flop is connected to a three-state buffer (or inverter) controlled by an output-enable signal marked in the diagram as *OE*. The output of the flip-flop is fed back into one of the inputs of the programmable AND gates to provide the present-state condition for the sequential circuit. A typical SPLD has from 8 to 10 macrocells within

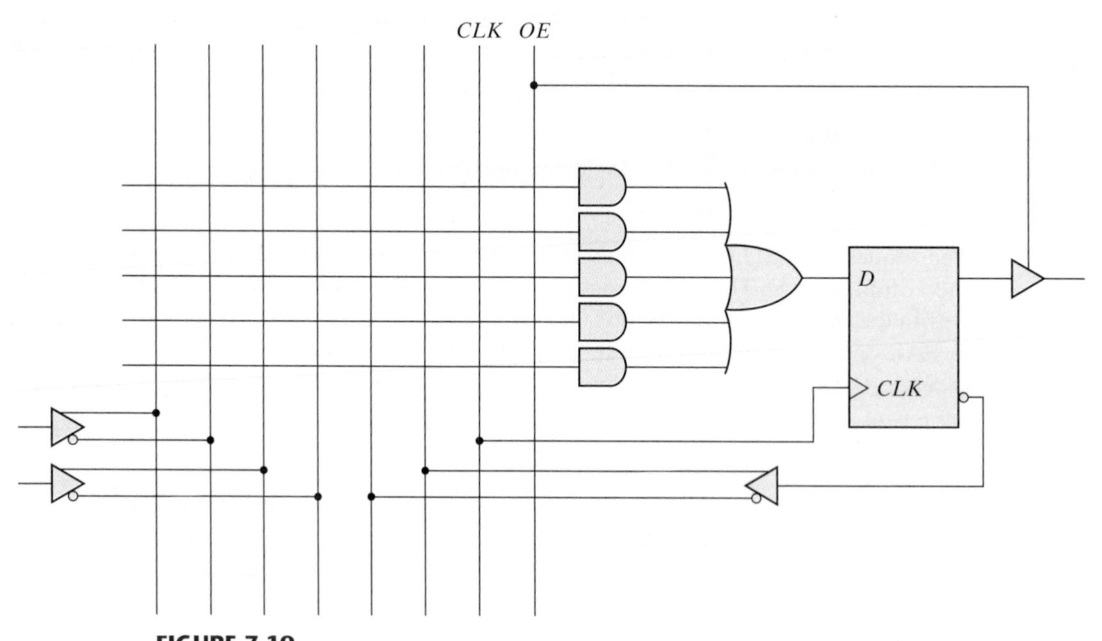

FIGURE 7.19
Basic macrocell logic

one IC package. All the flip-flops are connected to the common *CLK* input, and all three-state buffers are controlled by the *OE* input.

In addition to programming the AND array, a macrocell may have other programming features. Typical programming options include the ability to either use or bypass the flip-flop, the selection of clock edge polarity, the selection of preset and clear for the register, and the selection of the true value or complement of an output. An XOR gate is used to program a true/complement condition. Multiplexers select between two or four distinct paths by programming the selection inputs.

The design of a digital system using PLDs often requires the connection of several devices to produce the complete specification. For this type of application, it is more economical to use a complex programmable logic device (CPLD), which is a collection of individual PLDs on a single integrated circuit. A programmable interconnection structure allows the PLDs to be connected to each other in the same way that can be done with individual PLDs.

Figure 7.20 shows the general configuration of a CPLD. The device consists of multiple PLDs interconnected through a programmable switch matrix. The input–output (I/O) blocks provide the connections to the IC pins. Each I/O pin is driven by a three-state buffer and can be programmed to act as input or output. The switch matrix receives inputs from the I/O block and directs them to the individual macrocells. Similarly, selected outputs from macrocells are sent to the outputs as needed. Each PLD typically contains from 8 to 16 macrocells, usually fully connected. If a macrocell has unused product terms, they can be used by other nearby macrocells. In some cases the macrocell flip-flop is programmed to act as a *D*, *JK*, or *T* flip-flop.

Different manufacturers have taken different approaches to the general architecture of CPLDs. Areas in which they differ include the individual PLDs (sometimes called

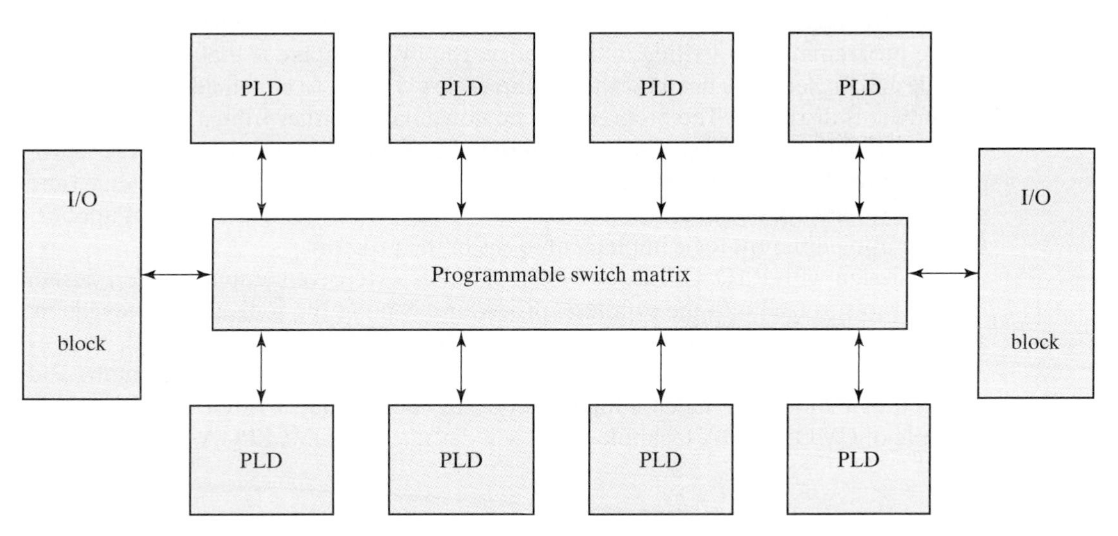

FIGURE 7.20
General CPLD configuration

function blocks), the type of macrocells, the I/O blocks, and the programmable interconnection structure. The best way to investigate a vendor-specific device is to look at the manufacturer's literature.

The basic component used in VLSI design is the *gate array,* which consists of a pattern of gates, fabricated in an area of silicon, that is repeated thousands of times until the entire chip is covered with gates. Arrays of one thousand to several hundred thousand gates are fabricated within a single IC chip, depending on the technology used. The design with gate arrays requires that the customer provide the manufacturer the desired interconnection pattern. The first few levels of the fabrication process are common and independent of the final logic function. Additional fabrication steps are required to interconnect the gates according to the specifications given by the designer.

A field-programmable gate array (FPGA) is a VLSI circuit that can be programmed at the user's location. A typical FPGA consists of an array of millions of logic blocks, surrounded by programmable input and output blocks and connected together via programmable interconnections. There is a wide variety of internal configurations within this group of devices. The performance of each type of device depends on the circuit contained in its logic blocks and the efficiency of its programmed interconnections.

A typical FPGA logic block consists of lookup tables, multiplexers, gates, and flip-flops. A lookup table is a truth table stored in an SRAM and provides the combinational circuit functions for the logic block. These functions are realized from the lookup table, in the same way that combinational circuit functions are implemented with ROM, as described in Section 7.5. For example, a 16×2 SRAM can store the truth table of a combinational circuit that has four inputs and two outputs. The combinational logic section, along with a number of programmable multiplexers, is used to configure the input equations for the flip-flop and the output of the logic block.

The advantage of using RAM instead of ROM to store the truth table is that the table can be programmed by writing into memory. The disadvantage is that the memory is volatile and presents the need for the lookup table's content to be reloaded in the event that power is disrupted. The program can be downloaded either from a host computer or from an onboard PROM. The program remains in SRAM until the FPGA is reprogrammed or the power is turned off. The device must be reprogrammed every time power is turned on. The ability to reprogram the FPGA can serve a variety of applications by using different logic implementations in the program.

The design with PLD, CPLD, or FPGA requires extensive computer-aided design (CAD) tools to facilitate the synthesis procedure. Among the tools that are available are schematic entry packages and hardware description languages (HDLs), such as ABEL, VHDL, and Verilog. Synthesis tools are available that allocate, configure, and connect logic blocks to match a high-level design description written in HDL. As an example of CMOS FPGA technology, we will discuss the Xilinx FPGA.[1]

---

[1] See www.Altera.com for an alternative CMOS FPGA architecture.

Xilinx FPGAs

Xilinx launched the world's first commercial FPGA in 1985, with the vintage XC2000 device family.[2] The XC3000 and XC4000 families soon followed, setting the stage for today's Spartan™, and Virtex™ device families. Each evolution of devices brought improvements in density, performance, power consumption, voltage levels, pin counts, and functionality. For example, the Spartan family of devices initially offered a maximum of 40K system gates, but today's Spartan-6 offers 150,000 logic cells plus 4.8Mb block RAM.

Basic Xilinx Architecture

The basic architecture of Spartan and earlier device families consists of an array of configurable logic blocks (CLBs), a variety of local and global routing resources, and input–output (I/O) blocks (IOBs), programmable I/O buffers, and an SRAM-based configuration memory, as shown in Fig. 7.21.

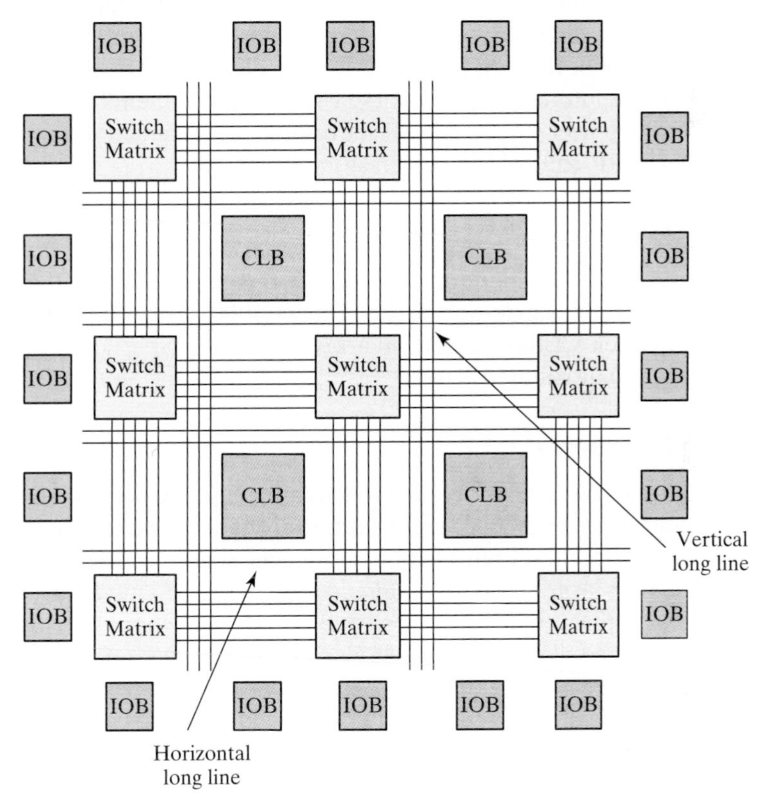

FIGURE 7.21
Basic architecture of Xilinx Spartan and predecessor devices

---

[2] See www.Xilinx.com for detailed, up-to-date information about Xilinx products.

Configurable Logic Block (CLB)

Each CLB consists of a programmable lookup table, multiplexers, registers, and paths for control signals, as shown in Fig. 7.22. Two of the function generators (F and G) of the lookup table can generate any arbitrary function of four inputs, and the third (H) can generate any Boolean function of three inputs. The H-function block can get its inputs from the F and G lookup tables or from external inputs. The three function generators can be programmed to generate (1) three different functions of three independent sets of variables (two with four inputs and one with three inputs—one function must be registered within the CLB), (2) an arbitrary function of five variables, (3) an arbitrary function of four variables together with some functions of six variables, and (4) some functions of nine variables.

Each CLB has two storage devices that can be configured as edge-triggered flip-flops with a common clock, or, in the XC4000X, they can be configured as flip-flops or as transparent latches with a common clock (programmed for either edge and separately invertible) and an enable. The storage elements can get their inputs from the function generators or from the D_{in} input. The other element can get an external input from the *H1* input. The function generators can also drive two outputs (X and Y) directly and independently of the outputs of the storage elements. All of these outputs can be connected to the interconnect network. The storage elements are driven by a global set/reset during power-up; the global set/reset is programmed to match the programming of the local S/R control for a given storage element.

Distributed RAM

The three function generators within a CLB can be used as either a 16×2 dual-port RAM or a 32×1 single-port RAM. The XC4000 devices do not have block RAM, but a group of their CLBs can form an array of memory. Spartan devices have block RAM in addition to distributed RAM.

Interconnect Resources

A grid of switch matrices overlays the architecture of CLBs to provide general-purpose interconnect for branching and routing throughout the device. The interconnect has three types of general-purpose interconnects: single-length lines, double-length lines, and long lines. A grid of horizontal and vertical single-length lines connects an array of switch boxes that provide a reduced number of connections between signal paths within each box, not a full crossbar switch. Each CLB has a pair of three-state buffers that can drive signals onto the nearest horizontal lines above or below the CLB.

Direct (dedicated) interconnect lines provide routing between adjacent vertical and horizontal CLBs in the same column or row. These are relatively high speed local connections through metal, but are not as fast as a hardwired metal connection because of the delay incurred by routing the signal paths through the transmission gates that configure the path. Direct interconnect lines do not use the switch matrices, thus eliminating the delay incurred on paths going through a matrix.[3]

---

[3] See Xilinx documentation for the pin-out conventions to establish local interconnects between CLBs.

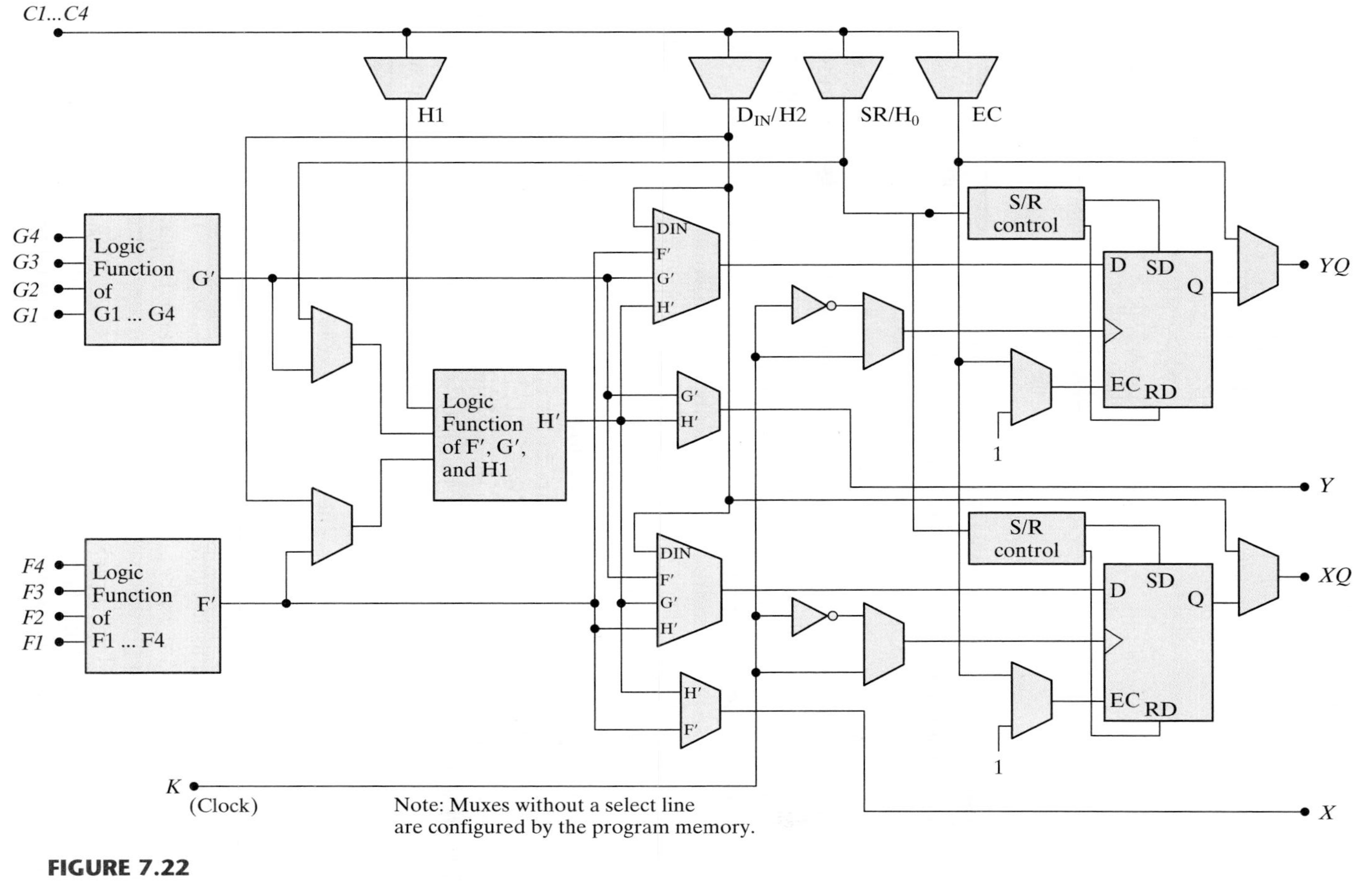

FIGURE 7.22
CLB architecture

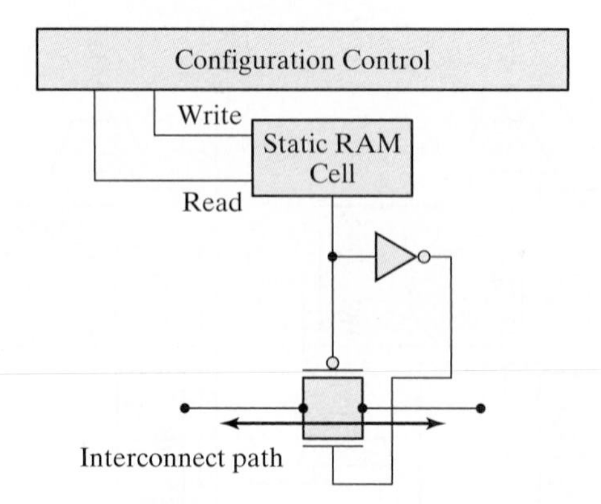

FIGURE 7.23
RAM cell controlling a PIP transmission gate

Double-length lines traverse the distance of two CLBs before entering a switch matrix, skipping every other CLB. These lines provide a more efficient implementation of intermediate-length connections by eliminating a switch matrix from the path, thereby reducing the delay of the path.

Long lines span the entire array vertically and horizontally. They drive low-skew, high-fan-out control signals. Long vertical lines have a programmable splitter that segments the lines and allows two independent routing channels spanning one-half of the array, but located in the same column. The routing resources are exploited automatically by the routing software. There are eight low-skew global buffers for clock distribution.

The signals that drive long lines are buffered. Long lines can be driven by adjacent CLBs or IOBs and may connect to three-state buffers that are available to CLBs. Long lines provide three-state buses within the architecture and implement wired-AND logic. Each horizontal long line is driven by a three-state buffer and can be programmed to connect to a pull-up resistor, which pulls the line to a logical 1 if no driver is asserted on the line.

The programmable interconnect resources of the device connect CLBs and IOBs, either directly or through switch boxes. These resources consist of a grid of two layers of metal segments and programmable interconnect points (PIPs) within switch boxes. A PIP is a CMOS transmission gate whose state (on or off) is determined by the content of a static RAM cell in the programmable memory, as shown in Fig. 7.23. The connection is established when the transmission gate is on (i.e., when a 1 is applied at the gate of the n-channel transistor), and a 0 is applied at the gate of the p-channel transistor. Thus, the device can be reprogrammed simply by changing the contents of the controlling memory cell.

The architecture of a PIP-based interconnection in a switch box is shown in Fig. 7.24, which shows possible signal paths through a PIP. The configuration of CMOS transmission gates determines the connection between a horizontal line and the opposite horizontal line and between the vertical lines at the connection. Each switch matrix PIP requires six pass transistors to establish full connectivity.

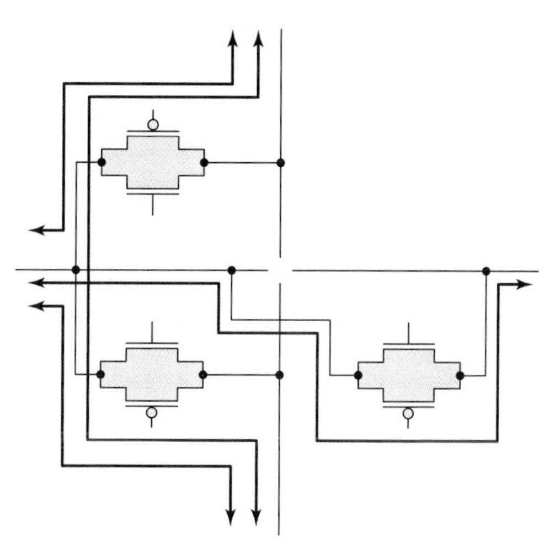

FIGURE 7.24
Circuit for a programmable PIP

I/O Block (IOB)

Each programmable I/O pin has a programmable IOB having buffers for compatibility with TTL and CMOS signal levels. Figure 7.25 shows a simplified schematic for a programmable IOB. It can be used as an input, an output, or a bidirectional port. An IOB that is configured as an input can have direct, latched, or registered input. In an output configuration, the IOB has direct or registered output. The output buffer of an IOB has skew and slew control. The registers available to the input and output path of an IOB are driven by separate, invertible clocks. There is a global set/reset.

Internal delay elements compensate for the delay induced when a clock signal passes through a global buffer before reaching an IOB. This strategy eliminates the hold condition on the data at an external pin. The three-state output of an IOB puts the output buffer in a high-impedance state. The output and the enable for the output can be inverted. The slew rate of the output buffer can be controlled to minimize transients on the power bus when noncritical signals are switched. The IOB pin can be programmed for pull-up or pull-down to prevent needless power consumption and noise.

The devices have embedded logic to support the IEEE 1149.1 (JTAG) boundary scan standard. There is an on-chip test access port (TAP) controller, and the I/O cells can be configured as a shift register. Under testing, the device can be checked to verify that all the pins on a PC board are connected and operate properly by creating a serial chain of all of the I/O pins of the chips on the board. A master three-state control signal puts all of the IOBs in high-impedance mode for board testing.

Enhancements

Spartan chips can accommodate embedded soft cores, and their on-chip distributed, dual-port, synchronous RAM (SelectRAM) can be used to implement first-in, first-out register

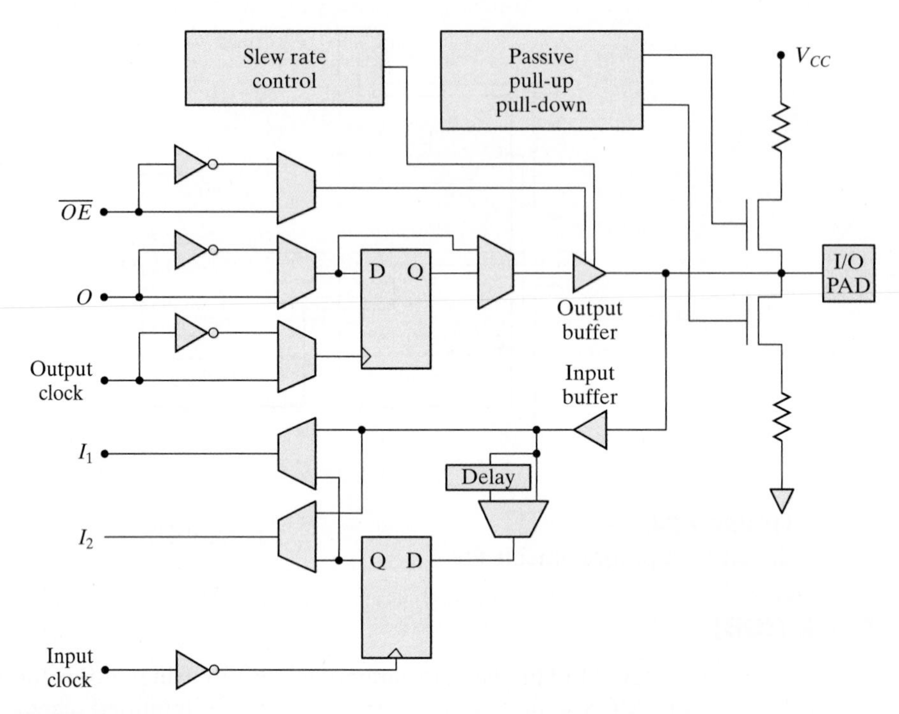

FIGURE 7.25
XC4000 series IOB

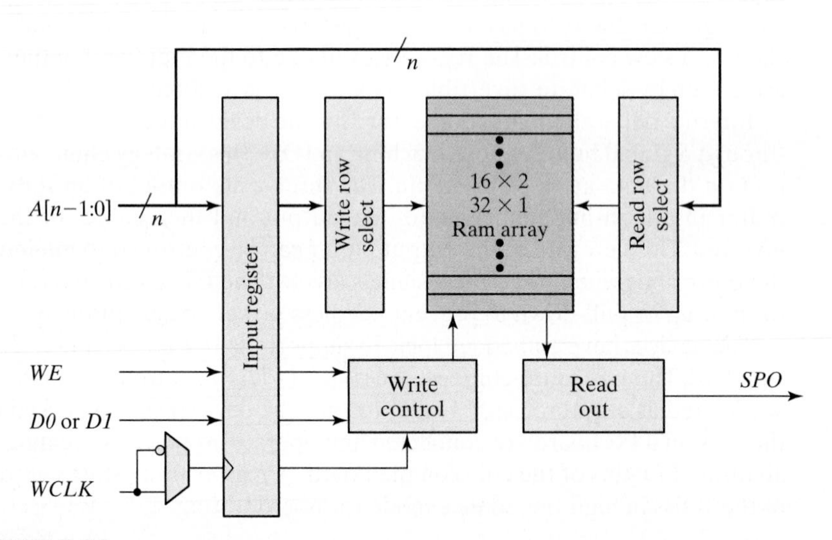

FIGURE 7.26
Distributed RAM cell formed from a lookup table

files (FIFOs), shift registers, and scratchpad memories. The blocks can be cascaded to any width and depth and located anywhere in the part, but their use reduces the CLBs available for logic. Figure 7.26 displays the structure of the on-chip RAM that is formed by

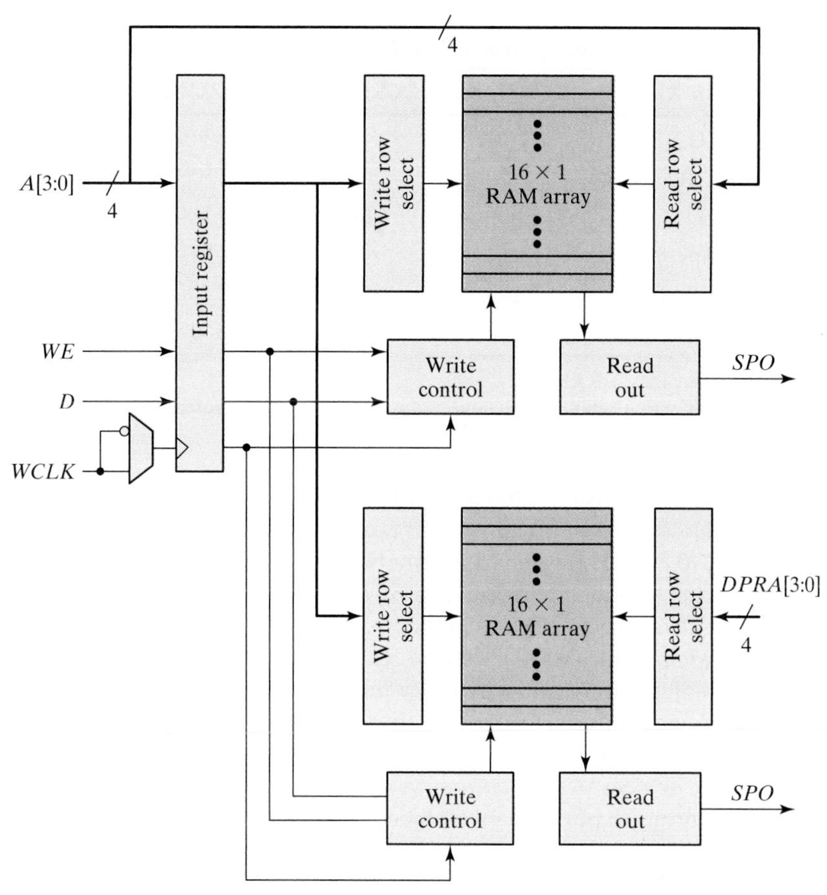

FIGURE 7.27
Spartan dual-port RAM

programming a lookup table to implement a single-port RAM with synchronous write and asynchronous read. Each CLB can be programmed as a 16 × 2 or 32 × 1 memory.

Dual-port RAMs are emulated in a Spartan device by the structure shown in Fig. 7.27, which has a single (common) write port and two asynchronous read ports. A CLB can form a memory having a maximum size of 16 × 1.

Xilinx Spartan XL FPGAs

Spartan XL chips are a further enhancement of Spartan chips, offering higher speed and density (40,000 system gates, approximately 6,000 of which are usable) and on-chip, distributed SelectRAM memory.[4] The lookup tables of the devices can implement 2^{2^n} different functions of n inputs.

[4] The maximum number of logic gates for a Xilinx FPGA is an estimate of the maximum number of logic gates that could be realized in a design consisting of only logic functions (no memory). Logic capacity is expressed in terms of the number of two-input NAND gates that would be required to implement the same number and type of logic functions (Xilinx App. Note).

Table 7.7
Attributes of the Xilinx Spartan XL Device Family

Spartan XL	XCS05/XL	XCS10/XL	XCS20/XL	XCS30/XL	XCS40/XL
System Gates[1]	2K–5K	3K–10K	7K–20K	10K–30K	13K–40K
Logic Cells[2]	238	466	950	1,368	1,862
Max Logic Gates	3,000	5,000	10,000	13,000	20,000
Flip-Flops	360	616	1,120	1,536	2,016
Max RAM Bits	3,200	6,272	12,800	18,432	25,088
Max Avail I/O	77	112	160	192	224

[1] 20–30% of CLBs as RAM.
[2] 1 Logic cell = four-input lookup table + flip-flop.

The XL series is targeted for applications for which low cost, low power, low packaging, and low test cost are important factors constraining the design. Spartan XL devices offer up to 80-MHz system performance, depending on the number of cascaded lookup tables, which reduce performance by introducing longer paths. Table 7.7 presents significant attributes of devices in the Spartan XL family.

The architecture of the Spartan XL and earlier devices consists of an array of CLB tiles mingled within an array of switch matrices, surrounded by a perimeter of IOBs. These devices support only distributed memory, whose use reduces the number of CLBs that could be used for logic. The relatively small amount of on-chip memory limits the devices to applications in which operations with off-chip memory devices do not compromise performance objectives. Beginning with the Spartan II series, Xilinx supported configurable embedded block memory, as well as distributed memory in a new architecture.

Xilinx Spartan II FPGAs

Aside from improvements in speed (200-MHz I/O switching frequency), density (up to 200,000 system gates) and operating voltage (2.5 V), four other features distinguish the Spartan II devices from the Spartan devices: (1) on-chip block memory, (2) a novel architecture, (3) support for multiple I/O standards, and (4) delay locked loops (DLLs).[5]

The Spartan II device family, manufactured in 0.22/0.18-μm CMOS technology with six layers of metal for interconnect, incorporates configurable block memory in addition to the distributed memory of the previous generations of devices, and the block memory does not reduce the amount of logic or distributed memory that is available for the

[5] Spartan II devices do not support low-voltage differential signaling (LVDS) or low-voltage positive emitter-coupled logic (LVPECL) I/O standards.

application. A large on-chip memory can improve system performance by eliminating or reducing the need to access off-chip storage.

Reliable clock distribution is the key to the synchronous operation of high-speed digital circuits. If the clock signal arrives at different times at different parts of a circuit, the device may fail to operate correctly. Clock skew reduces the available time budget of a circuit by lengthening the setup time at registers. It can also shorten the effective hold-time margin of a flip-flop in a shift register and cause the register to shift incorrectly. At high clock frequencies (shorter clock periods), the effect of skew is more significant because it represents a larger fraction of the clock cycle time. Buffered clock trees are commonly used to minimize clock skew in FPGAs. Xilinx provides all-digital DLLs for clock synchronization or management in high-speed circuits. DLLs eliminate the clock distribution delay and provide frequency multipliers, frequency dividers, and clock mirrors.

Spartan II devices are suitable for applications such as implementing the glue logic of a video capture system and the glue logic of an ISDN modem. Device attributes are summarized in Table 7.8, and the evolution of technology in the Spartan series is evident in the data in Table 7.9.

Table 7.8
Spartan II Device Attributes

Spartan II FPGAs	XC2S15	XC2S30	XC2S50	XC2S100	XC2S150	XC2S200
System Gates[1]	6K–15K	13K–30K	23K–50K	37K–100K	52K–150K	71K–200K
Logic Cells[2]	432	972	1,728	2,700	3,888	5,292
Block RAM Bits	16,384	24,576	32,768	40,960	49,152	57,344
Max Avail I/O	86	132	176	196	260	284

[1] 20–30% of CLBs as RAM.
[2] 1 Logic cell = four-input lookup table + flip-flop.

Table 7.9
Comparison of the Spartan Device Families

Part	Spartan	Spartan XL	Spartan II
Architecture	XC4000 Based	XC4000 Based	Virtex Based
Max # System Gates	5K–40K	5K–40K	15K–200K
Memory	Distributed RAM	Distributed RAM	Block + Distributed
I/O Performance	80 MHz	100 MHz	200 MHz
I/O Standards	4	4	16
Core Voltage	5 V	3.3 V	2.5 V
DLLs	No	No	Yes

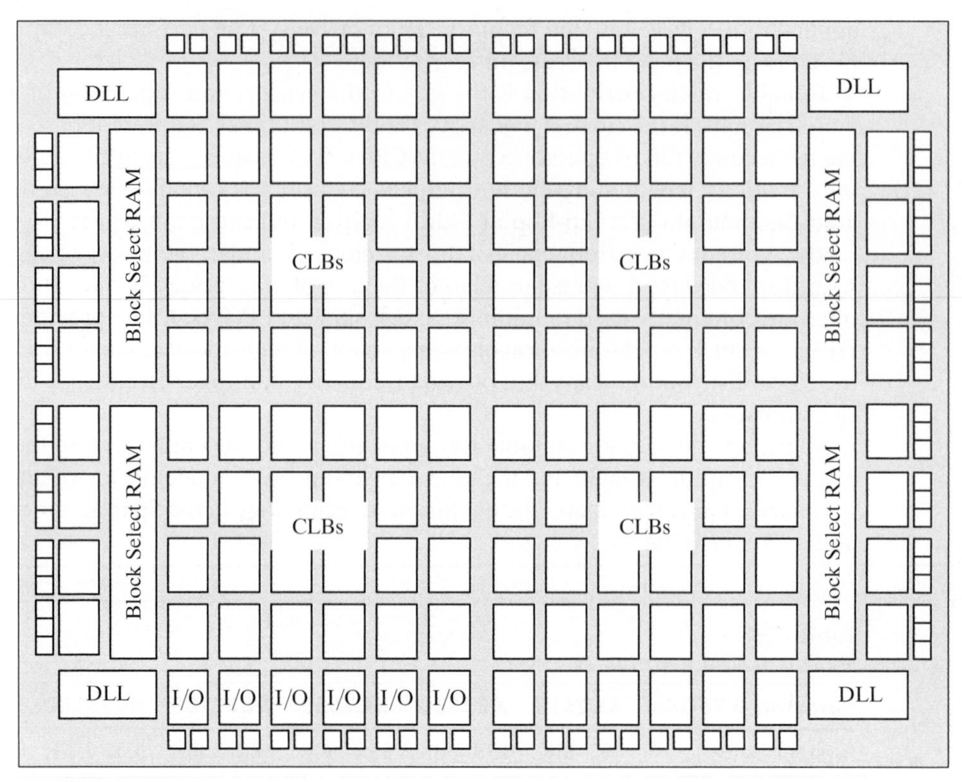

FIGURE 7.28
Spartan II architecture

The top-level tiled architecture of the Spartan II device, shown in Fig. 7.28, marks a new organization structure of the Xilinx parts. Each of four quadrants of CLBs is supported by a DLL and is flanked by a 4,096-bit block[6] of RAM, and the periphery of the chip is lined with IOBs.

Each CLB contains four logic cells, organized as a pair of slices. Each logic cell, shown in Fig. 7.29, has a four-input lookup table, logic for carry and control, and a *D*-type flip-flop. The CLB contains additional logic for configuring functions of five or six inputs.

The Spartan II part family provides the flexibility and capacity of an on-chip block RAM; in addition, each lookup table can be configured as a 16×1 RAM (distributed), and the pair of lookup tables in a logic cell can be configured as a 16×2 bit RAM or a 32×1 bit RAM.

The IOBs of the Spartan II family are individually programmable to support the reference, output voltage, and termination voltages of a variety of high-speed memory

[6] Parts are available with up to 14 blocks (56K bits).

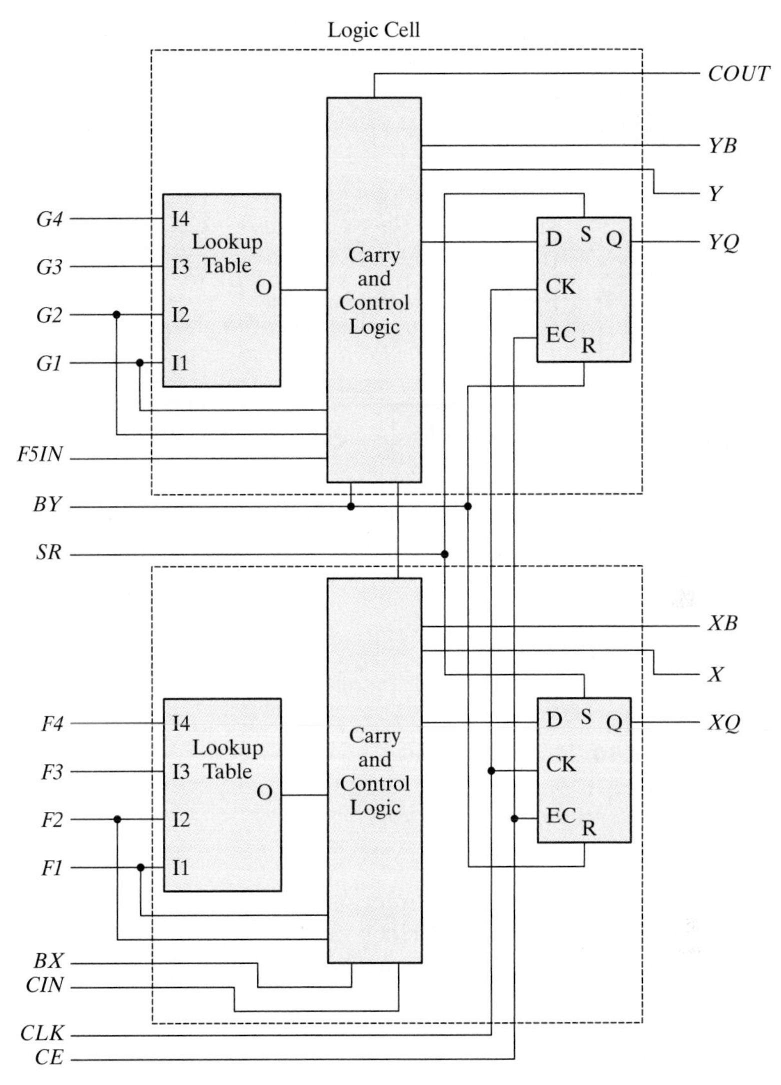

FIGURE 7.29
Spartan II CLB slice

and bus standards. (See Fig. 7.30.) Each IOB has three registers that can function as *D*-type flip-flops or as level-sensitive latches. One register (*TFF*) can be used to register the signal that (synchronously) controls the programmable output buffer. A second register (*OFF*) can be programmed to register a signal from the internal logic. (Alternatively, a signal from the internal logic can pass directly to the output buffer.) The third device can register the signal coming from the I/O pad. (Alternatively, this

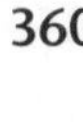

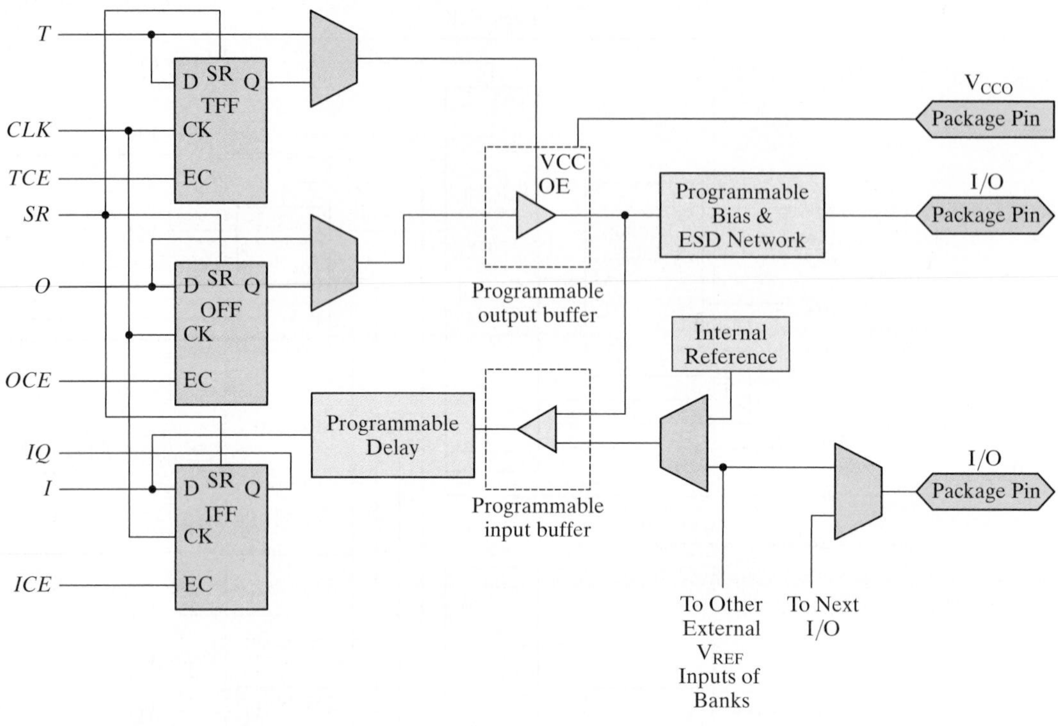

FIGURE 7.30
Spartan II IOB

signal can pass directly to the internal logic.) A common clock drives each register, but each has an independent clock enable. A programmable delay element on the input path can be used to eliminate the pad-to-pad hold time.

Xilinx Virtex FPGAs

The Virtex device series[7] is the leading edge of Xilinx technology. This family of devices addresses four key factors that influence the solution to complex system-level and system-on-chip designs: (1) the level of integration, (2) the amount of embedded memory, (3) performance (timing), and (4) subsystem interfaces. The family targets applications requiring a balance of high-performance logic, serial connectivity, signal processing, and embedded processing (e.g., wireless communications). Process rules

[7] Virtex, Virtex-II, II Platform, II-Pro/Pro X, and Virtex-5 Multi-Platform FPGA.

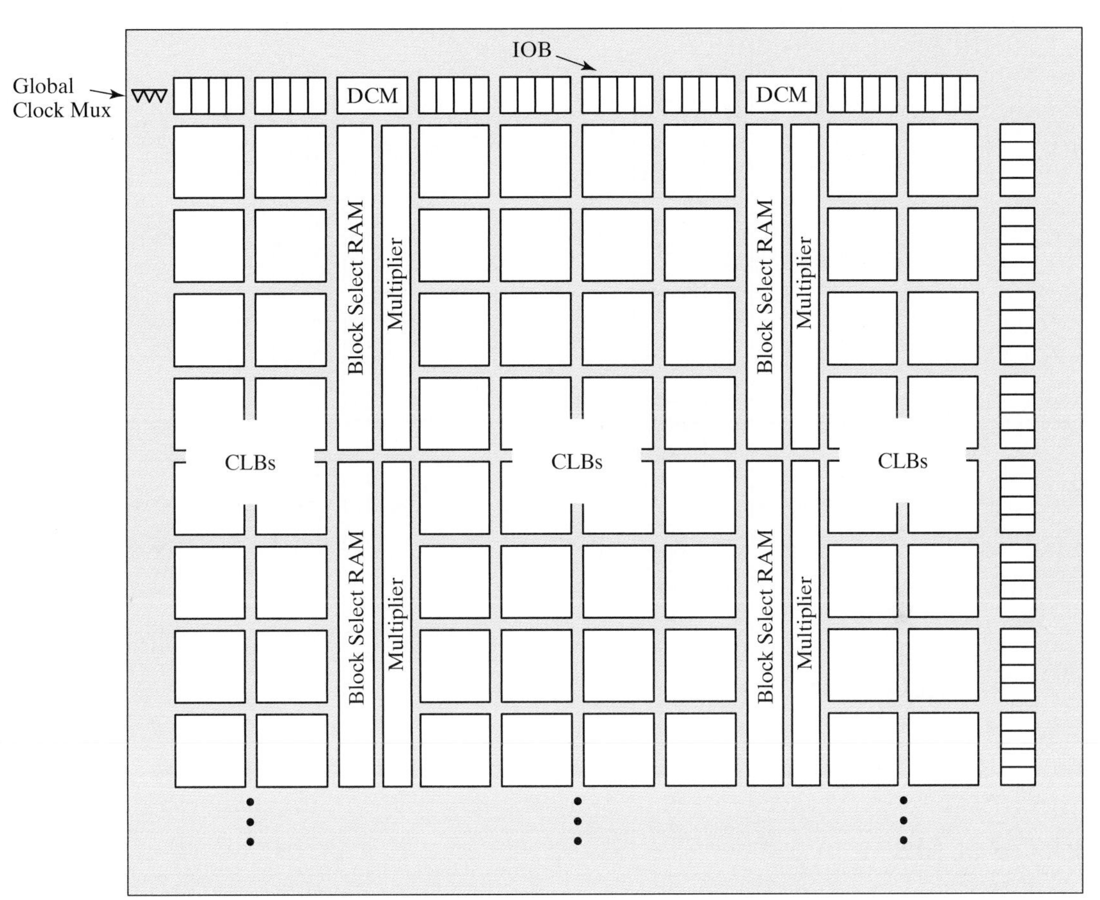

FIGURE 7.31
Virtex II overall architecture

for leading-edge Virtex parts stand at 65 nm, with a 1-V operating voltage. The rules allow up to 330,000 logic cells and over 200,000 internal flip-flops with clock enable, together with over 10 Mb of block RAM, and 550-MHz clock technology packed into a single die.

The Virtex family incorporates physical (electrical) and protocol support for 20 different I/O standards, including LVDS and LVPECL, with individually programmable pins. Up to 12 digital clock managers provide support for frequency synthesis and phase shifting in synchronous applications requiring multiple clock domains and high-frequency I/O. The Virtex architecture is shown in Fig. 7.31, and its IOB is shown in Fig. 7.32.

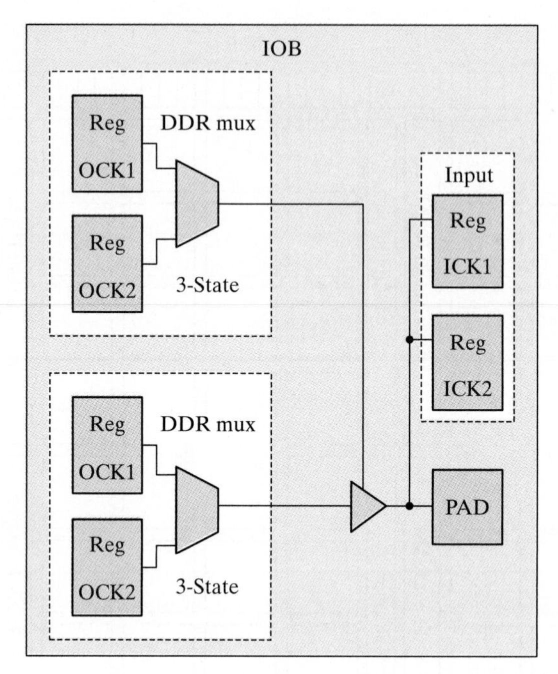

FIGURE 7.32
Virtex IOB block

PROBLEMS

Answers to problems marked with * appear at the end of the book.

7.1 The memory units that follow are specified by the number of words times the number of bits per word. How many address lines and input–output data lines are needed in each case?
(a) 16K × 8 (b) 512K × 16
(c) 32M × 32 (d) 8G × 16

7.2* Give the number of bytes stored in the memories listed in Problem 7.1.

7.3* Word number 875 in the memory shown in Fig. 7.3 contains the binary equivalent of 46,654. List the 10-bit address and the 16-bit memory content of the word.

7.4 Show the memory cycle timing waveforms for the write and read operations. Assume a CPU clock of 150 MHz and a memory cycle time of 20 ns.

7.5 Write a test bench for the ROM described in Example 7.1. The test program stores binary 7 in address 5 and binary 5 in address 7. Then the two addresses are read to verify their stored contents.

7.6 Enclose the 4 × 4 RAM of Fig. 7.6 in a block diagram showing all inputs and outputs. Assuming three-state outputs, construct an 8 × 12 memory using four 4 × 4 RAM units.

7.7* A 64K × 8 memory uses coincident decoding by splitting the internal decoder into X-selection and Y-selection.
(a) What is the size of each decoder, and how many AND gates are required for decoding the address?
(b) Determine the X and Y selection lines that are enabled when the input address is the binary equivalent of 36,952.

7.8* (a) How many 64K × 8 RAM chips are needed to provide a memory capacity of 512K bytes?
(b) How many lines of the address must be used to access 512K bytes? How many of these lines are connected to the address inputs of all chips?
(c) How many lines must be decoded for the chip select inputs? Specify the size of the decoder.

7.9 A DRAM chip uses two-dimensional address multiplexing. It has 15 common address pins, with the row address having one bit more than the column address. What is the capacity of the memory?

7.10* Given the 10-bit data word 0110101011, generate the 14-bit composite word for the Hamming code that corrects single errors and detects double errors.

7.11* Obtain the 13-bit Hamming code word for the 9-bit data word 110110110.

7.12* A 13-bit Hamming code word containing 9 bits of data and 4 parity bits is read from memory. What was the original 9-bit data word that was written into memory if the 13-bit word read out is as follows:
(a) 0 1110 0101 0100
(b) 1 1110 1010 0111
(c) 1 0101 0011 1010
(d) 1 0111 0111 0110
(e) 0 0001 0001 0101

7.13* How many parity check bits must be included with the data word to achieve single-error correction and double-error detection when the data word contains
(a) 25 bits.
(b) 55 bits.
(c) 100 bits.

7.14 It is necessary to formulate the Hamming code for five data bits, D_3, D_5, D_6, and D_9, together with three parity bits, P_1, P_2, P_4, and P_8.
(a)* Evaluate the 9-bit composite code word for the data word 01101.
(b) Evaluate four check bits, C_8, C_4, C_2, and C_1, assuming no error.
(c) Assume an error in bit D_6 during writing into memory. Show how the error in the bit is detected and corrected.
(d) Add parity bit P_{10} to include double-error detection in the code. Assume that errors occurred in bits P_1 and D_3. Show how the double error is detected.

7.15 Using 1K × 8 ROM chips with an enable input, construct a 4K × 8 ROM with four chips and a decoder.

7.16* A ROM chip of 8,192 × 8 bits has three chip select inputs and operates from a 5-V power supply. How many pins are needed for the integrated circuit package? Draw a block diagram, and label all input and output terminals in the ROM.

7.17 The 32 × 6 ROM, together with the 2^0 line, as shown in Fig. P7.17, converts a six-bit binary number to its corresponding two-digit BCD number. For example, binary 100001 converts to BCD 011 0011 (decimal 33). Specify the truth table for the ROM.

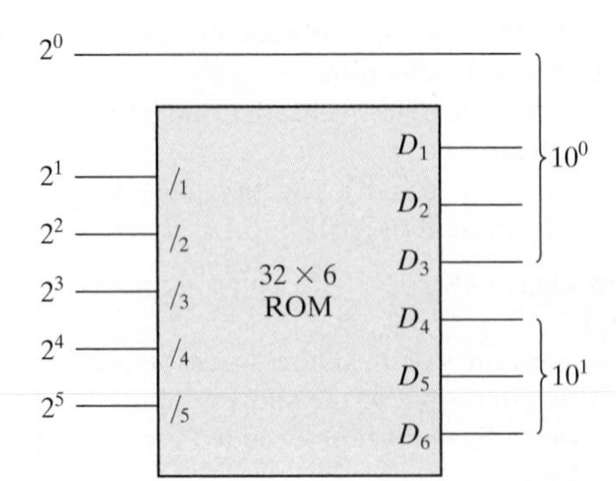

FIGURE P7.17

7.18* Specify the size of a ROM (number of words and number of bits per word) that will accommodate the truth table for the following combinational circuit components:
(a) a binary multiplier that multiplies two 5-bit binary words,
(b) a 5-bit adder–subtractor,
(c) a quadruple four-to-one-line multiplexer with common select and enable inputs, and
(d) a BCD-to-seven-segment decoder with an enable input.

7.19 Tabulate the PLA programming table for the four Boolean functions listed below. Minimize the numbers of product terms.

$$A(x, y, z) = \Sigma(0, 2, 3, 7)$$
$$B(x, y, z) = \Sigma(1, 2, 4, 5, 6)$$
$$C(x, y, z) = \Sigma(0, 1, 5, 7)$$
$$D(x, y, z) = \Sigma(0, 2, 3, 4, 6)$$

7.20 Tabulate the truth table for an 16 × 4 ROM that implements the Boolean functions

$$A(w, x, y, z) = \Sigma(1, 3, 5, 7, 9, 11, 13, 15)$$
$$B(w, x, y, z) = \Sigma(0, 2, 4, 6, 8, 10, 12, 14)$$
$$C(w, x, y, z) = \Sigma(1, 2, 5, 6, 7, 8, 13, 14)$$
$$D(w, x, y, z) = \Sigma(0, 3, 4, 7, 8, 11, 12, 15)$$

Considering now the ROM as a memory. Specify the memory contents at addresses 1 and 4.

7.21 Derive the PLA programming table for the combinational circuit that squares a three-bit number. Minimize the number of product terms. (See Fig. 7.12 for the equivalent ROM implementation.)

7.22 Derive the ROM programming table for the combinational circuit that squares a 4-bit number. Minimize the number of product terms.

7.23 List the PLA programming table for the BCD-to-excess-3-code converter whose Boolean functions are simplified in Fig. 4.3.

7.24 Repeat Problem 7.23, using a PAL.

7.25* The following is a truth table of a three-input, four-output combinational circuit:

Inputs			Outputs			
x	*y*	*z*	*A*	*B*	*C*	*D*
0	0	0	0	0	1	0
0	0	1	1	1	0	1
0	1	0	1	1	1	0
0	1	1	0	0	1	1
1	0	0	1	1	1	1
1	0	1	1	1	1	1
1	1	0	1	1	0	0
1	1	1	0	1	1	1

Tabulate the PAL programming table for the circuit, and mark the fuse map in a PAL diagram similar to the one shown in Fig. 7.17.

7.26 Using the registered macrocell of Fig. 7.19, show the fuse map for a sequential circuit with two inputs x and y and one flip-flop A described by the input equation

$$D_A = (x \oplus y \oplus A)'$$

7.27 Modify the PAL diagram of Fig. 7.16 by including three clocked D-type flip-flops between the OR gates and the outputs, as in Fig. 7.19. The diagram should conform with the block diagram of a sequential circuit. The modification will require three additional buffer–inverter gates and six vertical lines for the flip-flop outputs to be connected to the AND array through programmable connections. Using the modified registered PAL diagram, show the fuse map that will implement a three-bit binary counter with an output carry.

7.28 Draw a PLA circuit to implement the functions

$$F_1 = AB + B'C + A'B'C'$$

$$F_2 = B'C + A'C' + ABC$$

7.29 Develop the programming table for the PLA described in Problem 7.26.

REFERENCES

1. Hamming, R. W. 1950. Error Detecting and Error Correcting Codes. *Bell Syst. Tech. J.* 29: 147–160.
2. Kitson, B. 1984. *Programmable Array Logic Handbook*. Sunnyvale, CA: Advanced Micro Devices.
3. Lin, S. and D. J. Costello, Jr. 2004. *Error Control Coding*. 2nd ed. Englewood Cliffs, NJ: Prentice-Hall.
4. *Memory Components Handbook*. 1986. Santa Clara, CA: Intel.
5. Nelson, V. P., H. T. Nagle, J. D. Irwin, and B. D. Carroll. 1995. *Digital Logic Circuit Analysis and Design*. Upper Saddle River, NJ: Prentice Hall.

6. *The Programmable Logic Data Book*, 2nd ed. 1994. San Jose, CA: Xilinx, Inc.
7. Tocci, R. J. and N. S. Widmer. 2004. *Digital Systems Principles and Applications*, 9th ed. Upper Saddle River, NJ: Prentice Hall.
8. Trimberger, S. M. 1994. *Field Programmable Gate Array Technology*. Boston: Kluwer Academic Publishers.
9. Wakerly, J. F. 2006. *Digital Design: Principles and Practices*, 4th ed. Upper Saddle River, NJ: Prentice Hall.

WEB SEARCH TOPICS

FPGA
Gate array
Programmable array logic
Programmable logic data book
RAM
ROM

Chapter 8

Design at the Register Transfer Level

8.1 INTRODUCTION

The behavior of many digital systems depends on the history of their inputs, and the conditions that determine their future actions depend on the results of previous actions. Such systems are said to have "memory." A digital system is a sequential logic system constructed with flip-flops and gates. Sequential circuits can be specified by means of state tables as shown in Chapter 5. To specify a large digital system with a state table is very difficult, because the number of states would be enormous. To overcome this difficulty, digital systems are designed via a modular approach. The system is partitioned into subsystems, each of which performs some function. The modules are constructed from such digital devices as registers, decoders, multiplexers, arithmetic elements, and control logic. The various modules are interconnected with datapaths and control signals to form a digital system. In this chapter, we will introduce a design methodology for describing and designing large, complex digital systems.

8.2 REGISTER TRANSFER LEVEL NOTATION

The modules of a digital system are best defined by a set of registers and the operations that are performed on the binary information stored in them. Examples of register operations are *shift*, *count*, *clear*, and *load*. Registers are assumed to be the basic components of the digital system. The information flow and processing performed on the data stored in the registers are referred to as *register transfer operations*. We'll see subsequently how a hardware description language (HDL) includes operators that correspond to the register transfer operations of a digital system.

A digital system is represented at the *register transfer level* (RTL) when it is specified by the following three components:

1. The set of registers in the system.
2. The operations that are performed on the data stored in the registers.
3. The control that supervises the sequence of operations in the system.

A register is a connected group of flip-flops that stores binary information and has the capability of performing one or more elementary operations. A register can load new information or shift the information to the right or the left. A counter is a register that increments a number by a fixed value (e.g., 1). A flip-flop is a one-bit register that can be set, cleared, or complemented. In fact, the flip-flops and associated gates of any sequential circuit can be called registers by this definition.

The operations executed on the information stored in registers are elementary operations that are performed in parallel on the bits of a data word during one clock cycle. The data produced by the operation may replace the binary information that was in the register before the operation executed. Alternatively, the result may be transferred to another register (i.e., an operation on a register may leave its contents unchanged). The digital circuits introduced in Chapter 6 are registers that implement elementary operations. A counter with a parallel load is able to perform the increment-by-one and load operations. A bidirectional shift register is able to perform the shift-right and shift-left operations by shifting its contents by one or more bits in a specified direction.

The operations in a digital system are controlled by signals that sequence the operations in a prescribed manner. Certain conditions that depend on results of previous operations may determine the sequence of future operations. The outputs of the control logic of a digital system are binary variables that initiate the various operations in the system's registers.

Information transfer from one register to another is designated in symbolic form by means of a replacement operator. The statement

$$R2 \leftarrow R1$$

denotes a transfer of the contents of register $R1$ into register $R2$—that is, a replacement of the contents of register $R2$ by the contents of register $R1$. For example, an eight-bit register $R2$ holding the value 01011010 could have its contents replaced by $R1$ holding the value 10100101. By definition, the contents of the source register $R1$ do not change after the transfer. They are merely copied to $R1$. The arrow symbolizes the transfer and its direction; it points from the register whose contents are being transferred and towards the register that will receive the contents. A control signal would determine when the operation actually executes.

The controller in a digital system is a finite state machine (see Chapter 5) whose outputs are the control signals governing the register operations. In synchronous machines, the operations are synchronized by the system clock. For example, register $R2$ might be synchronized to have its contents replaced at the positive edge of the clock.

A statement that specifies a register transfer operation implies that a datapath (i.e., a set of circuit connections) is available from the outputs of the source register to the inputs of the destination register and that the destination register has a parallel load capability. Data can be transferred serially between registers, too, by repeatedly shifting their contents along a single wire, one bit at a time. Normally, we want a register transfer operation to occur, not with every clock cycle, but only under a predetermined condition. A conditional statement governing a register transfer operation is symbolized with an if–then statement such as

$$\text{If } (T1 = 1) \text{ then } (R2 \leftarrow R1)$$

where $T1$ is a control signal generated in the control section. Note that the clock is not included as a variable in the register transfer statements. It is assumed that all transfers occur at a clock-edge transition (i.e., a transition from 0 to 1 or from 1 to 0). Although a control condition such as $T1$ may become true before the clock transition, the actual transfer does not occur until the clock transition does. The transfers are initiated and synchronized by the action of the clock signal, but the actual transition of the outputs (in a physical system) does not result in instantaneous transitions at the outputs of the registers. Propagation delays depend on the physical characteristics of the transistors implementing the flip-flops of the register and the wires connecting devices. There is always a delay, however small, between a cause and its effect in a physical system.

A comma may be used to separate two or more operations that are executed at the same time (concurrently). Consider the statement

$$\text{If } (T3 = 1) \text{ then } (R2 \leftarrow R1,\ R1 \leftarrow R2)$$

This statement specifies an operation that exchanges the contents of two registers; moreover, the operation in both registers is triggered by the same clock edge, provided that $T3 = 1$. This simultaneous (concurrent) operation is possible with registers that have edge-triggered flip-flops controlled by a common clock (synchronizing signal). Other examples of register transfers are as follows:

$R1 \leftarrow R1 + R2$	Add contents of $R2$ to $R1$ ($R1$ gets $R1 + R2$)
$R3 \leftarrow R3 + 1$	Increment $R3$ by 1 (count upwards)
$R4 \leftarrow \text{shr } R4$	Shift right $R4$
$R5 \leftarrow 0$	Clear $R5$ to 0

In hardware, addition is done with a binary parallel adder, incrementing is done with a counter, and the shift operation is implemented with a shift register. The type of operations most often encountered in digital systems can be classified into four categories:

1. Transfer operations, which transfer (i.e., copy) data from one register to another.
2. Arithmetic operations, which perform arithmetic (e.g., multiplication) on data in registers.
3. Logic operations, which perform bit manipulation (e.g., logical OR) of nonnumeric data in registers.
4. Shift operations, which shift data between registers.

The transfer operation does not change the information content of the data being moved from the source register to the destination register unless the source and destination are the same. The other three operations change the information content during the transfer. The register transfer notation and the symbols used to represent the various register transfer operations are not standardized. In this text, we employ two types of notation. The notation introduced in this section will be used informally to specify and explain digital systems at the register transfer level. The next section introduces the RTL symbols used in the Verilog HDL, which is standardized.

8.3 REGISTER TRANSFER LEVEL IN HDL

Digital systems can be described at the register transfer level by means of a HDL. In the Verilog HDL, descriptions of RTL operations use a combination of behavioral and dataflow constructs and are employed to specify the register operations and the combinational logic functions implemented by hardware. Register transfers are specified by means of procedural assignment statements within an *edge-sensitive cyclic behavior.* Combinational circuit functions are specified at the RTL level by means of continuous assignment statements or by procedural assignment statements within a *level-sensitive cyclic behavior.* The symbol used to designate a register transfer is either an equals sign (=) or an arrow (<=); the symbol used to specify a combinational circuit function is an equals sign. Synchronization with the clock is represented by associating with an **always** statement an event control expression in which sensitivity to the clock event is qualified by **posedge** or **negedge**. The *always* keyword indicates that the associated block of statements will be executed repeatedly, for the life of the simulation. The @ operator and the event control expression preceding the block of statements synchronize the execution of the statements to the clock event.

The following examples show the various ways to specify a register transfer operation in Verilog:

```
(a) assign S = A + B;              // Continuous assignment for addition operation
(b) always @ (A, B)                // Level-sensitive cyclic behavior
      S = A + B;                   // Combinational logic for addition operation
(c) always @ (negedge clock)       // Edge-sensitive cyclic behavior
      begin
       RA = RA + RB;               // Blocking procedural assignment for addition
       RD = RA;                    // Register transfer operation
      end
(d) always @ (negedge clock)       // Edge-sensitive cyclic behavior
      begin
       RA <= RA + RB;              // Nonblocking procedural assignment for addition
       RD <= RA;                   // Register transfer operation
      end
```

Continuous assignments (e.g., **assign** S = A + B;) are used to represent and specify combinational logic circuits. In simulation, a continuous assignment statement executes when the expression on the right-hand side changes. The effect of execution is immediate.

(The variable on the left-hand side is updated.) Similarly, a level-sensitive cyclic behavior (e.g., **always** @ (A, B)) executes during simulation when a change is detected by its event control expression (sensitivity list). The effect of assignments made by the = operator is immediate. The continuous assignment statement (**assign** S = A + B) describes a binary adder with inputs *A* and *B* and output *S*. The target operand in a continuous assignment statement (*S* in this case) cannot be a register data type, but must be a type of net, for example, **wire**. The procedural assignment made in the level-sensitive cyclic behavior in the second example shows an alternative way of specifying a combinational circuit for addition. Within the cyclic behavior, the mechanism of the sensitivity list ensures that the output, *S*, will be updated whenever *A*, or *B*, or both change.

There are two kinds of procedural assignments: *blocking* and *nonblocking*. The two are distinguished by their symbols and by their operation. *Blocking* assignments use the equals symbol (=) as the assignment operator, and *nonblocking* assignments use the left arrow (<=) as the operator. Blocking assignment statements are executed *sequentially* in the order that they are listed in a sequential block; when they execute, they have an immediate effect on the contents of memory before the next statement can be executed. Nonblocking assignments are made *concurrently*. This feature is implemented by evaluating the expression on the right-hand side of each statement in the list of statements before making the assignment to their left-hand sides. Consequently, *there is no interaction between the result of any assignment and the evaluation of an expression affecting another assignment*. Also, the statements associated with an edge-sensitive cyclic behavior do not execute until the indicated edge condition occurs. Consider (c) in the example given above. In the list of blocking procedural assignment, the first statement transfers the sum ($RA + RB$) to RA, and the second statement transfers the new value of RA into RD. The value in RA after the clock event is the sum of the values in RA and RB immediately before the clock event. At the completion of the operation, both RA and RD have the same value. In the nonblocking procedural assignment ((d) above), the two assignments are performed concurrently, so that RD receives the original value of RA. The activity in both examples is launched by the clock undergoing a falling edge transition.

The registers in a system are clocked simultaneously (concurrently). The *D*-input of each flip-flop determines the value that will be assigned to its output, independently of the input to any other flip-flop. To ensure synchronous operations in RTL design, and to ensure a match between an HDL model and the circuit synthesized from the model, it is necessary that nonblocking procedural assignments be used for all variables that are assigned a value within an edge-sensitive cyclic behavior (**always** clocked). The nonblocking assignment that appears in an edge-sensitive cyclic behavior models the behavior of the hardware of a synchronous sequential circuit accurately. In general, the blocking assignment operator (=) is used in a procedural assignment statement only when it is necessary to specify a sequential ordering of multiple assignment statements.

HDL Operators

The Verilog HDL operators and their symbols used in RTL design are listed in Table 8.1. The arithmetic, logic, and shift operators describe register transfer operations. The

Table 8.1
Verilog 2001 HDL Operators

<table>
<tr><th>Operator Type</th><th>Symbol</th><th>Operation Performed</th></tr>
<tr><td>Arithmetic</td><td>+</td><td>addition</td></tr>
<tr><td></td><td>–</td><td>subtraction</td></tr>
<tr><td></td><td>*</td><td>multiplication</td></tr>
<tr><td></td><td>/</td><td>division</td></tr>
<tr><td></td><td>%</td><td>modulus</td></tr>
<tr><td></td><td>**</td><td>exponentiation</td></tr>
<tr><td>Bitwise or Reduction</td><td>~</td><td>negation (complement)</td></tr>
<tr><td></td><td>&</td><td>AND</td></tr>
<tr><td></td><td>|</td><td>OR</td></tr>
<tr><td></td><td>^</td><td>exclusive-OR (XOR)</td></tr>
<tr><td>Logical</td><td>!</td><td>negation</td></tr>
<tr><td></td><td>&&</td><td>AND</td></tr>
<tr><td></td><td>||</td><td>OR</td></tr>
<tr><td>Shift</td><td>>></td><td>logical right shift</td></tr>
<tr><td></td><td><<</td><td>logical left shift</td></tr>
<tr><td></td><td>>>></td><td>arithmetic right shift</td></tr>
<tr><td></td><td><<<</td><td>arithmetic left shift</td></tr>
<tr><td></td><td>{ , }</td><td>concatenation</td></tr>
<tr><td>Relational</td><td>></td><td>greater than</td></tr>
<tr><td></td><td><</td><td>less than</td></tr>
<tr><td></td><td>==</td><td>equality</td></tr>
<tr><td></td><td>!=</td><td>inequality</td></tr>
<tr><td></td><td>===</td><td>case equality</td></tr>
<tr><td></td><td>!==</td><td>case inequality</td></tr>
<tr><td></td><td>>=</td><td>greater than or equal</td></tr>
<tr><td></td><td><=</td><td>less than or equal</td></tr>
</table>

logical and relational operators specify control conditions and have Boolean expressions as their arguments.

The operands of the arithmetic operators are numbers. The +, –, *, and / operators form the sum, difference, product, and quotient, respectively, of a pair of operands. The exponentiation operator (**) was added to the language in 2001 and forms a double-precision floating-point value from a base and exponent having a real, integer, or signed value. Negative numbers are represented in 2's-complement form. The modulus operator produces the remainder from the division of two numbers. For example, 14 % 3 evaluates to 2.

There are two types of binary operators for binary words: bitwise and reduction. The bitwise operators perform a bit-by-bit operation on two vector operands to form a vector result. They take each bit in one operand and perform the operation with the corresponding bit in the other operand. Negation (~) is a unary operator; it complements the bits of a single vector operand to form a vector result. The reduction operators are also unary, acting on a single operand and producing a scalar (one-bit) result. They operate pairwise on the bits of a word, from right to left, and yield a one-bit result. For example, the reduction NOR (~|) results in 0 with operand 00101 and in 1 with operand 00000. The result of applying the NOR operation on the first two bits is used with the third bit, and so forth. Negation is not used as a reduction operator—its operation on a vector produces a vector. Truth tables for the bitwise operators acting on a pair of scalar operands are the same as those listed in Table 4.9 in Section 4.12 for the corresponding Verilog primitive (e.g., the **and** primitive and the **&** bitwise operator have the same truth table). The output of an AND gate with two scalar inputs is the same as the result produced by operating on the two bits with the **&** operator.

The logical and relational operators are used to form Boolean expressions and can take variables or expressions as operands. (*Note*: A variable is also an expression.) Used basically for determining true or false conditions, the logical and relational operators evaluate to 1 if the condition expressed is true and to 0 if the condition is false. If the condition is ambiguous, they evaluate to x. An operand that is a variable evaluates to 0 if the value of the variable is equal to zero and to 1 if the value is not equal to zero. For example, if $A = 1010$ and $B = 0000$, then the expression A has the Boolean value 1 (the number in question is not equal to 0) and the expression B has the Boolean value 0. Results of other operations with these values are as follows:

```
A && B = 0          // Logical AND:          (1010) && (0000) = 0
A & B = 0000        // Bitwise AND:          (1010) & (1010) = (0000)
A || B = 1          // Logical OR:           (1010) || (0000) = 1
A | B = 1010        // Bitwise OR:           (1010) | (0000) = (1010)
!A = 0              // Logical negation      !(1010) = !(1) = 0
~A = 0101           // Bitwise negation      ~(1010) = (0101)
!B = 1              // Logical negation      !(0000) = !(0) = 1
~B = 1111           // Bitwise negation      ~(0000) = 1111
(A > B) = 1         // is greater than
(A == B) = 0        // identity (equality)
```

The relational operators === and !== test for bitwise equality (identity) and inequality in Verilog's four-valued logic system. For example, if $A = 0xx0$ and $B = 0xx0$, the test $A === B$ would evaluate to true, but the test $A == B$ would evaluate to x.

Verilog 2001 has logical and arithmetic shift operators. The logical shift operators shift a vector operand to the right or the left by a specified number of bits. The vacated bit positions are filled with zeros. For example, if $R = 11010$, then the statement

```
R = R >> 1;
```

shifts R to the right one position. The value of R that results from the logical right-shift operation (11010) >> 1 is 01101. In contrast, the arithmetic right-shift operator fills the vacated cell (the most significant bit (MSB)) with its original contents when the word is shifted to the right. The arithmetic left-shift operator fills the vacated cell with a 0 when the word is shifted to the left. The arithmetic right-shift operator is used when the sign extension of a number is important. If $R = 11010$, then the statement

```
R >>> 1;
```

produces the result $R = 11101$; if $R = 01101$, it produces the result $R = 00110$. There is no distinction between the logical left-shift and the arithmetic left-shift operators.

The concatenation operator provides a mechanism for appending multiple operands. It can be used to specify a shift, including the bits transferred into the vacant positions. This aspect of its operation was shown in HDL Example 6.1 for the shift register.

Expressions are evaluated from left to right, and their operators associate from left to right (with the exception of the conditional operator) according to the precedence shown in Table 8.2. For example, in the expression $A + B - C$, the value of B is added to A, and then C is subtracted from the result. In the expression $A + B/C$, the value of B is divided by C, and then the result is added to A because the division operator (/) has a higher precedence than the addition operator (+). Use parentheses to establish precedence. For example, the expression $(A + B)/C$ is not the same as the expression $A + B/C$.

Loop Statements

Verilog HDL has four types of loops that execute procedural statements repeatedly: *repeat, forever, while*, and *for*. All looping statements must appear inside an **initial** or **always** block.

The **repeat** loop executes the associated statements a specified number of times. The following is an example that was used previously:

```
initial
 begin
  clock = 1'b0;
  repeat (16)
   #5 clock = ~ clock;
 end
```

This code toggles the clock 16 times and produces eight clock cycles with a cycle time of 10 time units.

Table 8.2
Verilog Operator Precedence

<table>
<tr><td>+ – ! ~ & ~& | ~| ^ ~^ ^~ (unary)</td><td>Highest precedence</td></tr>
<tr><td>**</td><td rowspan="12">↓</td></tr>
<tr><td>* / %</td></tr>
<tr><td>+ – (binary)</td></tr>
<tr><td><< >> <<< >>></td></tr>
<tr><td>< <= > >=</td></tr>
<tr><td>== != === !==</td></tr>
<tr><td>& (binary)</td></tr>
<tr><td>^ ^~ ~^ (binary)</td></tr>
<tr><td>| (binary)</td></tr>
<tr><td>&&</td></tr>
<tr><td>||</td></tr>
<tr><td>?: (conditional operator)</td></tr>
<tr><td>{ } { { } }</td><td>Lowest precedence</td></tr>
</table>

The **forever** loop causes unconditional, repetitive execution of a procedural statement or a block of procedural statements. For example, the following loop produces a continuous clock having a cycle time of 20 time units:

```
initial
 begin
  clock = 1'b0;
  forever
   #10 clock = ~ clock;
 end
```

The **while** loop executes a statement or a block of statements repeatedly while an expression is true. If the expression is false to begin with, the statement is never executed. The following example illustrates the use of the **while** loop:

```
integer count;
 initial
  begin
   count = 0;
   while (count < 64)
    #5 count = count + 1;
  end
```

The value of count is incremented from 0 to 63. Each increment is delayed by five time units, and the loop exits at the count of 64.

In dealing with looping statements, it is sometimes convenient to use the **integer** data type to index the loop. Integers are declared with the keyword **integer**, as in the previous example. Although it is possible to use a **reg** variable to index a loop, sometimes it is more convenient to declare an integer variable, rather than a **reg**, for counting purposes. Variables declared as data type **reg** are stored as unsigned numbers. Those declared as data type **integer** are store as signed numbers in 2's-complement format. The default width of an integer is a minimum of 32 bits.

The **for** loop is a compact way to express the operations implied by a list of statements whose variables are indexed. The **for** loop contains three parts separated by two semicolons:

- An initial condition.
- An expression to check for the terminating condition.
- An assignment to change the control variable.

The following is an example of a **for** loop:

```
for (j = 0; j < 8; j = j + 1)
  begin
   // procedural statements go here
  end
```

The **for** loop statement repeats the execution of the procedural statements eight times. The control variable is j, the initial condition is $j = 0$, and the loop is repeated as long as j is less than 8. After each execution of the loop statement, the value of j is incremented by 1.

A description of a two-to-four-line decoder using a **for** loop is shown in HDL Example 8.1. Since output Y is evaluated in a procedural statement, it must be declared as type **reg**. The control variable for the loop is the **integer** k. When the loop is expanded (unrolled), we get the following four conditions (IN and Y are in binary, and the index for Y is in decimal):

if IN = 00 then $Y(0)$ = 1; else $Y(0)$ = 0;

if IN = 01 then $Y(1)$ = 1; else $Y(1)$ = 0;

if IN = 10 then $Y(2)$ = 1; else $Y(2)$ = 0;

if IN = 11 then $Y(3)$ = 1; else $Y(3)$ = 0;

HDL Example 8.1 (Decoder)

```
// Description of 2 x 4 decoder using a for loop statement
module decoder (IN, Y);
 input       [1: 0] IN;       // Two binary inputs
 output      [3: 0]  Y;       // Four binary outputs
 reg         [3: 0]  Y;
 integer             k;       // Control (index) variable for loop
```

```
  always @ (IN)
   for (k = 0; k <= 3; k = k + 1)
   if (IN == k) Y[k] = 1;
   else Y[k] = 0;
endmodule
```

Logic Synthesis

Logic synthesis is the automatic process by which a computer-based program (i.e., a synthesis tool) transforms an HDL model of a logic circuit into an optimized netlist of gates that perform the operations specified by the source code. There are various target technologies that implement the synthesized design in hardware. The effective use of an HDL description requires that designers adopt a vendor-specific style suitable for the particular synthesis tools. The type of ICs that implement the design may be an application-specific integrated circuit (ASIC), a programmable logic device (PLD), or a field-programmable gate array (FPGA). Logic synthesis is widely used in industry to design and implement large circuits efficiently, correctly, and rapidly.

Logic synthesis tools interpret the source code of the HDL and translate it into an optimized gate structure, accomplishing (correctly) all of the work that would be done by manual methods using Karnaugh maps. Designs written in Verilog or a comparable language for the purpose of logic synthesis tend to be at the register transfer level. This is because the HDL constructs used in an RTL description can be converted into a gate-level description in a straightforward manner. The following examples discuss how a logic synthesizer can interpret an HDL construct and convert it into a gate structure.

The continuous assignment (**assign**) statement is used to describe combinational circuits. In an HDL, it represents a Boolean equation for a logic circuit. A continuous assignment with a Boolean expression for the right-hand side of the assignment statement is synthesized into the corresponding gate circuit implementing the expression. An expression with an addition operator (+) is interpreted as a binary adder using full-adder circuits. An expression with a subtraction operator (−) is converted into a gate-level subtractor consisting of full adders and exclusive-OR gates (Fig. 4.13). A statement with a conditional operator such as

```
assign Y = S ? In_1 : In_0;
```

translates into a two-to-one-line multiplexer with control input *S* and data inputs *In_1* and *In_0*. A statement with multiple conditional operators specifies a larger multiplexer.

A cyclic behavior (**always** . . .) may imply a combinational or sequential circuit, depending on whether the event control expression is level sensitive or edge sensitive. A synthesis tool will interpret as combinational logic a level-sensitive cyclic behavior whose event control expression is sensitive to every variable that is referenced within the behavior (e.g., by the variable's appearing in the right-hand side of an assignment statement). The event control expression in a description of combinational logic may not be sensitive to an edge of any signal. For example,

```
always @ (In_1 or In_0 or S)        // Alternative: (In_1, In_0, S)
 if (S) Y = In_1;
 else Y = In_0;
```

translates into a two-to-one-line multiplexer. As an alternative, the **case** statement may be used to imply large multiplexers. The **casex** statement treats the logic values x and z as don't-cares when they appear in either the case expression or a case item.

An edge-sensitive cyclic behavior (e.g., **always @ (posedge** clock)) specifies a synchronous (clocked) sequential circuit. The implementation of the corresponding circuit consists of *D* flip-flops and the gates that implement the synchronous register transfer operations specified by the statements associated with the event control expression. Examples of such circuits are registers and counters. A sequential circuit description with a **case** statement translates into a control circuit with *D* flip-flops and gates that form the inputs to the flip-flops. Thus, each statement in an RTL description is interpreted by the synthesizer and assigned to a corresponding gate and flip-flop circuit. For synthesizable sequential circuits, the event control expression must be sensitive to the positive or the negative edge of the clock (synchronizing signal), but not to both.

A simplified flowchart of the process used by industry to design digital systems is shown in Fig. 8.1. The RTL description of the HDL design is simulated and checked for proper operation. Its operational features must match those given in the specification for the behavior of the circuit. The test bench provides the stimulus signals to the simulator. If the result of the simulation is not satisfactory, the HDL description is corrected and checked again. After the simulation run shows a valid design, the RTL description is ready to be compiled by the logic synthesizer. All errors (syntax and functional) in the description must be eliminated before synthesis. The synthesis tool generates a netlist equivalent to a gate-level description of the design as it is represented by the model. If the model fails to express the functionality of the specification, the circuit will fail to do so also. The gate-level circuit is simulated with the same set of stimuli used to check the RTL design. If any corrections are needed, the process is repeated until a satisfactory simulation is achieved. The results of the two simulations are compared to see if they match. If they do not, the designer must change the RTL description to correct any errors in the design. Then the description is compiled again by the logic synthesizer to generate a new gate-level description. Once the designer is satisfied with the results of all simulation tests, the design of the circuit is ready for physical implementation in a technology. In practice, additional testing will be performed to verify that the timing specifications of the circuit can be met in the chosen hardware technology. That issue is not within the scope of this text.

Logic synthesis provides several advantages to the designer. It takes less time to write an HDL description and synthesize a gate-level realization than it does to develop the circuit by manual entry from schematic diagrams. The ease of changing the description facilitates exploration of design alternatives. It is faster, easier, less expensive, and less risky to check the validity of the design by simulation than it is to produce a hardware

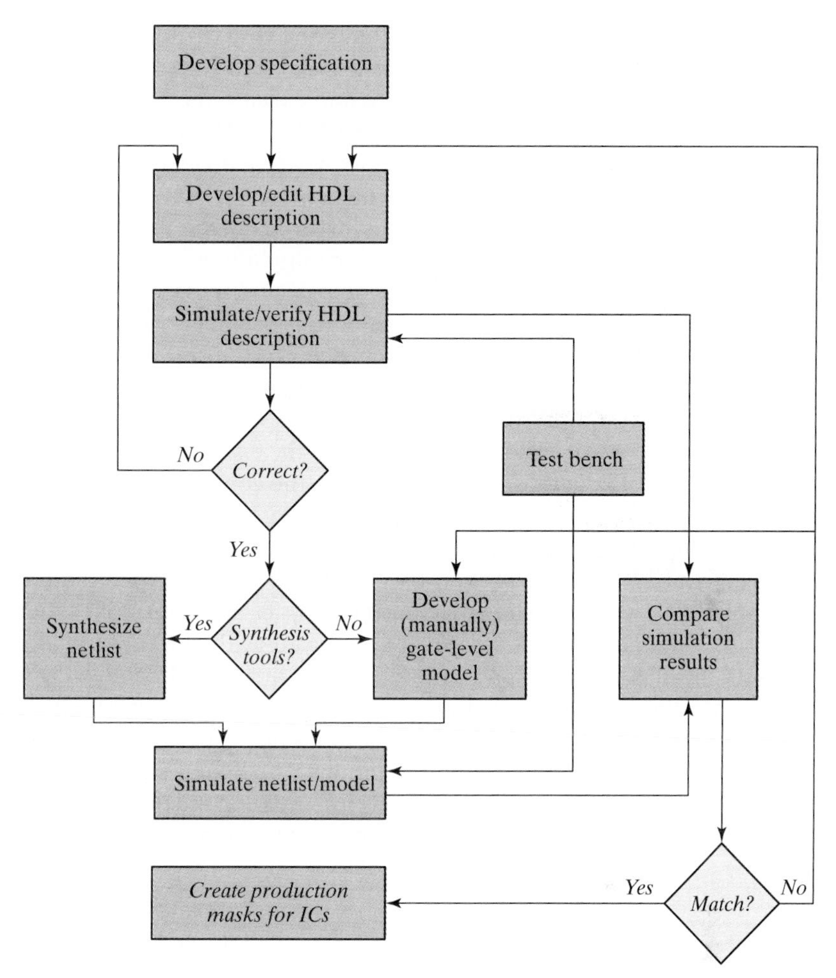

FIGURE 8.1
A simplified flowchart for HDL-based modeling, verification, and synthesis

prototype for evaluation. A schematic and the database for fabricating the integrated circuit can be generated automatically by synthesis tools. The HDL model can be compiled by different tools into different technologies (e.g., ASIC cells or FPGAs), providing multiple returns on the investment to create the model.

8.4 ALGORITHMIC STATE MACHINES (ASMs)

The binary information stored in a digital system can be classified as either data or control information. Data are discrete elements of information (binary words) that are manipulated by performing arithmetic, logic, shift, and other similar data-processing operations. These operations are implemented with digital hardware components such as adders, decoders, multiplexers, counters, and shift registers. Control information provides command signals that coordinate and execute the various operations in the data section of the machine in order to accomplish the desired data-processing tasks.

The design of the logic of a digital system can be divided into two distinct efforts. One part is concerned with designing the digital circuits that perform the data-processing operations. The other part is concerned with designing the control circuits that determine the sequence in which the various manipulations of data are performed.

The relationship between the control logic and the data-processing operations in a digital system is shown in Fig. 8.2. The data-processing path, commonly referred to as the *datapath unit*, manipulates data in registers according to the system's requirements. The *control unit* issues a sequence of commands to the datapath unit. Note that an internal feedback path from the datapath unit to the control unit provides status conditions that the control unit uses together with the external (primary) inputs to determine the sequence of control signals (outputs of the control unit) that direct the operation of the datapath unit. We'll see later that understanding how to model this feedback relationship with an HDL is very important.

The control logic that generates the signals for sequencing the operations in the datapath unit is a finite state machine (FSM), i.e., a synchronous sequential circuit. The control commands for the system are produced by the FSM as functions of the primary inputs, the status signals, and the state of the machine. In a given state, the outputs of the controller are the inputs to the datapath unit and determine the operations that it will execute. Depending on status conditions and other external inputs, the FSM goes to its next state to initiate other operations. The digital circuits that act as the control logic provide a time sequence of signals for initiating the operations in the datapath and also determine the next state of the control subsystem itself.

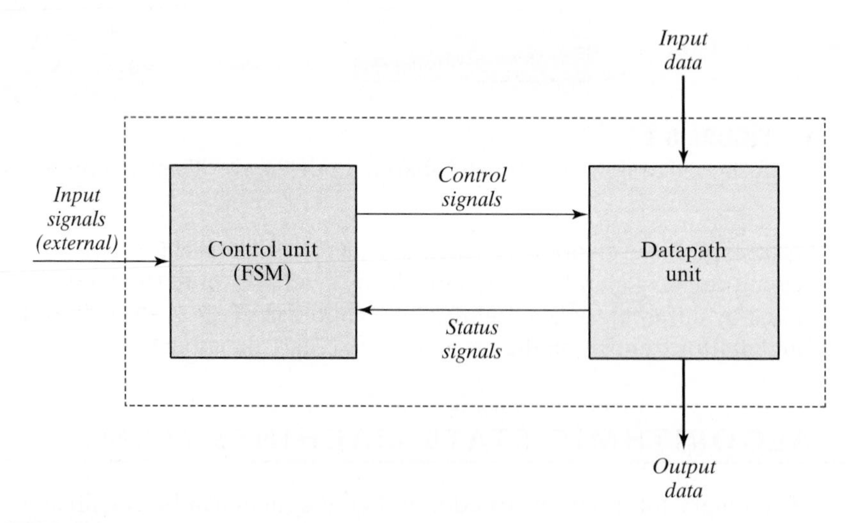

FIGURE 8.2
Control and datapath interaction

The control sequence and datapath tasks of a digital system are specified by means of a hardware algorithm. An algorithm consists of a finite number of procedural steps that specify how to obtain a solution to a problem. A hardware algorithm is a procedure for solving the problem with a given piece of equipment. The most challenging and creative part of digital design is the formulation of hardware algorithms for achieving required objectives. The goal is to implement the algorithms in silicon as an integrated circuit.

A flowchart is a convenient way to specify the sequence of procedural steps and decision paths for an algorithm. A flowchart for a hardware algorithm translates the verbal instructions to an information diagram that enumerates the sequence of operations together with the conditions necessary for their execution. An *algorithmic state machine* (ASM) chart is a flowchart that has been developed to specifically define algorithms for execution on digital hardware. A *state machine* is another term for a sequential circuit, which is the basic structure of a digital system.

ASM Chart

An ASM chart resembles a conventional flowchart, but is interpreted somewhat differently. A conventional flowchart describes the procedural steps and decision paths of an algorithm in a sequential manner, without taking into consideration their time relationship. The ASM chart describes the sequence of events, i.e., the ordering of events in time, as well as the timing relationship between the states of a sequential controller and the events that occur while going from one state to the next (i.e., the events that are synchronous with changes in the state). The chart is adapted to specify accurately the control sequence and datapath operations in a digital system, taking into consideration the constraints of digital hardware.

An ASM chart is composed of three basic elements: the state box, the decision box, and the conditional box. The boxes themselves are connected by directed edges indicating the sequential precedence and evolution of the states as the machine operates. There are various ways to attach information to an ASM chart. In one, a state in the control sequence is indicated by a state box, as shown in Fig. 8.3(a). The shape of the state box is a rectangle within which are written register operations or the names of output signals that the control generates while being in the indicated state. The state is given a symbolic name, which is placed within the upper left corner of the box. The binary code assigned to the state is placed at the upper right corner. (The state symbol and code can be placed

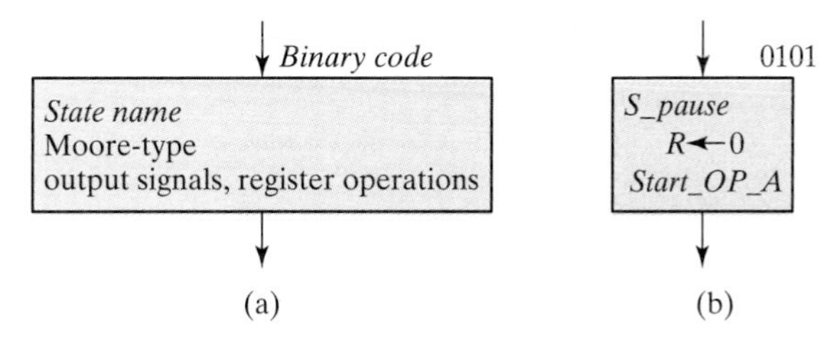

FIGURE 8.3
ASM chart state box

in other places as well.) Figure 8.3(b) gives an example of a state box. The state has the symbolic name *S_pause*, and the binary code assigned to it is 0101. Inside the box is written the register operation $R \leftarrow 0$, which indicates that register R is to be cleared to 0. The name *Start_OP_A* inside the box indicates, for example, a Moore-type output signal that is asserted while the machine is in state *S_pause* and that launches a certain operation in the datapath unit.

The style of state box shown in Fig. 8.3(b) is sometimes used in ASM charts, but it can lead to confusion about when the register operation $R \leftarrow 0$ is to execute. Although the operation is written inside the state box, it actually occurs when the machine makes a transition from *S_pause* to its next state. In fact, writing the register operation within the state box is a way (albeit possibly confusing) to indicate that the controller must assert a signal that will cause the register operation to occur when the machine changes state. Later we'll introduce a chart and notation that are more suited to digital design and that will eliminate any ambiguity about the register operations controlled by a state machine.

The decision box of an ASM chart describes the effect of an input (i.e., a primary, or external, input or a status, or internal, signal) on the control subsystem. The box is diamond shaped and has two or more exit paths, as shown in Fig. 8.4. The input condition to be tested is written inside the box. One or the other exit path is taken, depending on the evaluation of the condition. In the binary case, one path is taken if the condition is true and another when the condition is false. When an input condition is assigned a binary value, the two paths are indicated by 1 and 0, respectively.

The state and decision boxes of an ASM chart are similar to those used in conventional flowcharts. The third element, the conditional box, is unique to the ASM chart. The shape of the conditional box is shown in Fig. 8.5(a). Its rounded corners differentiate it from the state box. The input path to the conditional box must come from one of the exit paths of a decision box. The outputs listed inside the conditional box are generated as Mealy-type signals during a given state; the register operations listed in the conditional box are associated with a transition from the state. Figure 8.5(b) shows an example with a conditional box. The control generates the output signal *Start* while in state *S_1* and checks the status of input *Flag*. If *Flag* = 1, then R is cleared to 0; otherwise, R remains unchanged. In either case, the next state is *S_2*. A register operation is

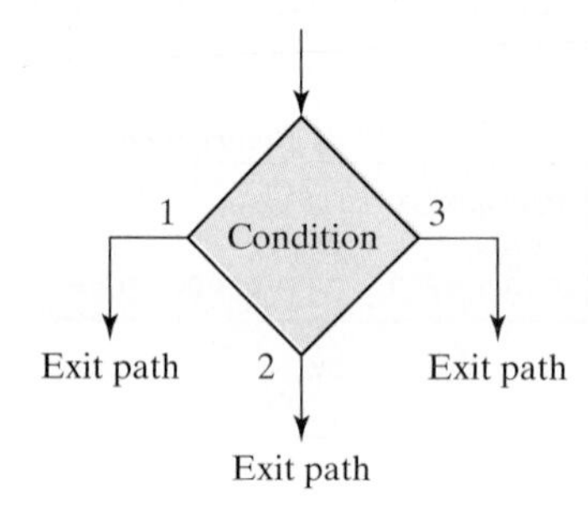

FIGURE 8.4
ASM chart decision box

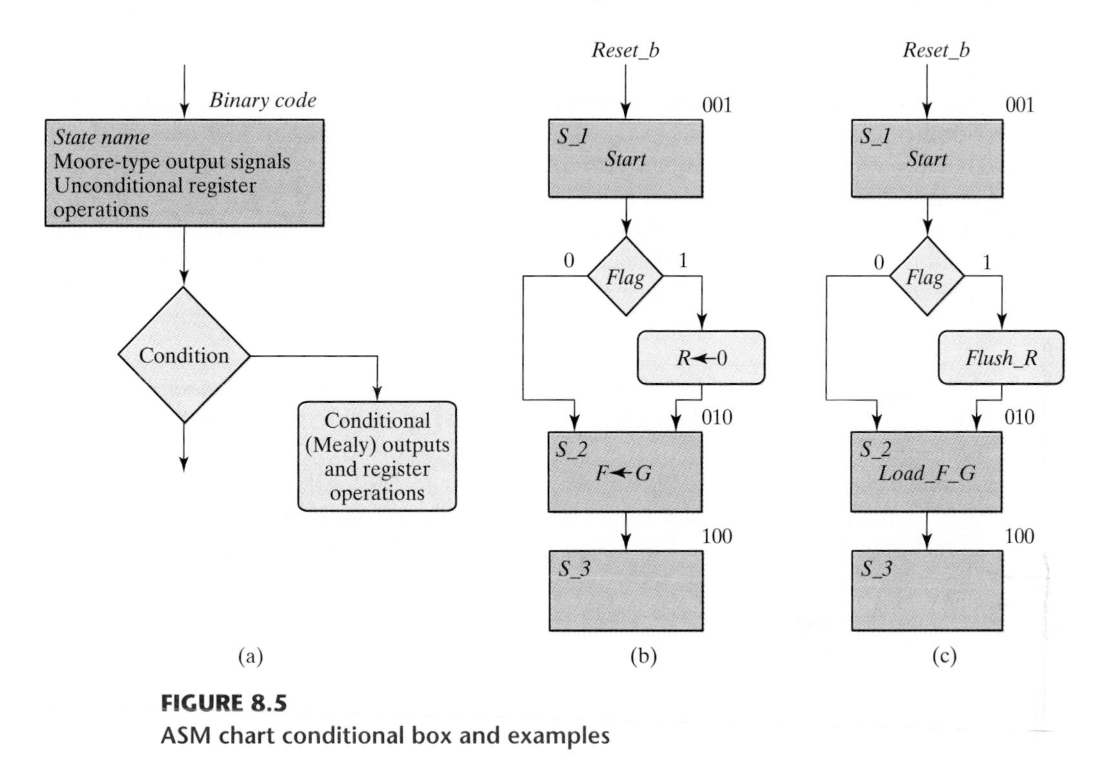

FIGURE 8.5
ASM chart conditional box and examples

associated with *S_2*. We again note that *this style of chart can be a source of confusion*, because the state machine does not execute the indicated register operation $R \leftarrow 0$ when it is in *S_1* or the operation $F \leftarrow G$when it is in *S_2*. The notation actually indicates that when the controller is in *S_1*, it must assert a Mealy-type signal that will cause the register operation $R \leftarrow 0$ to execute in the datapath unit[1], subject to the condition that *Flag* = 0. Likewise, in state *S_2*, the controller must generate a Moore-type output signal that causes the register operation $F \leftarrow G$ to execute in the datapath unit. The operations in the datapath unit are synchronized to the clock edge that causes the state to move from *S_1* to *S_2* and from *S_2* to *S_3*, respectively. Thus, *the control signal generated in a given state affects the operation of a register in the datapath when the next clock transition occurs.* The result of the operation is apparent in the next state.

The ASM chart in Fig. 8.5(b) mixes descriptions of the datapath and the controller. An ASM chart for only the controller is shown in Fig. 8.5(c), in which the register operations are omitted. In their place are the control signals that must be generated by the control unit to launch the operations of the datapath unit. This chart is useful for describing the controller, but it does not contain adequate information about the datapath. (We'll address this issue later.)

[1] If the path came from a state box the asserted signals would be moore type signals, dependent on only the state, and should be listed within the box.

ASM Block

An ASM block is a structure consisting of one state box and all the decision and conditional boxes connected to its exit path. An ASM block has one entrance and any number of exit paths represented by the structure of the decision boxes. An ASM chart consists of one or more interconnected blocks. An example of an ASM block is given in Fig. 8.6. Associated with state *S_0* are two decision boxes and one conditional box. The diagram distinguishes the block with dashed lines around the entire structure, but this is not usually done, since the ASM chart uniquely defines each block from its structure. A state box without any decision or conditional boxes constitutes a simple block.

Each block in the ASM chart describes the state of the system during one clock-pulse interval (i.e., the interval between two successive active edges of the clock). The operations within the state and conditional boxes in Fig. 8.6(a) are initiated by a common clock pulse when the state of the controller transitions from *S_0* to its next state. The same clock pulse transfers the system controller to one of the next states, *S_1, S_2*, or *S_3*, as dictated by the binary values of E and F. The ASM chart for the controller alone is shown in Fig. 8.6(b). The Moore-type signal *incr_A* is asserted unconditionally while the machine is in *S_0*; the Mealy-type signal *Clear_R* is generated conditionally when the state is *S_0* and E is asserted. In general, the Moore-type outputs of the controller are generated unconditionally and are indicated within a state box; the Mealy-type outputs are generated conditionally and are indicated in the conditional boxes connected to the edges that leave a decision box.

The ASM chart is similar to a state transition diagram. Each state block is equivalent to a state in a sequential circuit. The decision box is equivalent to the binary information

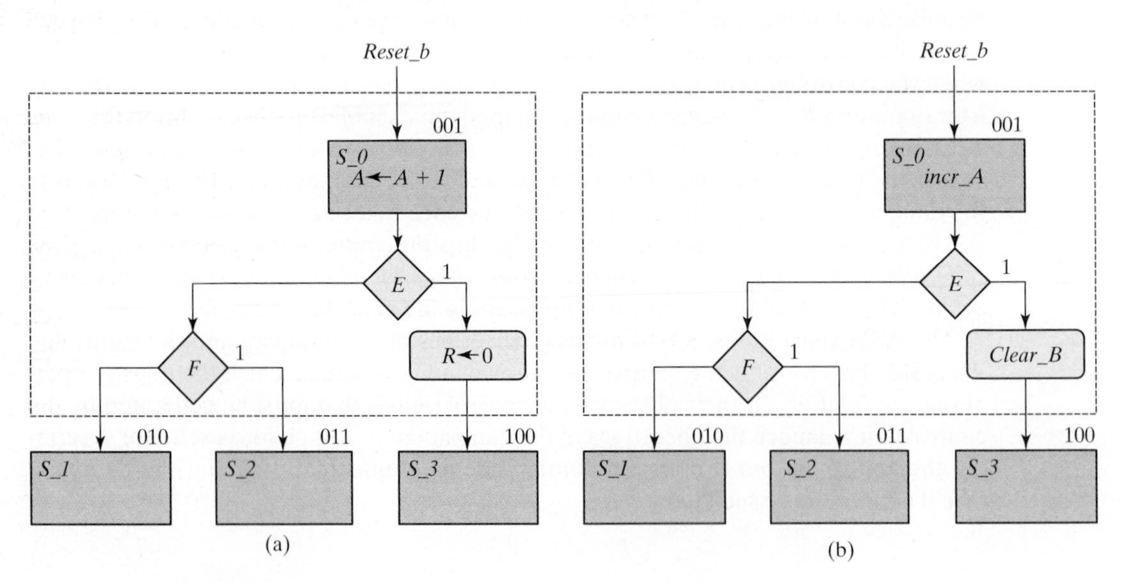

FIGURE 8.6
ASM blocks

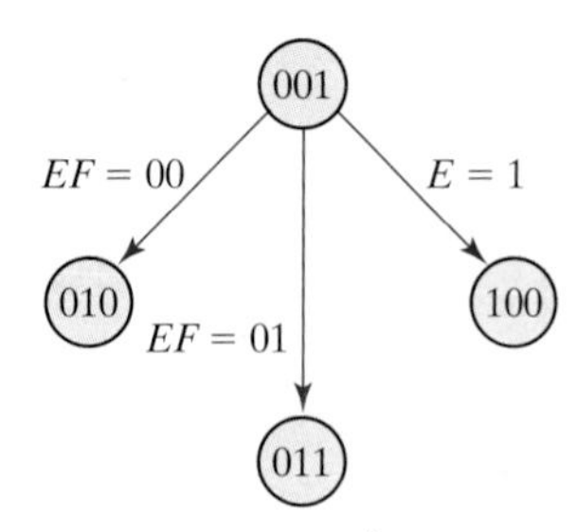

FIGURE 8.7
State diagram equivalent to the ASM chart of Fig. 8.6

written along the directed lines that connect two states in a state diagram. As a consequence, it is sometimes convenient to convert the chart into a state diagram and then use sequential circuit procedures to design the control logic. As an illustration, the ASM chart of Fig. 8.6 is drawn as a state diagram (outputs are omitted) in Fig. 8.7. The states are symbolized by circles, with their binary values written inside. The directed lines indicate the conditions that determine the next state. The unconditional and conditional operations that must be performed in the datapath unit are not indicated in the state diagram.

Simplifications

A binary decision box of an ASM chart can be simplified by labeling only the edge corresponding to the asserted decision variable and leaving the other edge without a label. A further simplification is to omit the edges corresponding to the state transitions that occur when a reset condition is asserted. Output signals that are not asserted are not shown on the chart; the presence of the name of an output signal indicates that it is asserted.

Timing Considerations

The timing for all registers and flip-flops in a digital system is controlled by a master-clock generator. The clock pulses are applied not only to the registers of the datapath, but also to all the flip-flops in the state machine implementing the control unit. Inputs are also synchronized to the clock, because they are normally generated as outputs of another circuit that uses the same clock signals. If the input signal changes at an arbitrary time independently of the clock, we call it an asynchronous input. Asynchronous inputs may cause a variety of problems. To simplify the design, we will assume that all inputs are synchronized with the clock and change state in response to an edge transition.

The major difference between a conventional flowchart and an ASM chart is in interpreting the time relationship among the various operations. For example, if Fig. 8.6 were a conventional flowchart, then the operations listed would be considered to follow one after another in sequence: First register A is incremented, and only then is E evaluated. If $E = 1$, then register R is cleared and control goes to state S_3. Otherwise (if $E = 0$), the next step is to evaluate F and go to state S_1 or S_2. In contrast, an ASM chart considers the entire block as one unit. All the register operations that are specified within

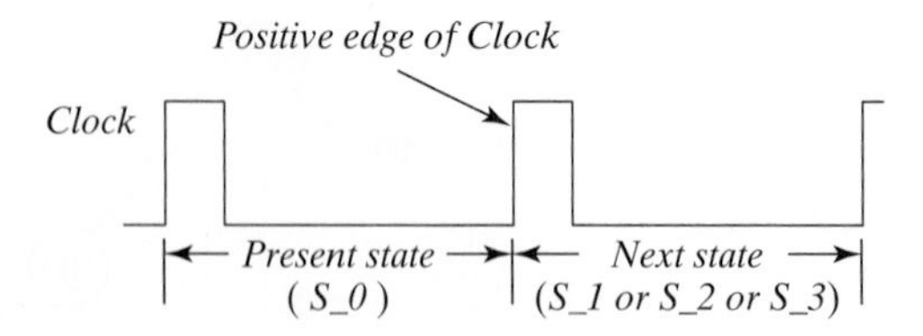

FIGURE 8.8
Transition between states

the block must occur in synchronism at the edge transition of the same clock pulse while the system changes from *S_0* to the next state. This sequence of events is presented pictorially in Fig. 8.8. In this illustration, we assume positive-edge triggering of all flip-flops. An asserted asynchronous reset signal (*reset_b*) transfers the control circuit into state *S_0*. While in state *S_0*, the control circuits check inputs E and F and generate appropriate signals accordingly. If *reset_b* is not asserted, the following operations occur simultaneously at the next positive edge of the clock:

1. Register A is incremented.
2. If $E = 1$, register R is cleared.
3. Control transfers to the next state, as specified in Fig. 8.7.

Note that the two operations in the datapath and the change of state in the control logic occur at the same time. Note also that the ASM chart in Fig. 8.6(a) indicates the register operations that must occur in the datapath unit, but does not indicate the control signal that is to be formed by the control unit. Conversely, the chart in Fig. 8.6(b) indicates the control signals, but not the datapath operations. We will now present an ASMD chart to provide the clarity and complete information needed by logic designers.

ASMD Chart

Algorithmic state machine and datapath (ASMD) charts were developed to clarify the information displayed by ASM charts and to provide an effective tool for designing a control unit for a given datapath unit. An ASMD chart differs from an ASM chart in three important ways: (1) An ASMD chart does not list register operations within a state box, (2) the edges of an ASMD chart are annotated with register operations that are concurrent with the state transition indicated by the edge, and (3) an ASMD chart includes conditional boxes identifying the signals which control the register operations that annotate the edges of the chart. Thus, *an ASMD chart associates register operations with state transitions rather than with states*; it also associates register operations with the signals that cause them. Consequently, an ASMD chart represents a partition of a complex digital machine into its datapath and control units and clearly indicates the relationship between them. There is no room for confusion about the timing of register operations or about the signals that launch them.

Designers form an ASMD chart in a three-step process that creates an annotated and completely specified ASM chart for the controller of a datapath unit.

The steps to form an ASMD chart are:

1. Form an ASM chart showing only the states of the controller and the input signals[2] that cause state transitions,
2. Convert the ASM chart into an ASMD chart by annotating the edges of the ASM chart to indicate the concurrent register operations of the datapath unit (i.e., register operations that are concurrent with a state transition), and
3. Modify the ASMD chart to identify the control signals that are generated by the controller and that cause the indicated operations in the datapath unit.

The ASMD chart produced by this process clearly and completely specifies the finite state machine of the controller, identifies the registers operations of the datapath unit, identifies signals reporting the status of the datapath to the controller, and links register operations to the signals that control them.

One important use of a state machine is to control register operations on a datapath in a sequential machine that has been partitioned into a controller and a datapath. An ASMD chart links the ASM chart of the controller to the datapath it controls in a manner that serves as a universal model representing all synchronous digital hardware design. ASMD charts help clarify the design of a sequential machine by separating the design of its datapath from the design of the controller, while maintaining a clear relationship between the two units. Register operations that occur concurrently with state transitions are annotated on a path of the chart, rather than in state boxes or in conditional boxes on the path, because these registers are not part of the controller. The outputs generated by the controller are the signals that control the registers of the datapath and cause the register operations annotated on the ASMD chart.

8.5 DESIGN EXAMPLE (ASMD CHART)

We will now present a simple example demonstrating the use of the ASMD chart and the register transfer representation. We start from the initial specifications of a system and proceed with the development of an appropriate ASMD chart from which the digital hardware is then designed.

The datapath unit is to consist of two *JK* flip-flops *E* and *F*, and one four-bit binary counter *A[3:0]*. The individual flip-flops in *A* are denoted by A_3, A_2, A_1, and A_0, with A_3 holding the most significant bit of the count. A signal, *Start*, initiates the system's operation by clearing the counter *A* and flip-flop *F*. At each subsequent clock pulse, the counter is incremented by 1 until the operations stop. Counter bits A_2 and A_3 determine the sequence of operations:

If $A_2 = 0$, *E* is cleared to 0 and the count continues.

If $A_2 = 1$, *E* is set to 1; then, if $A_3 = 0$, the count continues, but if $A_3 = 1$, *F* is set to 1 on the next clock pulse and the system stops counting.

---

[2] In general, the inputs to the control unit are external (primary) inputs and status signals that originate in the datapath unit.

Then, if *Start* = 0, the system remains in the initial state, but if *Start* = 1, the operation cycle repeats.

A block diagram of the system's architecture is shown in Fig. 8.9(a), with (1) the registers of the datapath unit, (2) the external (primary) input signals, (3) the status signals fed back from the datapath unit to the control unit, and (4) the control signals generated by the control unit and input to the datapath unit. Note that the names of the control signals clearly indicate the operations that they cause to be executed in the datapath unit. For example, *clr_A_F* clears registers *A* and *F*. The name of the signal *reset_b* (alternatively, *reset_bar*) indicates that the reset action is active low. The internal details of each unit are not shown.

ASMD Chart

An ASMD chart for the system is shown in Fig. 8.9(b) for asynchronous reset action and in Fig. 8.9(c) for synchronous reset action. The chart shows the state transitions of the controller and the datapath operations associated with those transitions. The chart is not in its final form, for it does not identify the control signals generated by the controller. The nonblocking Verilog operator (<=) is shown instead of the arrow ($\leftarrow$) for register transfer operations because we will ultimately use the ASMD chart to write a Verilog description of the system.

When the reset action is synchronous, the transition to the reset state is synchronous with the clock. This transition is shown for *S_idle* in the diagram, but *all other synchronous reset paths are omitted for clarity*. The system remains in the reset state, *S_idle*, until *Start* is asserted. When that happens (i.e., *Start* = 1), the state moves to *S_1*. *At the next clock edge*, depending on the values of A_2 and A_3 (decoded in a priority order), the state returns to *S_1* or goes to *S_2*. From *S_2*, it moves unconditionally to *S_idle*, where it awaits another assertion of *Start*.

The edges of the chart represent the state transitions that occur at the active (i.e., synchronizing) edge of the clock (e.g., the rising edge) and are annotated with the register operations that are to occur in the datapath. With *Start* asserted in *S_idle*, the state will transition to *S_1* and the registers *A* and *F* will be cleared. Note that, on the one hand, if a register operation is annotated on the *edge* leaving a state box, the operation occurs unconditionally and will be controlled by a Moore-type signal. For example, register *A* is incremented at every clock edge that occurs while the machine is in the state *S_1*. On the other hand, the register operation setting register *E* annotates the edge leaving the *decision box* for A_2. The signal controlling the operation will be a Mealy-type signal asserted when the system is in state *S_1* and A_2 has the value 1. Likewise, the control signal clearing *A* and *F* is asserted conditionally: The system is in state *S_idle* and *Start* is asserted.

In addition to showing that the counter is incremented in state *S_1*, the annotated paths show that other operations occur conditionally with the same clock edge:

Either *E* is cleared and control stays in state *S_1* ($A_2 = 0$) or

E is set and control stays in state *S_1* ($A_2A_3 = 10$) or

E is set and control goes to state *S_2* ($A_2A_3 = 11$).

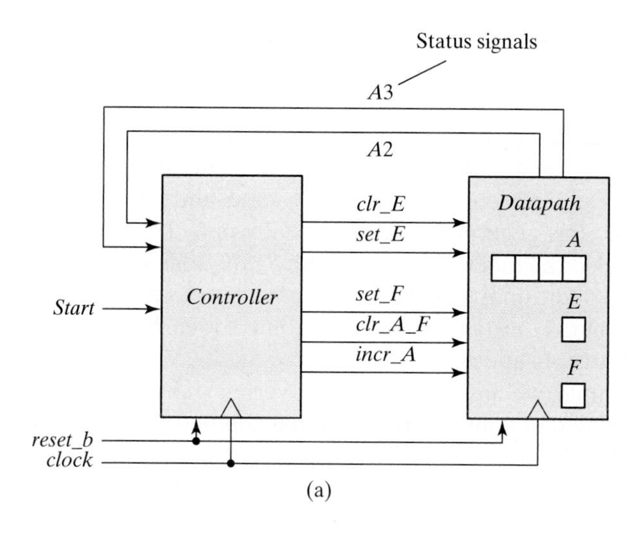

(a)

Note: A3 denotes A[3],
A2 denotes A[2],
<= denotes nonblocking assignment
reset_b denotes active-low reset condition

(b)

(c)

(d)

FIGURE 8.9
(a) Block diagram for design example
(b) ASMD chart for controller state transitions, asynchronous reset
(c) ASMD chart for controller state transitions, synchronous reset
(d) ASMD chart for a completely specified controller, asynchronous reset

When control is in state *S_2*, a Moore-type control signal must be asserted to set flip-flop *F* to 1, and the state returns to *S_idle* at the next active edge of the clock.

The third and final step in creating the ASMD chart is to insert conditional boxes for the signals generated by the controller or to insert Moore-type signals in the state boxes, as shown in Fig. 8.9(d). The signal *clr_A_F* is generated conditionally in state *S_idle*, *incr_A* is generated unconditionally in *S_1*, *clr_E* and *set_E* are generated conditionally in *S_1*, and *set_F* is generated unconditionally in *S_2*. The ASM chart has three states and three blocks. The block associated with *S_idle* consists of the state box, one decision box, and one conditional box. The block associated with *S_2* consists of only the state box. In addition to *clock* and *reset_b*, the control logic has one external input, *Start*, and two status inputs, A_2 and A_3.

In this example, we have shown how a verbal (text) description (specification) of a design is translated into an ASMD chart that completely describes the controller for the datapath, indicating the control signals and their associated register operations. This design example does not necessarily have a practical application, and in general, depending on the interpretation, the ASMD chart produced by the three-step design process for the controller may be simplified and formulated differently. However, once the ASMD chart is established, the procedure for designing the circuit is straightforward. *In practice, designers use the ASMD chart to write Verilog models of the controller and the datapath and then synthesize a circuit directly from the Verilog description.* We will first design the system manually and then write the HDL description, keeping synthesis as an optional step for those who have access to synthesis tools.

Timing Sequence

Every block in an ASMD chart specifies the signals which control the operations that are to be initiated by one common clock pulse. The control signals specified within the state and conditional boxes in the block are formed while the controller is in the indicated state, and the annotated operations occur in the datapath unit when the state makes a transition along an edge that exits the state. The change from one state to the next is performed in the control logic. In order to appreciate the timing relationship involved, we will list the step-by-step sequence of operations after each clock edge, beginning with an assertion of the signal *Start* until the system returns to the reset (initial) state, *S_idle*.

Table 8.3 shows the binary values of the counter and the two flip-flops after every clock pulse. The table also shows separately the status of A_2 and A_3, as well as the present state of the controller. We start with state *S_1* right after the input signal *Start* has caused the counter and flip-flop *F* to be cleared. We will assume that the machine had been running before it entered *S_idle*, instead of entering it from a reset condition. Therefore, the value of *E* is assumed to be 1, because *E* is set to 1 when the machine enters *S_2*, before moving to *S_idle* (as shown at the bottom of the table), and because *E* does not change during the transition from *S_idle* to *S_1*. The system stays in state *S_1* during the next 13 clock pulses. Each pulse increments the counter and either clears or sets *E*. Note the relationship between the time at which A_2 becomes a 1 and the time at

Table 8.3
Sequence of Operations for Design Example

Counter				Flip-Flops			
A_3	A_2	A_1	A_0	E	F	**Conditions**	**State**
0	0	0	0	1	0	$A_2 = 0, A_3 = 0$	*S_1*
0	0	0	1	0	0		
0	0	1	0	0	0		
0	0	1	1	0	0		
0	1	0	0	0	0	$A_2 = 1, A_3 = 0$	
0	1	0	1	1	0		
0	1	1	0	1	0		
0	1	1	1	1	0		
1	0	0	0	1	0	$A_2 = 0, A_3 = 1$	
1	0	0	1	0	0		
1	0	1	0	0	0		
1	0	1	1	0	0		
1	1	0	0	0	0	$A_2 = 1, A_3 = 1$	
1	1	0	1	1	0		*S_2*
1	1	0	1	1	1		*S_idle*

which E is set to 1. When $A = (A_3\ A_2\ A_1\ A_0)\ 0011$, the next (4th) clock pulse increments the counter to 0100, but that same clock edge sees the value of A_2 as 0, so E remains cleared. The next (5th) pulse changes the counter from 0100 to 0101, and because A_2 is equal to 1 *before* the clock pulse arrives, E is set to 1. Similarly, E is cleared to 0 not when the count goes from 0111 to 1000, but when it goes from 1000 to 1001, which is when A_2 is 0 in the *present* value of the counter.

When the count reaches 1100, both A_2 and A_3 are equal to 1. The next clock edge increments A by 1, sets E to 1, and transfers control to state *S_2*. Control stays in *S_2* for only one clock period. The clock edge associated with the path leaving *S_2* sets flip-flop F to 1 and transfers control to state *S_idle*. The system stays in the initial state *S_idle* as long as *Start* is equal to 0.

From an observation of Table 8.3, it may seem that the operations performed on E are delayed by one clock pulse. This is the difference between an ASMD chart and a conventional flowchart. If Fig. 8.9(d) were a conventional flowchart, we would assume that A is first incremented and the incremented value would have been used to check the status of A_2. The operations that are performed in the digital hardware, as specified by a block in the ASMD chart, occur during the same clock cycle and not in a sequence of operations following each other in time, as is the usual interpretation in a conventional flowchart. Thus, the value of A_2 to be considered in the decision box is taken from the value of the counter in the present state and before it is incremented. This is because the decision box for E belongs with the same block as state *S_1*. The digital circuits in the control unit generate the signals for all the operations specified in the present block *prior to the arrival of the next clock pulse*. The next clock edge executes all the operations in the registers and flip-flops, including the flip-flops in the controller that determine the next state, using the present values of the output signals of the controller. Thus, the signals that control the operations in the datapath unit are formed in the controller in the clock cycle (control state) *preceding* the clock edge at which the operations execute.

Controller and Datapath Hardware Design

The ASMD chart provides all the information needed to design the digital system—the datapath and the controller. The actual boundary between the hardware of the controller and that of the datapath can be arbitrary, but we advocate, first, that the datapath unit contain only the hardware associated with its operations and the logic required, perhaps, to form status signals used by the controller, and, second, that the control unit contain all of the logic required to generate the signals that control the operations of the datapath unit. The requirements for the design of the datapath are indicated by the control signals inside the state and conditional boxes of the ASMD chart and are specified by the annotations of the edges indicating datapath operations. The control logic is determined from the decision boxes and the required state transitions. The hardware configuration of the datapath and controller is shown in Fig. 8.10.

Note that the input signals of the control unit are the external (primary) inputs (*Start*, *reset_b*, and *clock*) and the status signals from the datapath (A_2 and A_3). The status signals provide information about the present condition of the datapath. This information, together with the primary inputs and information about the present state of the machine, is used to form the output of the controller and the value of the next state. The outputs of the controller are inputs to the datapath and determine which operations will be executed when the clock undergoes a transition. Note, also, that the state of the control unit is not an output of the control unit.

The control subsystem is shown in Fig. 8.10 with only its inputs and outputs, with names matching those of the ASMD chart. The detailed design of the controller is considered subsequently. The datapath unit consists of a four-bit binary counter and two *JK* flip-flops. The counter is similar to the one shown in Fig. 6.12, except that additional internal gates are required for the synchronous clear operation. The counter is incremented with every clock pulse when the controller state is *S_1*. It is cleared only when control is at state *S_idle* and *Start* is equal to 1. The logic for the signal *clr_A_F* will be included in the controller and requires an AND gate to guarantee that both conditions are present. Similarly, we can anticipate that the controller will use AND gates to form signals *set_E* and *clr_E*. Depending on whether the controller is in state *S_1* and whether A_2 is asserted, *set_F* controls flip-flop F and is asserted unconditionally during state *S_2*. Note that all flip-flops and registers, including the flip-flops in the control unit, use a common clock.

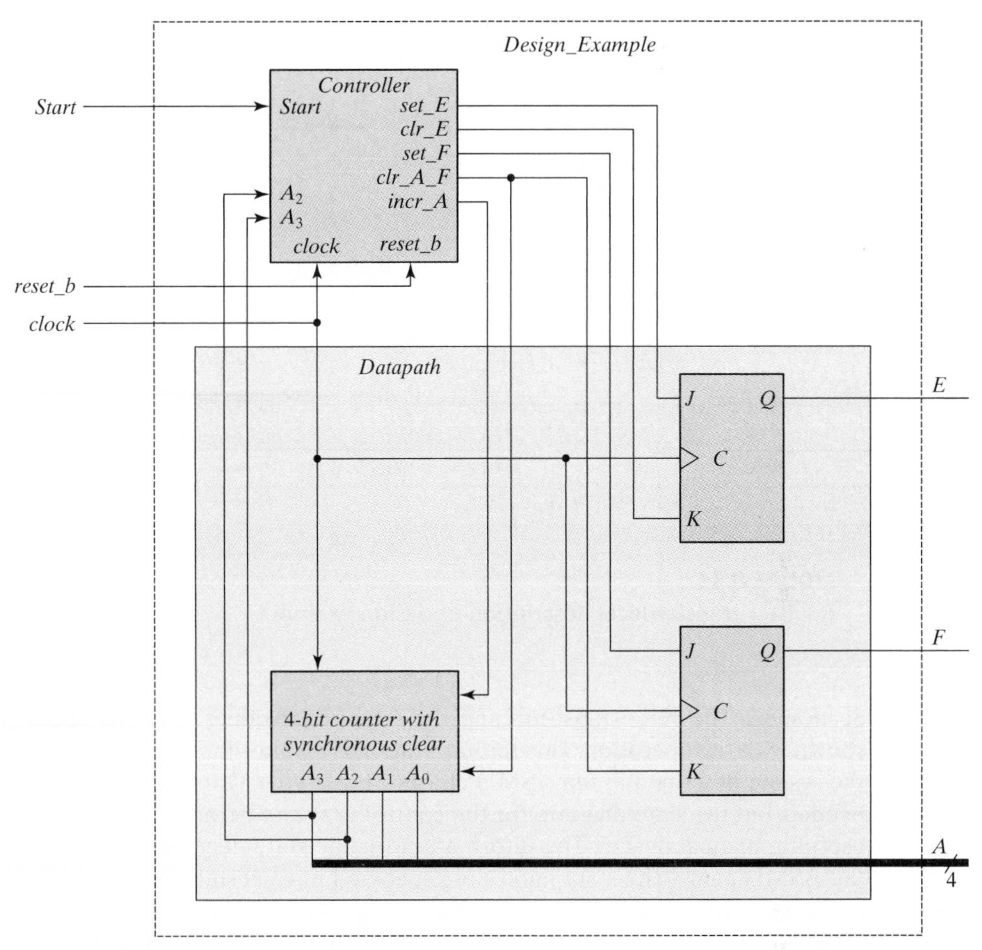

FIGURE 8.10
Datapath and controller for design example

Register Transfer Representation

A digital system is represented at the register transfer level by specifying the registers in the system, the operations performed, and the control sequence. The register operations and control information can be specified with an ASMD chart. It is convenient to separate the control logic from the register operations of the datapath. The ASMD chart provides this separation and a clear sequence of steps to design a controller for a datapath. The control information and register transfer operations can also be represented separately, as shown in Fig. 8.11. The state diagram specifies the control sequence, and the register operations are represented by the register transfer notation introduced in

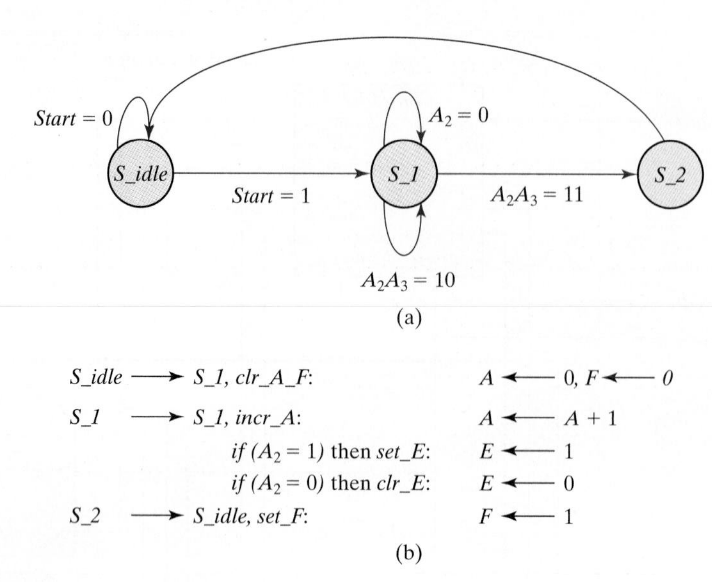

FIGURE 8.11
Register transfer-level description of design example

Section 8.2. The state transition and the signal controlling the register operation are shown with the operation. This representation is an alternative to the representation of the system described in the ASMD chart of Fig. 8.9(d). Only the ASMD chart is really needed, but the state diagram for the controller is an alternative representation that is useful in manual design. The information for the state diagram is taken directly from the ASMD chart. The state names are specified in each state box. The conditions that cause a change of state are specified inside the diamond-shaped decision boxes of the ASMD chart and are used to annotate the state diagram. The directed lines between states and the condition associated with each follow the same path as in the ASMD chart. The register transfer operations for each of the three states are listed following the name of the state. They are taken from the state boxes or the annotated edges of the ASMD chart.

State Table

The state diagram can be converted into a state table from which the sequential circuit of the controller can be designed. First, we must assign binary values to each state in the ASMD chart. For n flip-flops in the control sequential circuit, the ASMD chart can accommodate up to 2^n states. A chart with 3 or 4 states requires a sequential circuit with two flip-flops. With 5 to 8 states, there is a need for three flip-flops. Each combination of flip-flop values represents a binary number for one of the states.

A *state table* for a controller is a list of present states and inputs and their corresponding next states and outputs. In most cases, there are many don't-care input conditions

Table 8.4
State Table for the Controller of Fig. 8.10

	Present State		Inputs			Next State		Outputs				
Present-State Symbol	G_1	G_0	*Start*	A_2	A_3	G_1	G_0	*set_E*	*clr_E*	*set_F*	*clr_A_F*	*incr_A*
S_idle	0	0	0	X	X	0	0	0	0	0	0	0
S_idle	0	0	1	X	X	0	1	0	0	0	1	0
S_1	0	1	X	0	X	0	1	0	1	0	0	1
S_1	0	1	X	1	0	0	1	1	0	0	0	1
S_1	0	1	X	1	1	1	1	1	0	0	0	1
S_2	1	1	X	X	X	0	0	0	0	1	0	0

that must be included, so it is advisable to arrange the state table to take those conditions into consideration. We assign the following binary values to the three states: *S_idle* = 00, *S_1* = 01, and *S_2* = 11. Binary state 10 is not used and will be treated as a don't-care condition. The state table corresponding to the state diagram is shown in Table 8.4. Two flip-flops are needed, and they are labeled G_1 and G_0. There are three inputs and five outputs. The inputs are taken from the conditions in the decision boxes. The outputs depend on the inputs and the present state of the control. Note that there is a row in the table for each possible transition between states. Initial state 00 goes to state 01 or stays in 00, depending on the value of input *Start*. The other two inputs are marked with don't-care X's, as they do not determine the next state in this case. While the system is in binary state 00 with *Start* = 1, the control unit provides an output labeled *clr_A_F* to initiate the required register operations. The transition from binary state 01 depends on inputs A_2 and A_3. The system goes to binary state 11 only if $A_2A_3 = 11$; otherwise, it remains in binary state 01. Finally, binary state 11 goes to 00 independently of the input variables.

Control Logic

The procedure for designing a sequential circuit starting from a state table was presented in Chapter 5. If this procedure is applied to Table 8.4, we need to use five-variable maps to simplify the input equations. This is because there are five variables listed under the present-state and input columns of the table. Instead of using maps to simplify the input equations, we can obtain them directly from the state table by inspection. To design the sequential circuit of the controller with D flip-flops, it is necessary to go over the next-state columns in the state table and derive all the conditions that must set each flip-flop to 1. From Table 8.4, we note that the next-state column of G_1 has a single 1 in the fifth row. The D input of flip-flop G_1 must be equal to 1 during present state *S_1* when both inputs A_2 and A_3 are equal to 1. This condition is expressed with the *D* flip-flop input equation

$$D_{G1} = S\_1\, A_2 A_3$$

Similarly, the next-state column of G_0 has four 1's, and the condition for setting this flip-flop is

$$D_{G0} = Start\, S\_idle + S\_1$$

To derive the five output functions, we can exploit the fact that binary state 10 is not used, which simplifies the equation for *clr_A_F* and enables us to obtain the following simplified set of output equations:

$$set\_E = S\_1A_2$$
$$clr\_E = S\_1A_2'$$
$$set\_F = S\_2$$
$$clr\_A\_F = Start\, S\_idle$$
$$incr\_A = S\_1$$

The logic diagram showing the internal detail of the controller of Fig. 8.10 is drawn in Fig. 8.12. Note that although we derived the output equations from Table 8.4, they can also be obtained directly by inspection of Fig. 8.9(d). This simple example illustrates the

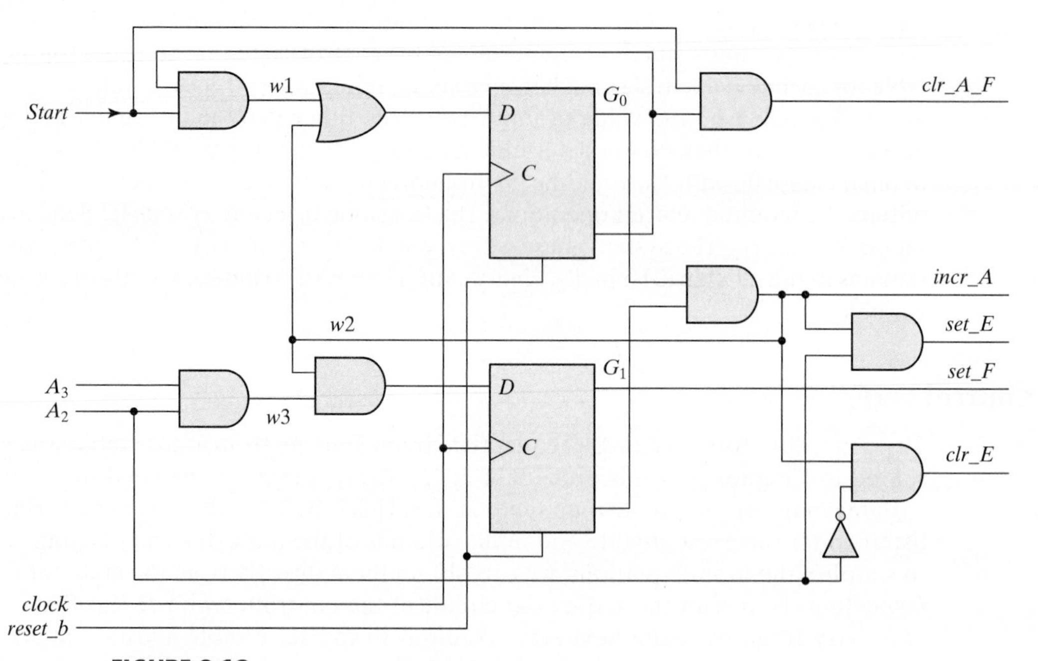

FIGURE 8.12
Logic diagram of the control unit for Fig. 8.10

manual design of a controller for a datapath, using an ASMD chart as a starting point. The fact that synthesis tools automatically execute these steps should be appreciated.

8.6 HDL DESCRIPTION OF DESIGN EXAMPLE

In previous chapters, we gave examples of HDL descriptions of combinational circuits, sequential circuits, and various standard components such as multiplexers, counters, and registers. We are now in a position to incorporate these components into the description of a specific design. As mentioned previously, a design can be described either at the structural or behavioral level. Behavioral descriptions may be classified as being either at the register transfer level or at an abstract algorithmic level. Consequently, we now consider three levels of design: structural description, RTL description, and algorithmic-based behavioral description.

The *structural* description is the lowest and most detailed level. The digital system is specified in terms of the physical components and their interconnection. The various components may include gates, flip-flops, and standard circuits such as multiplexers and counters. The design is hierarchically decomposed into functional units, and each unit is described by an HDL module. A top-level module combines the entire system by instantiating all the lower level modules. This style of description requires that the designer have sufficient experience not only to understand the functionality of the system, but also to implement it by selecting and connecting other functional elements.

The *RTL* description specifies the digital system in terms of the registers, the operations performed, and the control that sequences the operations. This type of description simplifies the design process because it consists of procedural statements that determine the relationship between the various operations of the design without reference to any specific structure. The RTL description implies a certain hardware configuration among the registers, allowing the designer to create a design that can be synthesized automatically, rather than manually, into standard digital components.

The *algorithmic-based behavioral* description is the most abstract level, describing the function of the design in a procedural, algorithmic form similar to a programming language. It does not provide any detail on how the design is to be implemented with hardware. The algorithmic-based behavioral description is most appropriate for simulating complex systems in order to verify design ideas and explore trade-offs. Descriptions at this level are accessible to nontechnical users who understand programming languages. Some algorithms, however, might not be synthesizable.

We will now illustrate the RTL and structural descriptions by using the design example of the previous section. The design example will serve as a model of coding style for future examples and will exploit alternative syntax options supported by revisions to the Verilog language. (An algorithmic-based description is illustrated in Section 8.9.)

RTL Description

The block diagram in Fig. 8.10 describes the design example. An HDL description of the design example can be written as a single RTL description in a Verilog module or as a top-level module having instantiations of separate modules for the controller and the datapath. The former option simply ignores the boundaries between the functional units; the modules in the latter option establish the boundaries shown in Fig. 8.9(a) and Fig. 8.10. We advocate the second option, because, in general, it distinguishes more clearly between the controller and the datapath. This choice also allows one to easily substitute alternative controllers for a given datapath (e.g., replace an RTL model by a structural model). The RTL description of the design example is shown in HDL Example 8.2. The description follows the ASMD chart of Fig. 8.9(d), which contains a complete description of the controller, the datapath, and the interface between them (i.e., the outputs of the controller and the status signals). Likewise, our description has three modules: *Design_Example_RTL, Controller_RTL*, and *Datapath_RTL*. The descriptions of the controller and the datapath units are taken directly from Fig. 8.9(d). *Design_Example_RTL* declares the input and output ports of the module and instantiates *Controller_RTL* and *Datapath_RTL*. At this stage of the description, it is important to remember to declare *A* as a vector. Failure to do so will produce *port mismatch* errors when the descriptions are compiled together. Note that the status signals *A[2]* and *A[3]*, but not *A[0]* and *A[1]*, are passed to the controller. The primary (external) inputs to the controller are *Start, clock* (to synchronize the system), and *reset_b*. The active-low input signal *reset_b* is needed to initialize the state of the controller to *S_idle*. Without that signal, the controller could not be placed in a known initial state.

The controller is described by three cyclic (**always**) behaviors. An edge-sensitive behavior updates the state at the positive edge of the clock, depending on whether a reset condition is asserted. Two level-sensitive behaviors describe the combinational logic for the next state and the outputs of the controller, as specified by the ASMD chart. Notice that the description includes default assignments to all of the outputs (e.g., *set_E* = 0). This approach allows the code of the **case** logic to be simplified by expressing only explicit assertions of the variables (i.e., values are assigned by exception). The approach also ensures that every path through the assignment logic assigns a value to every variable. Thus, a synthesis tool will interpret the logic to be combinational; failure to assign a value to every variable on every path of logic implies the need for a transparent latch (memory) to implement the logic. Synthesis tools will provide the latch, wasting silicon area.

The three states of the controller are given symbolic names and are encoded into binary values. Only three of the possible two-bit patterns are used, so the case statement for the next-state logic includes a **default** assignment to handle the possibility that one of the three assigned codes is not detected. The alternative is to allow the hardware to make an arbitrary assignment to the next state (*next_state* = *2'bx*;). Also, the first statement of the next-state logic assigns *next_state* = *S_idle* to guarantee that the next state is assigned in every thread of the logic. This is a precaution against accidentally forgetting to make an assignment to the next state in every thread of the logic, with the result that the description implies the need for memory, which a synthesis tool will implement with a transparent latch.

The description of *Datapath_RTL* is written by testing for an assertion of each control signal from *Controller_RTL*. The register transfer operations are displayed in the ASMD chart (Fig. 8.9(d)). Note that nonblocking assignments are used (with symbol <=) for the register transfer operations. This ensures that the register operations and state transitions are concurrent, a feature that is especially crucial during control state *S_1*. In this state, *A* is incremented by 1 and the value of *A2* (*A[2]*) is checked to determine the operation to execute at register *E* at the next clock. To accomplish a valid synchronous design, it is necessary to ensure that *A[2]* is checked before *A* is incremented. If blocking assignments were used, one would have to place the two statements that check *E* first and the *A* statement that increments last. However, by using nonblocking assignments, we accomplish the required synchronization without being concerned about the order in which the statements are listed. The counter *A* in *Datapath_RTL* is cleared synchronously because *clr_A_F* is synchronized to the clock.

The cyclic behaviors of the controller and the datapath interact in a chain reaction: At the active edge of the clock, the state and datapath registers are updated. A change in the state, a primary input, or a status input causes the level-sensitive behaviors of the controller to update the value of the next state and the outputs. The updated values are used at the next active edge of the clock to determine the state transition and the updates of the datapath.

Note that the manual method of design developed (1) a block diagram (Fig. 8.9(a)) showing the interface between the datapath and the controller, (2) an ASMD chart for the system (Fig. 8.9(d)), (3) the logic equations for the inputs to the flip-flops of the controller, and (4) a circuit that implements the controller (Fig. 8.12). In contrast, an RTL model describes the state transitions of the controller and the operations of the datapath as a step toward automatically synthesizing the circuit that implements them. The descriptions of the datapath and controller are derived directly from the ASMD chart in both cases.

HDL Example 8.2

```
// RTL description of design example (see Fig. 8.11)
module Design_Example_RTL (A, E, F, Start, clock, reset_b);
 // Specify ports of the top-level module of the design
 // See block diagram, Fig. 8.10
 output [3: 0] A;
 output        E, F;
 input         Start, clock, reset_b;
 // Instantiate controller and datapath units
 Controller_RTL M0 (set_E, clr_E, set_F, clr_A_F, incr_A, A[2], A[3], Start, clock, reset_b);
 Datapath_RTL M1 (A, E, F, set_E, clr_E, set_F, clr_A_F, incr_A, clock);
endmodule
module Controller_RTL (set_E, clr_E, set_F, clr_A_F, incr_A, A2, A3, Start, clock, reset_b);
  output reg    set_E, clr_E, set_F, clr_A_F, incr_A;
  input         Start, A2, A3, clock, reset_b;
```

```
  reg [1: 0]         state, next_state;
  parameter          S_idle = 2'b00, S_1 = 2'b01, S_2 = 2'b11;     // State codes
  always @ (posedge clock, negedge reset_b)       // State transitions (edge sensitive)
  if (reset_b == 0) state <= S_idle;
  else state <= next_state;
// Code next-state logic directly from ASMD chart (Fig. 8.9d)
always @ (state, Start, A2, A3) begin                     // Next-state logic (level sensitive)
 next_state = S_idle;
 case (state)
   S_idle:          if (Start) next_state = S_1; else next_state = S_idle;
   S_1:             if (A2 & A3) next_state = S_2; else next_state = S_1;
   S_2:             next_state = S_idle;
   default:         next_state = S_idle;
 endcase
end
// Code output logic directly from ASMD chart (Fig. 8.9d)
always @ (state, Start, A2) begin
 set_E   = 0;   // default assignments; assign by exception
 clr_E   = 0;
 set_F   = 0;
 clr_A_F = 0;
 incr_A  = 0;
 case (state)
  S_idle:          if (Start) clr_A_F = 1;
  S_1:             begin incr_A = 1; if (A2) set_E = 1; else clr_E = 1; end
  S_2:             set_F = 1;
 endcase
end
endmodule
module Datapath_RTL (A, E, F, set_E, clr_E, set_F, clr_A_F, incr_A, clock);
 output reg [3: 0] A;                         // register for counter
 output reg        E, F;                      // flags
 input             set_E, clr_E, set_F, clr_A_F, incr_A, clock;
 // Code register transfer operations directly from ASMD chart (Fig. 8.9(d))
 always @ (posedge clock) begin
  if (set_E)                                  E <= 1;
  if (clr_E)                                  E <= 0;
  if (set_F)                                  F <= 1;
  if (clr_A_F)                                begin A <= 0; F <= 0; end
  if (incr_A)                                 A <= A + 1;
 end
endmodule
```

Testing the Design Description

The sequence of operations for the design example was investigated in the previous section. Table 8.3 shows the values of *E* and *F* while register *A* is incremented. It is instructive to devise a test that checks the circuit to verify the validity of the HDL description. The test bench in HDL Example 8.3 provides such a module. (The procedure for writing test benches is explained in Section 4.12.) The test module generates signals for *Start, clock*, and *reset_b*, and checks the results obtained from registers *A, E*, and *F*. Initially, the *reset_b* signal is set to 0 to initialize the controller, and *Start* and *clock* are set to 0. At time $t = 5$, the *reset_b* signal is de-asserted by setting it to 1, the *Start* input is asserted by setting it to 1, and the clock is then repeated for 16 cycles. The **$monitor** statement displays the values of *A, E*, and *F* every 10 ns. The output of the simulation is listed in the example under the simulation log. Initially, at time $t = 0$, the values of the registers are unknown, so they are marked with the symbol **x**. The first positive clock transition, at time = 10, clears *A* and *F*, but does not affect *E*, so *E* is unknown at this time. The rest of the table is identical to Table 8.3. Note that since *Start* is still equal to 1 at time = 160, the last entry in the table shows that *A* and *F* are cleared to 0, and *E* does not change and remains at 1. This occurs during the second transition, from *S_idle* to *S_1*.

HDL Example 8.3

```
// Test bench for design example
'timescale 1 ns / 1 ps
module t_Design_Example_RTL;
 reg          Start, clock, reset_b;
 wire [3: 0] A;
 wire         E, F;
 // Instantiate design example
 Design_Example_RTL M0 (A, E, F, Start, clock, reset_b);
 // Describe stimulus waveforms
 initial #500 $finish;               // Stopwatch
 initial
  begin
   reset_b = 0;
   Start = 0;
   clock = 0;
   #5 reset_b = 1; Start = 1;
   repeat (32)
   begin
    #5 clock = ~ clock;      // Clock generator
   end
```

```
    end
  initial
    $monitor ("A = %b E = %b F = %b time = %0d", A, E, F, $time);
endmodule
Simulation log:
A = xxxx E = x F = x time = 0
A = 0000 E = x F = 0 time = 10
A = 0001 E = 0 F = 0 time = 20
A = 0010 E = 0 F = 0 time = 30
A = 0011 E = 0 F = 0 time = 40
A = 0100 E = 0 F = 0 time = 50
A = 0101 E = 1 F = 0 time = 60
A = 0110 E = 1 F = 0 time = 70
A = 0111 E = 1 F = 0 time = 80
A = 1000 E = 1 F = 0 time = 90
A = 1001 E = 0 F = 0 time = 100
A = 1010 E = 0 F = 0 time = 110
A = 1011 E = 0 F = 0 time = 120
A = 1100 E = 0 F = 0 time = 130
A = 1101 E = 1 F = 0 time = 140
A = 1101 E = 1 F = 1 time = 150
A = 0000 E = 1 F = 0 time = 160
```

Waveforms produced by a simulation of *Design_Example_RTL* with the test bench are shown in Fig. 8.13. Numerical values are shown in hexadecimal format. The results are annotated to call attention to the relationship between a control signal and the operation that it causes to execute. For example, the controller asserts *set_E* for one clock cycle *before* the clock edge at which *E* is set to 1. Likewise, *set_F* asserts during the clock cycle before the edge at which *F* is set to 1. Also, *clr_A_F* is formed in the cycle before *A* and *F* are cleared. A more thorough verification of *Design_Example_RTL* would confirm that the machine recovers from a reset on the fly (i.e., a reset that is asserted randomly after the machine is operating). Note that the signals in the output of the simulation have been listed in groups showing (1) *clock* and *reset_b*, (2) *Start* and the status inputs, (3) the state, (4) the control signals, and (5) the datapath registers. *It is strongly recommended that the state always be displayed*, because this information is essential for verifying that the machine is operating correctly and for debugging its description when it is not. For the chosen binary state code, *S_idle* $= 00_2 = 0_H$, *S_1*$= 01_2 = 1_H$, and *S_2* $= 11_2 = 3_H$.

Structural Description

The RTL description of a design consists of procedural statements that determine the functional behavior of the digital circuit. This type of description can be compiled by

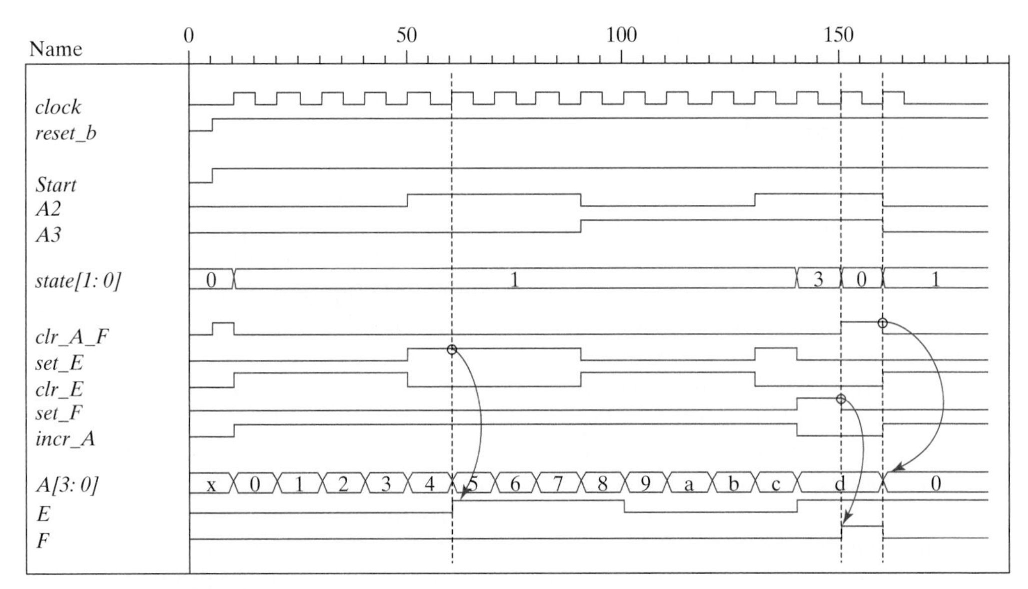

FIGURE 8.13
Simulation results for *Design_Example_RTL*

HDL synthesis tools, from which it is possible to obtain the equivalent gate-level circuit of the design. It is also possible to describe the design by its structure rather than its function. A structural description of a design consists of instantiations of components that define the circuit elements and their interconnections. In this regard, a structural description is equivalent to a schematic diagram or a block diagram of the circuit. Contemporary design practice relies heavily on RTL descriptions, but we will present a structural description here to contrast the two approaches.

For convenience, the circuit is again decomposed into two parts: the controller and the datapath. The block diagram of Fig. 8.10 shows the high-level partition between these units, and Fig. 8.12 provides additional underlying structural details of the controller. The structure of the datapath is evident in Fig. 8.10 and consists of the flip-flops and the four-bit counter with synchronous clear. The top level of the Verilog description replaces *Design_Example_RTL*, *Controller_RTL*, and *Datapath_RTL* by *Design_Example_STR*, *Controller_STR*, and *Datapath_STR*, respectively. The descriptions of *Controller_STR* and *Datapath_STR* will be structural.

HDL Example 8.4 presents the structural description of the design example. It consists of a nested hierarchy of modules and gates describing (1) the top-level module, *Design_Example_STR*, (2) the modules describing the controller and the datapath, (3) the modules describing the flip-flops and counters, and (4) gates implementing the logic of the controller. For simplicity, the counter and flip-flops are described by RTL models.

The top-level module (see Fig. 8.10) encapsulates the entire design by (1) instantiating the controller and the datapath modules, (2) declaring the primary (external) input signals,

(3) declaring the output signals, (4) declaring the control signals generated by the controller and connected to the datapath unit, and (5) declaring the status signals generated by the datapath unit and connected to the controller. The port list is identical to the list used in the RTL description. The outputs are declared as **wire** type here because they serve merely to connect the outputs of the datapath module to the outputs of the top-level module, with their logic value being determined within the datapath module.

The control module describes the circuit of Fig. 8.12. The outputs of the two flip-flops *G1* and *G0* are declared as **wire** data type. *G1* and *G0* cannot be declared as **reg** data type because they are outputs of an instantiated *D* flip-flop. *DG1* and *DG0* are undeclared identifiers, i.e., implicit wires. The name of a variable is local to the module or procedural block in which it is declared. Nets may not be declared within a procedural block (e.g., **begin** . . . **end**). The rule to remember is that a variable must be a declared register type (e.g., **reg**) if and only if its value is assigned by a procedural statement (i.e., a blocking or nonblocking assignment statement within a procedural block in cyclic or single-pass behavior or in the output of a sequential UDP). The instantiated gates specify the combinational part of the circuit. There are two flip-flop input equations and three output equations. The outputs of the flip-flops *G1* and *G0* and the input equations *DG1* and *DG0* replace output *Q* and input *D* in the instantiated flip-flops. The *D* flip-flop is then described in the next module. The structure of the datapath unit has direct inputs to the *JK* flip-flops. Note the correspondence between the modules of the HDL description and the structures in Figs. 8.9, 8.10, and 8.12.

HDL Example 8.4

```
// Structural description of design example (Figs. 8.9(a), 8.12)
module Design_Example_STR
 ( output     [3: 0]   A,                              // V 2001 port syntax
   output              E, F,
   input               Start, clock, reset_b
 );

  Controller_STR M0 (clr_A_F, set_E, clr_E, set_F, incr_A, Start, A[2], A[3], clock,
   reset_b );
  Datapath_STR M1 (A, E, F, clr_A_F, set_E, clr_E, set_F, incr_A, clock);
endmodule

module Controller_STR
( output clr_A_F, set_E, clr_E, set_F, incr_A,
  input Start, A2, A3, clock, reset_b
);

  wire             G0, G1;
  parameter        S_idle = 2'b00, S_1 = 2'b01, S_2 = 2'b11;
  wire             w1, w2, w3;
```

```
 not (G0_b, G0);
 not (G1_b, G1);
 buf (incr_A, w2);
 buf (set_F, G1);
 not (A2_b, A2);
 or (D_G0, w1, w2);
 and (w1, Start, G0_b);
 and (clr_A_F, G0_b, Start);
 and (w2, G0, G1_b);
 and (set_E, w2, A2);
 and (clr_E, w2, A2_b);
 and (D_G1, w3, w2);
 and (w3, A2, A3);
 D_flip_flop_AR M0 (G0, D_G0, clock, reset_b);
 D_flip_flop_AR M1 (G1, D_G1, clock, reset_b);
endmodule

// datapath unit

module Datapath_STR
( output [3: 0]       A,
  output              E, F,
  input               clr_A_F, set_E, clr_E, set_F, incr_A, clock
);
 JK_flip_flop_2 M0 (E, E_b, set_E, clr_E, clock);
 JK_flip_flop_2 M1 (F, F_b, set_F, clr_A_F, clock);
 Counter_4      M2 (A, incr_A, clr_A_F, clock);
endmodule

// Counter with synchronous clear

module Counter_4 (output reg [3: 0] A, input incr, clear, clock);
 always @ (posedge clock)
  if (clear)      A <= 0; else if (incr) A <= A + 1;
endmodule

module D_flip_flop_AR (Q, D, CLK, RST);
 output           Q;
 input            D, CLK, RST;
 reg              Q;

 always @ (posedge CLK, negedge RST)
  if (RST == 0) Q <= 1'b0;
  else Q <= D;
endmodule

// Description of JK flip-flop
```

```
module JK_flip_flop_2 (Q, Q_not, J, K, CLK);
 output       Q, Q_not;
 input        J, K, CLK;
 reg          Q;
 assign       Q_not = ~Q;
 always @ (posedge CLK)
  case ({J, K})
   2'b00:       Q <= Q;
   2'b01:       Q <= 1'b0;
   2'b10:       Q <= 1'b1;
   2'b11:       Q <= ~Q;
  endcase
endmodule

module t_Design_Example_STR;
 reg Start, clock, reset_b;
 wire [3: 0] A;
 wire      E, F;

 // Instantiate design example

Design_Example_STR M0 (A, E, F, Start, clock, reset_b);

// Describe stimulus waveforms

initial #500 $finish;              // Stopwatch
initial
 begin
  reset_b = 0;
  Start = 0;
  clock = 0;
  #5 reset_b = 1; Start = 1;
  repeat (32)
   begin
    #5 clock = ~ clock;            // Clock generator
   end
 end
initial
 $monitor ("A = %b E = %b F = %b time = %0d", A, E, F, $time);
endmodule
```

The structural description was tested with the test bench that verified the RTL description to produce the results shown in Fig. 8.13. The only change necessary is the replacement of the instantiation of the example from *Design_Example_RTL* by *Design_Example_STR*. The simulation results for *Design_Example_STR* matched those for *Design_Example_RTL*. However, a comparison of the two descriptions indicates that the RTL style is easier to write and will lead to results faster if synthesis tools are available to automatically synthesize the registers, the combinational logic, and their interconnections.

8.7 SEQUENTIAL BINARY MULTIPLIER

This section introduces a second design example. It presents a hardware algorithm for binary multiplication, proposes the register configuration for its implementation, and then shows how to use an ASMD chart to design its datapath and its controller.

The system we will examine multiplies two unsigned binary numbers. The hardware developed in Section 4.7 to execute multiplication resulted in a combinational circuit multiplier with many adders and AND gates, and requires large area of silicon as an integrated circuit. In contrast, in this section, a more efficient hardware algorithm results in a sequential multiplier that uses only one adder and a shift register. The savings in hardware and silicon area come about from a trade-off in the space (hardware)–time domain. A parallel adder uses more hardware, but forms its result in one cycle of the clock; a sequential adder uses less hardware, but takes multiple clock cycles to form its result.

The multiplication of two binary numbers is done with paper and pencil by successive (i.e., sequential) additions and shifting. The process is best illustrated with a numerical example. Let us multiply the two binary numbers 10111 and 10011:

```
 23        10111  multiplican
 19        10011  multiplier
 --        -----
           10111
          10111
         00000
        00000
       10111
---    ---------
437    110110101  product
```

The process consists of successively adding and shifting copies of the multiplicand. Successive bits of the multiplier are examined, least significant bit first. If the multiplier bit is 1, the multiplicand is copied down; otherwise, 0's are copied down. The numbers copied in successive lines are shifted one position to the left from the previous number. Finally, the numbers are added and their sum forms the product. The product obtained from the multiplication of two binary numbers of n bits each can have up to $2n$ bits. It is apparent that the operations of addition and shifting are executed by the algorithm.

When the multiplication process is implemented with digital hardware, it is convenient to change the process slightly. First, we note that, in the context of synthesizing a sequential machine, the add-and-shift algorithm for binary multiplication can be executed in a single clock cycle or over multiple clock cycles. A choice to form the product in the time span of a single clock cycle will synthesize the circuit of a parallel multiplier like the one discussed in Section 4.7. On the other hand, an RTL model of the algorithm adds shifted copies of the multiplicand to an accumulated partial product. The values of the multiplier, multiplicand, and partial product are stored in registers, and the operations of shifting and adding their contents are executed under the control of a state machine. Among the many possibilities for distributing the effort of multiplication over multiple clock cycles, we will consider that in which only one partial product is formed and accumulated in a single cycle of the clock. (One alternative would be to use additional hardware to form and accumulate two partial products in a clock cycle, but this would require more logic gates and either faster circuits or a slower clock.) Instead of providing digital circuits to store and add simultaneously as many binary numbers as there are 1's in the multiplier, it is less expensive to provide only the hardware needed to sum two binary numbers and accumulate the partial products in a register. Second, instead of shifting the multiplicand to the left, the partial product being formed is shifted to the right. This leaves the partial product and the multiplicand in the required relative positions. Third, when the corresponding bit of the multiplier is 0, there is no need to add all 0's to the partial product, since doing so will not alter its resulting value.

Register Configuration

A block diagram for the sequential binary multiplier is shown in Fig. 8.14(a), and the register configuration of the datapath is shown in Fig. 8.14(b). The multiplicand is stored in register B, the multiplier is stored in register Q, and the partial product is formed in register A and stored in A and Q. A parallel adder adds the contents of register B to register A. The C flip-flop stores the carry after the addition. The counter P is initially set to hold a binary number equal to the number of bits in the multiplier. This counter is decremented after the formation of each partial product. When the content of the counter reaches zero, the product is formed in the double register A and Q, and the process stops. The control logic stays in an initial state until *Start* becomes 1. The system then performs the multiplication. The sum of A and B forms the n most significant bits of the partial product, which is transferred to A. The output carry from the addition, whether 0 or 1, is transferred to C. Both the partial product in A and the multiplier in Q are shifted to the right. The least significant bit of A is shifted into the most significant position of Q, the carry from C is shifted into the most significant position of A, and 0 is shifted into C. After the shift-right operation, one bit of the partial product is transferred into Q while the multiplier bits in Q are shifted one position to the right. In this manner, the least significant bit of register Q, designated by *Q[0]*, holds the bit of the multiplier that must be inspected next. The control logic determines whether to add or not on the basis of this input bit. The control logic also receives a signal, *Zero*, from a circuit that checks counter P for zero. *Q[0]* and *Zero* are status inputs for the control unit. The input signal *Start* is an external control input. The outputs of the control logic launch the required operations in the registers of the datapath unit.

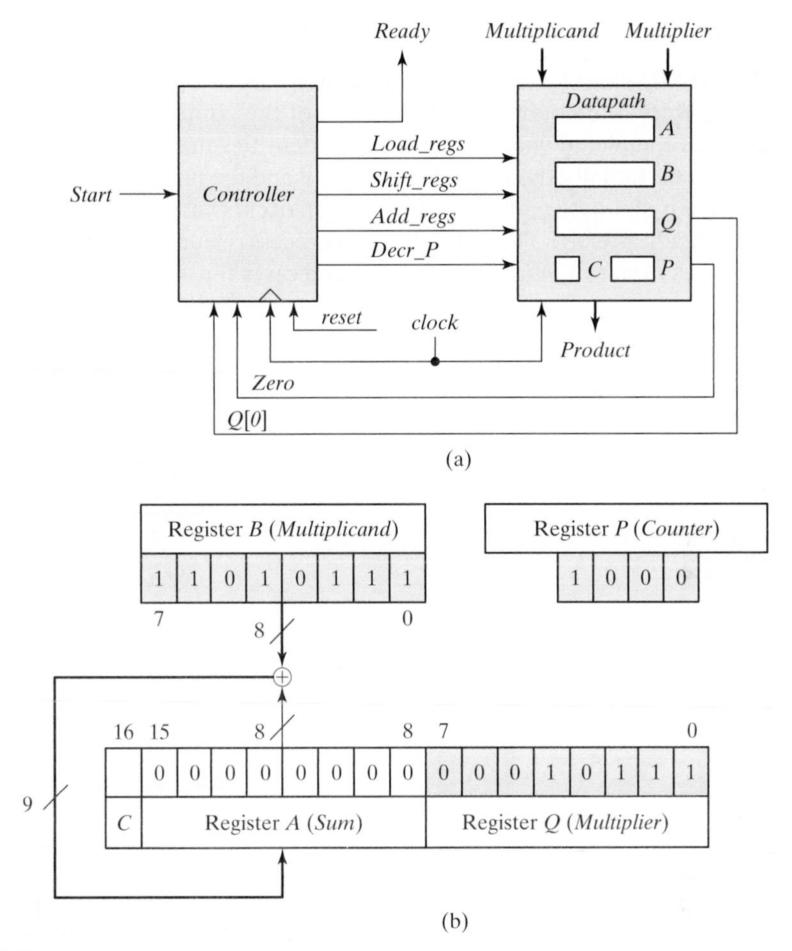

FIGURE 8.14
(a) Block diagram and (b) datapath of a binary multiplier

The interface between the controller and the datapath consists of the status signals and the output signals of the controller. The control signals govern the synchronous register operations of the datapath. Signal *Load_regs* loads the internal registers of the datapath, *Shift_regs* causes the shift register to shift, *Add_regs* forms the sum of the multiplicand and register *A*, and *Decr_P* decrements the counter. The controller also forms output *Ready* to signal to the host environment that the machine is ready to multiply. The contents of the register holding the product vary during execution, so it is useful to have a signal indicating that its contents are valid. Note, again, that the state of the control is not an interface signal between the control unit and the datapath. Only the signals needed to control the datapath are included in the interface. Putting the state in the interface would require a decoder in the datapath, and would require a wider and more active bus than the control signals alone. Not good.

ASMD Chart

The ASMD chart for the binary multiplier is shown in Fig. 8.15. The intermediate form in Fig. 8.15(a) annotates the ASM chart of the controller with the register operations, and the completed chart in Fig. 8.15(b) identifies the Moore and Mealy outputs of the controller. Initially, the multiplicand is in B and the multiplier in Q. As long as the circuit is in the initial state and $Start = 0$, no action occurs and the system remains in state *S_idle* with *Ready* asserted. The multiplication process is launched when $Start = 1$. Then, (1) control goes to state *S_add*, (2) register A and carry flip-flop C are cleared to 0, (3) registers

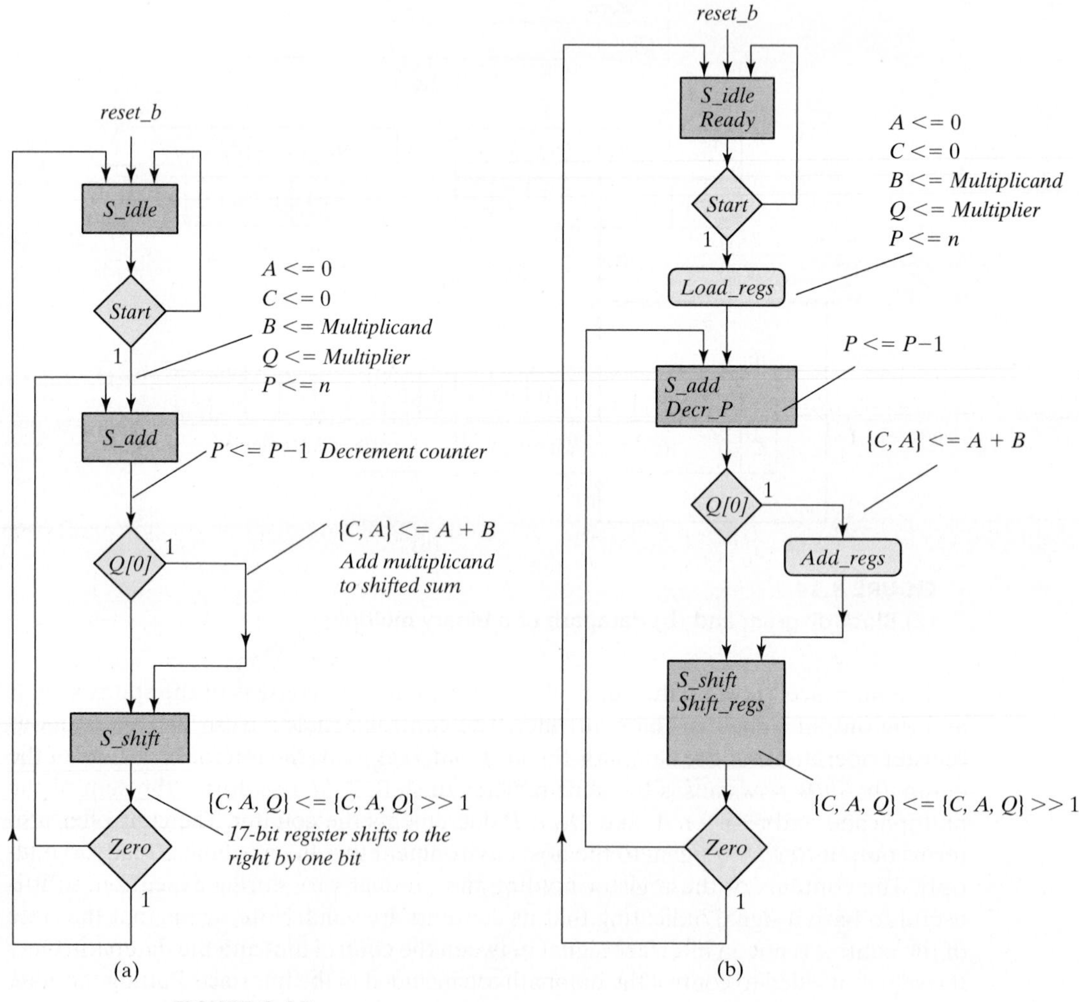

FIGURE 8.15
ASMD chart for binary multiplier

B and Q are loaded with the multiplicand and the multiplier, respectively, and (4) the sequence counter P is set to a binary number n, equal to the number of bits in the multiplier. In state *S_add*, the multiplier bit in *Q[0]* is checked, and if it is equal to 1, the multiplicand in B is added to the partial product in A. The carry from the addition is transferred to C. The partial product in A and C is left unchanged if $Q[0] = 0$. The counter P is decremented by 1 regardless of the value of *Q[0]*, so *Decr_P* is formed in state *S_add* as a Moore output of the controller. In both cases, the next state is *S_shift*. Registers C, A, and Q are combined into one composite register CAQ, denoted by the concatenation $\{C, A, Q\}$, and its contents are shifted once to the right to obtain a new partial product. This shift operation is symbolized in the flowchart with the Verilog logical right-shift operator, >>. It is equivalent to the following statement in register transfer notation:

$$\text{Shift right } CAQ, C \leftarrow 0$$

In terms of individual register symbols, the shift operation can be described by the following register operations:

$$A \leftarrow \text{shr } A, A_{n-1} \leftarrow C$$
$$Q \leftarrow \text{shr } Q, Q_{n-1} \leftarrow A_0$$
$$C \leftarrow 0$$

Both registers A and Q are shifted right. The leftmost bit of A, designated by A_{n-1}, receives the carry from C. The leftmost bit of Q, Q_{n-1}, receives the bit from the rightmost position of A in A_0, and C is reset to 0. In essence, this is a long shift of the composite register CAQ with 0 inserted into the serial input, which is at C.

The value in counter P is checked after the formation of each partial product. If the contents of P are different from zero, status bit *Zero* is set equal to 0 and the process is repeated to form a new partial product. The process stops when the counter reaches 0 and the controller's status input *Zero* is equal to 1. Note that the partial product formed in A is shifted into Q one bit at a time and eventually replaces the multiplier. The final product is available in A and Q, with A holding the most significant bits and Q the least significant bits of the product.

The previous numerical example is repeated in Table 8.5 to clarify the multiplication process. The procedure follows the steps outlined in the ASMD chart. The data shown in the table can be compared with simulation results.

The type of registers needed for the data processor subsystem can be derived from the register operations listed in the ASMD chart. Register A is a shift register with parallel load to accept the sum from the adder and must have a synchronous clear capability to reset the register to 0. Register Q is a shift register. The counter P is a binary down counter with a facility to parallel load a binary constant. The C flip-flop must be designed to accept the input carry and have a synchronous clear. Registers B and Q need a parallel load capability in order to receive the multiplicand and multiplier prior to the start of the multiplication process.

Table 8.5
Numerical Example For Binary Multiplier

Multiplicand $B = 10111_2 = 17_H = 23_{10}$	**Multiplier $Q = 10011_2 = 13_H = 19_{10}$**			
	C	**A**	**Q**	**P**
Multiplier in Q	0	00000	10011	101
$Q_0 = 1$; add B		10111		
First partial product	0	10111		100
Shift right CAQ	0	01011	11001	
$Q_0 = 1$; add B		10111		
Second partial product	1	00010		011
Shift right CAQ	0	10001	01100	
$Q_0 = 0$; shift right CAQ	0	01000	10110	010
$Q_0 = 0$; shift right CAQ	0	00100	01011	001
$Q_0 = 1$; add B		10111		
Fifth partial product	0	11011		
Shift right CAQ	0	01101	10101	000
Final product in $AQ = 0110110101_2 = 1b5_H$				

8.8 CONTROL LOGIC

The design of a digital system can be divided into two parts: the design of the register transfers in the datapath unit and the design of the control logic of the control unit. The control logic is a finite state machine; its Mealy- and Moore-type outputs control the operations of the datapath. The inputs to the control unit are the primary (external) inputs and the internal status signals fed back from the datapath to the controller. The design of the system can be synthesized from an RTL description derived from the ASMD chart. Alternatively, a manual design must derive the logic governing the inputs to the flip-flops holding the state of the controller. The information needed to form the state diagram of the controller is already contained in the ASMD chart, since the rectangular blocks that designate state boxes are the states of the sequential circuit. The diamond-shaped blocks that designate decision boxes determine the logical conditions for the next state transition in the state diagram and assertions of the conditional outputs.

As an example, the control state diagram for the binary multiplier developed in the previous section is shown in Fig. 8.16(a). The information for the diagram is taken directly from the ASMD chart of Fig. 8.15. The three states *S_idle* through *S_shift* are taken from the rectangular state boxes. The inputs *Start* and *Zero* are taken from the diamond-shaped decision boxes. The register transfer operations for each of the three states are listed in Fig. 8.16(b) and are taken from the corresponding state and conditional boxes in the ASMD chart. Establishing the state transitions is the initial focus, so the outputs of the controller are not shown.

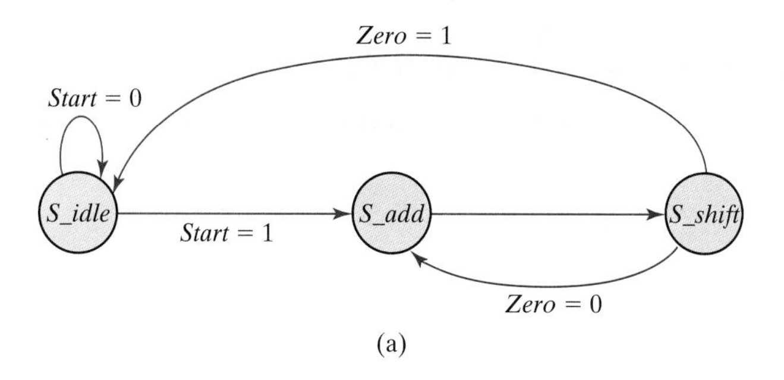

(a)

State Transition		Register Operations
From	To	
S_idle		Initial state
S_idle	*S_add*	$A <= 0, C <= 0, P <= dp\_width$
S_add	*S_shift*	$P <= P - 1$ if ($Q[0]$) then ($A <= A + B, C <= C_{out}$)
S_shift		shift right $\{CAQ\}, C <= 0$

(b)

FIGURE 8.16
Control specifications for binary multiplier

We must execute two steps when implementing the control logic: (1) establish the required sequence of states, and (2) provide signals to control the register operations. The sequence of states is specified in the ASMD chart or the state diagram. The signals for controlling the operations in the registers are specified in the register transfer statements annotated on the ASMD chart or listed in tabular format. For the multiplier, these signals are *Load_regs* (for parallel loading the registers in the datapath unit), *Decr_P* (for decrementing the counter), *Add_regs* (for adding the multiplicand and the partial product), and *Shift_regs* (for shifting register *CAQ*). The block diagram of the control unit is shown in Fig. 8.14(a). The inputs to the controller are *Start, Q[0]*, and *Zero*, and the outputs are *Ready, Load_regs, Decr_P, Add_regs*, and *Shift_regs*, as specified in the ASMD chart. We note that *Q[0]* affects only the output of the controller, not its state transitions. The machine transitions from *S_add* to *S_shift* unconditionally.

An important step in the design is the assignment of coded binary values to the states. The simplest assignment is the sequence of binary numbers, as shown in Table 8.6. Another assignment is the Gray code, according to which only one bit changes when going from one number to the next. A state assignment often used in control design is the *one-hot* assignment. This assignment uses as many bits and flip-flops as there are states in the circuit. At any given time, only one bit is equal to 1 (the one that is hot)

Table 8.6
State Assignment for Control

State	Binary	Gray Code	One-Hot
S_idle	00	00	001
S_add	01	01	010
S_shift	10	11	100

while all others are kept at 0 (all cold). This type of assignment uses a flip-flop for each state. Indeed, one-hot encoding uses more flip-flops than other types of coding, but it usually leads to simpler decoding logic for the next state and the output of the machine. Because the decoding logic does not become more complex as states are added to the machine, the speed at which the machine can operate is not limited by the time required to decode the state.

Since the controller is a sequential circuit, it can be designed manually by the sequential logic procedure outlined in Chapter 5. However, in most cases this method is difficult to carry out manually because of the large number of states and inputs that a typical control circuit may have. As a consequence, it is necessary to use specialized methods for control logic design that may be considered as variations of the classical sequential logic method. We will now present two such design procedures. One uses a sequence register and decoder, and the other uses one flip-flop per state. The method will be presented for a small circuit, but it applies to larger circuits as well. Of course, the need for these methods is eliminated if one has software that automatically synthesizes the circuit from an HDL description.

Sequence Register and Decoder

The sequence-register-and-decoder (manual) method, as the name implies, uses a register for the control states and a decoder to provide an output corresponding to each of the states. (The decoder is not needed if a one-hot code is used.) A register with n flip-flops can have up to 2^n states, and an n-to-2^n-line decoder has up to 2^n outputs. An n-bit sequence register is essentially a circuit with n flip-flops, together with the associated gates that effect their state transitions.

The ASMD chart and the state diagram for the controller of the binary multiplier have three states and two inputs. (There is no need to consider *Q[0]*.) To implement the design with a sequence register and decoder, we need two flip-flops for the register and a two-to-four-line decoder. The outputs of the decoder will form the Moore-type outputs of the controller directly. The Mealy-type outputs will be formed from the Moore outputs and the inputs.

The state table for the finite state machine of the controller is shown in Table 8.7. It is derived directly from the ASMD chart of Fig. 8.15(b) or the state diagram of Fig. 8.16(a). We designate the two flip-flops as G_1 and G_0 and assign the binary states 00, 01, and 10 to *S_idle, S_add*, and *S_shift*, respectively. Note that the input columns have don't-care entries whenever the input variable is not used to determine the next

Table 8.7
State Table for Control Circuit

	Present State		Inputs			Next State						
Present-State Symbol	G_1	G_0	*Start*	*Q[0]*	*Zero*	G_1	G_0	*Ready*	*Load_regs*	*Decr_P*	*Add_regs*	*Shift_regs*
S_idle	0	0	0	X	X	0	0	1	0	0	0	0
S_idle	0	0	1	X	X	0	1	1	1	0	0	0
S_add	0	1	X	0	X	1	0	0	0	1	0	0
S_add	0	1	X	1	X	1	0	0	0	1	1	0
S_shift	1	0	X	X	0	0	1	0	0	0	0	1
S_shift	1	0	X	X	1	0	0	0	0	0	0	1

state. The outputs of the control circuit are designated by the names given in the ASMD chart. The particular Moore-type output variable that is equal to 1 at any given time is determined from the equivalent binary value of the present state. Those output variables are shaded in Table 8.7. Thus, when the present state is $G_1G_0 = 00$, output *Ready* must be equal to 1, while the other outputs remain at 0. Since the Moore-type outputs are a function of only the present state, they can be generated with a decoder circuit having the two inputs G_1 and G_0 and using three of the decoder outputs T_0 through T_2, as shown in Fig. 8.17(a), which does not include the wiring for the state feedback.

The state machine of the controller can be designed from the state table by means of the classical procedure presented in Chapter 5. This example has a small number of states and inputs, so we could use maps to simplify the Boolean functions. In most control logic applications, the number of states and inputs is much larger. In general, the application of the classical method requires an excessive amount of work to obtain the simplified input equations for the flip-flops and is prone to error. The design can be simplified if we take into consideration the fact that the decoder outputs are available for use in the design. Instead of using flip-flop outputs as the present-state conditions, *we use the outputs of the decoder to indicate the present-state condition of the sequential circuit.* Moreover, instead of using maps to simplify the flip-flop equations, we can obtain them directly by inspection of the state table. For example, from the next-state conditions in the state table, we find that the next state of G_1 is equal to 1 when the present state is *S_add* and is equal to 0 when the present state is *S_idle* or *S_shift*. These conditions can be specified by the equation

$$D_{G1} = T_1$$

where D_{G1} is the D input of flip-flop G_1. Similarly, the D input of G_0 is

$$D_{G0} = T_0 Start + T_2 Zero'$$

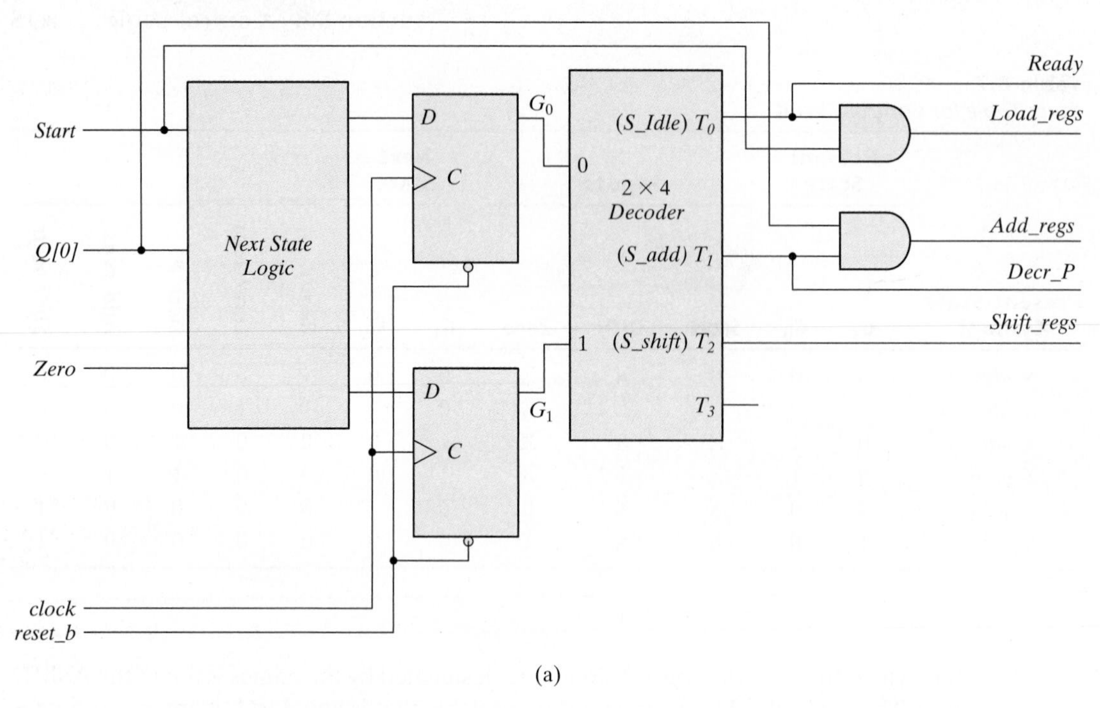

(a)

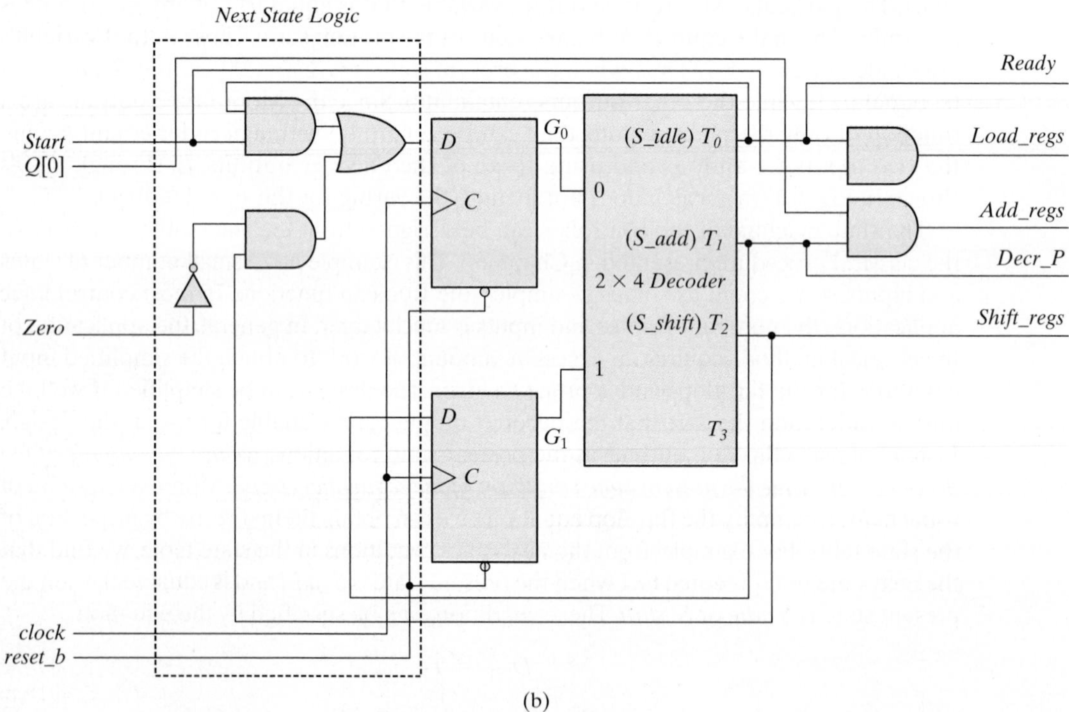

(b)

FIGURE 8.17
Logic diagram of control for binary multiplier using a sequence register and decoder

When deriving input equations by inspection from the state table, we cannot be sure that the Boolean functions have been simplified in the best possible way. (Synthesis tools take care of this detail automatically.) In general, it is advisable to analyze the circuit to ensure that the equations derived do indeed produce the required state transitions.

The logic diagram of the control circuit is drawn in Fig. 8.17(b). It consists of a register with two flip-flops G_1 and G_0 and a 2×4 decoder. The outputs of the decoder are used to generate the inputs to the next-state logic as well as the control outputs. The outputs of the controller should be connected to the datapath to activate the required register operations.

One-Hot Design (One Flip-Flop per State)

Another method of control logic design is the one-hot assignment, which results in a sequential circuit with one flip-flop per state. Only one of the flip-flops contains a 1 at any time; all others are reset to 0. The single 1 propagates from one flip-flop to another under the control of decision logic. In such a configuration, each flip-flop represents a state that is present only when the control bit is transferred to it.

This method uses the maximum number of flip-flops for the sequential circuit. For example, a sequential circuit with 12 states requires a minimum of four flip-flops. By contrast, with the method of one flip-flop per state, the circuit requires 12 flip-flops, one for each state. At first glance, it may seem that this method would increase system cost, since more flip-flops are used. But the method offers some advantages that may not be apparent. One advantage is the simplicity with which the logic can be designed by inspection of the ASMD chart or the state diagram. No state or excitation tables are needed if *D*-type flip-flops are employed. The one-hot method offers a savings in design effort, an increase in operational simplicity, and a possible decrease in the total number of gates, since a decoder is not needed.

The design procedure for a one-hot state assignment will be demonstrated by obtaining the control circuit specified by the state diagram of Fig. 8.16(a). Since there are three states in the state diagram, we choose three D flip-flops and label their outputs G_0, G_1, and G_2, *corresponding to S_idle, S_add*, and *S_shift*, respectively. The input equations for setting each flip-flop to 1 are determined from the present state and the input conditions along the corresponding directed lines going into the state. For example, D_{G0}, the input to flip-flop G_0, is set to 1 if the machine is in state G_0 and *Start* is not asserted, or if the machine is in state G_2 and *Zero* is asserted. These conditions are specified by the input equation:

$$D_{G0} = G_0 Start' + G_2 Zero$$

In fact, the condition for setting a flip-flop to 1 is obtained directly from the state diagram, from the condition specified in the directed lines going into the corresponding flip-flop state ANDed with the previous flip-flop state. If there is more than one directed line going into a state, all conditions must be ORed. Using this procedure for the other three flip-flops, we obtain the remaining input equations:

$$D_{G1} = G_0 Start + G_2 Zero'$$
$$D_{G2} = G_1$$

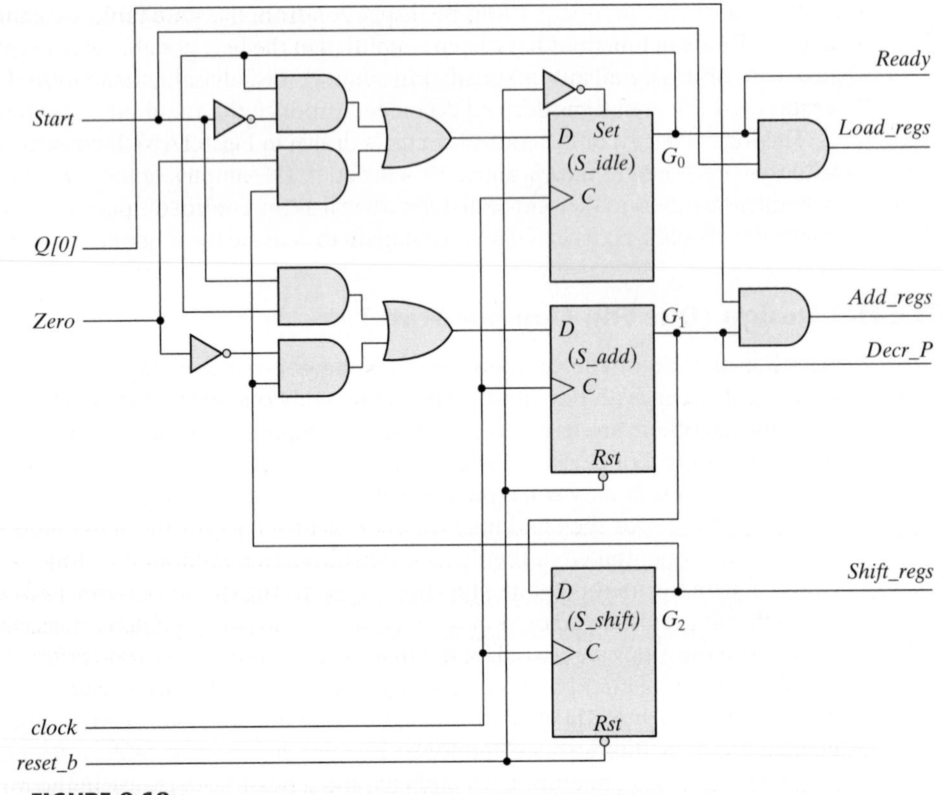

FIGURE 8.18
Logic diagram for one-hot state controller

The logic diagram of the one-hot controller (with one flip-flop per state) is shown in Fig. 8.18. The circuit consists of three D flip-flops labeled G_0 through G_2, together with the associated gates specified by the input equations. Initially, flip-flop G_0 must be set to 1 and all other flip-flops must be reset to 0, so that the flip-flop representing the initial state is enabled. This can be done by using an asynchronous preset on flip-flop G_0 and an asynchronous clear for the other flip-flops. Once started, the controller with one flip-flop per state will propagate from one state to the other in the proper manner. Only one flip-flop will be set to 1 with each clock edge; all others are reset to 0, because their D inputs are equal to 0.

8.9 HDL DESCRIPTION OF BINARY MULTIPLIER

A second example of an HDL description of an RTL design is given in HDL Example 8.5, the binary multiplier designed in Section 8.7. For simplicity, the entire description is "flattened" and encapsulated in one module. Comments will identify the controller and the datapath. The first part of the description declares all of the inputs and outputs as specified in the block diagram of Fig. 8.14(a). The machine will be parameterized for a five-bit datapath to enable a comparison between its simulation data and the result of the multiplication with the numerical example listed in Table 8.5. The same model can be used for a datapath having a different size merely by changing the value of the parameters. The second part of the description declares all registers in the controller and the datapath, as well as the one-hot encoding of the states. The third part specifies implicit combinational logic (continuous assignment statements) for the concatenated register *CAQ*, the *Zero* status signal, and the *Ready* output signal. The continuous assignments for *Zero* and *Ready* are accomplished by assigning a Boolean expression to their **wire** declarations. The next section describes the control unit, using a single edge-sensitive cyclic behavior to describe the state transitions, and a level-sensitive cyclic behavior to describe the combinational logic for the next state and the outputs. Again, note that default assignments are made to *next_state, Load_regs, Decr_P, Add_regs*, and *Shift_regs*. The subsequent logic of the case statement assigns their value by exception. The state transitions and the output logic are written directly from the ASMD chart of Fig. 8.15(b).

The datapath unit describes the register operations within a separate edge-sensitive cyclic behavior.[3] (For clarity, separate cyclic behaviors are used; we do not mix the description of the datapath with the description of the controller.) Each control input is decoded and is used to specify the associated operations. The addition and subtraction operations will be implemented in hardware by combinational logic. Signal *Load_regs* causes the counter and the other registers to be loaded with their initial values, etc. Because the controller and datapath have been partitioned into separate units, the control signals completely specify the behavior of the datapath; explicit information about the state of the controller is not needed and is not made available to the datapath unit.

The next-state logic of the controller includes a default case item to direct a synthesis tool to map any of the unused codes to *S_idle*. The default case item and the default assignments preceding the **case** statement ensure that the machine will recover if it somehow enters an unused state. They also prevent unintentional synthesis of latches. (Remember, a synthesis tool will synthesize latches when what was intended to be combinational logic in fact fails to completely specify the input–output function of the logic.)

HDL Example 8.5 (Sequential Multiplier)

```
module Sequential_Binary_Multiplier (Product, Ready, Multiplicand, Multiplier, Start,
clock, reset_b);
// Default configuration: five-bit datapath
 parameter                          dp_width = 5;     // Set to width of datapath
 output      [2*dp_width -1: 0]     Product;
 output                             Ready;
 input       [dp_width -1: 0]       Multiplicand, Multiplier;
 input                              Start, clock, reset_b;
```

[3]The width of the datapath here is *dp-width*.

```
parameter                               BC_size = 3;      // Size of bit counter
parameter                               S_idle = 3'b001, // one-hot code
                                        S_add = 3'b010,
                                        S_shift = 3'b100;
reg     [2: 0]                          state, next_state;
reg     [dp_width -1: 0]          A, B, Q;           // Sized for datapath
reg                               C;
reg     [BC_size -1: 0]           P;
reg                               Load_regs, Decr_P, Add_regs, Shift_regs;

// Miscellaneous combinational logic

assign                            Product = {A, Q};
wire                              Zero = (P == 0);          // counter is zero
                                  // Zero = ~|P;             // alternative
wire                              Ready = (state == S_idle); // controller status
// control unit
always @ (posedge clock, negedge reset_b)
  if (~reset_b) state <= S_idle; else state <= next_state;

always @ (state, Start, Q[0], Zero) begin
  next_state = S_idle;
  Load_regs = 0;
  Decr_P = 0;
  Add_regs = 0;
  Shift_regs = 0;
  case (state)
    S_idle:     begin if (Start) next_state = S_add; Load_regs = 1; end
    S_add:      begin next_state = S_shift; Decr_P = 1; if (Q[0]) Add_regs = 1; end
    S_shift:    begin Shift_regs = 1; if (Zero) next_state = S_idle;
                else next_state = S_add; end
    default:    next_state = S_idle;
  endcase
end
// datapath unit
always @ (posedge clock) begin
  if (Load_regs) begin
    P <= dp_width;
    A <= 0;
    C <= 0;
    B <= Multiplicand;
    Q <= Multiplier;
  end
  if (Add_regs) {C, A} <= A + B;
```

```
    if (Shift_regs) {C, A, Q} <= {C, A, Q} >> 1;
    if (Decr_P) P <= P -1;
  end
endmodule
```

Testing the Multiplier

HDL Example 8.6 shows a test bench for testing the multiplier. The inputs and outputs are the same as those shown in the block diagram of Fig. 8.14(a). It is naive to conclude that an HDL description of a system is correct on the basis of the output it generates under the application of a few input signals. A more strategic approach to testing and verification exploits the partition of the design into its datapath and control unit. This partition supports separate verification of the controller and the datapath. A separate test bench can be developed to verify that the datapath executes each operation and generates status signals correctly. After the datapath unit is verified, the next step is to verify that each control signal is formed correctly by the control unit. A separate test bench can verify that the control unit exhibits the complete functionality specified by the ASMD chart (i.e., that it makes the correct state transitions and asserts its outputs in response to the external inputs and the status signals).

A verified control unit and a verified datapath unit together do not guarantee that the system will operate correctly. The final step in the design process is to integrate the verified models within a parent module and verify the functionality of the overall machine. The interface between the controller and the datapath must be examined in order to verify that the ports are connected correctly. For example, a mismatch in the listed order of signals may not be detected by the compiler. After the datapath unit and the control unit have been verified, a third test bench should verify the specified functionality of the complete system. In practice, this requires writing a comprehensive test plan identifying that functionality. For example, the test plan would identify the need to verify that the sequential multiplier asserts the signal *Ready* in state *S_idle*. The exercise to write a test plan is not academic: The quality and scope of the test plan determine the worth of the verification effort. The test plan guides the development of the test bench and increases the likelihood that the final design will match its specification.

Testing and verifying an HDL model usually requires access to more information than the inputs and outputs of the machine. Knowledge of the state of the control unit, the control signals, the status signals, and the internal registers of the datapath might all be necessary for debugging. Fortunately, Verilog provides a mechanism to hierarchically de-reference identifiers so that any variable at any level of the design hierarchy can be visible to the test bench. Procedural statements can display the information required to support efforts to debug the machine. Simulators use this mechanism to display waveforms of any variable in the design hierarchy. To use the mechanism, we reference the variable by its hierarchical path name. For example, the register *P* within the datapath unit is not an output port of the multiplier, but it can be referenced as *M0.P*. The hierarchical path name consists of the sequence of module identifiers or block names, separated by periods and specifying the location of the variable in the design hierarchy. We also note that simulators commonly have a graphical user interface that displays all levels of the hierarchy of a design.

The first test bench in HDL Example 8.6 uses the system task **$strobe** to display the result of the computations. This task is similar to the **$display** and **$monitor** tasks explained in Section 4.12. The **$strobe** system task provides a synchronization mechanism to ensure that data are displayed only after all assignments in a given time step are executed. This is very useful in synchronous sequential circuits, where the time step begins at a clock edge and multiple assignments may occur at the same time step of simulation. When the system is synchronized to the positive edge of the clock, using **$strobe** after the ***always @ (posedge** clock)* statement ensures that the display shows values of the signal after the clock pulse.

The test bench module *t_Sequential_Binary_Multiplier* in HDL Example 8.6 instantiates the module *Sequential Binary_Multiplier* of HDL Example 8.5. Both modules must be included as source files when simulating the multiplier with a Verilog HDL simulator. The result of this simulation displays a simulation log with numbers identical to the ones in Table 8.5. The code includes a second test bench to exhaustively multiply five-bit values of the multiplicand and the multiplier. Waveforms for a sample of simulation results are shown in Fig. 8.19. The numerical values of *Multiplicand, Multiplier*, and *Product* are displayed in decimal and hexadecimal formats. Insight can be gained by studying the displayed waveforms of the control state, the control signals, the status signals, and the register operations. Enhancements to the multiplier and its test bench are considered in the problems at the end of this chapter. In this example, $19_{10} \times 23_{10} = 437_{10}$, and $17_H + 0b_H = 02_H$ with C = 1. Note the need for the carry bit.

HDL Example 8.6

```
// Test bench for the binary multiplier
module t_Sequential_Binary_Multiplier;
 parameter                        dp_width = 5;          // Set to width of datapath
 wire          [2*dp_width -1: 0] Product;               // Output from multiplier
 wire                             Ready;
 reg           [dp_width -1: 0]   Multiplicand, Multiplier; // Inputs to multiplier
 reg                              Start, clock, reset_b;
// Instantiate multiplier
Sequential_Binary_Multiplier M0 (Product, Ready, Multiplicand, Multiplier, Start, clock,
 reset_b);
 // Generate stimulus waveforms
initial #200 $finish;
initial
  begin
   Start = 0;
```

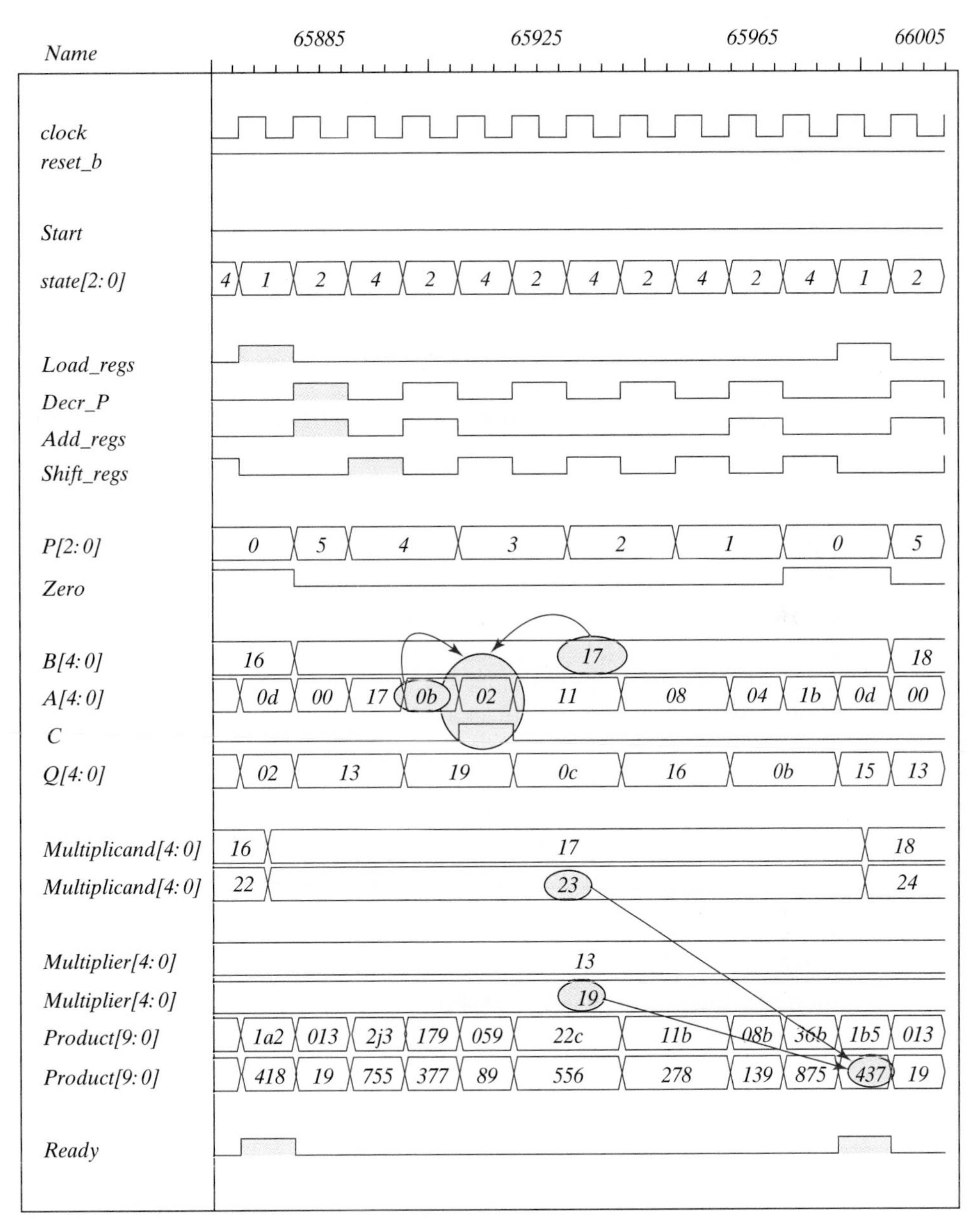

FIGURE 8.19
Simulation waveforms for one-hot state controller

```
    reset_b = 0;
    #2 Start = 1; reset_b = 1;
    Multiplicand = 5'b10111;      Multiplier = 5'b10011;
    #10 Start = 0;
  end
 initial
  begin
   clock = 0;
   repeat (26) #5 clock = ~clock;
  end
 // Display results and compare with Table 8.5
 always @ (posedge clock)
  $strobe ("C=%b A=%b Q=%b P=%b time=%0d",M0.C,M0.A,M0.Q,M0.P, $time);
endmodule
```

```
Simulation log:
C=0 A=00000  Q=10011  P=101 time=5
C=0 A=10111  Q=10011  P=100 time=15
C=0 A=01011  Q=11001  P=100 time=25
C=1 A=00010  Q=11001  P=011 time=35
C=0 A=10001  Q=01100  P=011 time=45
C=0 A=10001  Q=01100  P=010 time=55
C=0 A=01000  Q=10110  P=010 time=65
C=0 A=01000  Q=10110  P=001 time=75
C=0 A=00100  Q=01011  P=001 time=85
C=0 A=11011  Q=01011  P=000 time=95
C=0 A=01101  Q=10101  P=000 time=105
C=0 A=01101  Q=10101  P=000 time=115
C=0 A=01101  Q=10101  P=000 time=125
```

```
/* Test bench for exhaustive simulation
module t_Sequential_Binary_Multiplier;
 parameter                            dp_width = 5;          // Width of datapath
 wire          [2 * dp_width -1: 0]   Product;
 wire                                 Ready;
 reg           [dp_width -1: 0]       Multiplicand, Multiplier;
 reg                                  Start, clock, reset_b;
  Sequential_Binary_Multiplier M0 (Product, Ready, Multiplicand, Multiplier, Start, clock,
   reset_b);
 initial #1030000 $finish;
 initial begin clock = 0; #5 forever #5 clock = ~clock; end
 initial fork
  reset_b = 1;
  #2 reset_b = 0;
  #3 reset_b = 1;
 join
```

```
  initial begin #5 Start = 1; end
  initial begin
   #5 Multiplicand = 0;
   Multiplier = 0;
   repeat (32) #10 begin Multiplier = Multiplier + 1;
    repeat (32) @ (posedge M0.Ready) 5 Multiplicand = Multiplicand + 1;
   end
  end
endmodule
*/
```

Behavioral Description of a Parallel Multiplier

Structural modeling implicitly specifies the functionality of a digital machine by prescribing an interconnection of gate-level hardware units. In this form of modeling, a synthesis tool performs Boolean optimization and translates the HDL description of a circuit into a netlist of gates in a particular technology, e.g., CMOS. Hardware design at this level often requires cleverness and accrued experience. It is the most tedious and detailed form of modeling. In contrast, behavioral RTL modeling specifies functionality abstractly, in terms of HDL operators. The RTL model does not specify a gate-level implementation of the registers or the logic to control the operations that manipulate their contents—those tasks are accomplished by a synthesis tool. RTL modeling implicitly schedules operations by explicitly assigning them to clock cycles. The most abstract form of behavioral modeling describes only an algorithm, without any reference to a physical implementation, a set of resources, or a schedule for their use. Thus, algorithmic modeling allows a designer to explore trade-offs in the space (hardware) and time domains, trading processing speed for hardware complexity.

HDL Example 8.7 presents an RTL model and an algorithmic model of a binary multiplier. Both use a level-sensitive cyclic behavior. The RTL model expresses the functionality of a multiplier in a single statement. A synthesis tool will associate with the multiplication operator a gate-level circuit equivalent to that shown in Section 4.7. In simulation, when either the multiplier or the multiplicand changes, the product will be updated. The time required to form the product will depend on the propagation delays of the gates available in the library of standard cells used by the synthesis tool. The second model is an algorithmic description of the multiplier. A synthesis tool will unroll the loop of the algorithm and infer the need for a gate-level circuit equivalent to that shown in Section 4.7.

Be aware that a synthesis tool may not be able to synthesize a given algorithmic description, even though the associated HDL model will simulate and produce correct results. One difficulty is that the sequence of operations implied by an algorithm might not be physically realizable in a single clock cycle. It then becomes necessary to distribute the operations over multiple clock cycles. A tool for synthesizing RTL logic will not be able to automatically accomplish the required distribution of effort, but a tool that

HDL Example 8.7

```
// Behavioral (RTL) description of a parallel multiplier (n = 8)
module Mult (Product, Multiplicand, Multiplier);
  input [7: 0]      Multiplicand, Multiplier;
  output reg [15: 0] Product;
  always @ (Multiplicand, Multiplier)
   Product = Multiplicand * Multiplier;
endmodule
module Algorithmic_Binary_Multiplier #(parameter dp_width = 5) (
  output [2*dp_width -1: 0] Product, input [dp_width -1: 0] Multiplicand, Multiplier);
  reg [dp_width -1: 0]            A, B, Q;          // Sized for datapath
  reg                             C;
  integer                         k;
  assign                          Product = {C, A, Q};
  always @ (Multiplier, Multiplicand) begin
    Q = Multiplier;
    B = Multiplicand;
    C = 0;
    A = 0;
    for (k = 0; k <= dp_width -1; k = k + 1) begin
    if (Q[0]) {C, A} = A + B;
    {C, A, Q} = {C, A, Q} >> 1;
  end
 end
endmodule
module t_Algorithmic_Binary_Multiplier;
 parameter                      dp_width = 5;     // Width of datapath
 wire [2* dp_width -1: 0]       Product;
 reg [dp_width -1: 0]           Multiplicand, Multiplier;
 integer                        Exp_Value;
 reg                            Error;
  Algorithmic_Binary_Multiplier M0 (Product, Multiplicand, Multiplier);
 // Error detection
  initial # 1030000 finish;
  always @ (Product) begin
   Exp_Value = Multiplier * Multiplicand;
   // Exp_Value = Multiplier * Multiplicand +1; // Inject error to confirm detection
   Error = Exp_Value ^ Product;
  end
// Generate multiplier and multiplicand exhaustively for 5 bit operands
 initial begin
  #5 Multiplicand = 0;
  Multiplier = 0;
```

```
    repeat (32) #10 begin Multiplier = Multiplier + 1;
     repeat (32) #5 Multiplicand = Multiplicand + 1;
    end
  end
endmodule
```

is designed to synthesize algorithms should be successful. In effect, a behavioral synthesis tool would have to allocate the registers and adders to implement multiplication. If only a single adder is to be shared by all of the operations that form a partial sum, the activity must be distributed over multiple clock cycles and in the correct sequence, ultimately leading to the sequential binary multiplier for which we have explicitly designed the controller for its datapath. Behavioral synthesis tools require a different and more sophisticated style of modeling and are not within the scope of this text.

8.10 DESIGN WITH MULTIPLEXERS

The register-and-decoder scheme for the design of a controller has three parts: the flip-flops that hold the binary state value, the decoder that generates the control outputs, and the gates that determine the next-state and output signals. In Section 4.11, it was shown that a combinational circuit can be implemented with multiplexers instead of individual gates. Replacing the gates with multiplexers results in a regular pattern of three levels of components. The first level consists of multiplexers that determine the next state of the register. The second level contains a register that holds the present binary state. The third level has a decoder that asserts a unique output line for each control state. These three components are predefined standard cells in many integrated circuits.

Consider, for example, the ASM chart of Fig. 8.20, consisting of four states and four control inputs. We are interested in only the control signals governing the state sequence. These signals are independent of the register operations of the datapath, so the edges of the graph are not annotated with datapath register operations, and the graph does not identify the output signals of the controller. The binary assignment for each state is indicated at the upper right corner of the state boxes. The decision boxes specify the state transitions as a function of the four control inputs: w, x, y, and z. The three-level control implementation, shown in Fig. 8.21, consists of two multiplexers, MUX1 and MUX2; a register with two flip-flops, $G_1$and G_0; and a decoder with four outputs—d_0, d_1, d_2, and d_3, corresponding to *S_0, S_1, S_2*, and *S_3*, respectively. The outputs of the state-register flip-flops are applied to the decoder inputs and also to the select inputs of the multiplexers. In this way, the present state of the register is used to select one of the inputs from each multiplexer. The outputs of the multiplexers are then applied to the D inputs of G_1 and G_0. The purpose of each multiplexer is to produce an input to its corresponding flip-flop equal to the binary value of that bit of the next-state vector. The inputs of the multiplexers

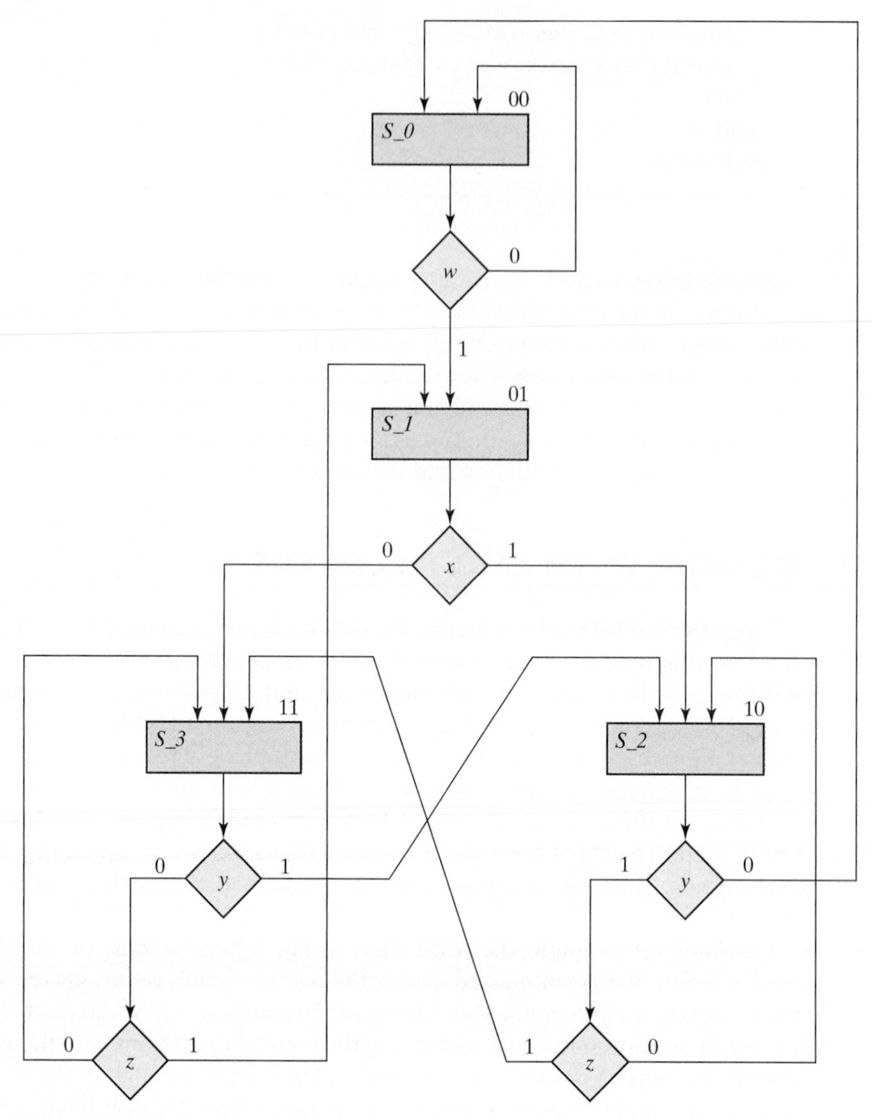

FIGURE 8.20
Example of ASM chart with four control inputs

are determined from the decision boxes and state transitions given in the ASM chart. For example, state 00 stays at 00 or goes to 01, depending on the value of input w. Since the next state of G_1 is 0 in either case, we place a signal equivalent to logic 0 in MUX1 input 0. The next state of G_0 is 0 if $w = 0$ and 1 if $w = 1$. Since the next state of G_0 is equal to w, we apply control input w to MUX2 input 0. This means that when the select inputs of the multiplexers are equal to present state 00, the outputs of the multiplexers provide the binary value that is transferred to the register at the next clock pulse.

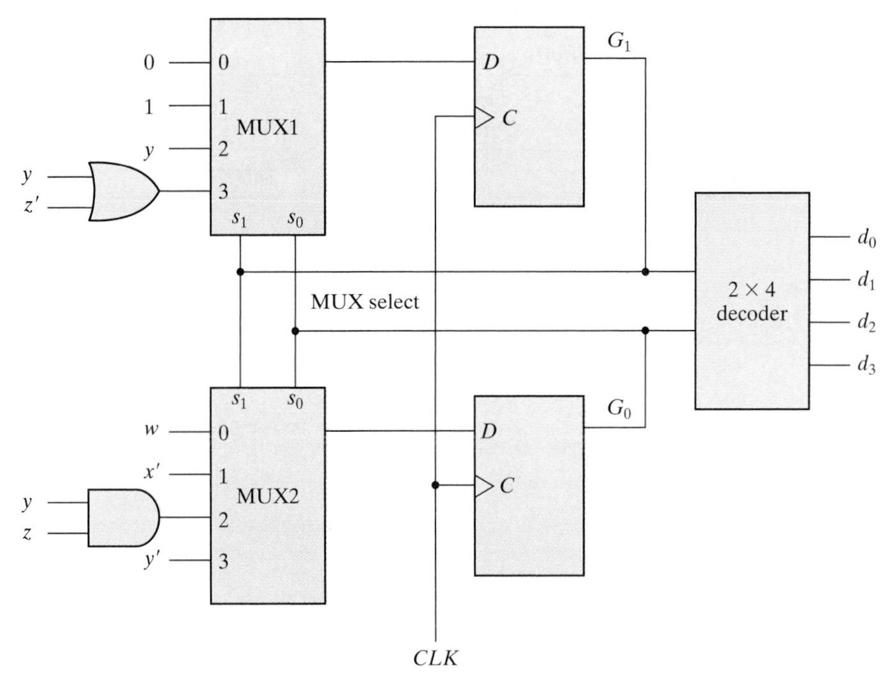

FIGURE 8.21
Control implementation with multiplexers

To facilitate the evaluation of the multiplexer inputs, we prepare a table showing the input conditions for each possible state transition in the ASM chart. Table 8.8 gives this information for the ASM chart of Fig. 8.20. There are two transitions from present state 00 or 01 and three from present state 10 or 11. The sets of transitions are separated by horizontal lines across the table. The input conditions listed in the table are obtained from the decision boxes in the ASM chart. For example, from Fig. 8.20, we note that present state 01 will go to next state 10 if $x = 1$ or to next state 11 if $x = 0$. In the table, we mark these input conditions as x and x', respectively. The two columns under "multiplexer inputs" in the table specify the input values that must be applied to MUX1 and MUX2. The multiplexer input for each present state is determined from the input conditions when the next state of the flip-flop is equal to 1. Thus, after present state 01, the next state of G_1 is always equal to 1 and the next state of G_0 is equal to the complement of x. Therefore, the input of MUX1 is made equal to 1 and that of MUX2 to x' when the present state of the register is 01. As another example, after present state 10, the next state of G_1 must be equal to 1 if the input conditions are yz' or yz. When these two Boolean terms are ORed together and then simplified, we obtain the single binary variable y, as indicated in the table. The next state of G_0 is equal to 1 if the input conditions are $yz = 11$. If the next state of G_1 remains at 0 after a given present state, we place a 0 in the multiplexer input, as shown in present state 00 for

Table 8.8
Multiplexer Input Conditions

Present State		Next State		Input Condition	Inputs	
G_1	G_0	G_1	G_0	s	MUX1	MUX2
0	0	0	0	w'		
0	0	0	1	w	0	w
0	1	1	0	x		
0	1	1	1	x'	1	x'
1	0	0	0	y'		
1	0	1	0	yz'		
1	0	1	1	yz	$yz' + yz = y$	yz
1	1	0	1	$y'z$		
1	1	1	0	y		
1	1	1	1	$y'z'$	$y + y'z' = y + z'$	$y'z + y'z' = y'$

MUX1. If the next state of G_1 is always 1, we place a 1 in the multiplexer input, as shown in present state 01 for MUX1. The other entries for MUX1 and MUX2 are derived in a similar manner. The multiplexer inputs from the table are then used in the control implementation of Fig. 8.21. Note that if the next state of a flip-flop is a function of two or more control variables, the multiplexer may require one or more gates in its input. Otherwise, the multiplexer input is equal to the control variable, the complement of the control variable, 0, or 1.

Design Example: Count the Number of Ones in a Register

We will demonstrate the multiplexer implementation of the logic for a control unit by means of a design example—a system that is to count the number of 1's in a word of data. The example will also demonstrate the formulation of the ASMD chart and the implementation of the datapath subsystem.

From among various alternatives, we will consider a ones counter consisting of two registers *R1* and *R2*, and a flip-flop *E*. (A more efficient implementation is considered in the problems at the end of the chapter.) The system counts the number of 1's in the number loaded into register *R1* and sets register *R2* to that number. For example, if the binary number loaded into *R1* is 10111001, the circuit counts the five 1's in *R1* and sets register *R2* to the binary count 101. This is done by shifting each bit from register *R1* one at a time into flip-flop *E*. The value in *E* is checked by the control, and each time it is equal to 1, register *R2* is incremented by 1.

The block diagram of the datapath and controller are shown in Fig. 8.22(a). The datapath contains registers *R1*, *R2*, and *E*, as well as logic to shift the leftmost bit of *R1* into *E*. The unit also contains logic (a NOR gate to detect whether *R1* is 0, but that

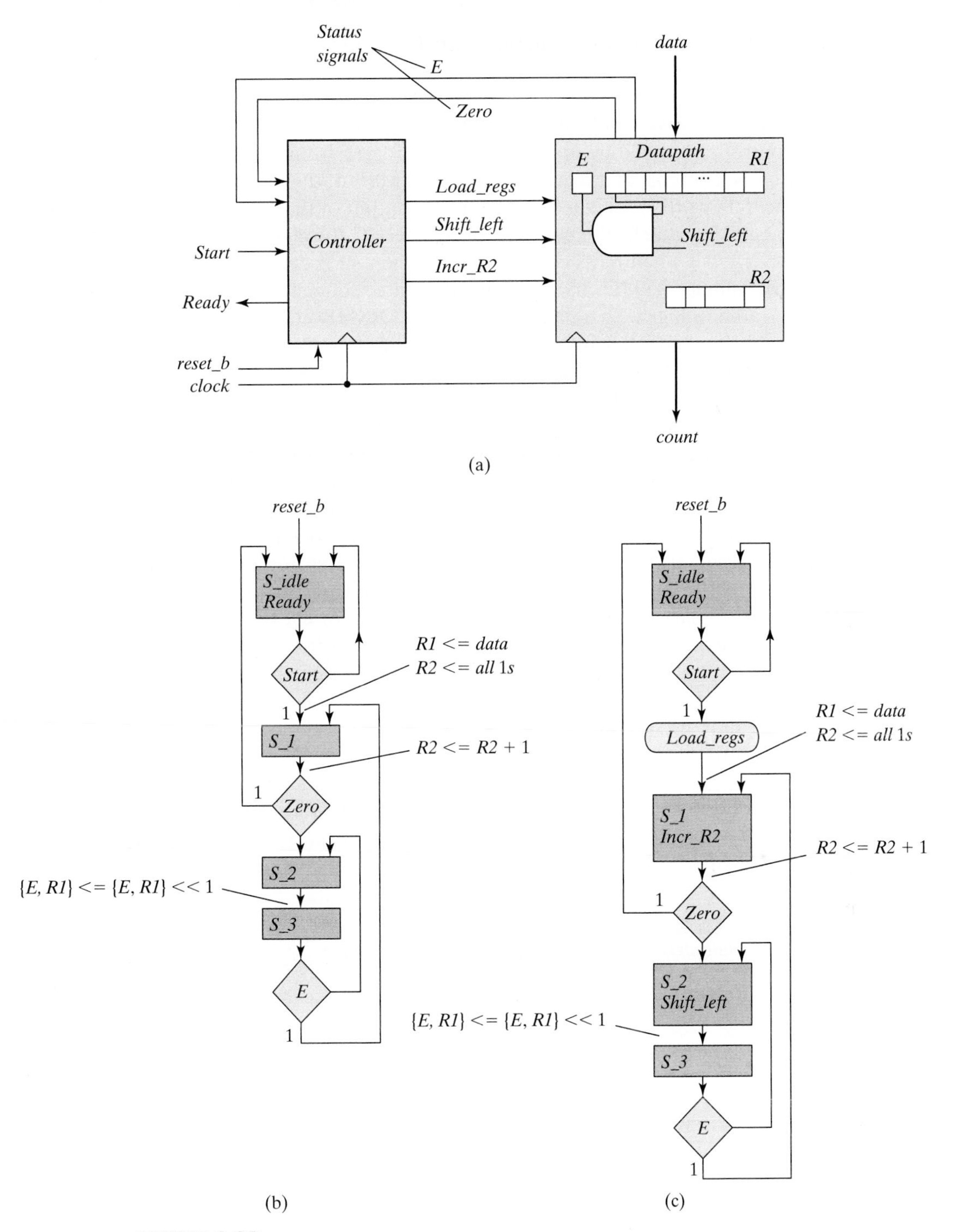

FIGURE 8.22
Block diagram and ASMD chart for count-of-ones circuit

detail is omitted in the figure). The external input signal *Start* launches the operation of the machine; *Ready* indicates the status of the machine to the external environment. The controller has status input signals *E* and *Zero* from the datapath. These signals indicate the contents of a register holding the MSB of the data word and the condition that the data word is 0, respectively. *E* is the output of the flip-flop. *Zero* is the output of a circuit that checks the contents of register *R1* for all 0's. The circuit produces an output *Zero* = 1 when *R1* is equal to 0 (i.e., when *R1* is empty of 1's).

A preliminary ASMD chart showing the state sequence and the register operations is illustrated in Fig. 8.22(b), and the complete ASMD chart in Fig. 8.22(c). Asserting *Start* with the controller in *S_idle* transfers the state to *S_1*, concurrently loads register *R1* with the binary data word, and fills the cells of *R2* with 1's. Note that incrementing a number with all 1's in a counter register produces a number with all 0's. Thus, the first transition from *S_1* to *S_2* will clear *R2*. Subsequent transitions will have *R2* holding a count of the bits of data that have been processed. The content of *R1*, as indicated by *Zero*, will also be examined in *S_1*. If *R1* is empty, *Zero* = 1, and the state returns to *S_idle*, where it asserts *Ready*. In state *S_1, Incr_R2* is asserted to cause the datapath unit to increment *R2* at each clock pulse. If *R1* is not empty of 1's, then *Zero* = 0, indicating that there are some 1's stored in the register. The number in *R1* is shifted and its leftmost bit is transferred into *E*. This is done as many times as necessary, until a 1 is transferred into *E*. For every 1 detected in *E*, register *R2* is incremented and register *R1* is checked again for more 1's. The major loop is repeated until all the 1's in *R1* are counted. Note that the state box of *S_3* has no register operations, but the block associated with it contains the decision box for *E*. Note also that the serial input to shift register *R1* must be equal to 0 because we don't want to shift external 1's into *R1*. The register *R1* in Fig. 8.22(a) is a shift register. Register *R2* is a counter with parallel load. The multiplexer input conditions for the control are determined from Table 8.9. The input conditions are obtained from the ASMD chart for each possible binary state transition. The four states are assigned

Table 8.9
Multiplexer Input Conditions for Design Example

Present State		Next State		Input Conditions	Multiplexer Inputs	
G_1	G_0	G_1	G_0		MUX1	MUX2
0	0	0	0	*Start'*		
0	0	0	1	*Start*	0	*Start*
0	1	0	0	*Zero*		
0	1	1	0	*Zero'*	*Zero'*	0
1	0	1	1	*None*	1	1
1	1	1	0	*E'*		
1	1	0	1	*E*	*E'*	*E*

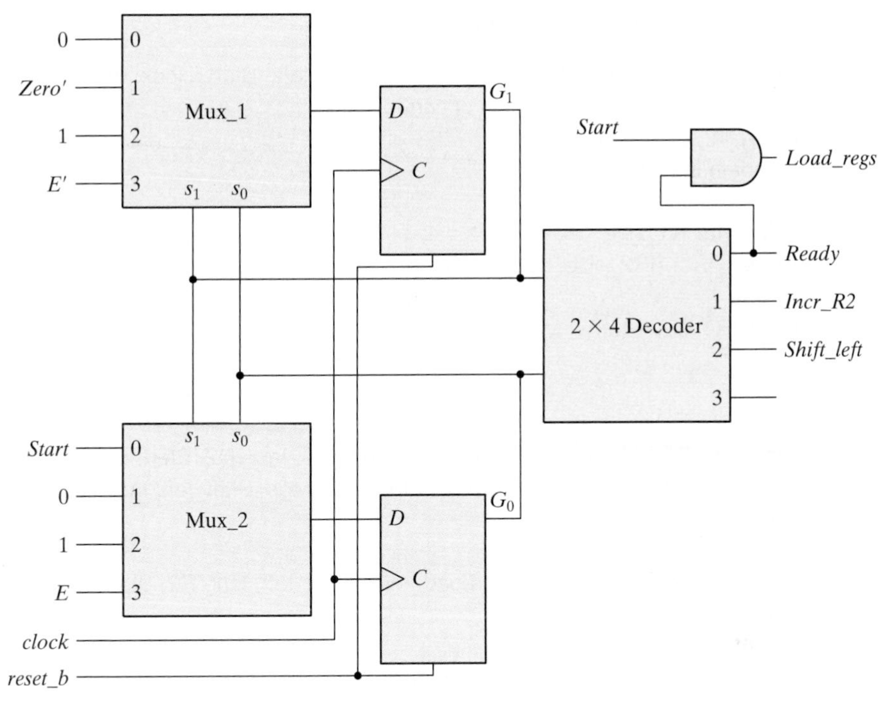

FIGURE 8.23
Control implementation for count-of-ones circuit

binary values 00 through 11. The transition from present state 00 depends on *Start*. The transition from present state 01 depends on *Zero*, and the transition from present state 11 on *E*. Present state 10 goes to next state 11 unconditionally. The values under MUX1 and MUX2 in the table are determined from the Boolean input conditions for the next state of G_1 and G_0, respectively.

The control implementation of the design example is shown in Fig. 8.23. This is a three-level implementation, with the multiplexers in the first level. The inputs to the multiplexers are obtained from Table 8.9. The Verilog description in HDL Example 8.8 instantiates structural models of the controller and the datapath. The listing of code includes the lower level modules implementing their structures. Note that the datapath unit does not have a reset signal to clear the registers, but the models for the flip-flop, shift register, and counter have an active-low reset. This illustrates the use of Verilog data type **supply1** to hardwire those ports to logic value 1 in their instantiation within *Datapath_STR*. Note also that the test bench uses hierarchical de-referencing to access the state of the controller to make the debug and verification tasks easier, without having to alter the module ports to provide access to the internal signals. Another detail to observe is that the serial input to the shift register is hardwired to 0. The lower level models are described behaviorally for simplicity.

HDL Example 8.8 (Ones Counter)

```
module Count_Ones_STR_STR (count, Ready, data, Start, clock, reset_b);
// Mux – decoder implementation of control logic
// controller is structural
// datapath is structural

 parameter R1_size = 8, R2_size = 4;
 output      [R2_size -1: 0]   count;
 output                        Ready;
 input [R1_size -1: 0]         data;
 input                         Start, clock, reset_b;
 wire                          Load_regs, Shift_left, Incr_R2, Zero, E;

 Controller_STR M0 (Ready, Load_regs, Shift_left, Incr_R2, Start, E, Zero, clock, reset_b);
 Datapath_STR M1 (count, E, Zero, data, Load_regs, Shift_left, Incr_R2, clock);
endmodule

module Controller_STR (Ready, Load_regs, Shift_left, Incr_R2, Start, E, Zero, clock,
 reset_b);
 output       Ready;
 output       Load_regs, Shift_left, Incr_R2;
 input        Start;
 input        E, Zero;
 input        clock, reset_b;
 supply0      GND;
 supply1      PWR;
 parameter    S0 = 2'b00, S1 = 2'b01, S2 = 2'b10, S3 = 2'b11; // Binary code
 wire                                   Load_regs, Shift_left, Incr_R2;
 wire                                   G0, G0_b, D_in0, D_in1, G1, G1_b;
 wire                                   Zero_b = ~Zero;
 wire                                   E_b = ~E;
 wire [1: 0]                            select = {G1, G0};
 wire [0: 3]                            Decoder_out;
 assign                                 Ready = ~Decoder_out[0];
 assign                                 Incr_R2 = ~Decoder_out[1];
 assign                                 Shift_left = ~Decoder_out[2];
 and                                    (Load_regs, Ready, Start);
 mux_4x1_beh             Mux_1          (D_in1, GND, Zero_b, PWR, E_b, select);
 mux_4x1_beh             Mux_0          (D_in0, Start, GND, PWR, E, select);
 D_flip_flop_AR_b        M1             (G1, G1_b, D_in1, clock, reset_b);
 D_flip_flop_AR_b        M0             (G0, G0_b, D_in0, clock, reset_b);
 decoder_2x4_df          M2             (Decoder_out, G1, G0, GND);

endmodule
```

```
module Datapath_STR (count, E, Zero, data, Load_regs, Shift_left, Incr_R2, clock);
 parameter                         R1_size = 8, R2_size = 4;
 output [R2_size -1: 0]  count;
 output                            E, Zero;
 input   [R1_size -1: 0]           data;
 input                             Load_regs, Shift_left, Incr_R2, clock;
 wire    [R1_size -1: 0]           R1;
 wire                              Zero;
 supply0                           Gnd;
 supply1                           Pwr;
 assign Zero = (R1 == 0);          // implicit combinational logic
 Shift_Reg         M1              (R1, data, Gnd, Shift_left, Load_regs, clock, Pwr);
 Counter           M2              (count, Load_regs, Incr_R2, clock, Pwr);
 D_flip_flop_AR    M3              (E, w1, clock, Pwr);
 and                               (w1, R1[R1_size - 1], Shift_left);
endmodule

module Shift_Reg (R1, data, SI_0, Shift_left, Load_regs, clock, reset_b);
 parameter R1_size = 8;
 output     [R1_size -1: 0]     R1;
 input      [R1_size -1: 0]     data;
 input                          SI_0, Shift_left, Load_regs;
 input                          clock, reset_b;
 reg        [R1_size -1: 0]     R1;
 always @ (posedge clock, negedge reset_b)
  if (reset_b == 0) R1 <= 0;
  else begin
   if (Load_regs) R1 <= data; else
    if (Shift_left) R1 <= {R1[R1_size -2: 0], SI_0}; end
endmodule
module Counter (R2, Load_regs, Incr_R2, clock, reset_b);
 parameter                     R2_size = 4;
 output    [R2_size -1: 0]     R2;
 input                         Load_regs, Incr_R2;
 input                         clock, reset_b;
 reg       [R2_size -1: 0]     R2;
 always @ (posedge clock, negedge reset_b)
  if (reset_b == 0) R2 <= 0;
  else if (Load_regs) R2 <= {R2_size {1'b1}};         // Fill with 1
   else if (Incr_R2 == 1) R2 <= R2 + 1;
endmodule
module D_flip_flop_AR (Q, D, CLK, RST_b);
 output        Q;
 input         D, CLK, RST_b;
```

```
 reg            Q;
 always @ (posedge CLK, negedge RST_b)
  if (RST_b == 0) Q <= 1'b0;
  else Q <= D;
endmodule
module D_flip_flop_AR_b (Q, Q_b, D, CLK, RST_b);
 output         Q, Q_b;
 input          D, CLK, RST_b;
 reg            Q;
 assign         Q_b = ~Q;
 always @ (posedge CLK, negedge RST_b)
  if (RST_b == 0) Q <= 1'b0;
  else Q <= D;
endmodule
// Behavioral description of four-to-one line multiplexer
// Verilog 2005 port syntax
module mux_4x1_beh
(output reg        m_out,
 input             in_0, in_1, in_2, in_3,
 input [1: 0]      select
);
 always @ (in_0, in_1, in_2, in_3, select)     // Verilog 2005 syntax
  case (select)
   2'b00:      m_out = in_0;
   2'b01:      m_out = in_1;
   2'b10:      m_out = in_2;
   2'b11:      m_out = in_3;
  endcase
 endmodule

// Dataflow description of two-to-four-line decoder
// See Fig. 4.19. Note: The figure uses symbol E, but the
// Verilog model uses enable to indicate functionality clearly.
module decoder_2x4_df (D, A, B, enable);
 output        [0: 3]   D;
 input                  A, B;
 input                  enable;

 assign        D[0] = !(!A && !B && !enable),
               D[1] = !(!A && B && !enable),
               D[2] = !(A && !B && !enable),
               D[3] = !(A && B && !enable);
 endmodule

module t_Count_Ones;
```

```
parameter R1_size = 8, R2_size = 4;
wire          [R2_size -1: 0]    R2;
wire          [R2_size -1: 0]    count;
wire                             Ready;
reg           [R1_size -1: 0]    data;
reg                              Start, clock, reset_b;
wire          [1: 0]             state;              // Use only for debug
assign state = {M0.M0.G1, M0.M0.G0};
Count_Ones_STR_STR M0 (count, Ready, data, Start, clock, reset_b);
initial #650 $finish;
initial begin clock = 0; #5 forever #5 clock = ~clock; end
initial fork
  #1 reset_b = 1;
  #3 reset_b = 0;
  #4 reset_b = 1;
  #27 reset_b = 0;
  #29 reset_b = 1;
  #355 reset_b = 0;
  #365 reset_b = 1;
  #4 data = 8'Hff;
  #145 data = 8'haa;
  # 25 Start = 1;
  # 35 Start = 0;
  #55 Start = 1;
  #65 Start = 0;
  #395 Start = 1;
  #405 Start = 0;
 join
endmodule
```

Testing the Ones Counter

The test bench in HDL Example 8.8 was used to produce the simulation results in Fig. 8.24. Annotations have been added for clarification. In Fig. 8.24(a), *reset_b* is toggled low at $t = 3$ to drive the controller into *S_idle*, but with *Start* not yet having an assigned value. (The default is x.) Consequently, the controller enters an unknown state (the shaded waveform) at the next clock, and its outputs are unknown.[4] When *reset_b* is asserted (low) again at $t = 27$, the state enters *S_idle*. Then, with $Start = 1$ at the first clock after *reset_b* is de-asserted, (1) the controller enters *S_1*, (2) *Load_regs* causes *R1* to be set to the value of *data*, namely, 8′Hff, and (3) *R2* is filled with 1's. At the next clock, *R2*

[4] Remember, this simulation is in Verilog's four-valued logic system. In actual hardware, the values will be 0 or 1. Without a known applied value for the inputs, the next state and outputs will be undetermined, even after the reset signal has been applied.

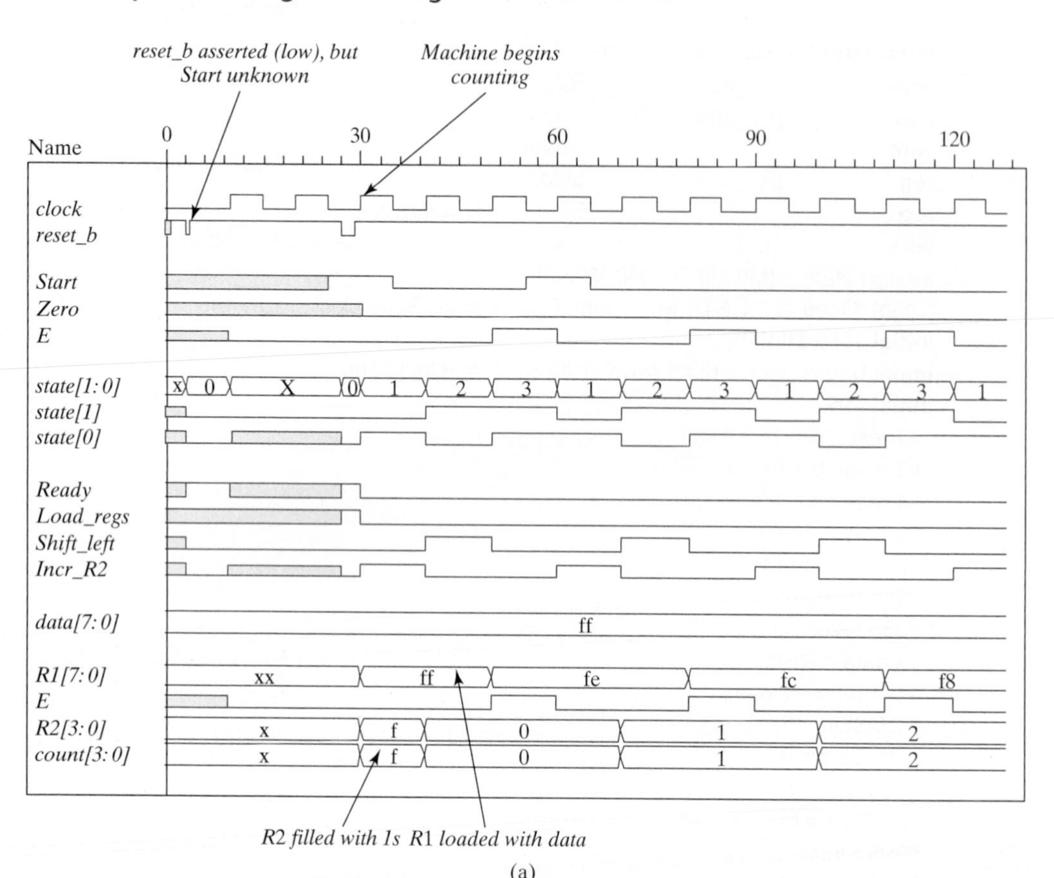

(a)

FIGURE 8.24
Simulation waveforms for count-of-ones circuit

starts counting from 0. *Shift_left* is asserted while the controller is in state *S_2*, and *incr_R2* is asserted while the controller is in state *S_1*. Notice that *R2* is incremented in the next cycle after *incr_R2* is asserted. No output is asserted in state *S_3*. The counting sequence continues in Fig. 8.24(b) until *Zero* is asserted, with *E* holding the last 1 of the data word. The next clock produces *count* = 8, and *state* returns to *S_idle*. (Additional testing is addressed in the problems at the end of the chapter.)

8.11 RACE-FREE DESIGN (SOFTWARE RACE CONDITIONS)

Once a circuit has been synthesized, either manually or with tools, it is necessary to verify that the simulation results produced by the HDL behavioral model match those of the netlist of the gates (standard cells) of the physical circuit. It is important to resolve any

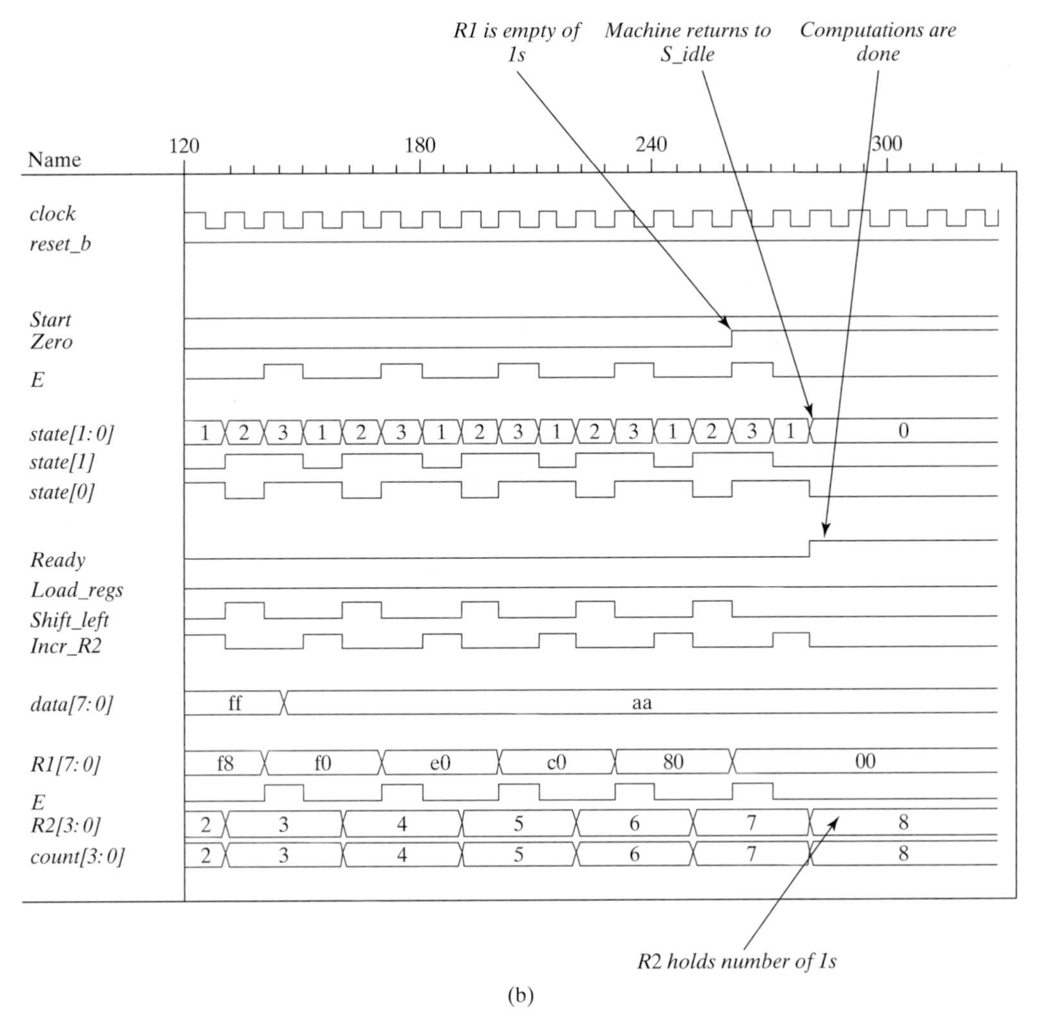

(b)

Figure 8.24 (Continued)

mismatch, because the behavioral model was presumed to be correct. There are various potential sources of mismatch between the results of a simulation, but we will consider one that typically happens in HDL-based design methodology. Three realities contribute to the potential problem: (1) a physical feedback path exists between a datapath unit and a control unit whose inputs include status signals fed back from the datapath unit; (2) blocked procedural assignments execute immediately, and behavioral models simulate with 0 propagation delays, effectively creating immediate changes in the outputs of combinational logic when its inputs change (i.e., changes in the inputs and the outputs are scheduled in the same time step of the simulation); and (3) the order in which a simulator executes multiple blocked assignments to the same variable at a given time step of the simulation is indeterminate (i.e., unpredictable).

Now consider a sequential machine with an HDL model in which all assignments are made with the blocked assignment operator. At a clock pulse, the register operations in the datapath, the state transitions in the controller, the updates of the next state and output logic of the controller, and the updates to the status signals in the datapath are all scheduled to occur at the same time step of the simulation. Which executes first? Suppose that when a clock pulse occurs, the state of the controller changes before the register operations execute. The change in the state could change the outputs of the control unit. The new values of the outputs would be used by the datapath when it finally executes its assignments at that same clock pulse. The result might not be the same as it would have been if the datapath had executed its assignments before the control unit updated its state and outputs. Conversely, suppose that when the clock pulse occurs, the datapath unit executes its operations and updates its status signals first. The updated status signals could cause a change in the value of the next state of the controller, which would be used to update the state. The result could differ from that which would result if the state had been updated before the edge-sensitive operations in the datapath executed. In either case, the timing of register transfer operations and state transitions in the different representations of the system might not match. *Failing to detect a mismatch can have disastrous consequences for the user of the design. Finding the source of the mismatch can be very time-consuming and costly. It is better to* avoid the mismatch *by following a strict discipline in your design*. Fortunately, there is a solution to this dilemma.

A designer can eliminate the *software race conditions* just described by observing the rule of modeling combinational logic with blocked assignments and modeling state transitions and edge-sensitive register operations with nonblocking assignments. A software race cannot happen if nonblocking operators are used as shown in all of the examples in this text, because the sampling mechanism of the nonblocking operator breaks the feedback path between a state transition or edge-sensitive datapath operation and the combinational logic that forms the next state or inputs to the registers in the datapath unit. The mechanism does this because simulators evaluate the expressions on the right-hand side of their nonblocking assignment statements before any blocked assignments are made. Thus, the nonblocking assignments cannot be affected by the results of the blocked assignments. This matches the hardware reality. Always use the blocking operator to model combinational logic, and use the nonblocking operator to model edge-sensitive register operations and state transitions.

It also might appear that the physical structure of a datapath and the controller together create a physical (i.e., hardware), race condition, because the status signals are fed back to the controller and the outputs of the controller are fed forward to the datapath. However, timing analysis can verify that a change in the output of the controller will not propagate through the datapath logic and then through the input logic of the controller in time to have an effect on the output of the controller until the next clock pulse. The state cannot update until the next edge of the clock, even though the status signals update the value of the next state. The flip-flop cuts the feedback path between clock cycles. In practice, timing analysis verifies that the circuit will operate at the specified clock frequency, or it identifies signal paths whose propagation delays are problematic. Remember, the design must implement the correct logic and operate at the speed prescribed by the clock.

8.12 LATCH-FREE DESIGN (WHY WASTE SILICON?)

Continuous assignments model combinational logic implicitly. A feedback-free continuous assignment will synthesize to combinational logic, and the input–output relationship of the logic is automatically sensitive to all of the inputs of the circuit. In simulation, the simulator monitors the right-hand sides of all continuous assignments, detects a change in any of the referenced variables, and updates the left-hand side of an affected assignment statement. Unlike a continuous assignment, a cyclic behavior is not necessarily completely sensitive to all of the variables that are referenced by its assignment statements. If a level-sensitive cyclic behavior is used to describe combinational logic, it is essential that the sensitivity list include every variable that is referenced on the right-hand side of an assignment statement in the behavior. If the list is incomplete, the logic described by the behavior will be synthesized with latches at the outputs of the logic. This implementation wastes silicon area and may have a mismatch between the simulation of the behavioral model and the synthesized circuit. These difficulties can be avoided by ensuring that the sensitivity list is complete, but, in large circuits, it is easy to fail to include every referenced variable in the sensitivity list of a level-sensitive cyclic behavior. Consequently, Verilog 2001 included a new operator to reduce the risk of accidentally synthesizing latches.

In Verilog 2001, the tokens @ and * can be combined as @* or @(*) and are used without a sensitivity list to indicate that execution of the associated statement is sensitive to every variable that is referenced on the right-hand side of an assignment statement in the logic. In effect, the operator @* indicates that the logic is to be interpreted

---

HDL Example 8.9

The following level-sensitive cyclic behavior will synthesize a two-channel multiplexer:

```
module mux_2_V2001 (output reg [31: 0] y, input [31: 0] a, b, input sel);
 always @*
 y = sel ? a: b;
endmodule
```

The cyclic behavior has an implicit sensitivity list consisting of *a*, *b*, and *sel*.

---

and synthesized as level-sensitive combinational logic; the logic has an implicit sensitivity list composed of all of the variables that are referenced by the procedural assignments. Using the @* operator will prevent accidental synthesis of latches.

8.13 OTHER LANGUAGE FEATURES

The examples in this text have used only those features of the Verilog HDL that are appropriate for an introductory course in logic design. Verilog 2001 contains features that are very useful to designers, but which are not considered here. Among them are multidimensional arrays, variable part selects, array bit and part selects, signed reg, net, and port declarations, and local parameters. These enhancements are treated in more advanced texts using Verilog 2001 and Verilog 2005.

PROBLEMS

Answers to problems marked with * appear at the end of the book.

8.1* Explain in words and write HDL statements for the operations specified by the following register transfer notation:
(a) $R2 \leftarrow R2 + 1, R1 \leftarrow R$
(b) $R3 \leftarrow R3 - 1$
(c) If ($S_1 = 1$) then ($R0 \leftarrow R1$) else if ($S_2 = 1$) then ($R0 \leftarrow R2$)

8.2 A logic circuit with active-low synchronous reset has two control inputs x and y. If x is 1 and y is 0, register R is incremented by 1 and control goes to a second state. If x is 0 and y is 1, register R is cleared to zero and control goes from the initial state to a third state. Otherwise, control stays in the initial state. Draw (1) a block diagram showing the controller, datapath unit (with internal registers), and signals, and (2) the portion of an ASMD chart starting from an initial state.

8.3 Draw the ASMD charts for the following state transitions:
(a) If $x = 1$, control goes from state S_1 to state S_2; if $x = 0$, generate a conditional operation $R <= R + 2$ and go from S_1 to S_2.
(b) If $x = 1$, control goes from S_1 to S_2 and then to S_3; if $x = 0$, control goes from S_1 to S_3.
(c) Start from state S_1; then if $xy = 11$, go to S_2; if $xy = 01$ go to S_3; and if $xy = 10$, go to S_1; otherwise, go to S_3.

8.4 Show the eight exit paths in an ASM block emanating from the decision boxes that check the eight possible binary values of three control variables x, y, and z.

8.5 Explain how the ASM and ASMD charts differ from a conventional flowchart. Using Fig. 8.5 as an illustration, show the difference in interpretation. Explain the difference between and ASM chart and an ASMD chart. In your own words, discuss the use and merit of using an ASMD chart.

8.6 Construct a block diagram and an ASMD chart for a digital system that counts the number of people in a room. The one door through which people enter the room has a photocell that changes a signal x from 1 to 0 while the light is interrupted. They leave the room from a second door with a similar photocell that changes a signal y from 1 to 0 while the light is interrupted. The datapath circuit consists of an up–down counter with a display that shows how many people are in the room.

8.7* Draw a block diagram and an ASMD chart for a circuit with two eight-bit registers RA and RB that receive two unsigned binary numbers. The circuit performs the subtraction operation

$$RA \leftarrow RA - RB$$

Use the method for subtraction described in Section 1.5, and set a borrow flip-flop to 1 if the answer is negative. Write and verify an HDL model of the circuit.

8.8* Design a digital circuit with three 16-bit registers AR, BR, and CR that perform the following operations:

(a) Transfer two 16-bit signed numbers (in 2's-complement representation) to AR and BR.
(b) If the number in AR is negative, divide the number in AR by 2 and transfer the result to register CR.
(c) If the number in AR is positive but nonzero, multiply the number in BR by 2 and transfer the result to register CR.
(d) If the number in AR is zero, clear register CR to 0.
(e) Write and verify a behavioral model of the circuit.

8.9* Design the controller whose state diagram is given by Fig. 8.11(a). Use one flip-flop per state (a one-hot assignment). Write, simulate, verify, and compare RTL and structural models of the controller.

8.10 The state diagram of a control unit is shown in Fig. P8.10. It has four states and two inputs x and y. Draw the equivalent ASM chart. Write and verify a Verilog model of the controller.

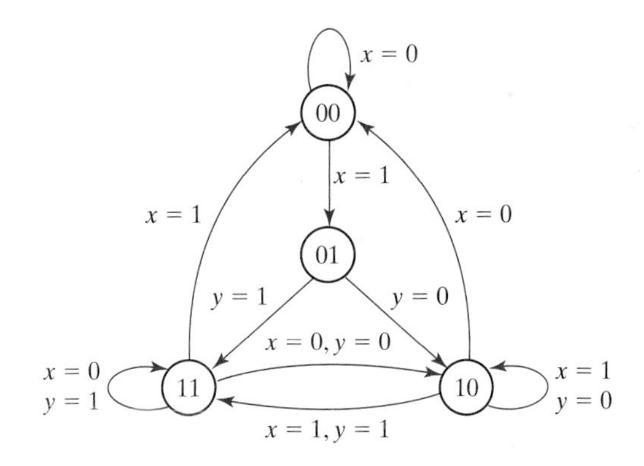

FIGURE P8.10
Control state diagram for Problems 8.10 and 8.11

8.11* Design the controller whose state diagram is shown in Fig. P8.10. Use D flip-flops.

8.12 Design the four-bit counter with synchronous clear specified in Fig. 8.10. Repeat for asynchronous clear.

8.13 Simulate *Design_Example_STR* (see HDL Example 8.4), and verify that its behavior matches that of the RTL description. Obtain state information by displaying *G0* and *G1* as a concatenated vector for the state.

8.14 What, if any, are the consequences of the machine in *Design_Example_RTL* (see HDL Example 8.2) entering an unused state?

8.15 *Simulate Design_Example_RTL* in HDL Example 8.2, and verify that it recovers from an unexpected reset condition during its operation, i.e., a "running reset" or a "reset on-the-fly."

8.16* Develop a block diagram and an ASMD chart for a digital circuit that multiplies two binary numbers by the repeated-addition method. For example, to multiply 5×4, the digital system evaluates the product by adding the multiplicand four times: $5 + 5 + 5 + 5 = 20$. Design the circuit. Let the multiplicand be in register *BR*, the multiplier in register *AR*, and the product in register *PR*. An adder circuit adds the contents of *BR* to *PR*. A zero-detection signal indicates whether *AR* is 0. Write and verify a Verilog behavioral model of the circuit.

8.17* Prove that the multiplication of two n-bit numbers gives a product of length less than or equal to $2n$ bits.

8.18* In Fig. 8.14, the *Q* register holds the multiplier and the *B* register holds the multiplicand. Assume that each number consists of 16 bits.

(a) How many bits can be expected in the product, and where is it available?

(b) How many bits are in the *P* counter, and what is the binary number loaded into it initially?

(c) Design the circuit that checks for zero in the *P* counter.

8.19 List the contents of registers *C, A, Q,* and *P* in a manner similar to Table 8.5 during the process of multiplying the two numbers 11011 (multiplicand) and 10111 (multiplier).

8.20* Determine the time it takes to process the multiplication operation in the binary multiplier described in Section 8.8. Assume that the *Q* register has n bits and the clock cycle is t ns.

8.21 Design the control circuit of the binary multiplier specified by the state diagram of Fig. 8.16, using multiplexers, a decoder, and a register.

8.22 Figure P8.22 shows an alternative ASMD chart for a sequential binary multiplier. Write and verify an RTL model of the system. Compare this design with that described by the ASMD chart in Fig. 8.15(b).

8.23 Figure P8.23 shows an alternative ASMD chart for a sequential binary multiplier. Write and verify an RTL model of the system. Compare this design with that described by the ASMD chart in Fig. 8.15(b).

8.24 The HDL description of a sequential binary multiplier given in HDL Example 8.5 encapsulates the descriptions of the controller and the datapath in a single Verilog module. Write and verify a model that encapsulates the controller and datapath in separate modules.

8.25 The sequential binary multiplier described by the ASMD chart in Fig. 8.15 does not consider whether the multiplicand or the shifted multiplier is 0. Therefore, it executes for a fixed number of clock cycles, independently of the data.

(a) Develop an ASMD chart for a more efficient multiplier that will terminate execution as soon as either word is found to be zero.

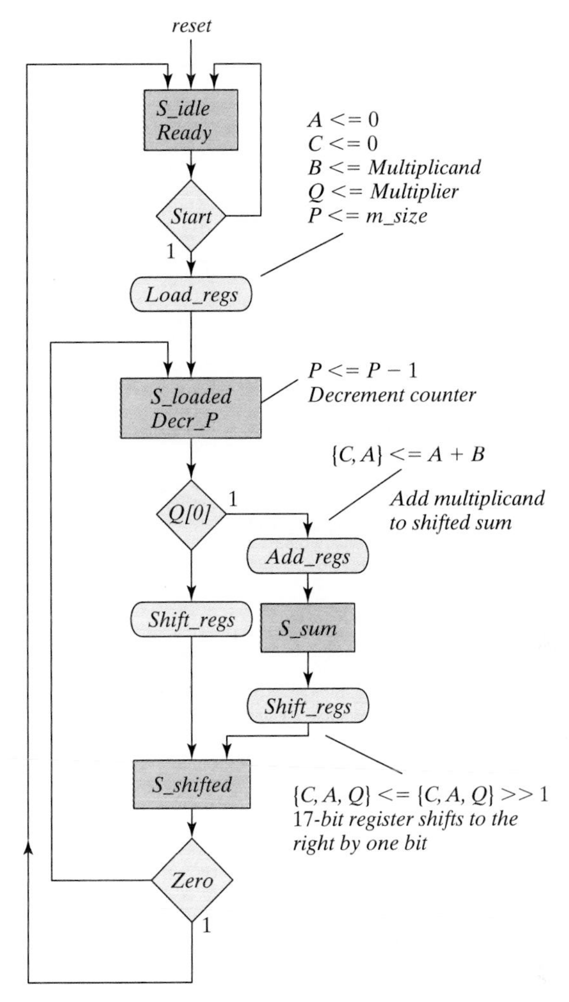

FIGURE P8.22
ASMD chart for Problem 8.22

(b) Write an HDL description of the circuit. The controller and datapath are to be encapsulated in separate Verilog modules.

(c) Write a test plan and a test bench, and verify the circuit.

8.26 Modify the ASMD chart of the sequential binary multiplier shown in Fig. 8.15 to add and shift in the same clock cycle. Write and verify an RTL description of the system.

8.27 The second test bench given in HDL Example 8.6 generates a product for all possible values of the multiplicand and multiplier. Verifying that each result is correct would not be practical, so modify the test bench to include a statement that forms the expected

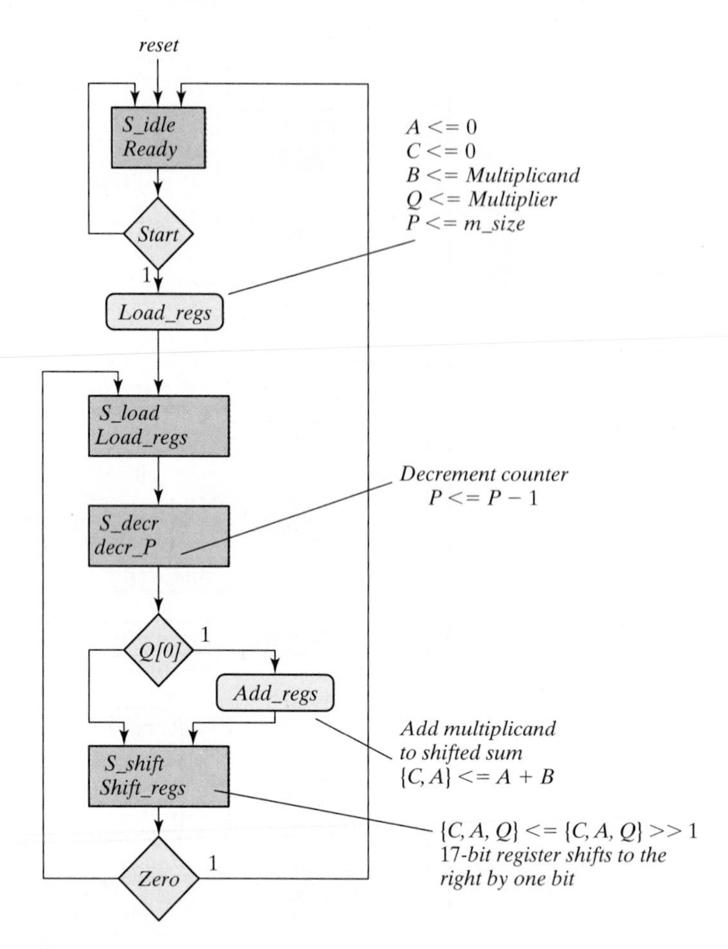

FIGURE P8.23
ASMD chart for Problem 8.23

product. Write additional statements to compare the result produced by the RTL description with the expected result. Your simulation is to produce an error signal indicating the result of the comparison. Repeat for the structural model of the multiplier.

8.28 Write the HDL structural description of the multiplier designed in Section 8.8. Use the block diagram of Fig. 8.14(a) and the control circuit of Fig. 8.18. Simulate the design and verify its functionality by using the test bench of HDL Example 8.6.

8.29 An incomplete ASMD chart for a finite state machine is shown in Fig. P8.29. The register operations are not specified, because we are interested only in designing the control logic.
(a) Draw the equivalent state diagram.
(b) Design the control unit with one flip-flop per state.
(c) List the state table for the control unit.

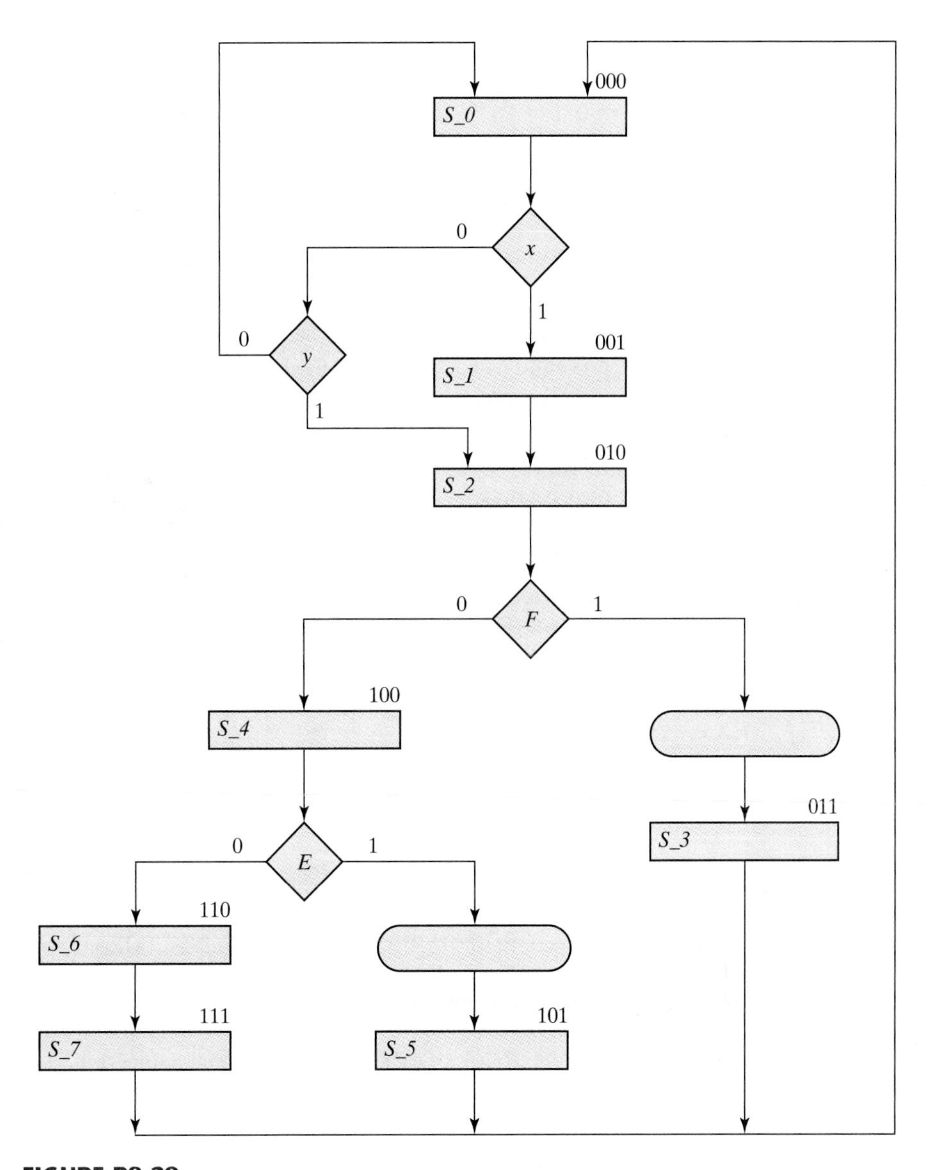

FIGURE P8.29
ASMD chart for Problem 8.29

(d) Design the control unit with three *D* flip-flops, a decoder, and gates.
(e) Derive a table showing the multiplexer input conditions for the control unit.
(f) Design the control unit with three multiplexers, a register with three flip-flops, and a 3×8 decoder.
(g) Using the results of (f), write and verify a structural model of the controller.
(h) Write and verify an RTL description of the controller.

8.30* What is the value of E in each HDL block, assuming that $RA = 1$?

(a)
```
RA = RA - 1;
if (RA == 0) E = 1;
else E = 0;
```

(b)
```
RA <= RA - 1;
if (RA == 0) E <= 1;
else E <= 0;
```

8.31* Using the Verilog HDL operators listed in Table 8.2, assume that $A = 4'b0110$, $B = 4'b0010$, and $C = 4'b0000$ and evaluate the result of the following operations:

```
A * B; A + B; A - B; ~C; A & B; A | B; A ^ B; & A; ~|C; A || B; A && C; |A; A < B; A > B;
A != B;
```

8.32 Consider the following always block:

```
always @ (posedge CLK)
  if (S1) R1 <= R1 + R2;
  else if (S2) R1 <= R1 + 1;
  else R1 <= R1;
```

Using a four-bit counter with parallel load for *R1* (as in Fig. 6.15) and a four-bit adder, draw a block diagram showing the connections of components and control signals for a possible synthesis of the block.

8.33 The multilevel case statement is often translated by a logic synthesizer into hardware multiplexers. How would you translate the following **case** block into hardware (assume registers of eight bits each)?

```
case (state)
  S0: R4 = R0;
  S1: R4 = R1;
  S2: R4 = R2;
  S3: R4 = R3;
endcase
```

8.34 The design of a circuit that counts the number of ones in a register is carried out in Section 8.10. The block diagram for the circuit is shown in Fig. 8.22(a), a complete ASMD chart for the circuit appears in Fig. 8.22(c), and structural HDL models of the datapath and controller are given in HDL Example 8.8. Using the operations and signal names indicated on the ASMD chart,

(a) Write *Datapath_BEH*, an RTL description of the datapath unit of the ones counter. Write a test plan specifying the functionality that will be tested, and write a test bench to implement the plan. Execute the test plan to verify the functionality of the datapath unit, and produce annotated simulation results relating the test plan to the waveforms produced in a simulation.

(b) Write *Controller_BEH*, an RTL description of the control unit of the ones counter. Write a test plan specifying the functionality that will be tested, and write a test bench to implement the plan. Execute the test plan to verify the functionality of the control unit, and produce annotated simulation results relating the test plan to the waveforms produced in a simulation.

(c) *Write Count_Ones_BEH_BEH*, a top-level module encapsulating and integrating *Controller_BEH* and *Datapath_BEH*. Write a test plan and a test bench, and verify the description. Produce annotated simulation results relating the test plan to the waveforms produced in a simulation.

(d) Write *Controller_BEH_1Hot*, an RTL description of a one-hot controller implementing the ASMD chart of Fig. 8.22(c). Write a test plan specifying the functionality that will be tested, and write a test bench to implement the plan. Execute the test plan and produce annotated simulation results relating the test plan to the waveforms produced in a simulation.

(e) Write *Count_Ones_BEH_1_Hot*, a top-level module encapsulating the module *Controller_BEH_1_Hot* and *Datapath_BEH*. Write a test plan and a test bench, and verify the description. Produce annotated simulation results relating the test plan to the waveforms produced in a simulation.

8.35 The HDL description and test bench for a circuit that counts the number of ones in a register are given in HDL Example 8.8. Modify the test bench and simulate the circuit to verify that the system operates correctly for the following patterns of data: 8′hff, 8′h0f, 8′hf0, 8′h00, 8′haa, 8′h0a, 8′ha0, 8′h55, 8′h05, 8′h50, 8′ha5, and 8′h5a.

8.36 The design of a circuit that counts the number of ones in a register is carried out in Section 8.10. The block diagram for the circuit is shown in Fig. 8.22(a), a complete ASMD chart for this circuit appears in Fig. 8.22(c), and structural HDL models of the datapath and controller are given in HDL Example 8.8. Using the operations and signal names indicated on the ASMD chart,

(a) Design the control logic, employing one flip-flop per state (a one-hot assignment). List the input equations for the four flip-flops.

(b) Write *Controller_Gates_1_Hot*, a gate-level HDL structural description of the circuit, using the control designed in part (a) and the signals shown in the block diagram of Fig. 8.22(a).

(c) Write a test plan and a test bench, and then verify the controller.

(d) Write *Count_Ones_Gates_1_Hot_STR*, a top-level module encapsulating and integrating instantiations of *Controller_Gates_1_Hot* and *Datapath_STR*. Write a test plan and a test bench to verify the description. Produce annotated simulation results relating the test plan to the waveforms produced in a simulation.

8.37 Compared with the circuit presented in HDL Example 8.8, a more efficient circuit that counts the number of ones in a data word is described by the block diagram and the partially completed ASMD chart in Fig. P8.37. This circuit accomplishes addition and shifting in the same clock cycle and adds the LSB of the data register to the counter register at every clock cycle.

(a) Complete the ASMD chart.

(b) Using the ASMD chart, write an RTL description of the circuit. A top-level Verilog module, *Count_of_ones_2_Beh* is to instantiate separate modules for the datapath and control units.

(c) Design the control logic, using one flip-flop per state (a one-hot assignment). List the input equations for the flip-flops.

(d) Write the HDL structural description of the circuit, using the controller designed in part (c) and the block diagram of Fig. P8.37(a).

(e) Write a test bench to test the circuit. Simulate the circuit to verify the operation described in both the RTL and the structural programs.

8.38 The addition of two signed binary numbers in the signed-magnitude representation follows the rules of ordinary arithmetic: If the two numbers have the same sign (both positive or both negative), the two magnitudes are added and the sum has the common sign; if the two numbers have opposite signs, the smaller magnitude is subtracted from the larger and

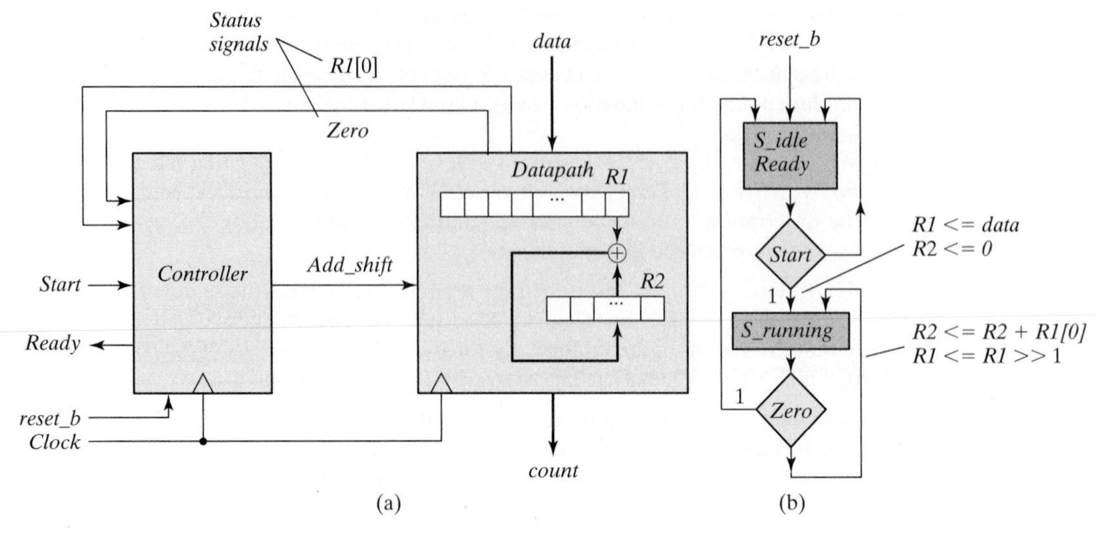

FIGURE P8.37
(a) Alternative circuit for a ones counter
(b) ASMD Chart for Problem 8.37

the result has the sign of the larger magnitude. Write an HDL behavioral description for adding two 8-bit signed numbers in signed-magnitude representation and verify. The leftmost bit of the number holds the sign and the other seven bits hold the magnitude.

8.39* For the circuit designed in Problem 8.16,

(a) Write and verify a structural HDL description of the circuit. The datapath and controller are to be described in separate units.

(b) Write and verify an RTL description of the circuit. The datapath and controller are to be described in separate units.

8.40 Modify the block diagram of the sequential multiplier given in Fig. 8.14(a) and the ASMD chart in Fig. 8.15(b) to describe a system that multiplies 32-bit words, but with 8-bit (byte-wide) external datapaths. The machine is to assert *Ready* in the (initial) reset state. When *Start* is asserted, the machine is to fetch the data bytes from a single 8-bit data bus in consecutive clock cycles (multiplicand bytes first, followed by multiplier bytes, least significant byte first) and store the data in datapath registers. *Got_Data* is to be asserted for one cycle of the clock when the transfer is complete. When *Run* is asserted, the product is to be formed sequentially. *Done_Product* is to be asserted for one clock cycle when the multiplication is complete. When a signal *Send_Data* is asserted, each byte of the product is to be placed on an 8-bit output bus for one clock cycle, in sequence, beginning with the least significant byte. The machine is to return to the initial state after the product has been transmitted. Consider safeguards, such as not attempting to send or receive data while the product is being formed. Consider also other features that might eliminate needless multiplication by 0. For example, do not continue to multiply if the shifted multiplier is empty of 1's.

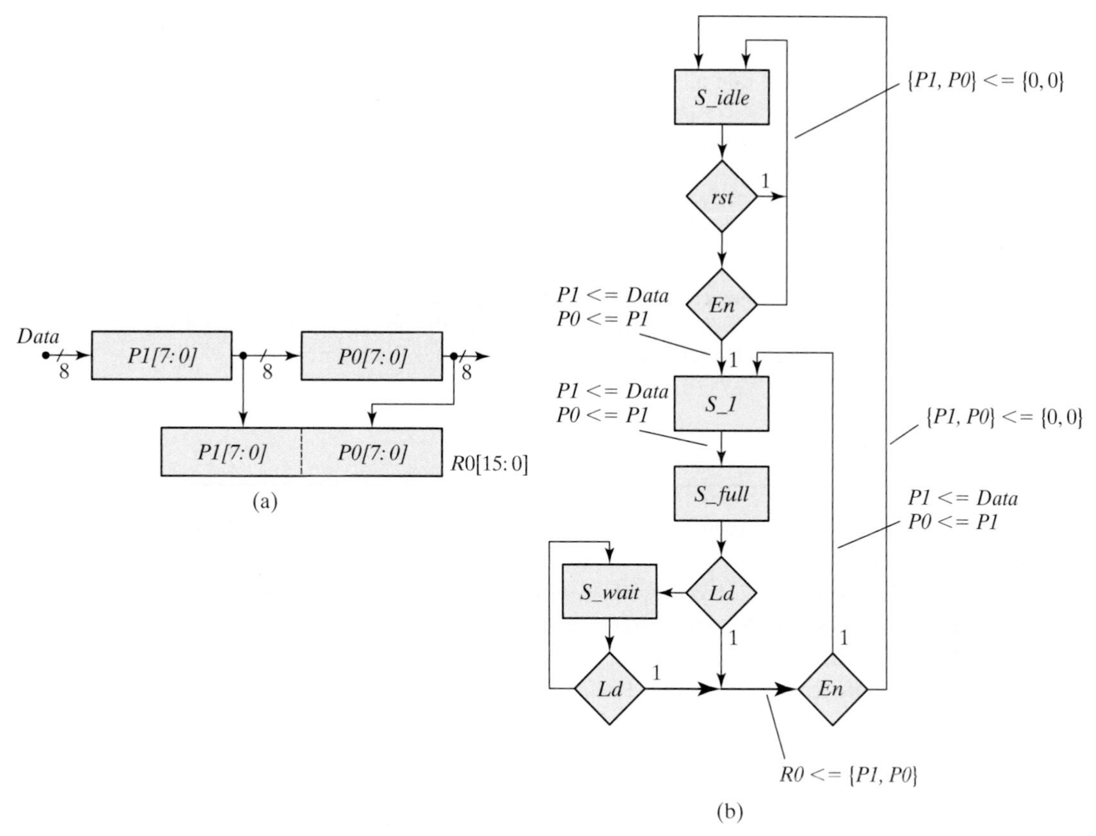

FIGURE P8.41
Two-stage pipeline register: Datapath unit and ASMD chart

8.41 The block diagram and partially completed ASMD chart in Fig. P8.41 describe the behavior of a two-stage pipeline that acts as a 2:1 decimator with a parallel input and output. Decimators are used in digital signal processors to move data from a datapath with a high clock rate to a datapath with a lower clock rate, converting data from a parallel format to a serial format in the process. In the datapath shown, entire words of data can be transferred into the pipeline at twice the rate at which the contents of the pipeline must be dumped into a holding register or consumed by some processor. The contents of the holding register *R0* can be shifted out serially, to accomplish an overall parallel-to-serial conversion of the data stream. The ASMD chart indicates that the machine has synchronous reset to *S_idle*, where it waits until *rst* is de-asserted and En is asserted. Note that synchronous transitions which would occur from the other states to *S_idle* under the action of *rst* are not shown. With *En* asserted, the machine transitions from *S_idle* to *S_1*, accompanied by concurrent register operations that load the MSByte of the pipe with *Data* and move the content of *P1* to the LSByte (*P0*). At the next clock, the state goes to *S_full*, and now the pipe is full. If *Ld* is asserted at the next clock, the machine moves to *S_1* while dumping the pipe into a holding register *R0*. If *Ld* is not asserted, the machine enters *S_wait* and remains there until *Ld* is asserted, at which time it dumps the pipe and returns to *S_1* or to *S_idle*, depending on whether *En* is asserted, too. The data rate at R_0 is one-half the rate at which data are supplied to the unit from an external datapath.

(a) Develop the complete ASMD chart.
(b) Using the ASMD chart developed in (a), write and verify an HDL model of the datapath.
(c) Write and verify a Verilog behavioral model of the control unit.
(d) Encapsulate the datapath and controller in a top-level module, and verify the integrated system.

8.42 The count-of-ones circuit described in Fig. 8.22 has a latency that is to be eliminated. It arises because the status signal *E* is formed as the output of a flip-flop into which the MSB of *R*1 is shifted. Develop a design that eliminates the latency.

REFERENCES

1. ARNOLD, M. G. 1999. *Verilog Digital Computer Design*. Upper Saddle River, NJ: Prentice Hall.
2. BHASKER, J. 1997. *A Verilog HDL Primer*. Allentown, PA: Star Galaxy Press.
3. BHASKER, J. 1998. *Verilog HDL Synthesis*. Allentown, PA: Star Galaxy Press.
4. CILETTI, M. D. 2003. *Modeling, Synthesis, and Rapid Prototyping with Verilog HDL*. Upper Saddle River, NJ: Prentice Hall.
5. CILETTI, M. D. 2010. *Advanced Digital Design with the Verilog HDL*. Upper Saddle River, NJ: Prentice Hall.
6. CLARE, C. R. 1971. *Designing Logic Systems Using State Machines*. New York: McGraw-Hill.
7. HAYES, J. P. 1993. *Introduction to Digital Logic Design*. Reading, MA: Addison-Wesley.
8. *IEEE Standard Hardware Description Language Based on the Verilog Hardware Description Language* (IEEE Std 1364-2005). 2005. New York: Institute of Electrical and Electronics Engineers.
9. MANO, M. M. 1993. *Computer System Architecture,* 3rd ed. Upper Saddle River, NJ: Prentice Hall.
10. MANO, M. M., and C. R. KIME. 2005. *Logic and Computer Design Fundamentals,* 3rd ed. Upper Saddle River, NJ: Prentice Hall.
11. PALNITKAR, S. 2003. *Verilog HDL: A Guide to Digital Design and Synthesis*. Mountain View, CA: SunSoft Press (a Prentice Hall Title).
12. SMITH, D. J. 1996. *HDL Chip Design*. Madison, AL: Doone Publications.
13. THOMAS, D. E., and P. R. MOORBY. 2002. *The Verilog Hardware Description Language,* 5th ed. Boston: Kluwer Academic Publishers.
14. WINKLER, D., and F. PROSSER. 1987. *The Art of Digital Design,* 2nd ed. Englewood Cliffs, NJ: Prentice-Hall.

WEB SEARCH TOPICS

Algorithmic state machine
Algorithmic state machine chart
Asynchronous circuit
Decimator
Digital control unit
Digital datapath unit
Mealy machine
Moore machine
Race condition

Chapter 9

Laboratory Experiments with Standard ICs and FPGAs

9.1 INTRODUCTION TO EXPERIMENTS

This chapter presents 17 laboratory experiments in digital circuits and logic design. The experiments give the student using this book hands-on experience. The digital circuits can be constructed by using standard integrated circuits (ICs) mounted on breadboards that are easily assembled in the laboratory. The experiments are ordered according to the material presented in the book. The last section consists of a number of supplements with suggestions for using the Verilog HDL to simulate and verify the functionality of the digital circuits presented in the experiments. If an FPGA prototyping board is available, the experiments can be implemented in an FPGA as an alternative to standard ICs.

A logic breadboard suitable for performing the experiments must have the following equipment:

1. Light-emitting diode (LED) indicator lamps.
2. Toggle switches to provide logic-1 and logic-0 signals.
3. Pulsers with push buttons and debounce circuits to generate single pulses.
4. A clock-pulse generator with at least two frequencies: a low frequency of about 1 pulse per second to observe slow changes in digital signals and a higher frequency for observing waveforms in an oscilloscope.
5. A power supply of 5 V.
6. Socket strips for mounting the ICs.
7. Solid hookup wires and a pair of wire strippers for cutting the wires.

Digital logic trainers that include the required equipment are available from several manufacturers. A digital logic trainer contains LED lamps, toggle switches, pulsers, a variable clock, a power supply, and IC socket strips. Some experiments may require additional switches, lamps, or IC socket strips. Extended breadboards with more solderless sockets and plug-in switches and lamps may be needed.

Additional equipment required is a dual-trace oscilloscope (for Experiments 1, 2, 8, and 15), a logic probe to be used for debugging, and a number of ICs. The ICs required for the experiments are of the TTL or CMOS series 7400.

The integrated circuits to be used in the experiments can be classified as small-scale integration (SSI) or medium-scale integration (MSI) circuits. SSI circuits contain individual gates or flip-flops, and MSI circuits perform specific digital functions. The eight SSI gate ICs needed for the experiments—two-input NAND, NOR, AND, OR, and XOR gates, inverters, and three-input and four-input NAND gates—are shown in Fig. 9.1. The pin assignments for the gates are indicated in the diagram. The pins are numbered from 1 to 14. Pin number 14 is marked V_{CC}, and pin number 7 is marked *GND* (ground). These are the supply terminals, which must be connected to a power supply of 5 V for proper operation of the circuit. Each IC is recognized by its identification number; for example, the two-input NAND gates are found inside the IC whose number is 7400.

Detailed descriptions of the MSI circuits can be found in data books published by the manufacturers. The best way to acquire experience with a commercial MSI circuit is to study its description in a data book that provides complete information on the internal, external, and electrical characteristics of integrated circuits. Various semiconductor companies publish data books for the 7400 series. The MSI circuits that are needed for the experiments are introduced and explained when they are used for the first time. The operation of the circuit is explained by referring to similar circuits in previous chapters. The information given in this chapter about the MSI circuits should be sufficient for performing the experiments adequately. Nevertheless, reference to a data book will always be preferable, as it gives more detailed description of the circuits.

We will now demonstrate the method of presentation of MSI circuits adopted here. To illustrate, we introduce the ripple counter IC, type 7493. This IC is used in Experiment 1 and in subsequent experiments to generate a sequence of binary numbers for verifying the operation of combinational circuits.

The information about the 7493 IC that is found in a data book is shown in Figs. 9.2(a) and (b). Part (a) shows a diagram of the internal logic circuit and its connection to external pins. All inputs and outputs are given symbolic letters and assigned to pin numbers. Part (b) shows the physical layout of the IC, together with its 14-pin assignment to signal names. Some of the pins are not used by the circuit and are marked as *NC* (no connection). The IC is inserted into a socket, and wires are connected to the various pins through the socket terminals. When drawing schematic diagrams in this chapter, we will show the IC in block diagram form, as in Fig. 9.2(c). The IC number (here, 7493) is written inside the block. All input terminals are placed on the left of the block and all output terminals on the right. The letter symbols of the signals, such as *A*, *R1*, and *QA*, are written inside the block, and the corresponding pin numbers, such as 14, 2, and 12, are written along the external lines. V_{CC}, and *GND* are the power terminals connected to pins 5 and 10. The size of the block may vary to accommodate all input

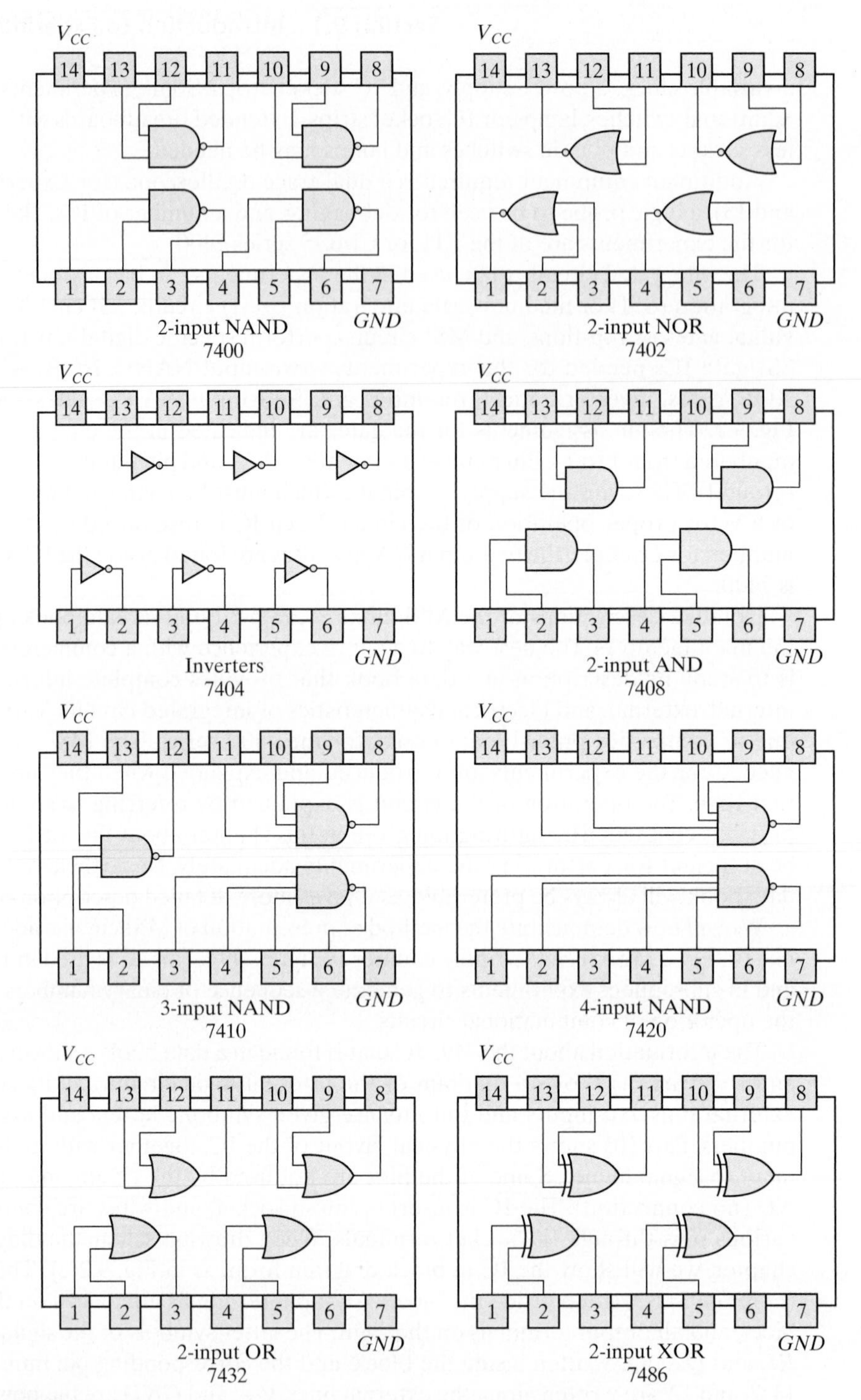

FIGURE 9.1
Digital gates in IC packages with identification numbers and pin assignments

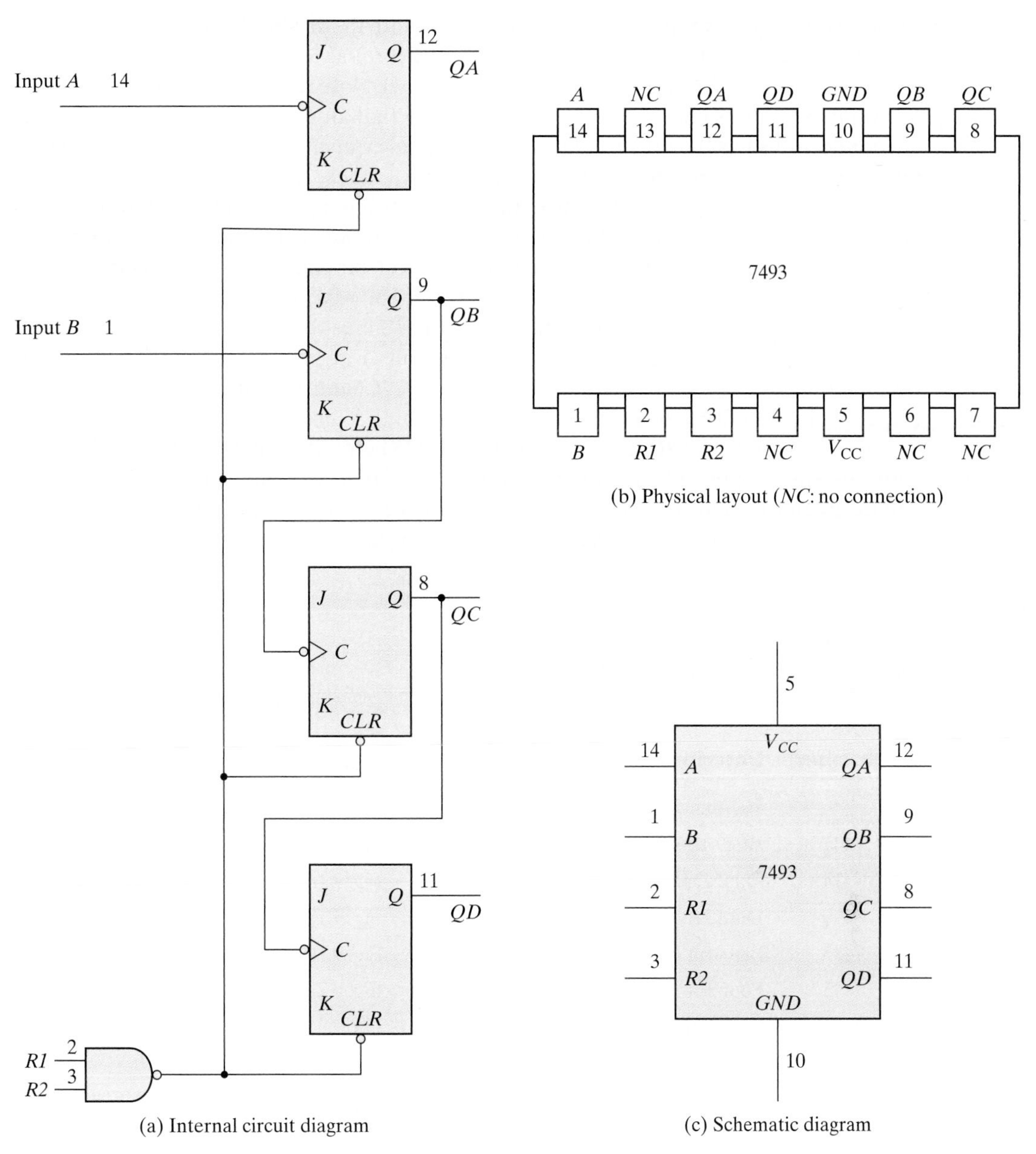

(a) Internal circuit diagram

(b) Physical layout (*NC*: no connection)

(c) Schematic diagram

FIGURE 9.2
IC type 7493 ripple counter

and output terminals. Inputs or outputs may sometimes be placed on the top or the bottom of the block for convenience.

The operation of the circuit is similar to the ripple counter shown in Fig. 6.8(a) with an asynchronous clear to each flip-flop. When input *R1* or *R2* or both are equal to logic 0 (ground), all asynchronous clears are equal to 1 and are disabled. To clear all four flip-flops to 0, the output of the NAND gate must be equal to 0. This is accomplished by having both inputs *R1* and *R2* at logic 1 (about 5 V). Note that the *J* and *K* inputs show no connections. It is characteristic of TTL circuits that an input terminal with no external connections has the effect of producing a signal equivalent to logic 1. Note also that output *QA* is not connected to input *B* internally.

The 7493 IC can operate as a three-bit counter using input *B* and flip-flops *QB, QC*, and *QD*. It can operate as a four-bit counter using input *A* if output *QA* is connected to input *B*. Therefore, to operate the circuit as a four-bit counter, it is necessary to have an external connection between pin 12 and pin 1. The reset inputs, *R1* and *R2*, at pins 2 and 3, respectively, must be grounded. Pins 5 and 10 must be connected to a 5-V power supply. The input pulses must be applied to input *A* at pin 14, and the four flip-flop outputs of the counter are taken from *QA, QB, QC*, and *QD* at pins 12, 9, 8, and 11, respectively, with *QA* being the least significant bit.

Figure 9.2(c) demonstrates the way that all MSI circuits will be symbolized graphically in this chapter. Only a block diagram similar to the one shown in this figure will be given for each IC. The letter symbols for the inputs and outputs in the IC block diagram will be according to the symbols used in the data book. The operation of the

Table 9.1
Integrated Circuits Required for the Experiments

		Graphic Symbol	
IC Number	**Description**	**In Chapter 9**	**In Chapter 10**
	Various gates	Fig. 9.1	Fig. 10.1
7447	BCD-to-seven-segment decoder	Fig. 9.8	—
7474	Dual *D*-type flip-flops	Fig. 9.13	Fig. 10.9(b))
7476	Dual *JK*-type flip-flops	Fig. 9.12	Fig. 10.9(a)
7483	Four-bit binary adder	Fig. 9.10	Fig. 10.2
7493	Four-bit ripple counter	Fig. 9.2	Fig. 10.13
74151	8×1 multiplexer	Fig. 9.9	Fig. 10.7(a)
74155	3×8 decoder	Fig. 9.7	Fig. 10.6
74157	Quadruple 2×1 multiplexers	Fig. 9.17	Fig. 10.7(b)
74161	Four-bit synchronous counter	Fig. 9.15	Fig. 10.14
74189	16×4 random-access memory	Fig. 9.18	Fig. 10.15
74194	Bidirectional shift register	Fig. 9.19	Fig. 10.12
74195	Four-bit shift register	Fig. 9.16	Fig. 10.11
7730	Seven-segment LED display	Fig. 9.8	—
72555	Timer (same as 555)	Fig. 9.21	—

circuit will be explained with reference to logic diagrams from previous chapters. The operation of the circuit will be specified by means of a truth table or a function table.

Other possible graphic symbols for the ICs are presented in Chapter 10. These are standard graphic symbols approved by the Institute of Electrical and Electronics Engineers and are given in IEEE Standard 91-1984. The standard graphic symbols for SSI gates have rectangular shapes, as shown in Fig. 10.1. The standard graphic symbol for the 7493 IC is shown in Fig. 10.13. This symbol can be substituted in place of the one shown in Fig. 9.2(c). The standard graphic symbols of the other ICs that are needed to run the experiments are presented in Chapter 10. They can be used to draw schematic diagrams of the logic circuits if the standard symbols are preferred.

Table 9.1 lists the ICs that are needed for the experiments, together with the numbers of the figures in which they are presented in this chapter. In addition, the table lists the numbers of the figures in Chapter 10 in which the equivalent standard graphic symbols are drawn.

The next 18 sections present 18 hardware experiments requiring the use of digital integrated circuits. Section 9.20 outlines HDL simulation experiments requiring a Verilog HDL compiler and simulator.

9.2 EXPERIMENT 1: BINARY AND DECIMAL NUMBERS

This experiment demonstrates the count sequence of binary numbers and the binary-coded decimal (BCD) representation. It serves as an introduction to the breadboard used in the laboratory and acquaints the student with the cathode-ray oscilloscope. Reference material from the text that may be useful to know while performing the experiment can be found in Section 1.2, on binary numbers, and Section 1.7, on BCD numbers.

Binary Count

IC type 7493 consists of four flip-flops, as shown in Fig. 9.2. They can be connected to count in binary or in BCD. Connect the IC to operate as a four-bit binary counter by wiring the external terminals, as shown in Fig. 9.3. This is done by connecting a wire from pin 12 (output *QA*) to pin 1 (input *B*). Input *A* at pin 14 is connected to a pulser that provides single pulses. The two reset inputs, *R1* and *R2*, are connected to ground. The four outputs go to four indicator lamps, with the low-order bit of the counter from *QA* connected to the rightmost indicator lamp. Do not forget to supply 5 V and ground to the IC. All connections should be made with the power supply in the off position.

Turn the power on and observe the four indicator lamps. The four-bit number in the output is incremented by 1 for every pulse generated in the push-button pulser. The count goes to binary 15 and then back to 0. Disconnect the input of the counter at pin 14 from the pulser, and connect it to a clock generator that produces a train of pulses at a low frequency of about 1 pulse per second. This will provide an automatic binary count. Note that the binary counter will be used in subsequent experiments to provide the input binary signals for testing combinational circuits.

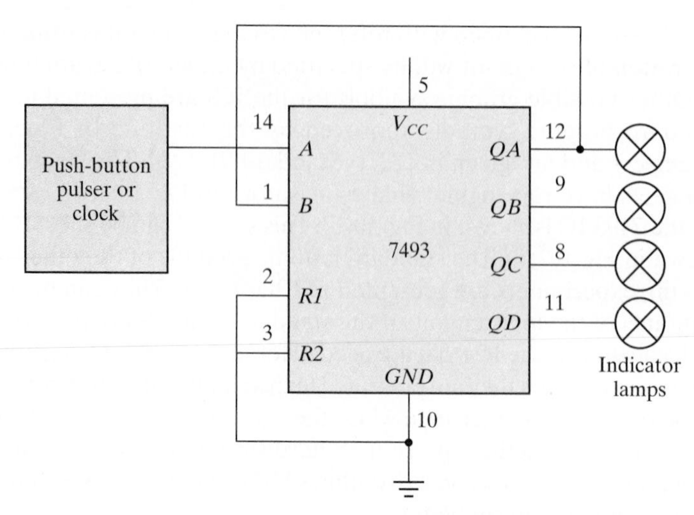

FIGURE 9.3
Binary counter

Oscilloscope Display

Increase the frequency of the clock to 10 kHz or higher and connect its output to an oscilloscope. Observe the clock output on the oscilloscope and sketch its waveform. Using a dual-trace oscilloscope, connect the output of *QA* to one channel and the output of the clock to the second channel. Note that the output of *QA* is complemented every time the clock pulse goes through a negative transition from 1 to 0. Note also that the clock frequency at the output of the first flip-flop is one-half that of the input clock frequency. Each flip-flop in turn divides its incoming frequency by 2. The four-bit counter divides the incoming frequency by 16 at output *QD*. Obtain a timing diagram showing the relationship of the clock to the four outputs of the counter. Make sure that you include at least 16 clock cycles. The way to proceed with a dual-trace oscilloscope is as follows: First, observe the clock pulses and *QA*, and record their timing waveforms. Then repeat by observing and recording the waveforms of *QA* together with *QB*, followed by the waveforms of *QB* with *QC* and then *QC* with *QD*. Your final result should be a diagram showing the relationship of the clock to the four outputs in one composite diagram having at least 16 clock cycles.

BCD Count

The BCD representation uses the binary numbers from 0000 to 1001 to represent the coded decimal digits from 0 to 9. IC type 7493 can be operated as a BCD counter by making the external connections shown in Fig. 9.4. Outputs *QB* and *QD* are connected to the two reset inputs, *R1* and *R2*. When both *R1* and *R2* are equal to 1, all four cells in the counter clear to 0 irrespective of the input pulse. The counter starts from 0, and every input pulse increments it by 1 until it reaches the count of 1001. The next pulse changes the ouput to 1010, making *QB* and *QD* equal to 1. This momentary output cannot be

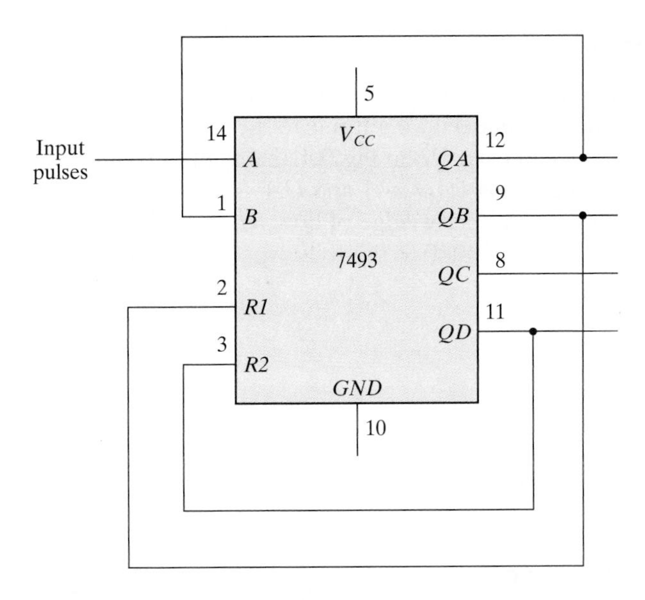

FIGURE 9.4
BCD counter

sustained, because the four cells immediately clear to 0, with the result that the output goes to 0000. Thus, the pulse after the count of 1001 changes the output to 0000, producing a BCD count.

Connect the IC to operate as a BCD counter. Connect the input to a pulser and the four outputs to indicator lamps. Verify that the count goes from 0000 to 1001.

Disconnect the input from the pulser and connect it to a clock generator. Observe the clock waveform and the four outputs on the oscilloscope. Obtain an accurate timing diagram showing the relationship between the clock and the four outputs. Make sure to include at least 10 clock cycles in the oscilloscope display and in the composite timing diagram.

Output Pattern

When the count pulses into the BCD counter are continuous, the counter keeps repeating the sequence from 0000 to 1001 and back to 0000. This means that each bit in the four outputs produces a fixed pattern of 1's and 0's that is repeated every 10 pulses. These patterns can be predicted from a list of the binary numbers from 0000 to 1001. The list will show that output *QA*, being the least significant bit, produces a pattern of alternate 1's and 0's. Output *QD*, being the most significant bit, produces a pattern of eight 0's followed by two 1's. Obtain the pattern for the other two outputs and then check all four patterns on the oscilloscope. This is done with a dual-trace oscilloscope by displaying the clock pulses in one channel and one of the output waveforms in the other channel. The pattern of 1's and 0's for the corresponding output is obtained by observing the output levels at the vertical positions where the pulses change from 1 to 0.

Other Counts

IC type 7493 can be connected to count from 0 to a variety of final counts. This is done by connecting one or two outputs to the reset inputs, *R1* and *R2*. Thus, if *R1* is connected to *QA* instead of to *QB* in Fig. 9.4, the resulting count will be from 0000 to 1000, which is 1 less than 1001 ($QD = 1$ and $QA = 1$).

Utilizing your knowledge of how *R1* and *R2* affect the final count, connect the 7493 IC to count from 0000 to the following final counts:

(a) 0101
(b) 0111
(c) 1011

Connect each circuit and verify its count sequence by applying pulses from the pulser and observing the output count in the indicator lamps. If the initial count starts with a value greater than the final count, keep applying input pulses until the output clears to 0.

9.3 EXPERIMENT 2: DIGITAL LOGIC GATES

In this experiment, you will investigate the logic behavior of various IC gates:

7400 quadruple two-input NAND gates
7402 quadruple two-input NOR gates
7404 hex inverters
7408 quadruple two-input AND gates
7432 quadruple two-input OR gates
7486 quadruple two-input XOR gates

The pin assignments to the various gates are shown in Fig. 9.1. "Quadruple" means that there are four gates within the package. The digital logic gates and their characteristics are discussed in Section 2.8. A NAND implementation is discussed in Section 3.7.

Truth Tables

Use one gate from each IC listed and obtain the truth table of the gate. The truth table is obtained by connecting the inputs of the gate to switches and the output to an indicator lamp. Compare your results with the truth tables listed in Fig. 2.5.

Waveforms

For each gate listed, obtain the input–output waveform of the gate. The waveforms are to be observed in the oscilloscope. Use the two low-order outputs of a binary counter (Fig. 9.3) to provide the inputs to the gate. As an example, the circuit and waveforms for the NAND gate are illustrated in Fig. 9.5. The oscilloscope display will repeat this waveform, but you should record only the nonrepetitive portion.

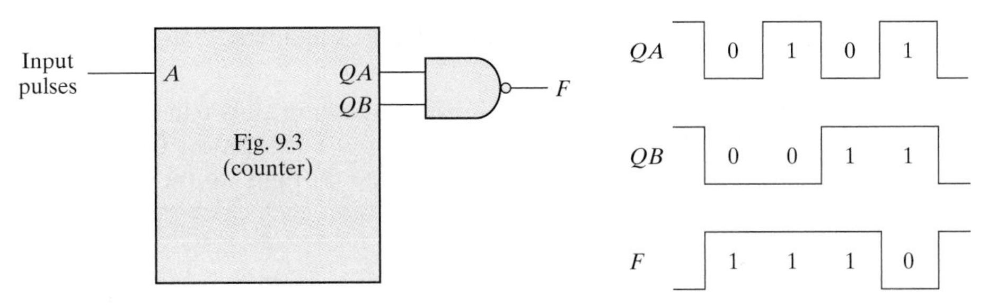

FIGURE 9.5
Waveforms for NAND gate

Propagation Delay

Connect the six inverters inside the 7404 IC in cascade. The output will be the same as the input, except that it will be delayed by the time it takes the signal to propagate through all six inverters. Apply clock pulses to the input of the first inverter. Using the oscilloscope, determine the delay from the input to the output of the sixth inverter during the upswing of the pulse and again during the downswing. This is done with a dual-trace oscilloscope by applying the input clock pulses to one of the channels and the output of the sixth inverter to the second channel. Set the time-base knob to the lowest time-per-division setting. The rise or fall time of the two pulses should appear on the screen. Divide the total delay by 6 to obtain an average propagation delay per inverter.

Universal NAND Gate

Using a single 7400 IC, connect a circuit that produces

(a) an inverter,
(b) a two-input AND,
(c) a two-input OR,
(d) a two-input NOR,
(e) a two-input XOR. (See Fig. 3.32.)

In each case, verify your circuit by checking its truth table.

NAND Circuit

Using a single 7400 IC, construct a circuit with NAND gates that implements the Boolean function

$$F = AB + CD$$

1. Draw the circuit diagram.
2. Obtain the truth table for F as a function of the four inputs.
3. Connect the circuit and verify the truth table.

4. Record the patterns of 1's and 0's for F as inputs A, B, C, and D go from binary 0 to binary 15.
5. Connect the four outputs of the binary counter shown in Fig. 9.3 to the four inputs of the NAND circuit. Connect the input clock pulses from the counter to one channel of a dual-trace oscilloscope and output F to the other channel. Observe and record the 1's and 0's pattern of F after each clock pulse, and compare it with the pattern recorded in step 4.

9.4 EXPERIMENT 3: SIMPLIFICATION OF BOOLEAN FUNCTIONS

This experiment demonstrates the relationship between a Boolean function and the corresponding logic diagram. The Boolean functions are simplified by using the map method, as discussed in Chapter 3. The logic diagrams are to be drawn with NAND gates, as explained in Section 3.7.

The gate ICs to be used for the logic diagrams must be those from Fig. 9.1 which contain the following NAND gates:

7400 two-input NAND

7404 inverter (one-input NAND)

7410 three-input NAND

7420 four-input NAND

If an input to a NAND gate is not used, it should not be left open, but instead should be connected to another input that is used. For example, if the circuit needs an inverter and there is an extra two-input gate available in a 7400 IC, then both inputs of the gate are to be connected together to form a single input for an inverter.

Logic Diagram

This part of the experiment starts with a given logic diagram from which we proceed to apply simplification procedures to reduce the number of gates and, possibly, the number of ICs. The logic diagram shown in Fig. 9.6 requires two ICs—a 7400 and a 7410. Note that the inverters for inputs x, y, and z are obtained from the remaining three gates in the 7400 IC. If the inverters were taken from a 7404 IC, the circuit would have required three ICs. Note also that, in drawing SSI circuits, the gates are not enclosed in blocks as is done with MSI circuits.

Assign pin numbers to all inputs and outputs of the gates, and connect the circuit with the x, y, and z inputs going to three switches and the output F to an indicator lamp. Test the circuit by obtaining its truth table.

Obtain the Boolean function of the circuit and simplify it, using the map method. Construct the simplified circuit without disconnecting the original circuit. Test both circuits by applying identical inputs to each and observing the separate outputs. Show that, for each of the eight possible input combinations, the two circuits have identical outputs. This will prove that the simplified circuit behaves exactly like the original circuit.

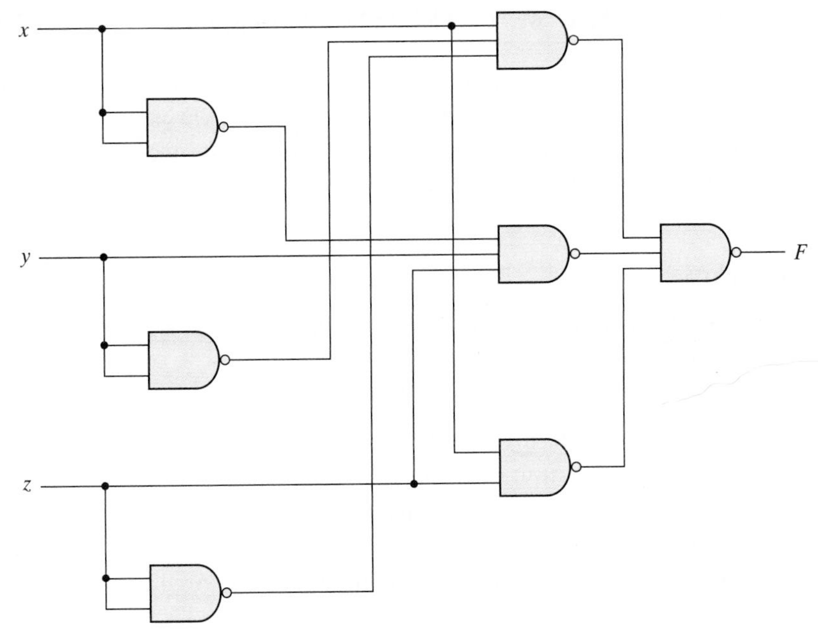

FIGURE 9.6
Logic diagram for Experiment 3

Boolean Functions

Consider two Boolean functions in sum-of-minterms form:

$$F_1(A, B, C, D) = (0, 1, 4, 5, 8, 9, 10, 12, 13)$$
$$F_2(A, B, C, D) = (3, 5, 7, 8, 10, 11, 13, 15)$$

Simplify these functions by means of maps. Obtain a composite logic diagram with four inputs, A, B, C, and D, and two outputs, F_1 and F_2. Implement the two functions together, using a minimum number of NAND ICs. Do not duplicate the same gate if the corresponding term is needed for both functions. Use any extra gates in existing ICs for inverters when possible. Connect the circuit and check its operation. The truth table for F_1 and F_2 obtained from the circuit should conform with the minterms listed.

Complement

Plot the following Boolean function in a map:

$$F = A'D + BD + B'C + AB'D$$

Combine the 1's in the map to obtain the simplified function for F in sum-of-products form. Then combine the 0's in the map to obtain the simplified function for F', also in sum-of-products form. Implement both F and F' with NAND gates, and connect the two circuits to the same input switches, but to separate output indicator lamps. Obtain the truth table of each circuit in the laboratory and show that they are the complements of each other.

9.5 EXPERIMENT 4: COMBINATIONAL CIRCUITS

In this experiment, you will design, construct, and test four combinational logic circuits. The first two circuits are to be constructed with NAND gates, the third with XOR gates, and the fourth with a decoder and NAND gates. Reference to a parity generator can be found in Section 3.9. Implementation with a decoder is discussed in Section 4.9.

Design Example

Design a combinational circuit with four inputs—A, B, C, and D—and one output, F. F is to be equal to 1 when $A = 1$, provided that $B = 0$, or when $B = 1$, provided that either C or D is also equal to 1. Otherwise, the output is to be equal to 0.

1. Obtain the truth table of the circuit.
2. Simplify the output function.
3. Draw the logic diagram of the circuit, using NAND gates with a minimum number of ICs.
4. Construct the circuit and test it for proper operation by verifying the given conditions.

Majority Logic

A majority logic is a digital circuit whose output is equal to 1 if the majority of the inputs are 1's. The output is 0 otherwise. Design and test a three-input majority circuit using NAND gates with a minimum number of ICs.

Parity Generator

Design, construct, and test a circuit that generates an even parity bit from four message bits. Use XOR gates. Adding one more XOR gate, expand the circuit so that it generates an odd parity bit also.

Decoder Implementation

A combinational circuit has three inputs—x, y, and z—and three outputs—F_1, F_2, and F_3. The simplified Boolean functions for the circuit are

$$F_1 = xz + x'y'z'$$
$$F_2 = x'y + xy'z'$$
$$F_3 = xy + x'y'z$$

Implement and test the combinational circuit, using a 74155 decoder IC and external NAND gates.

The block diagram of the decoder and its truth table are shown in Fig. 9.7. The 74155 can be connected as a dual 2 × 4 decoder or as a single 3 × 8 decoder. When a 3 × 8 decoder is desired, inputs *C1* and *C2*, as well as inputs *G1* and *G2*, must be connected together, as shown in the block diagram. The function of the circuit is similar to that illustrated in Fig. 4.18. *G* is the enable input and must be equal to 0 for proper operation. The eight outputs are labeled with symbols given in the data book. The 74155 uses NAND gates, with the result that the selected output goes to 0 while all other outputs remain at 1. The implementation with the decoder is as shown in Fig. 4.21, except that the OR gates must be replaced with external NAND gates when the 74155 is used.

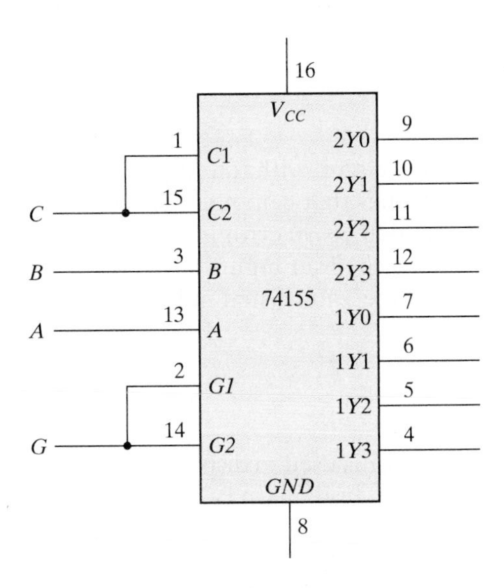

Truth table

Inputs				Outputs							
G	*C*	*B*	*A*	2*Y*0	2*Y*1	2*Y*2	2*Y*3	1*Y*0	1*Y*1	1*Y*2	1*Y*3
1	X	X	X	1	1	1	1	1	1	1	1
0	0	0	0	0	1	1	1	1	1	1	1
0	0	0	1	1	0	1	1	1	1	1	1
0	0	1	0	1	1	0	1	1	1	1	1
0	0	1	1	1	1	1	0	1	1	1	1
0	1	0	0	1	1	1	1	0	1	1	1
0	1	0	1	1	1	1	1	1	0	1	1
0	1	1	0	1	1	1	1	1	1	0	1
0	1	1	1	1	1	1	1	1	1	1	0

FIGURE 9.7
IC type 74155 connected as a 3 × 8 decoder

9.6 EXPERIMENT 5: CODE CONVERTERS

The conversion from one binary code to another is common in digital systems. In this experiment, you will design and construct three combinational-circuit converters. Code conversion is discussed in Section 4.4.

Gray Code to Binary

Design a combinational circuit with four inputs and four outputs that converts a four-bit Gray code number (Table 1.6) into the equivalent four-bit binary number. Implement the circuit with exclusive-OR gates. (This can be done with one 7486 IC.) Connect the circuit to four switches and four indicator lamps, and check for proper operation.

9's Complementer

Design a combinational circuit with four input lines that represent a decimal digit in BCD and four output lines that generate the 9's complement of the input digit. Provide a fifth output that detects an error in the input BCD number. This output should be equal to logic 1 when the four inputs have one of the unused combinations of the BCD code. Use any of the gates listed in Fig. 9.1, but minimize the total number of ICs used.

Seven-Segment Display

A seven-segment indicator is used to display any one of the decimal digits 0 through 9. Usually, the decimal digit is available in BCD. A BCD-to-seven-segment decoder accepts a decimal digit in BCD and generates the corresponding seven-segment code, as is shown pictorially in Problem 4.9.

Figure 9.8 shows the connections necessary between the decoder and the display. The 7447 IC is a BCD-to-seven-segment decoder/driver that has four inputs for the BCD digit. Input D is the most significant and input A the least significant. The four-bit BCD digit is converted to a seven-segment code with outputs a through g. The outputs of the 7447 are applied to the inputs of the 7730 (or equivalent) seven-segment display. This IC contains the seven light-emitting diode (LED) segments on top of the package. The input at pin 14 is the common anode (CA) for all the LEDs. A 47-Ω resistor to V_{CC} is needed in order to supply the proper current to the selected LED segments. Other equivalent seven-segment display ICs may have additional anode terminals and may require different resistor values.

Construct the circuit shown in Fig. 9.8. Apply the four-bit BCD digits through four switches, and observe the decimal display from 0 to 9. Inputs 1010 through 1111 have no meaning in BCD. Depending on the decoder, these values may cause either a blank or a meaningless pattern to be displayed. Observe and record the output patterns of the six unused input combinations.

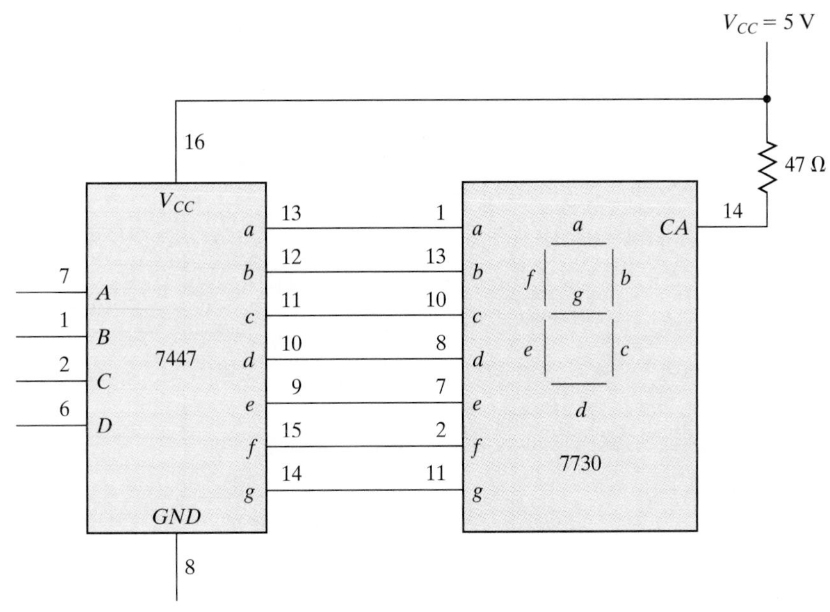

FIGURE 9.8
BCD-to-seven-segment decoder (7447) and seven-segment display (7730)

9.7 EXPERIMENT 6: DESIGN WITH MULTIPLEXERS

In this experiment, you will design a combinational circuit and implement it with multiplexers, as explained in Section 4.11. The multiplexer to be used is IC type 74151, shown in Fig. 9.9. The internal construction of the 74151 is similar to the diagram shown in Fig. 4.25, except that there are eight inputs instead of four. The eight inputs are designated *D0* through *D7*. The three selection lines—*C*, *B*, and *A*—select the particular input to be multiplexed and applied to the output. A strobe control *S* acts as an enable signal. The function table specifies the value of output *Y* as a function of the selection lines. Output *W* is the complement of *Y*. For proper operation, the strobe input *S* must be connected to ground.

Design Specifications

A small corporation has 10 shares of stock, and each share entitles its owner to one vote at a stockholder's meeting. The 10 shares of stock are owned by four people as follows:

Mr. W: 1 share

Mr. X: 2 shares

Mr. Y: 3 shares

Mrs. Z: 4 shares

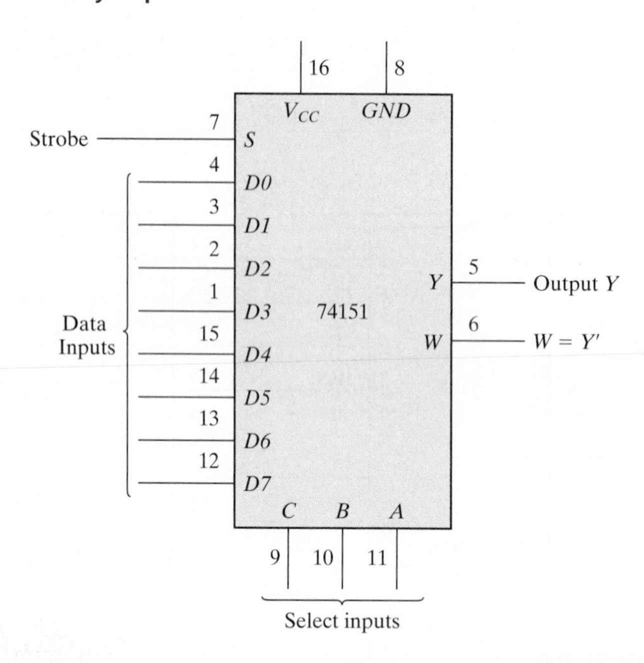

Function table

Strobe	Select			Output
S	C	B	A	Y
1	X	X	X	0
0	0	0	0	D0
0	0	0	1	D1
0	0	1	0	D2
0	0	1	1	D3
0	1	0	0	D4
0	1	0	1	D5
0	1	1	0	D6
0	1	1	1	D7

FIGURE 9.9
IC type 74151 38 × 1 multiplexer

Each of these persons has a switch to close when voting yes and to open when voting no for his or her shares.

It is necessary to design a circuit that displays the total number of shares that vote yes for each measure. Use a seven-segment display and a decoder, as shown in Fig. 9.8, to display the required number. If all shares vote no for a measure, the display should be blank. (Note that binary input 15 into the 7447 blanks out all seven segments.) If 10 shares vote yes for a measure, the display should show 0. Otherwise, the display shows a decimal number equal to the number of shares that vote yes. Use four 74151 multiplexers to design the combinational circuit that converts the inputs from the stock owners' switches into the BCD digit for the 7447. Do not use 5 V for logic 1. Use the output of an inverter whose input is grounded.

9.8 EXPERIMENT 7: ADDERS AND SUBTRACTORS

In this experiment, you will construct and test various adder and subtractor circuits. The subtractor circuit is then used to compare the relative magnitudes of two numbers. Adders are discussed in Section 4.3. Subtraction with 2's complement is explained in Section 1.6. A four-bit parallel adder–subtractor is shown in Fig. 4.13, and the comparison of two numbers is explained in Section 4.8.

Half Adder

Design, construct, and test a half-adder circuit using one XOR gate and two NAND gates.

Full Adder

Design, construct, and test a full-adder circuit using two ICs, 7486 and 7400.

Parallel Adder

IC type 7483 is a four-bit binary parallel adder. The pin assignment is shown in Fig. 9.10. The 2 four-bit input binary numbers are *A1* through *A4* and *B1* through *B4*. The four-bit sum is obtained from *S1* through *S4*. *C0* is the input carry and *C4* the output carry.

Test the four-bit binary adder 7483 by connecting the power supply and ground terminals. Then connect the four *A* inputs to a fixed binary number, such as 1001, and the *B* inputs and the input carry to five toggle switches. The five outputs are applied to

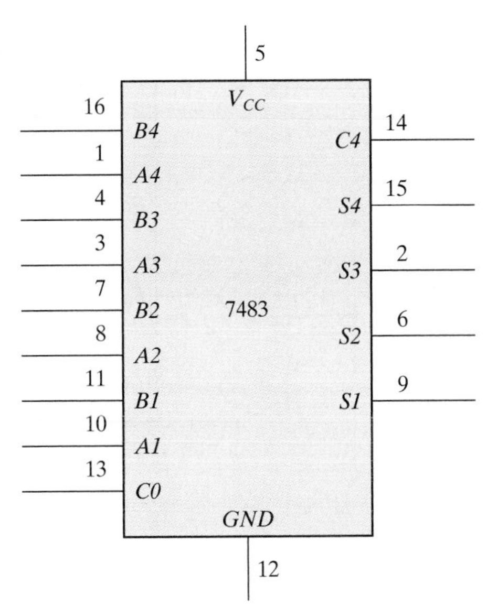

FIGURE 9.10
IC type 7483 four-bit binary adder

indicator lamps. Perform the addition of a few binary numbers and check that the output sum and output carry give the proper values. Show that when the input carry is equal to 1, it adds 1 to the output sum.

Adder–Subtractor

Two binary numbers can be subtracted by taking the 2's complement of the subtrahend and adding it to the minuend. The 2's complement can be obtained by taking the 1's complement and adding 1. To perform $A - B$, we complement the four bits of B, add them to the four bits of A, and add 1 through the input carry. This is done as shown in Fig. 9.11. The four XOR gates complement the bits of B when the mode select $M = 1$ (because $x \oplus 1 = x'$ and leave the bits of B unchanged when $M = 0$ (because $x \oplus 0 = x$). Thus, when the mode select M is equal to 1, the input carry $C0$ is equal to 1 and the sum output is A plus the 2's complement of B. When M is equal to 0, the input carry is equal to 0 and the sum generates $A + B$.

Connect the adder–subtractor circuit and test it for proper operation. Connect the four A inputs to a fixed binary number 1001 and the B inputs to switches. Perform

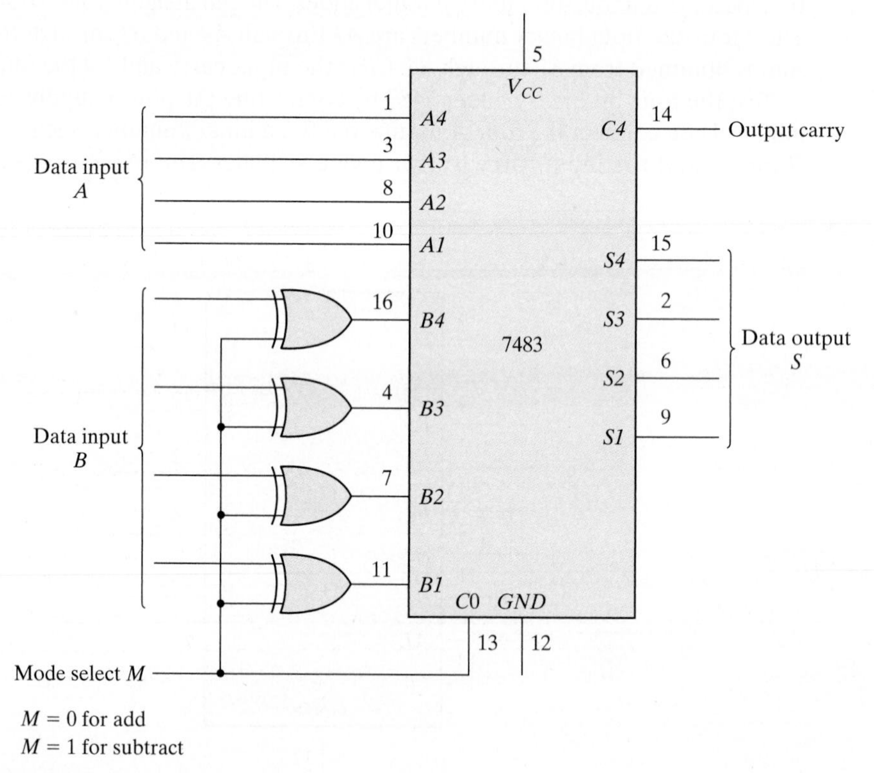

FIGURE 9.11
Four-bit adder–subtractor

the following operations and record the values of the output sum and the output carry $C4$:

$$9 + 5 \quad 9 - 5$$
$$9 + 9 \quad 9 - 9$$
$$9 + 15 \quad 9 - 15$$

Show that during addition, the output carry is equal to 1 when the sum exceeds 15. Also, show that when $A \geq B$, the subtraction operation gives the correct answer, $A - B$, and the output carry $C4$ is equal to 1, but when $A < B$, the subtraction gives the 2's complement of $B - A$ and the output carry is equal to 0.

Magnitude Comparator

The comparison of two numbers is an operation that determines whether one number is greater than, equal to, or less than the other number. Two numbers, A and B, can be compared by first subtracting $A - B$ as is done in Fig. 9.11. If the output in S is equal to zero, then $A = B$. The output carry from $C4$ determines the relative magnitudes of the numbers: When $C4 = 1, A \geq B$; when $C4 = 0, A < B$; and when $C4 = 1$ and $S \neq 0, A > B$.

It is necessary to supplement the subtractor circuit of Fig. 9.11 to provide the comparison logic. This is done with a combinational circuit that has five inputs—$S1$ through $S4$ and $C4$—and three outputs, designated by x, y, and z, so that

$$x = 1 \quad \text{if } A = B \quad (S = 0000)$$
$$y = 1 \quad \text{if } A < B \quad (C4 = 0)$$
$$z = 1 \quad \text{if } A > B \quad (C4 = 1 \text{ and } S \neq 0000)$$

The combinational circuit can be implemented with the 7404 and 7408 ICs.

Construct the comparator circuit and test its operation. Use at least two sets of numbers for A and B to check each of the outputs x, y, and z.

9.9 EXPERIMENT 8: FLIP-FLOPS

In this experiment, you will construct, test, and investigate the operation of various latches and flip-flops. The internal construction of latches and flip-flops can be found in Sections 5.3 and 5.4.

SR Latch

Construct an SR latch with two cross-coupled NAND gates. Connect the two inputs to switches and the two outputs to indicator lamps. Set the two switches to logic 1, and then momentarily turn each switch separately to the logic-0 position and back to 1. Obtain the function table of the circuit.

D Latch

Construct a D latch with four NAND gates (only one 7400 IC) and verify its function table.

Master–Slave Flip-Flop

Connect a master–slave D flip-flop using two D latches and an inverter. Connect the D input to a switch and the clock input to a pulser. Connect the output of the master latch to one indicator lamp and the output of the slave latch to another indicator lamp. Set the value of the input to the complement value of the output. Press the push button in the pulser and then release it to produce a single pulse. Observe that the master changes when the pulse goes positive and the slave follows the change when the pulse goes negative. Press the push button again a few times while observing the two indicator lamps. Explain the transfer sequence from input to master and from master to slave.

Disconnect the clock input from the pulser and connect it to a clock generator. Connect the complement output of the flip-flop to the D input. This causes the flip-flop to be complemented with each clock pulse. Using a dual-trace oscilloscope, observe the waveforms of the clock and the master and slave outputs. Verify that the delay between the master and the slave outputs is equal to the positive half of the clock cycle. Obtain a timing diagram showing the relationship between the clock waveform and the master and slave outputs.

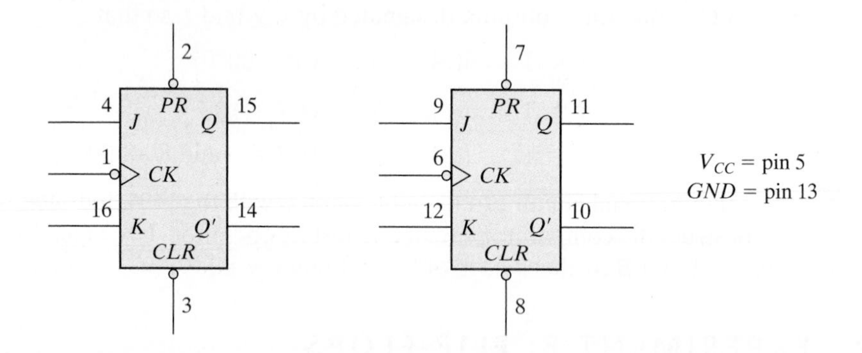

Function table

Inputs					Outputs	
Preset	Clear	Clock	J	K	Q	Q'
0	1	X	X	X	1	0
1	0	X	X	X	0	1
0	0	X	X	X	1	1
1	1	⎍	0	0	No change	
1	1	⎍	0	1	0	1
1	1	⎍	1	0	1	0
1	1	⎍	1	1	Toggle	

FIGURE 9.12
IC type 7476 dual JK master–slave flip-flops

Edge-Triggered Flip-Flop

Construct a D-type positive-edge-triggered flip-flop using six NAND gates. Connect the clock input to a pulser, the D input to a toggle switch, and the output Q to an indicator lamp. Set the value of D to the complement of Q. Show that the flip-flop output changes only in response to a positive transition of the clock pulse. Verify that the output does not change when the clock input is logic 1, when the clock goes through a negative transition, or when the clock input is logic 0. Continue changing the D input to correspond to the complement of the Q output at all times.

Disconnect the input from the pulser and connect it to the clock generator. Connect the complement output Q' to the D input. This causes the output to be complemented with each positive transition of the clock pulse. Using a dual-trace oscilloscope, observe and record the timing relationship between the input clock and the output Q. Show that the output changes in response to a positive edge transition.

IC Flip-Flops

IC type 7476 consists of two JK master–slave flip-flops with preset and clear. The pin assignment for each flip-flop is shown in Fig. 9.12. The function table specifies the circuit's operation. The first three entries in the table specify the operation of the asynchronous

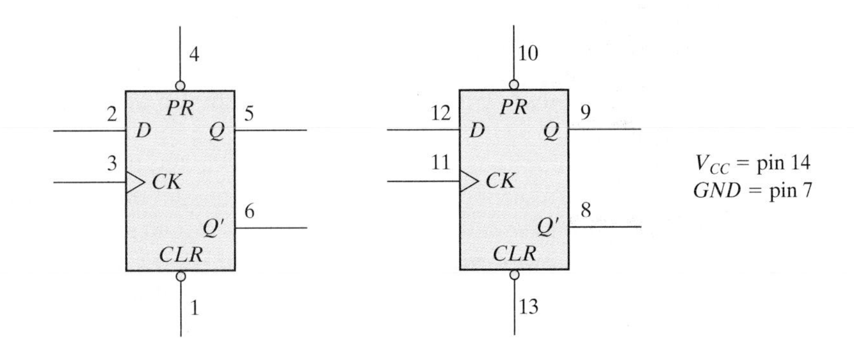

Function table

Inputs				Outputs	
Preset	Clear	Clock	D	Q	Q'
0	1	X	X	1	0
1	0	X	X	0	1
0	0	X	X	1	1
1	1	↑	0	0	1
1	1	↑	1	1	0
1	1	0	X	No change	

FIGURE 9.13
IC type 7474 dual D positive-edge-triggered flip-flops

preset and clear inputs. These inputs behave like a NAND SR latch and are independent of the clock or the J and K inputs. (The X's indicate don't-care conditions.) The last four entries in the function table specify the operation of the clock with both the preset and clear inputs maintained at logic 1. The clock value is shown as a single pulse. The positive transition of the pulse changes the master flip-flop, and the negative transition changes the slave flip-flop as well as the output of the circuit. With $J = K = 0$, the output does not change. The flip-flop toggles, or is complemented, when $J = K = 1$. Investigate the operation of one 7476 flip-flop and verify its function table.

IC type 7474 consists of two D positive-edge-triggered flip-flops with preset and clear. The pin assignment is shown in Fig. 9.13. The function table specifies the preset and clear operations and the clock's operation. The clock is shown with an upward arrow to indicate that it is a positive-edge-triggered flip-flop. Investigate the operation of one of the flip-flops and verify its function table.

9.10 EXPERIMENT 9: SEQUENTIAL CIRCUITS

In this experiment, you will design, construct, and test three synchronous sequential circuits. Use IC type 7476 (Fig. 9.12) or 7474 (Fig. 9.13). Choose any type of gate that will minimize the total number of ICs. The design of synchronous sequential circuits is covered in Section 5.7.

Up–Down Counter with Enable

Design, construct, and test a two-bit counter that counts up or down. An enable input E determines whether the counter is on or off. If $E = 0$, the counter is disabled and remains at its present count even though clock pulses are applied to the flip-flops. If $E = 1$, the counter is enabled and a second input, x, determines the direction of the count. If $x = 1$, the circuit counts upward with the sequence 00, 01, 10, 11, and the count repeats. If $x = 0$, the circuit counts downward with the sequence 11, 10, 01, 00, and the count repeats. Do not use E to disable the clock. Design the sequential circuit with E and x as inputs.

State Diagram

Design, construct, and test a sequential circuit whose state diagram is shown in Fig. 9.14. Designate the two flip-flops as A and B, the input as x, and the output as y.

Connect the output of the least significant flip-flop B to the input x, and predict the sequence of states and output that will occur with the application of clock pulses. Verify the state transition and output by testing the circuit.

Design of Counter

Design, construct, and test a counter that goes through the following sequence of binary states: 0, 1, 2, 3, 6, 7, 10, 11, 12, 13, 14, 15, and back to 0 to repeat. Note that binary states 4, 5, 8, and 9 are not used. The counter must be self-starting; that is, if the circuit starts from any one of the four invalid states, the count pulses must transfer the circuit to one of the valid states to continue the count correctly.

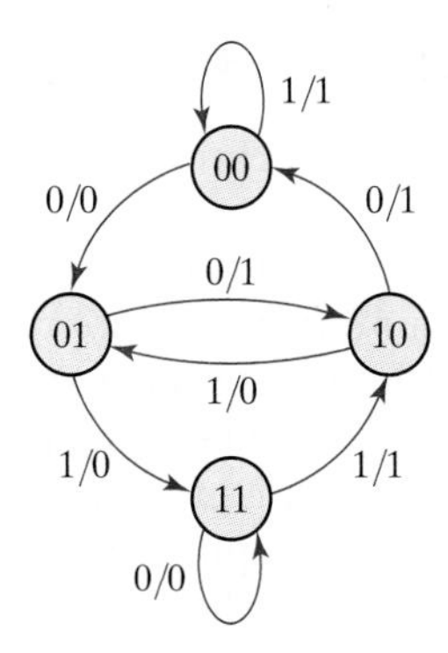

FIGURE 9.14
State diagram for Experiment 9

Check the circuit's operation for the required count sequence. Verify that the counter is self-starting. This is done by initializing the circuit to each unused state by means of the preset and clear inputs and then applying pulses to see whether the counter reaches one of the valid states.

9.11 EXPERIMENT 10: COUNTERS

In this experiment, you will construct and test various ripple and synchronous counter circuits. Ripple counters are discussed in Section 6.3 and synchronous counters are covered in Section 6.4.

Ripple Counter

Construct a four-bit binary ripple counter using two 7476 ICs (Fig. 9.12). Connect all asynchronous clear and preset inputs to logic 1. Connect the count-pulse input to a pulser and check the counter for proper operation.

Modify the counter so that it will count downward instead of upward. Check that each input pulse decrements the counter by 1.

Synchronous Counter

Construct a synchronous four-bit binary counter and check its operation. Use two 7476 ICs and one 7408 IC.

Decimal Counter

Design a synchronous BCD counter that counts from 0000 to 1001. Use two 7476 ICs and one 7408 IC. Test the counter for the proper sequence. Determine whether the counter is self-starting. This is done by initializing the counter to each of the six unused states by means of the preset and clear inputs. The application of pulses will transfer the counter to one of the valid states if the counter is self-starting.

Binary Counter with Parallel Load

IC type 74161 is a four-bit synchronous binary counter with parallel load and asynchronous clear. The internal logic is similar to that of the circuit shown in Fig. 6.14. The pin assignments to the inputs and outputs are shown in Fig. 9.15. When the load signal is enabled, the four data inputs are transferred into four internal flip-flops, QA through QD, with QD being the most significant bit. There are two count-enable inputs called P and T. Both must be equal to 1 for the counter to operate. The function table is similar to Table 6.6, with one exception: The load input in the 74161 is enabled when equal to 0. To load the input data, the clear input must be equal to 1 and the load input must be equal to 0. The two count inputs have don't-care conditions and may be equal to either 1 or 0. The internal flip-flops trigger on the positive transition of the clock pulse. The circuit functions as a counter when the load input is equal to 1 and both count inputs P and T are equal to 1. If either P or T goes to 0, the output

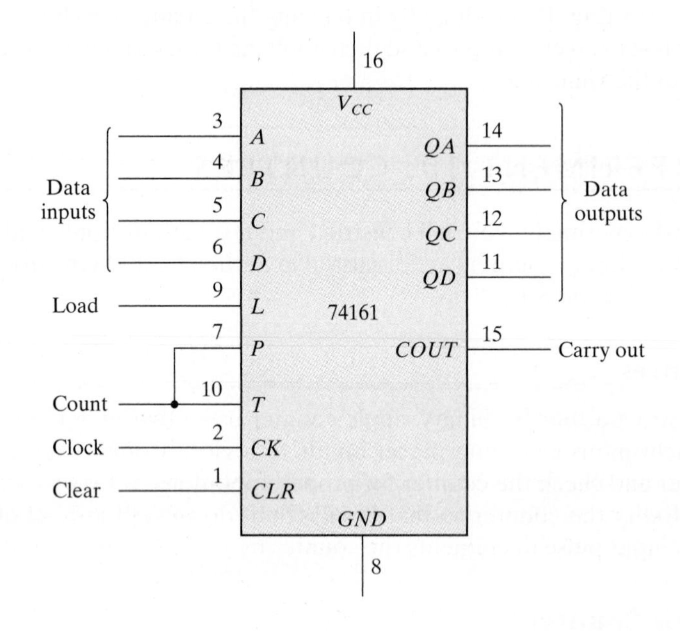

Function table

Clear	Clock	Load	Count	Function
0	X	X	X	Clear outputs to 0
1	↑	0	X	Load input data
1	↑	1	1	Count to next binary value
1	↑	1	0	No change in output

FIGURE 9.15
IC type 74161 binary counter with parallel load

does not change. The carry-out output is equal to 1 when all four data outputs are equal to 1. Perform an experiment to verify the operation of the 74161 IC according to the function table.

Show how the 74161 IC, together with a two-input NAND gate, can be made to operate as a synchronous BCD counter that counts from 0000 to 1001. Do not use the clear input. Use the NAND gate to detect the count of 1001, which then causes all 0's to be loaded into the counter.

9.12 EXPERIMENT 11: SHIFT REGISTERS

In this experiment, you will investigate the operation of shift registers. The IC to be used is the 74195 shift register with parallel load. Shift registers are explained in Section 6.2.

IC Shift Register

IC type 74195 is a four-bit shift register with parallel load and asynchronous clear. The pin assignments to the inputs and outputs are shown in Fig. 9.16. The single control line labeled SH/LD (shift/load) determines the synchronous operation of the register. When $SH/LD = 0$, the control input is in the load mode and the four data inputs are transferred into the four internal flip-flops, QA through QD. When $SH/LD = 1$, the control input is in the shift mode and the information in the register is shifted right from QA toward QD. The serial input into QA during the shift is determined from the J and $\overline{K}$ inputs. The two inputs behave like the J and the complement of K of a JK flip-flop. When both J and $\overline{K}$ are equal to 0, flip-flop QA is cleared to 0 after the shift. If both inputs are equal to 1, QA is set to 1 after the shift. The other two conditions for the J and $\overline{K}$ inputs provide a complement or no change in the output of flip-flop QA after the shift.

The function table for the 74195 shows the mode of operation of the register. When the clear input goes to 0, the four flip-flops clear to 0 asynchronously—that is, without the need of a clock. Synchronous operations are affected by a positive transition of the clock. To load the input data, SH/LD must be equal to 0 and a positive clock-pulse transition must occur. To shift right, SH/LD must be equal to 1. The J and $\overline{K}$ inputs must be connected together to form the serial input.

Perform an experiment that will verify the operation of the 74195 IC. Show that it performs all the operations listed in the function table. Include in your function table the two conditions for $J\overline{K} = 01$ and 10.

Ring Counter

A ring counter is a circular shift register with the signal from the serial output QD going into the serial input. Connect the J and $\overline{K}$ input together to form the serial input. Use the load condition to preset the ring counter to an initial value of 1000. Rotate the single bit with the shift condition and check the state of the register after each clock pulse.

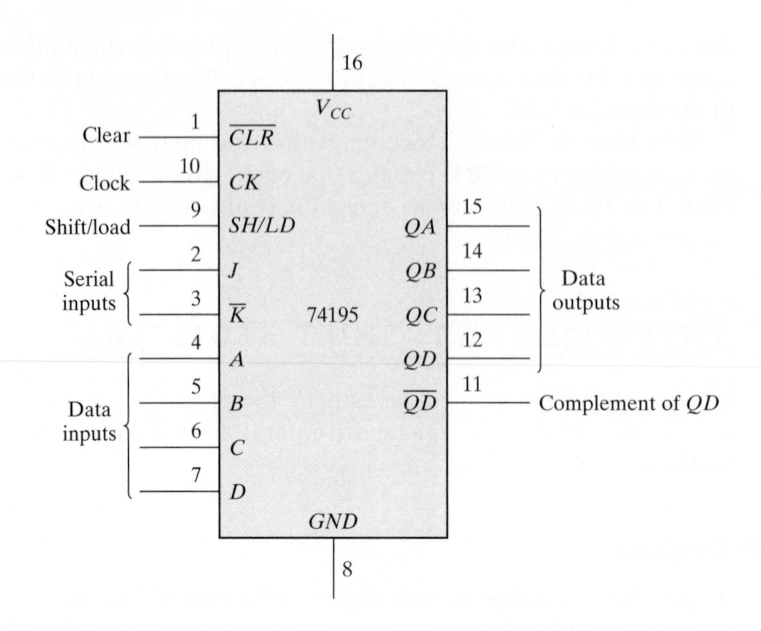

Function table

Clear	Shift/ load	Clock	J	$\overline{K}$	Serial input	Function
0	X	X	X	X	X	Asynchronous clear
1	X	0	X	X	X	No change in output
1	0	↑	X	X	X	Load input data
1	1	↑	0	0	0	Shift from QA toward QD, $QA = 0$
1	1	↑	1	1	1	Shift from QA toward QD, $QA = 1$

FIGURE 9.16
IC type 74195 shift register with parallel load

A switch-tail ring counter uses the complement output of QD for the serial input. Preset the switch-tail ring counter to 0000 and predict the sequence of states that will result from shifting. Verify your prediction by observing the state sequence after each shift.

Feedback Shift Register

A feedback shift register is a shift register whose serial input is connected to some function of selected register outputs. Connect a feedback shift register whose serial input is the exclusive-OR of outputs *QC* and *QD*. Predict the sequence of states of the register, starting from state 1000. Verify your prediction by observing the state sequence after each clock pulse.

Bidirectional Shift Register

The 74195 IC can shift only right from *QA* toward *QD*. It is possible to convert the register to a bidirectional shift register by using the load mode to obtain a shift-left operation (from *QD* toward *QA*). This is accomplished by connecting the output of each flip-flop to the input of the flip-flop on its left and using the load mode of the *SH/LD* input as a shift-left control. Input *D* becomes the serial input for the shift-left operation.

Connect the 74195 as a bidirectional shift register (without parallel load). Connect the serial input for shift right to a toggle switch. Construct the shift left as a ring counter by connecting the serial output *QA* to the serial input *D*. Clear the register and then check its operation by shifting a single 1 from the serial input switch. Shift right three more times and insert 0's from the serial input switch. Then rotate left with the shift-left (load) control. The single 1 should remain visible while shifting.

Bidirectional Shift Register with Parallel Load

The 74195 IC can be converted to a bidirectional shift register with parallel load in conjunction with a multiplexer circuit. We will use IC type 74157 for this purpose. The 74157 is a quadruple two-to-one-line multiplexer whose internal logic is shown in Fig. 4.26. The pin assignments to the inputs and outputs of the 74157 are shown in Fig. 9.17. Note that the enable input is called a strobe in the 74157.

Construct a bidirectional shift register with parallel load using the 74195 register and the 74157 multiplexer. The circuit should be able to perform the following operations:

1. Asynchronous clear
2. Shift right
3. Shift left
4. Parallel load
5. Synchronous clear

Derive a table for the five operations as a function of the clear, clock, and *SH/LD* inputs of the 74195 and the strobe and select inputs of the 74157. Connect the circuit and verify your function table. Use the parallel-load condition to provide an initial value to the register, and connect the serial outputs to the serial inputs of both shifts in order not to lose the binary information while shifting.

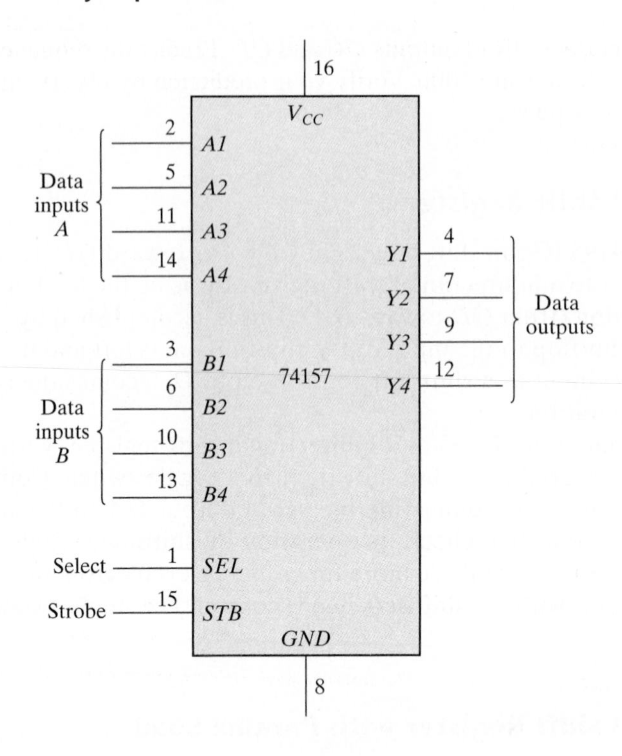

Function table

Strobe	Select	Data outputs *Y*
1	X	All 0's
0	0	Select data inputs *A*
0	1	Select data inputs *B*

FIGURE 9.17
IC type 74157 quadruple 2 × 1 multiplexers

9.13 EXPERIMENT 12: SERIAL ADDITION

In this experiment, you will construct and test a serial adder–subtractor circuit. Serial addition of two binary numbers can be done by means of shift registers and a full adder, as explained in Section 6.2.

Serial Adder

Starting from the diagram of Fig. 6.6, design and construct a four-bit serial adder using the following ICs: 74195 (two), 7408, 7486, and 7476. Provide a facility for register *B* to accept parallel data from four toggle switches, and connect its serial input to ground so that 0's are shifted into register *B* during the addition. Provide a toggle switch to clear the registers and the flip-flop. Another switch will be needed to specify whether register B is to accept parallel data or is to be shifted during the addition.

Testing the Adder

To test your serial adder, perform the binary addition $5 + 6 + 15 = 26$. This is done by first clearing the registers and the carry flip-flop. Parallel load the binary value 0101 into register B. Apply four pulses to add B to A serially, and check that the result in A is 0101. (Note that clock pulses for the 7476 must be as shown in Fig. 9.12.) Parallel load 0110 into B and add it to A serially. Check that A has the proper sum. Parallel load 1111 into B and add to A. Check that the value in A is 1010 and that the carry flip-flop is set.

Clear the registers and flip-flop and try a few other numbers to verify that your serial adder is functioning properly.

Serial Adder–Subtractor

If we follow the procedure used in Section 6.2 for the design of a serial subtractor (that subtracts $A - B$), we will find that the output difference is the same as the output sum, but that the input to the J and K of the borrow flip-flop needs the complement of QD (available in the 74195). Using the other two XOR gates from the 7486, convert the serial adder to a serial adder–subtractor with a mode control M. When $M = 0$, the circuit adds $A + B$. When $M = 1$, the circuit subtracts $A - B$ and the flip-flop holds the borrow instead of the carry.

Test the adder part of the circuit by repeating the operations recommended to ensure that the modification did not change the operation. Test the serial subtractor part by performing the subtraction $15 - 4 - 5 - 13 = -7$. Binary 15 can be transferred to register A by first clearing it to 0 and adding 15 from B. Check the intermediate results during the subtraction. Note that -7 will appear as the 2's complement of 7 with a borrow of 1 in the flip-flop.

9.14 EXPERIMENT 13: MEMORY UNIT

In this experiment, you will investigate the behavior of a random-access memory (RAM) unit and its storage capability. The RAM will be used to simulate a read-only memory (ROM). The ROM simulator will then be used to implement combinational circuits, as explained in Section 7.5. The memory unit is discussed in Sections 7.2 and 7.3.

IC RAM

IC type 74189 is a 16×4 RAM. The internal logic is similar to the circuit shown in Fig. 7.6 for a 4×4 RAM. The pin assignments to the inputs and outputs are shown in Fig. 9.18. The four address inputs select 1 of 16 words in the memory. The least significant bit of the address is A and the most significant is A_3. The chip select (CS) input must be equal to 0 to enable the memory. If CS is equal to 1, the memory is disabled and all four outputs are in a high-impedance state. The write enable (WE) input determines the type of operation, as indicated in the function table. The write operation is performed when $WE = 0$. This

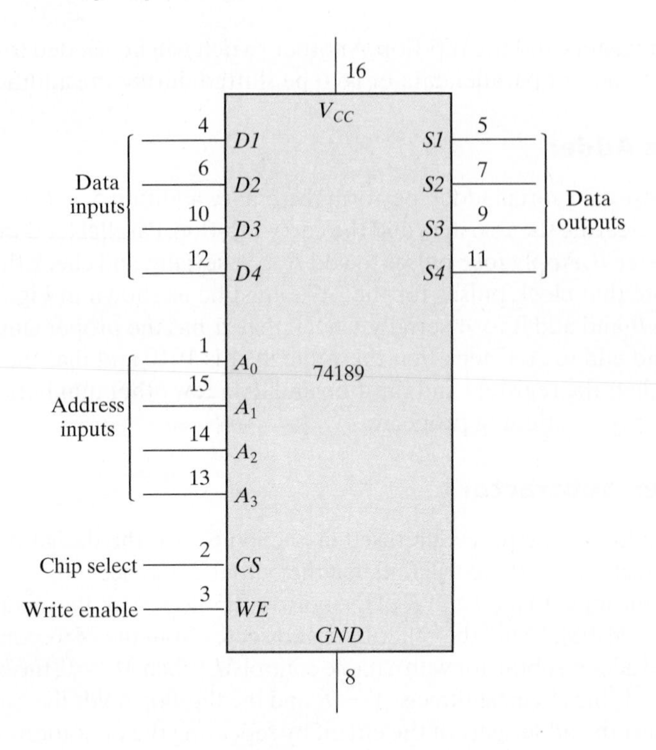

Function table

CS	*WE*	Operation	Data outputs
0	0	Write	High impedance
0	1	Read	Complement of selected word
1	X	Disable	High impedance

FIGURE 9.18
IC type 74189 16 × 4 RAM

operation is a transfer of the binary number from the data inputs into the selected word in memory. The read operation is performed when $WE = 1$. This operation transfers the complemented value stored in the selected word into the output data lines. The memory has three-state outputs to facilitate memory expansion.

Testing the RAM

Since the outputs of the 74189 produce the complemented values, we need to insert four inverters to change the outputs to their normal value. The RAM can be tested after making the following connections: Connect the address inputs to a binary counter using the 7493 IC (shown in Fig. 9.3). Connect the four data inputs to toggle switches and the data outputs to four 7404 inverters. Provide four indicator lamps for the address and four more for the outputs of the inverters. Connect input *CS* to ground and *WE* to a toggle switch (or a pulser that provides a negative pulse). Store a few words into the memory, and then read them to verify that the write and read operations are functioning properly. You must be careful when using the *WE* switch. Always leave the *WE* input in the read mode, unless you want to write into memory. The proper way to write is first to set the address in the counter and the inputs in the four toggle switches. Then, store the word in memory, flip the *WE* switch to the write position and return it to the read position. Be careful not to change the address or the inputs when *WE* is in the write mode.

ROM Simulator

A ROM simulator is obtained from a RAM operated in the read mode only. The pattern of 1's and 0's is first entered into the simulating RAM by placing the unit momentarily in the write mode. Simulation is achieved by placing the unit in the read mode and taking the address lines as inputs to the ROM. The ROM can then be used to implement any combinational circuit.

Implement a combinational circuit using the ROM simulator that converts a four-bit binary number to its equivalent Gray code as defined in Table 1.6. This is done as follows: Obtain the truth table of the code converter. Store the truth table into the 74189 memory by setting the address inputs to the binary value and the data inputs to the corresponding Gray code value. After all 16 entries of the table are written into memory, the ROM simulator is set by permanently connecting the *WE* line to logic 1. Check the code converter by applying the inputs to the address lines and verifying the correct outputs in the data output lines.

Memory Expansion

Expand the memory unit to a 32×4 RAM using two 74189 ICs. Use the *CS* inputs to select between the two ICs. Note that since the data outputs are three-stated, you can tie pairs of terminals together to obtain a logic OR operation between the two ICs. Test your circuit by using it as a ROM simulator that adds a three-bit number to a two-bit number to produce a four-bit sum. For example, if the input of the ROM is 10110, then the output is calculated to be $101 + 10 = 0111$. (The first three bits of the input represent 5, the last two bits represent 2, and the output sum is binary 7.) Use the counter to provide four bits of the address and a switch for the fifth bit of the address.

9.15 EXPERIMENT 14: LAMP HANDBALL

In this experiment, you will construct an electronic game of handball, using a single light to simulate the moving ball. The experiment demonstrates the application of a bidirectional shift register with parallel load. It also shows the operation of the asynchronous inputs of flip-flops. We will first introduce an IC that is needed for the experiment and then present the logic diagram of the simulated lamp handball game.

IC Type 74194

This is a four-bit bidirectional shift register with parallel load. The internal logic is similar to that shown in Fig. 6.7. The pin assignments to the inputs and outputs are shown in Fig. 9.19. The two mode-control inputs determine the type of operation, as specified in the function table.

Logic Diagram

The logic diagram of the electronic lamp handball game is shown in Fig. 9.20. It consists of two 74194 ICs, a dual *D* flip-flop 7474 IC, and three gate ICs: the 7400, 7404, and 7408. The ball is simulated by a moving light that is shifted left or right through the

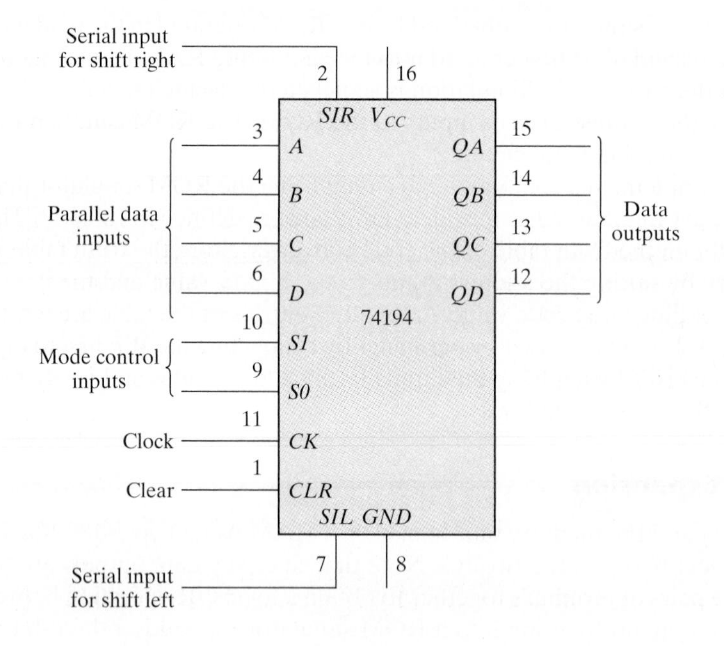

Function table

Clear	Clock	Mode *S1*	Mode *S0*	Function
0	X	X	X	Clear outputs to 0
1	↑	0	0	No change in output
1	↑	0	1	Shift right in the direction from *QA* to *QD*. *SIR* to *QA*
1	↑	1	0	Shift left in the direction from *QD* to *QA*. *SIL* to *QD*
1	↑	1	1	Parallel-load input data

FIGURE 9.19
IC type 74194 bidirectional shift register with parallel load

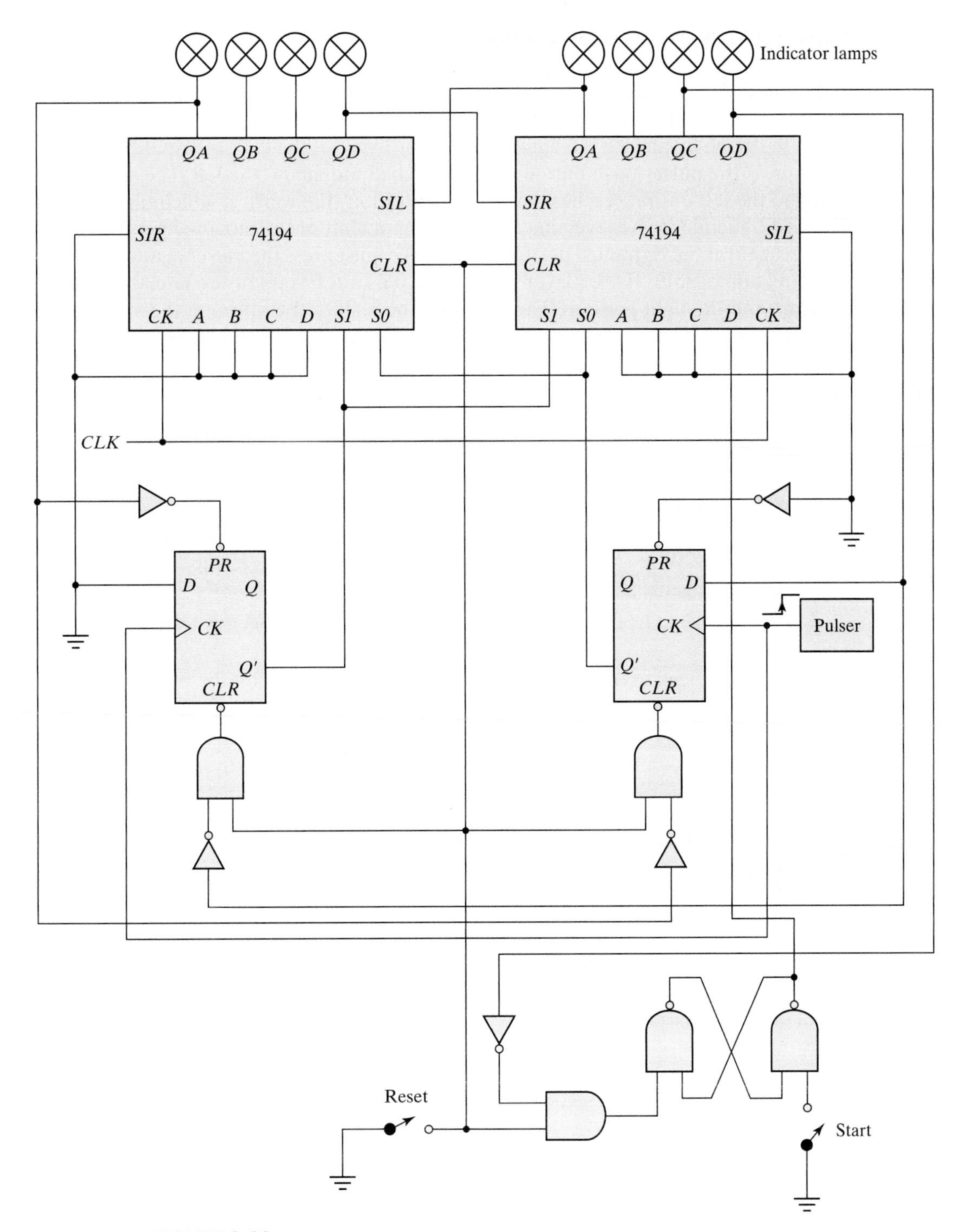

FIGURE 9.20
Lamp handball logic diagram

bidirectional shift register. The rate at which the light moves is determined by the frequency of the clock. The circuit is first initialized with the *reset* switch. The *start* switch starts the game by placing the ball (an indicator lamp) at the extreme right. The player must press the pulser push button to start the ball moving to the left. The single light shifts to the left until it reaches the leftmost position (the wall), at which time the ball returns to the player by reversing the direction of shift of the moving light. When the light is again at the rightmost position, the player must press the pulser again to reverse the direction of shift. If the player presses the pulser too soon or too late, the ball disappears and the light goes off. The game can be restarted by turning the start switch on and then off. The start switch must be open (logic 1) during the game.

Circuit Analysis

Prior to connecting the circuit, analyze the logic diagram to ensure that you understand how the circuit operates. In particular, try to answer the following questions:

1. What is the function of the reset switch?
2. How does the light in the rightmost position come on when the start switch is grounded? Why is it necessary to place the start switch in the logic-1 position before the game starts?
3. What happens to the two mode-control inputs, *S1* and *S0*, once the ball is set in motion?
4. What happens to the mode-control inputs and to the ball if the pulser is pressed while the ball is moving to the left? What happens if the ball is moving to the right, but has not yet reached the rightmost position?
5. If the ball has returned to the rightmost position, but the pulser has not yet been pressed, what is the state of the mode-control inputs if the pulser is pressed? What happens if it is not pressed?

Playing the Game

Wire the circuit of Fig. 9.20. Test the circuit for proper operation by playing the game. Note that the pulser must provide a positive-edge transition and that both the reset and start switches must be open (i.e., must be in the logic-1 state) during the game. Start with a low clock rate, and increase the clock frequency to make the handball game more challenging.

Counting the Number of Losses

Design a circuit that keeps score of the number of times the player loses while playing the game. Use a BCD-to-seven-segment decoder and a seven-segment display, as in Fig. 9.8, to display the count from 0 through 9. Counting is done with either the 7493 as a ripple decimal counter or the 74161 and a NAND gate as a synchronous decimal counter. The display should show 0 when the circuit is reset. Every time the ball disappears and the light goes off, the display should increase by 1. If the light stays on during the play, the number in the display should not change. The final design should be an automatic scoring circuit, with the decimal display incremented automatically each time the player loses when the light disappears.

Lamp Ping-Pong™

Modify the circuit of Fig. 9.20 so as to obtain a lamp Ping-Pong game. Two players can participate in this game, with each player having his or her own pulser. The player with the right pulser returns the ball when it is in the extreme right position, and the player with the left pulser returns the ball when it is in the extreme left position. The only modification required for the Ping-Pong game is a second pulser and a change of a few wires.

With a second start circuit, the game can be made to start by either one of the two players (i.e., either one serves). This addition is optional.

9.16 EXPERIMENT 15: CLOCK-PULSE GENERATOR

In this experiment, you will use an IC timer unit and connect it to produce clock pulses at a given frequency. The circuit requires the connection of two external resistors and two external capacitors. The cathode-ray oscilloscope is used to observe the waveforms and measure the frequency of the pulses.

IC Timer

IC type 72555 (or 555) is a precision timer circuit whose internal logic is shown in Fig. 9.21. (The resistors, R_A and R_B, and the two capacitors are not part of the IC.) The circuit consists of two voltage comparators, a flip-flop, and an internal transistor. The voltage division from $V_{CC} = 5$ V through the three internal resistors to ground produces $\frac{2}{3}$ and $\frac{1}{3}$ of V_{CC} (3.3 V and 1.7 V, respectively) into the fixed inputs of the comparators. When the threshold input at pin 6 goes above 3.3 V, the upper comparator resets the flip-flop and the output goes low to about 0 V. When the trigger input at pin 2 goes below 1.7 V, the lower comparator sets the flip-flop and the output goes high to about 5 V. When the output is low, Q' is high and the base–emitter junction of the transistor is forward biased. When the output is high, Q' is low and the transistor is cut off. (See Section 10.3.) The timer circuit is capable of producing accurate time delays controlled by an external *RC* circuit. In this experiment, the IC timer will be operated in the astable mode to produce clock pulses.

Circuit Operation

Figure 9.21 shows the external connections for astable operation of the circuit. Capacitor C charges through resistors R_A and R_B when the transistor is cut off and discharges through R_B when the transistor is forward biased and conducting. When the charging voltage across capacitor C reaches 3.3 V, the threshold input at pin 6 causes the flip-flop to reset and the transistor turns on. When the discharging voltage reaches 1.7 V, the trigger input at pin 2 causes the flip-flop to set and the transistor turns off. Thus, the output continually alternates

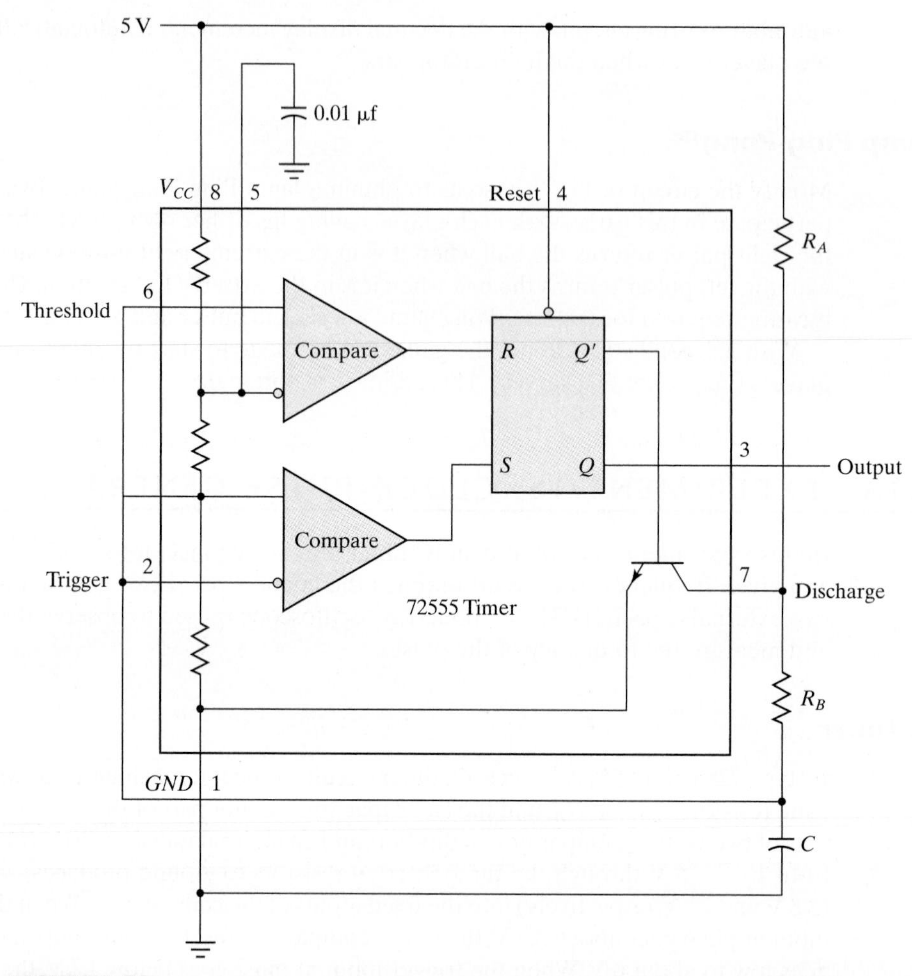

FIGURE 9.21
IC type 72555 timer connected as a clock-pulse generator

between two voltage levels at the output of the flip-flop. The output remains high for a duration equal to the charge time. This duration is determined from the equation

$$t_H = 0.693(R_A + R_B)C$$

The output remains low for a duration equal to the discharge time. This duration is determined from the equation

$$t_L = 0.693R_BC$$

Clock-Pulse Generator

Starting with a capacitor C of 0.001 μF calculate values for R_A and R_B to produce clock pulses, as shown in Fig. 9.22. The pulse width is 1 μs in the low level and repeats at a

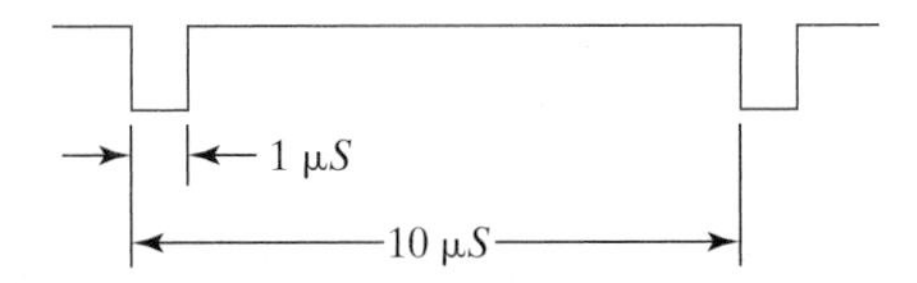

FIGURE 9.22
Output waveform for clock generator

frequency rate of 100 kHz (every 10 μs). Connect the circuit and check the output in the oscilloscope.

Observe the output across the capacitor C, and record its two levels to verify that they are between the trigger and threshold values.

Observe the waveform in the collector of the transistor at pin 7 and record all pertinent information. Explain the waveform by analyzing the circuit's action.

Connect a variable resistor (potentiometer) in series with R_A to produce a variable-frequency pulse generator. The low-level duration remains at 1 μs The frequency should range from 20 to 100 kHz.

Change the low-level pulses to high-level pulses with a 7404 inverter. This will produce positive pulses of 1 μs with a variable-frequency range.

9.17 EXPERIMENT 16: PARALLEL ADDER AND ACCUMULATOR

In this experiment, you will construct a four-bit parallel adder whose sum can be loaded into a register. The numbers to be added will be stored in a RAM. A set of binary numbers will be selected from memory and their sum will be accumulated in the register.

Block Diagram

Use the RAM circuit from the memory experiment of Section 9.14, a four-bit parallel adder, a four-bit shift register with parallel load, a carry flip-flop, and a multiplexer to construct the circuit. The block diagram and the ICs to be used are shown in Fig. 9.23. Information can be written into RAM from data in four switches or from the four-bit data available in the outputs of the register. The selection is done by means of a multiplexer. The data in RAM can be added to the contents of the register and the sum transferred back to the register.

Control of Register

Provide toggle switches to control the 74194 register and the 7476 carry flip-flop as follows:

(a) A LOAD condition transfers the sum to the register and the output carry to the flip-flop upon the application of a clock pulse.

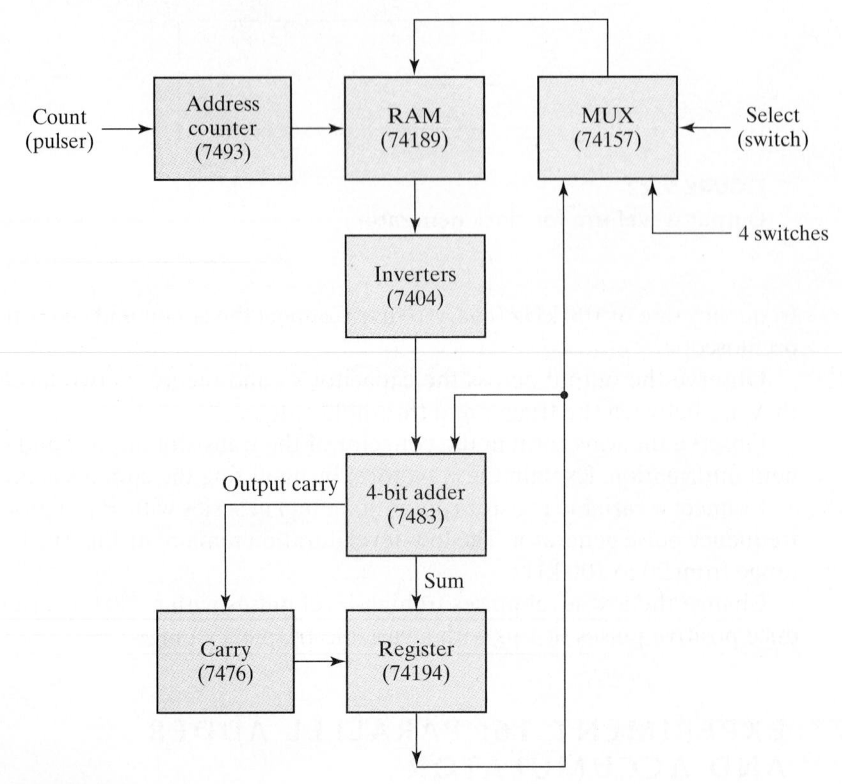

FIGURE 9.23
Block diagram of a parallel adder for Experiment 16

(b) A SHIFT condition shifts the register right with the carry from the carry flip-flop transferred into the leftmost position of the register upon the application of a clock pulse. The value in the carry flip-flop should not change during the shift.

(c) A NO-CHANGE condition leaves the contents of the register and flip-flop unchanged even when clock pulses are applied.

Carry Circuit

To conform with the preceding specifications, it is necessary to provide a circuit between the output carry from the adder and the J and K inputs of the 7476 flip-flop so that the output carry is transferred into the flip-flop (whether it is equal to 0 or 1) only when the LOAD condition is activated and a pulse is applied to the clock input of the flip-flop. The carry flip-flop should not change if the LOAD condition is disabled or the SHIFT condition is enabled.

Detailed Circuit

Draw a detailed diagram showing all the wiring between the ICs. Connect the circuit, and provide indicator lamps for the outputs of the register and carry flip-flop and for the address and output data of the RAM.

Checking the Circuit

Store the numbers 0110, 1110, 1101, 0101, and 0011 in RAM and then add them to the register one at a time. Start with a cleared register and flip-flop. Predict the values in the output of the register and carry after each addition in the following sum, and verify your results:

$$0110 + 1110 + 1101 + 0101 + 0011$$

Circuit Operation

Clear the register and the carry flip-flop to zero, and store the following four-bit numbers in RAM in the indicated addresses:

Address	Content
0	0110
3	1110
6	1101
9	0101
12	0011

Now perform the following four operations:

1. Add the contents of address 0 to the contents of the register, using the LOAD condition.
2. Store the sum from the register into RAM at address 1.
3. Shift right the contents of the register and carry with the SHIFT condition.
4. Store the shifted contents of the register at address 2 of RAM.

Check that the contents of the first three locations in RAM are as follows:

Address	Contents
0	0110
1	0110
2	0011

Repeat the foregoing four operations for each of the other four binary numbers stored in RAM. Use addresses 4, 7, 10, and 13 to store the sum from the register in step 2.

Use addresses 5, 8, 11, and 14 to store the shifted value from the register in step 4. Predict what the contents of RAM at addresses 0 through 14 would be, and check to verify your results.

9.18 EXPERIMENT 17: BINARY MULTIPLIER

In this experiment, you will design and construct a circuit that multiplies 2 four-bit unsigned numbers to produce an eight-bit product. An algorithm for multiplying two binary numbers is presented in Section 8.7. The algorithm implemented in this experiment differs from the one described in Figs. 8.14 and 8.15, by treating only a four-bit datapath and by incrementing, instead of decrementing, a bit counter.

Block Diagram

The ASMD chart and block diagram of the binary multiplier with those ICs recommended to be used are shown in Fig. 9.24(a) and (b). The multiplicand, B, is available from four switches instead of a register. The multiplier, Q, is obtained from another set of four switches. The product is displayed with eight indicator lamps. Counter P is initialized to 0 and then incremented after each partial product is formed. When the counter reaches the count of four, output *Done* becomes 1 and the multiplication operation terminates.

Control of Registers

The ASMD chart for the binary multiplier in Fig. 9.24(a) shows that the three registers and the carry flip-flop of the datapath unit are controlled with signals *Load_regs*, *Incr_P*, *Add_regs*, and *Shift_regs*. The external input signals of the control unit are *clock*, *reset_b* (active-low), and *Start*; another input to the control unit is the internal status signal, *Done*, which is formed by the datapath unit to indicate that the counter has reached a count of four, corresponding to the number of bits in the multiplier. *Load_regs* clears the product register (A) and the carry flip-flop (C), loads the multiplicand into register B, loads the multiplier into register Q, and clears the bit counter. *Incr_P* increments the bit counter concurrently with the accumulation of a partial product. *Add_regs* adds the multiplicand to A, if the least significant bit of the shifted multiplier (Q*[0]*) is 1. Flip-flop C accommodates a carry that results from the addition. The concatenated register *CAQ* is updated by storing the result of shifting its contents one bit to the right. *Shift_regs* shifts *CAQ* one bit to the right, which also clears flip-flop C.

The state diagram for the control unit is shown in Fig. 9.24(c). Note that it does not show the register operations of the datapath unit or the output signals that control them. That information is apparent in Fig. 9.24(d). Note that *Incr_P* and *Shift_regs* are generated unconditionally in states *S_add* and *S_shift*, respectively. *Load_regs* is generated under the condition that *Start* is asserted conditionally while the state is in *S_idle*; *Add_regs* is asserted conditionally in *S_add* if *Q[0]* = 1.

Multiplication Example

Before connecting the circuit, make sure that you understand the operation of the multiplier. To do this, construct a table similar to Table 8.5, but with $B = 1111$ for the multiplicand and $Q = 1011$ for the multiplier. Along with each comment listed on the left side of the table, specify the state.

Datapath Design

Draw a detailed diagram of the datapath part of the multiplier, showing all IC pin connections. Generate the four control signals with switches, and use them to provide the required control operations for the various registers. Connect the circuit and check that each component is functioning properly. With the control signals at 0, set the multiplicand switches to 1111 and the multiplier switches to 1011. Assert the control signals manually by means of the control switches, as specified by the state diagram of Fig. 9.24(c). Apply a single pulse while in each control state, and observe the outputs of registers A and Q and the values in C and P. Compare these outputs with the numbers in your numerical example to verify that the circuit is functioning properly. Note that IC type 74161 has master–slave flip-flops. To operate it manually, it is necessary that the single clock pulse be a negative pulse.

Design of Control

Design the control circuit specified by the state diagram. You can use any method of control implementation discussed in Section 8.8.

Choose the method that minimizes the number of ICs. Verify the operation of the control circuit prior to its connection to the datapath unit.

Checking the Multiplier

Connect the outputs of the control circuit to the datapath unit, and verify the total circuit operation by repeating the steps of multiplying 1111 by 1011. The single clock pulses should now sequence the control states as well. (Remove the manual switches.) The start signal (*Start*) can be generated with a switch that is on while the control is in state *S_idle*.

Generate the start signal (*Start*) with a pulser or any other short pulse, and operate the multiplier with continuous clock pulses from a clock generator. Pressing the pulser for *Start* should initiate the multiplication operation, and upon its completion, the product should be displayed in the A and Q registers. Note that the multiplication will be repeated as long as signal *Start* is enabled. Make sure that *Start* goes back to 0. Then set the switches to two other four-bit numbers and press *Start* again. The new product should appear at the outputs. Repeat the multiplication of a few numbers to verify the operation of the circuit.

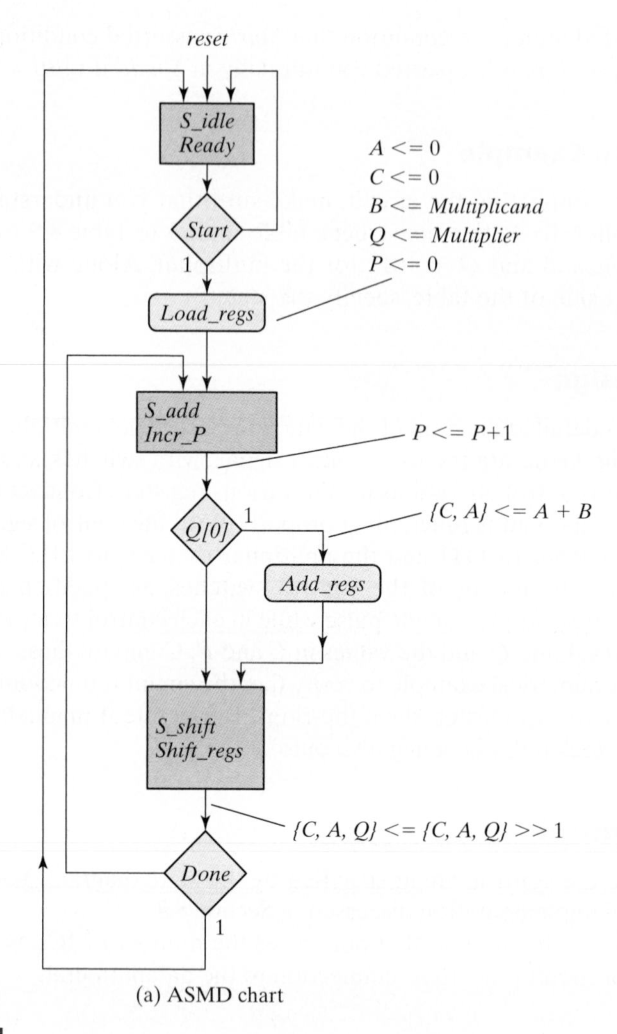

(a) ASMD chart

FIGURE 9.24
ASMD chart, block diagram of the datapath, control state diagram, and register operations of the binary multiplier circuit

9.19 VERILOG HDL SIMULATION EXPERIMENTS AND RAPID PROTOTYPING WITH FPGAS

Field programmable gate arrays (FPGAs) are used by industry to implement logic when the system is complex, the time-to-market is short, the performance (e.g., speed) of an FPGA is acceptable, and the volume of potential sales does not warrant the investment in a standard cell-based ASIC. Circuits can be rapidly prototyped into an FPGA using an

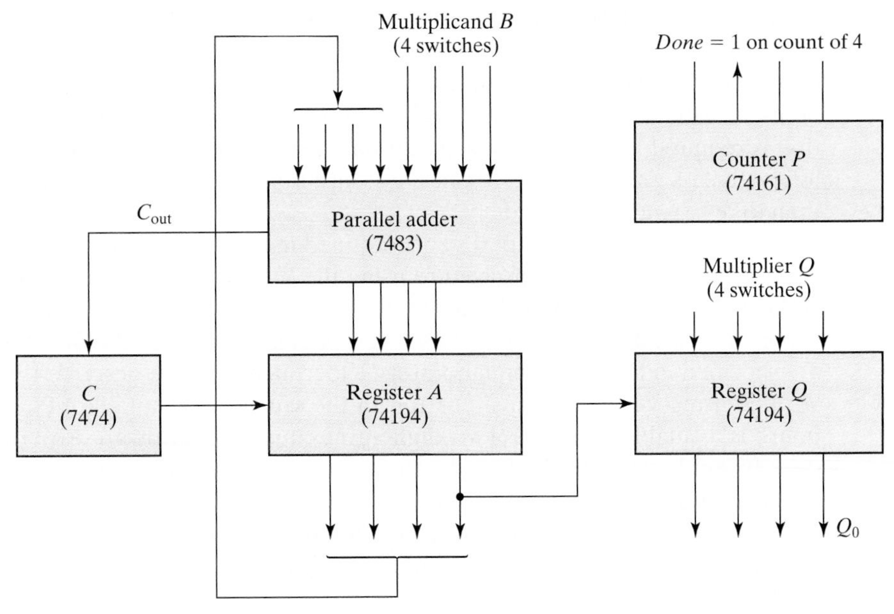

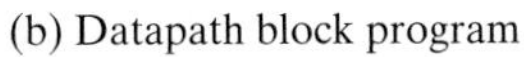
(b) Datapath block program

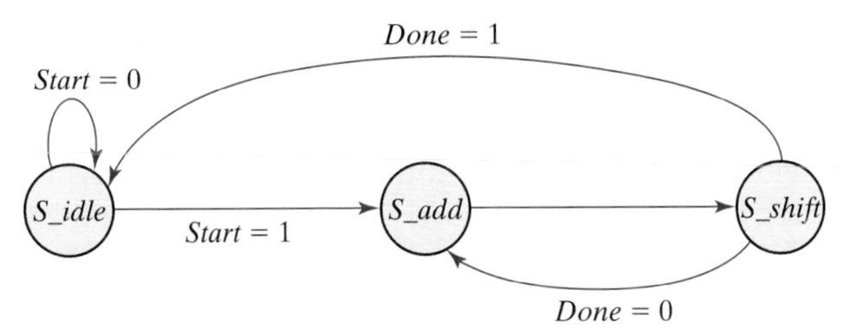

(c) Control state diagram

State Transition		Register Operations	Control signal
From	To		
S_idle		Initial state reached by reset action	
S_idle	*S_add*	$A <= 0, C <= 0, P <= 0$	*Load_regs*
S_add	*S_shift*	$P <= P + 1$ if ($Q[0]$) then ($A <= A + B, C <= C_{out}$)	*Incr_P* *Add_regs*
S_shift		shift right $\{CAQ\}, C <= 0$	*Shift_regs*

(d) Register operations

FIGURE 9.24
(*Continued*)

HDL. Once the HDL model is verified, the description is synthesized and mapped into the FPGA. FPGA vendors provide software tools for synthesizing the HDL description of a circuit into an optimized gate-level description and mapping (fitting) the resulting netlist into the resources of their FPGA. This process avoids the detailed assembly of ICs that is required by composing a circuit on a breadboard, and the process involves significantly less risk of failure, because it is easier and faster to edit an HDL description than to re-wire a breadboard.

Most of the hardware experiments outlined in this chapter can be supplemented by a corresponding software procedure using the Verilog hardware description language (HDL). A Verilog compiler and simulator are necessary for these supplements. The supplemental experiments have two levels of engagement. In the first, the circuits that are specified in the hands-on laboratory experiments can be described, simulated, and verified using Verilog and a simulator. In the second, if a suitable FPGA prototyping board is available (e.g., see www.digilentinc.com), the hardware experiments can be done by synthesizing the Verilog descriptions and implementing the circuits in an FPGA. Where appropriate, the identity of the individual (structural) hardware units (e.g., a 4-bit counter) can be preserved by encapsulating them in separate Verilog modules whose internal detail is described behaviorally or by a mixture of behavioral and structural models.

Prototyping a circuit with an FPGA requires synthesizing a Verilog description to produce a bit stream that can be downloaded to configure the internal resources (e.g., CLBS of a Xilinx FPGA) and connectivity of the FPGA. Three details require attention: (1) The pins of the prototyping board are connected to the pins of the FPGA, and the hardware implementation of the synthesized circuit requires that its input and output signals be associated with the pins of the prototyping board (this association is made using the synthesis tool provided by the vendor of the FPGA (such tools are available free)), (2) FPGA prototyping boards have a clock generator, but it will be necessary, in some cases, to implement a clock divider (in Verilog) to obtain an internal clock whose frequency is suitable for the experiment, and (3) inputs to an FPGA-based circuit can be made using switches and pushbuttons located on the prototyping board, but it might be necessary to implement a pulser circuit in software to control and observe the activity of a counter or a state machine (see the supplement to Experiment 1).

Supplement to Experiment 1 (Section 9.2)

The functionality of the counters specified in Experiment 1 can be described in Verilog and synthesized for implementation in an FPGA. Note that the circuit shown in Fig. 9.3 uses a push-button pulser or a clock to cause the count to increment in a circuit built with standard ICs. A software pulser circuit can be developed to work with a switch on the prototyping board of an FPGA so that the operation of the counters can be verified by visual inspection.

The software pulser has the ASM chart shown in Fig. 9.25, where the external input (*Pushed*) is obtained from a mechanical switch or pushbutton. This circuit asserts *Start* for one cycle of the clock and then waits for the switch to be opened (or the pushbutton

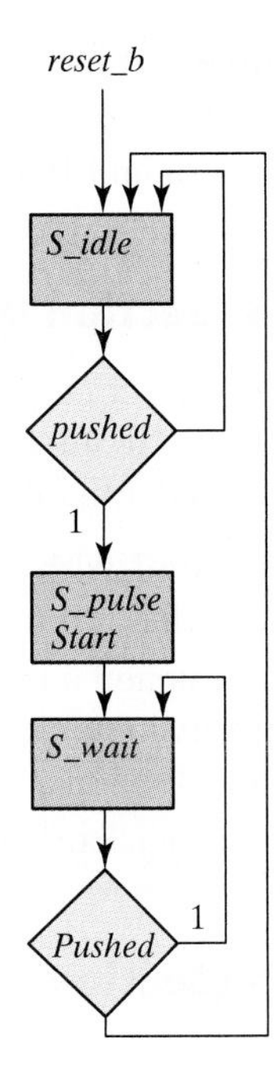

FIGURE 9.25
Pulser circuit for FPGA implementation of Experiment 1

to be released) to ensure that each action of the switch or pushbutton will produce only one pulse of *Start*. If the counter, or a state machine, is in the reset state (*S_idle*) when the switch is closed, the pulse will launch the activity of the counter or state machine. It will be necessary to open the switch (or release the pushbutton) before *Start* can be reasserted. Using the software pulser will allow each value of the count to be observed. If necessary, a simple synchronizer circuit can be used with *Pushed*.

Supplement to Experiment 2 (Section 9.3)

The various logic gates and their propagation delays were introduced in the hardware experiment. In Section 3.10, a simple circuit with gate delays was investigated. As an introduction to the laboratory Verilog program, compile the circuit described in HDL Example 3.3 and then run the simulator to verify the waveforms shown in Fig. 3.38.

Assign the following delays to the exclusive-OR circuit shown in Fig. 3.32(a): 10 ns for an inverter, 20 ns for an AND gate, and 30 ns for an OR gate. The input of the circuit goes from $xy = 00$ to $xy = 01$.

(a) Determine the signals at the output of each gate from $t = 0$ to $t = 50$ ns.

(b) Write the HDL description of the circuit including the delays.

(c) Write a stimulus module (similar to HDL Example 3.3) and simulate the circuit to verify the answer in part (a).

(d) Implement the circuit with an FPGA and test its operation.

Supplement to Experiment 4 (Section 9.5)

The operation of a combinational circuit is verified by checking the output and comparing it with the truth table for the circuit. HDL Example 4.10 (Section 4.12) demonstrates the procedure for obtaining the truth table of a combinational circuit by simulating it.

(a) In order to get acquainted with this procedure, compile and simulate HDL Example 4.10 and check the output truth table.

(b) In Experiment 4, you designed a majority logic circuit. Write the HDL gate-level description of the majority logic circuit together with a stimulus for displaying the truth table. Compile and simulate the circuit and check the output response.

(c) Implement the majority logic circuit units in an FPGA and test its operation.

Supplement to Experiment 5 (Section 9.6)

This experiment deals with code conversion. A BCD-to-excess-3 converter was designed in Section 4.4. Use the result of the design to check it with an HDL simulator.

(a) Write an HDL gate-level description of the circuit shown in Fig. 4.4.

(b) Write a dataflow description using the Boolean expressions listed in Fig. 4.3.

(c) Write an HDL behavioral description of a BCD-to-excess-3 converter.

(d) Write a test bench to simulate and test the BCD-to-excess-3 converter circuit in order to verify the truth table. Check all three circuits.

(e) Implement the behavioral description with an FPGA and test the operation of the circuit.

Supplement to Experiment 7 (Section 9.8)

A four-bit adder–subtractor is developed in this experiment. An adder–subtractor circuit is also developed in Section 4.5.

(a) Write the HDL behavioral description of the 7483 four-bit adder.

(b) Write a behavioral description of the adder–subtractor circuit shown in Fig. 9.11.

(c) Write the HDL hierarchical description of the four-bit adder–subtractor shown in Fig. 4.13 (including V). This can be done by instantiating a modified version of the four-bit adder described in HDL Example 4.2 (Section 4.12).

(d) Write an HDL test bench to simulate and test the circuits of part (c). Check and verify the values that cause an overflow with $V = 1$.

(e) Implement the circuit of part (c) with an FPGA and test its operation.

Supplement to Experiment 8 (Section 9.9)

The edge-triggered *D* flip-flop 7474 is shown in Fig. 9.13. The flip-flop has asynchronous preset and clear inputs.

(a) Write an HDL behavioral description of the 7474 *D* flip-flop, using only the *Q* output. (Note that when *Preset* = 0, *Q* goes to 1, and when *Preset* = 1 and *Clear* = 0, *Q* goes to 0. Thus, *Preset* takes precedence over *Clear*.)

(b) Write an HDL behavioral description of the 7474 *D* flip-flop, using both outputs. Label the second output *Q_not,* and note that this is not always the complement of Q. (When *Preset* = *Clear* = 0, both *Q* and *Q_not* go to 1.)

Supplement to Experiment 9 (Section 9.10)

In this hardware experiment, you are asked to design and test a sequential circuit whose state diagram is given by Fig. 9.14. This is a Mealy model sequential circuit similar to the one described in HDL Example 5.5 (Section 5.6).

(a) Write the HDL description of the state diagram of Fig. 9.14.

(b) Write the HDL structural description of the sequential circuit obtained from the design. (This is similar to HDL Example 5.7 in Section 5.6.)

(c) Figure 9.24(c) (Section 9.18) shows a control state diagram. Write the HDL description of the state diagram, using the one-hot binary assignment (see Table 5.9 in Section 5.7) and four outputs—T_0, T_1, T_2, and T_3.

(d) Write a behavioral model of the datapath unit, and verify that the interconnected control unit and datapath unit operate correctly.

(e) Implement the integrated circuit with an FPGA and test its operation.

Supplement to Experiment 10 (Section 9.11)

The synchronous counter with parallel load IC type 74161 is shown in Fig. 9.15. This circuit is similar to the one described in HDL Example 6.3 (Section 6.6), with two exceptions: The load input is enabled when equal to 0, and there are two inputs (*P* and *T*) that control the count. Write the HDL description of the 74161 IC. Implement the counter with an FPGA and test its operation.

Supplement to Experiment 11 (Section 9.12)

A bidirectional shift register with parallel load is designed in this experiment by using the 74195 and 74157 IC types.

(a) Write the HDL description of the 74195 shift register. Assume that inputs *J* and $\overline{K}$ are connected together to form the serial input.

(b) Write the HDL description of the 74157 multiplexer.

(c) Obtain the HDL description of the four-bit bidirectional shift register that has been designed in this experiment. (1) Write the structural description by instantiating the two ICs and specifying their interconnection, and (2) write the behavioral description of the circuit, using the function table that is derived in this design experiment.

(d) Implement the circuit with an FPGA and test its operation.

Supplement to Experiment 13 (Section 9.14)

This experiment investigates the operation of a random-access memory (RAM). The way a memory is described in HDL is explained in Section 7.2 in conjunction with HDL Example 7.1.

(a) Write the HDL description of IC type 74189 RAM, shown in Fig. 9.18.

(b) Test the operation of the memory by writing a stimulus program that stores binary 3 in address 0 and binary 1 in address 14. Then read the stored numbers from the two addresses to check whether the numbers were stored correctly.

(c) Implement the RAM with an FPGA and test its operation.

Supplement to Experiment 14 (Section 9.15)

(a) Write the HDL behavioral description of the 74194 bidirectional shift register with parallel load shown in Fig. 9.19.

(b) Implement the shift register with an FPGA and test its operation.

Supplement to Experiment 16 (Section 9.17)

A parallel adder with an accumulator register and a memory unit is shown in the block diagram of Fig. 9.23. Write the structural description of the circuit specified by the block diagram. The HDL structural description of this circuit can be obtained by instantiating the various components. An example of a structural description of a design can be found in HDL Example 8.4 in Section 8.6. First, it is necessary to write the behavioral description of each component. Use counter 74161 instead of 7493, and substitute the *D* flip-flop 7474 instead of the *JK* flip-flop 7476. The block diagram of the various components can be found from the list in Table 9.1. Write a test bench for each model, and then write a test bench to verify the entire design. Implement the circuit with an FPGA and test its operation.

Supplement to Experiment 17 (Section 9.18)

The block diagram of a four-bit binary multiplier is shown in Fig. 9.24. The multiplier can be described in one of two ways: (1) by using the register transfer level statements listed in part (b) of the figure or (2) by using the block diagram shown in part (a) of the figure. The description of the multiplier in terms of the register transfer level (RTL) format is carried out in HDL Example 8.5 (Section 8.7).

(a) Use the integrated circuit components specified in the block diagram to write the HDL structural description of the binary multiplier. The structural description is obtained by using the module description of each component and then instantiating all the components to show how they are interconnected. (See Section 8.5 for an example.) The HDL descriptions of the components may be available from the solutions to previous experiments. The 7483 is described with a solution to Experiment 7(a), the 7474 with Experiment 8(a), the 74161 with Experiment 10, and the 74194 with Experiment 14. The description of the control is available from a solution to Experiment 9(c). Be sure to verify each structural unit before attempting to verify the multiplier.

(b) Implement the binary multiplier with an FPGA. Use the pulser described in the supplement to Experiment 1.

Chapter 10

Standard Graphic Symbols

10.1 RECTANGULAR-SHAPE SYMBOLS

Digital components such as gates, decoders, multiplexers, and registers are available commercially in integrated circuits and are classified as SSI or MSI circuits. Standard graphic symbols have been developed for these and other components so that the user can recognize each function from the unique graphic symbol assigned to it. This standard, known as ANSI/IEEE Std. 91-1984, has been approved by industry, government, and professional organizations and is consistent with international standards.

The standard uses a rectangular-shape outline to represent each particular logic function. Within the outline, there is a general qualifying symbol denoting the logical operation performed by the unit. For example, the general qualifying symbol for a multiplexer is MUX. The size of the outline is arbitrary and can be either a square or a rectangular shape with an arbitrary length–width ratio. Input lines are placed on the left and output lines are placed on the right. If the direction of signal flow is reversed, it must be indicated by arrows.

The rectangular-shape graphic symbols for SSI gates are shown in Fig. 10.1. The qualifying symbol for the AND gate is the ampersand (&). The OR gate has the qualifying symbol that designates greater than or equal to 1, indicating that at least one input must be active for the output to be active. The symbol for the buffer gate is 1, showing that only one input is present. The exclusive-OR symbol designates the fact that only one input must be active for the output to be active. The inclusion of the logic negation small circle in the output converts the gates to their complement values. Although the rectangular-shape symbols for the gates are recommended, the standard also recognizes the distinctive-shape symbols for the gates shown in Fig. 2.5.

An example of an MSI standard graphic symbol is the four-bit parallel adder shown in Fig. 10.2. The qualifying symbol for an adder is the Greek letter Σ. The preferred

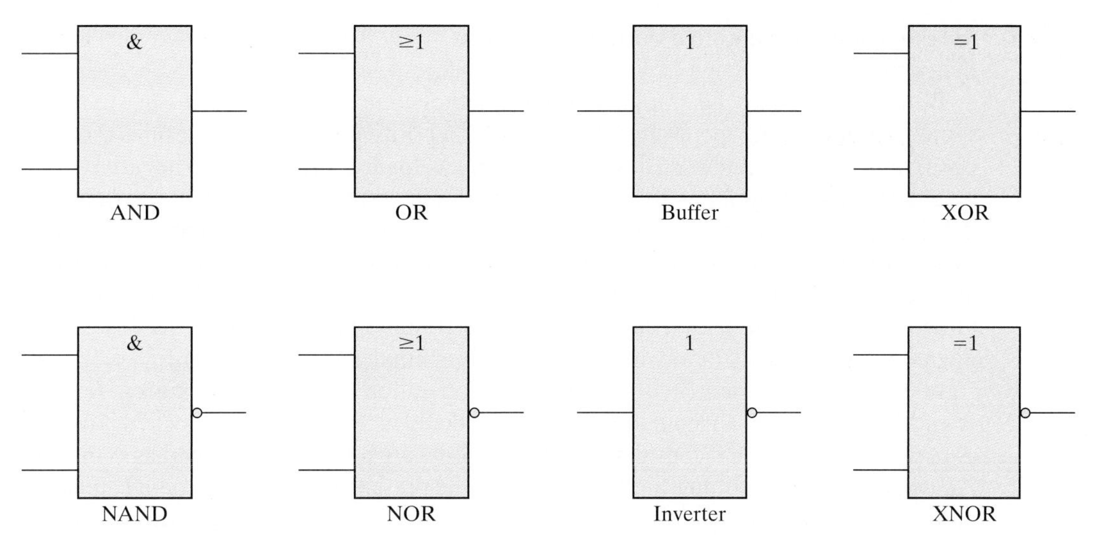

FIGURE 10.1
Rectangular-shape graphic symbols for gates

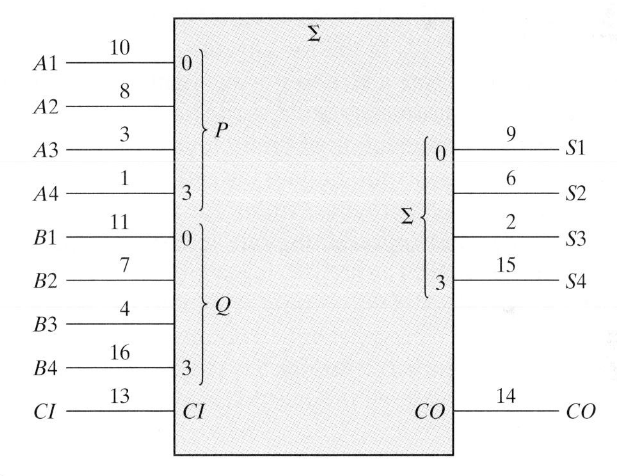

FIGURE 10.2
Standard graphic symbol for a four-bit parallel adder, IC type 7483

letters for the arithmetic operands are *P* and *Q*. The bit-grouping symbols in the two types of inputs and the sum output are the decimal equivalents of the weights of the bits to the power of 2. Thus, the input labeled 3 corresponds to the value of $2^3 = 8$. The input carry is designated by *CI* and the output carry by *CO*. When the digital component represented by the outline is also a commercial integrated circuit, it is customary to write the IC pin number along each input and output. Thus, IC type 7483 is a four-bit adder with look-ahead carry. It is enclosed in a package with 16 pins. The pin numbers for the nine inputs and five outputs are shown in Fig. 10.2. The other two pins are for the power supply.

Before introducing the graphic symbols of other components, it is necessary to review some terminology. As mentioned in Section 2.8, a positive-logic system defines the more positive of two signal levels (designated by H) as logic 1 and the more negative signal level (designated by L) as logic 0. Negative logic assumes the opposite assignment. A third alternative is to employ a mixed-logic convention, where the signals are considered entirely in terms of their H and L values. At any point in the circuit, the user is allowed to define the logic polarity by assigning logic 1 to either the H or L signal. The mixed-logic notation uses a small right-angle-triangle graphic symbol to designate a negative-logic polarity at any input or output terminal. (See Fig. 2.10(f).)

Integrated-circuit manufacturers specify the operation of integrated circuits in terms of H and L signals. When an input or output is considered in terms of positive logic, it is defined as *active high*. When it is considered in terms of negative logic, it is defined as *active low*. Active-low inputs or outputs are recognized by the presence of the small-triangle polarity-indicator symbol. When positive logic is used exclusively throughout the entire system, the small-triangle polarity symbol is equivalent to the small circle that designates negation. In this book, we have assumed positive logic throughout and employed the small circle when drawing logic diagrams. When an input or output line does not include the small circle, we define it to be active if it is logic 1. An input or output that includes the small-circle symbol is considered active if it is in the logic-0 state. However, we will use the small-triangle polarity symbol to indicate active-low assignment in all drawings that represent standard diagrams. This will conform with integrated-circuit data books, where the polarity symbol is usually employed. Note that the bottom four gates in Fig. 10.1 could have been drawn with a small triangle in the output lines instead of a small circle.

Another example of a graphic symbol for an MSI circuit is shown in Fig. 10.3. This is a 2-to-4-line decoder representing one-half of IC type 74155. Inputs are on the left and outputs on the right. The identifying symbol X/Y indicates that the circuit converts from code X to code Y. Data inputs A and B are assigned binary weights 1 and 2 equivalent to 2^0 and 2^1, respectively. The outputs are assigned numbers from 0 to 3, corresponding to outputs D_0 through D_3, respectively. The decoder has one active-low input E_1 and one active-high input E_2. These two inputs go through an internal AND

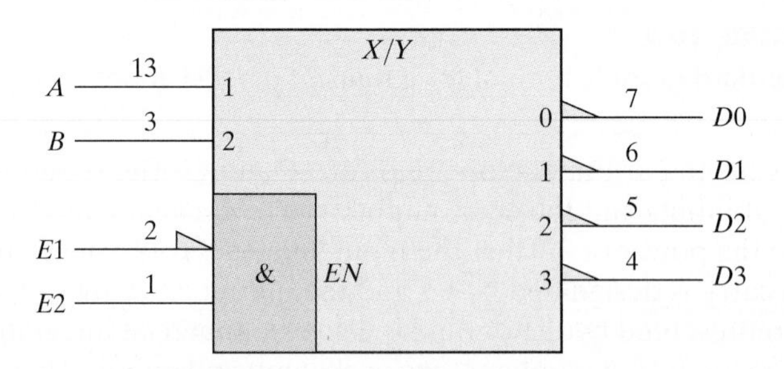

FIGURE 10.3
Standard graphic symbol for a 2-to-4-line decoder (one-half of IC type 74155)

gate to enable the decoder. The output of the AND gate is labeled *EN* (enable) and is activated when E_1 is at a low-level state and E_2 at a high-level state.

10.2 QUALIFYING SYMBOLS

The IEEE standard graphic symbols for logic functions provide a list of qualifying symbols to be used in conjunction with the outline. A qualifying symbol is added to the basic outline to designate the overall logic characteristics of the element or the physical characteristics of an input or output. Table 10.1 lists some of the general qualifying symbols specified in the standard. A general qualifying symbol defines the basic function performed by the device represented in the diagram. It is placed near the top center position of the rectangular-shape outline. The general qualifying symbols for the gates, decoder, and adder were shown in previous diagrams. The other symbols are self-explanatory and will be used later in diagrams representing the corresponding digital elements.

Some of the qualifying symbols associated with inputs and outputs are shown in Fig. 10.4. Symbols associated with inputs are placed on the left side of the column labeled *symbol*. Symbols associated with outputs are placed on the right side of the column. The active-low input or output symbol is the polarity indicator. As mentioned

Table 10.1
General Qualifying Symbols

Symbol	Description
&	AND gate or function
≥ 1	OR gate or function
1	Buffer gate or inverter
$= 1$	Exclusive-OR gate or function
2k	Even function or even parity element
2k + 1	Odd function or odd parity element
X/Y	Coder, decoder, or code converter
MUX	Multiplexer
DMUX	Demultiplexer
Σ	Adder
Π	Multiplier
COMP	Magnitude comparator
ALU	Arithmetic logic unit
SRG	Shift register
CTR	Counter
RCTR	Ripple counter
ROM	Read-only memory
RAM	Random-access memory

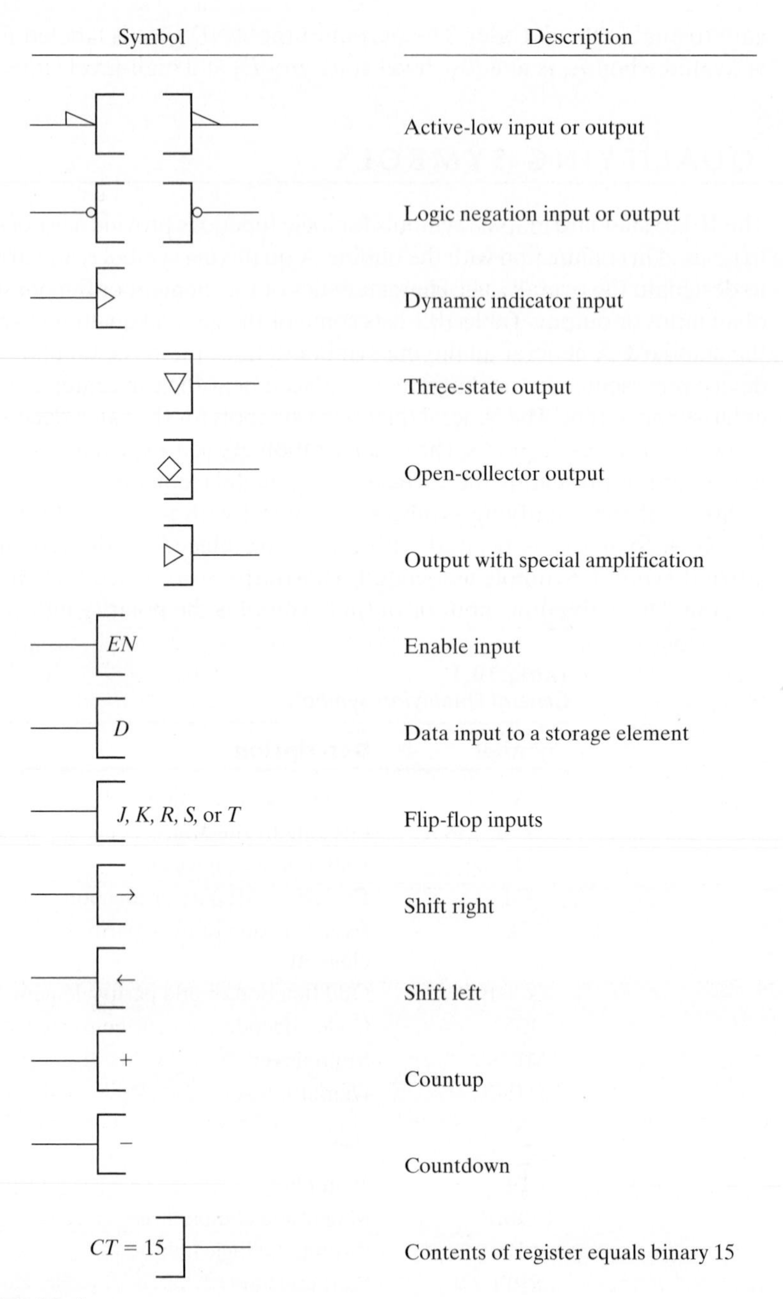

FIGURE 10.4
Qualifying symbols associated with inputs and outputs

previously, it is equivalent to the logic negation when positive logic is assumed. The dynamic input is associated with the clock input in flip-flop circuits. It indicates that the input is active on a transition from a low-to-high-level signal. The three-state output has a third high-impedance state, which has no logic significance. When the circuit is enabled, the output is in the normal 0 or 1 logic state, but when the circuit is disabled, the three-state output is in a high-impedance state. This state is equivalent to an open circuit.

The open-collector output has one state that exhibits a high-impedance condition. An externally connected resistor is sometimes required in order to produce the proper logic level. The diamond-shape symbol may have a bar on top (for high type) or on the bottom (for low type). The high or low type specifies the logic level when the output is not in the high-impedance state. For example, TTL-type integrated circuits have special outputs called open-collector outputs. These outputs are recognized by a diamond-shape symbol with a bar under it. This indicates that the output can be either in a high-impedance state or in a low-level state. When used as part of a distribution function, two or more open-collector NAND gates when connected to a common resistor perform a positive-logic AND function or a negative-logic OR function.

The output with special amplification is used in gates that provide special driving capabilities. Such gates are employed in components such as clock drivers or bus-oriented transmitters. The *EN* symbol designates an enable input. It has the effect of enabling all outputs when it is active. When the input marked with *EN* is inactive, all outputs are disabled. The symbols for flip-flop inputs have the usual meaning. The *D* input is also associated with other storage elements such as memory input.

The symbols for shift right and shift left are arrows pointing to the right or the left, respectively. The symbols for count-up and count-down counters are the plus and minus symbols, respectively. An output designated by $CT = 15$ will be active when the contents of the register reach the binary count of 15. When nonstandard information is shown inside the outline, it is enclosed in square brackets [like this].

10.3 DEPENDENCY NOTATION

The most important aspect of the standard logic symbols is the dependency notation. Dependency notation is used to provide the means of denoting the relationship between different inputs or outputs without actually showing all the elements and interconnections between them. We will first demonstrate the dependency notation with an example of the AND dependency and then define all the other symbols associated with this notation.

The AND dependency is represented with the letter *G* followed by a number. Any input or output in a diagram that is labeled with the number associated with *G* is considered to be ANDed with it. For example, if one input in the diagram has the label *G*1 and another input is labeled with the number 1, then the two inputs labeled *G*1 and 1 are considered to be ANDed together internally.

An example of AND dependency is shown in Fig. 10.5. In (a), we have a portion of a graphic symbol with two AND dependency labels, *G*1 and *G*2. There are two inputs labeled with the number 1 and one input labeled with the number 2. The equivalent

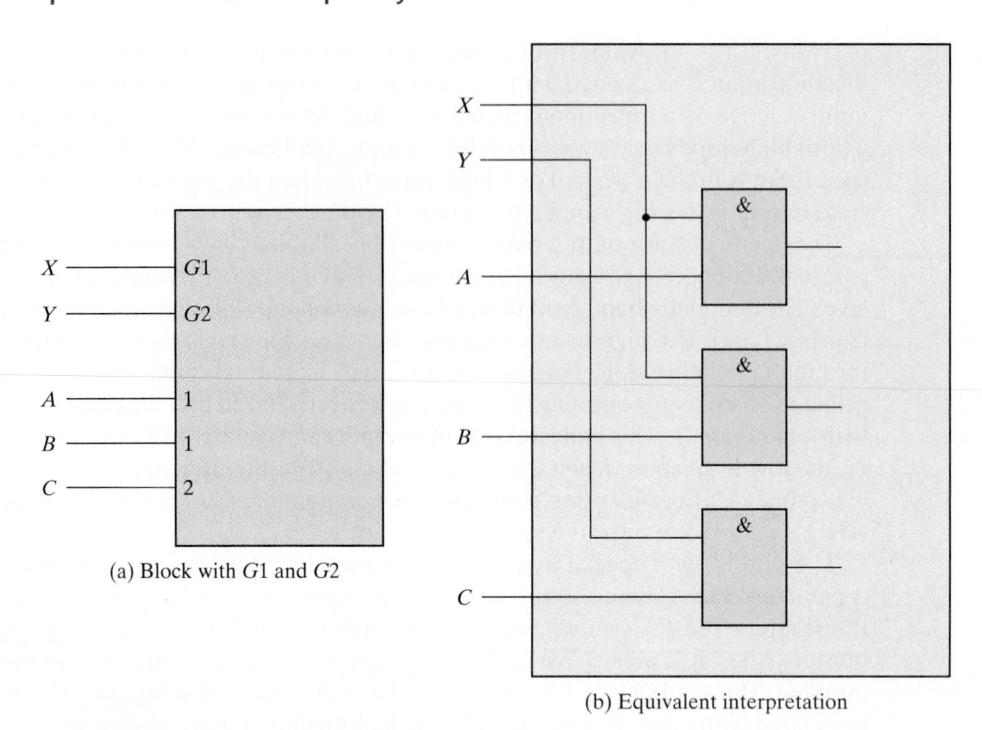

(a) Block with $G1$ and $G2$

(b) Equivalent interpretation

FIGURE 10.5
Example of G (AND) dependency

interpretation is shown in part (b) of the figure. Input X associated with $G1$ is considered to be ANDed with inputs A and B, which are labeled with a 1. Similarly, input Y is ANDed with input C to conform with the dependency between $G2$ and 2.

The standard defines 10 other dependencies. Each dependency is denoted by a letter symbol (except EN). The letter appears at the input or output and is followed by a number. Each input or output affected by that dependency is labeled with that same number. The 11 dependencies and their corresponding letter designation are as follows:

- *G* Denotes an AND (gate) relationship
- *V* Denotes an OR relationship
- *N* Denotes a negate (exclusive-OR) relationship
- *EN* Specifies an enable action
- *C* Identifies a control dependency
- *S* Specifies a setting action
- *R* Specifies a resetting action
- *M* Identifies a mode dependency
- *A* Identifies an address dependency

Z Indicates an internal interconnection

X Indicates a controlled transmission

The *V* and *N* dependencies are used to denote the Boolean relationships of OR and exclusive-OR similar to the *G* that denotes the Boolean AND. The *EN* dependency is similar to the qualifying symbol *EN* except that a number follows it (for example, *EN* 2). Only the outputs marked with that number are disabled when the input associated with *EN* is active.

The control dependency *C* is used to identify a clock input in a sequential element and to indicate which input is controlled by it. The set *S* and reset *R* dependencies are used to specify internal logic states of an *SR* flip-flop. The *C*, *S*, and *R* dependencies are explained in Section 10.5 in conjunction with the flip-flop circuit. The mode *M* dependency is used to identify inputs that select the mode of operation of the unit. The mode dependency is presented in Section 10.6 in conjunction with registers and counters. The address *A* dependency is used to identify the address input of a memory. It is introduced in Section 10.8 in conjunction with the memory unit.

The *Z* dependency is used to indicate interconnections inside the unit. It signifies the existence of internal logic connections between inputs, outputs, internal inputs, and internal outputs, in any combination. The *X* dependency is used to indicate the controlled transmission path in a CMOS transmission gate.

10.4 SYMBOLS FOR COMBINATIONAL ELEMENTS

The examples in this section and the rest of this chapter illustrate the use of the standard in representing various digital components with graphic symbols. The examples demonstrate actual commercial integrated circuits with the pin numbers included in the inputs and outputs. Most of the ICs presented in this chapter are included with the suggested experiments outlined in Chapter 9.

The graphic symbols for the adder and decoder were shown in Section 10.2. IC type 74155 can be connected as a 3×8 decoder, as shown in Fig. 10.6. (The truth table of this decoder is shown in Fig. 9.7.) There are two *C* and two *G* inputs in the IC. Each pair must be connected together as shown in the diagram. The enable input is active when in the low-level state. The outputs are all active low. The inputs are assigned binary weights 1, 2, and 4, equivalent to $2^0, 2^1$, and 2^2, respectively. The outputs are assigned numbers from 0 to 7. The sum of the weights of the inputs determines the output that is active. Thus, if the two input lines with weights 1 and 4 are activated, the total weight is $1 + 4 = 5$ and output 5 is activated. Of course, the *EN* input must be activated for any output to be active.

The decoder is a special case of a more general component referred to as a *coder*. A coder is a device that receives an input binary code on a number of inputs and produces a different binary code on a number of outputs. Instead of using the qualifying symbol *X/Y*, the coder can be specified by the code name. For example, the 3-to-8-line decoder of Fig. 10.6 can be symbolized with the name *BIN/OCT* since the circuit converts a 3-bit binary number into 8 octal values, 0 through 7.

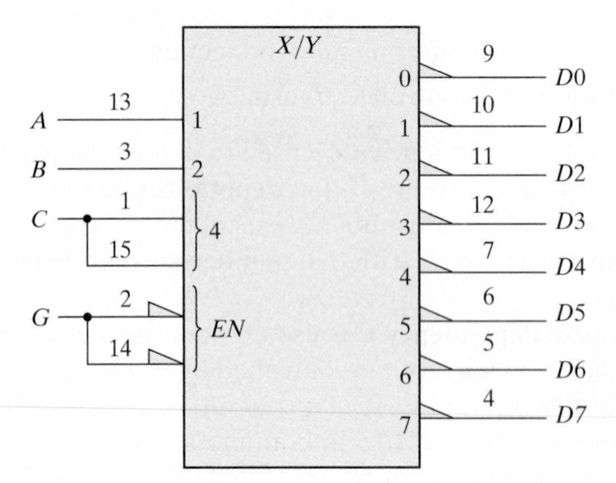

FIGURE 10.6
IC type 74155 connected as a 3 × 8 decoder

Before showing the graphic symbol for the multiplexer, it is necessary to show a variation of the AND dependency. The AND dependency is sometimes represented by a shorthand notation like $G\,\frac{0}{7}$. This symbol stands for eight AND dependency symbols from 0 to 7 as follows:

$$G0, G1, G2, G3, G4, G5, G6, G7$$

At any given time, only one out of the eight AND gates can be active. The active AND gate is determined from the inputs associated with the *G* symbol. These inputs are marked with weights equal to the powers of 2. For the eight AND gates just listed, the weights are 0, 1, and 2, corresponding to the numbers 2^0, 2^1, and 2^2, respectively. The AND gate that is active at any given time is determined from the sum of the weights of the active inputs. Thus, if inputs 0 and 2 are active, then the AND gate that is active has the number $2^0 + 2^2 = 5$. This makes *G*5 active and the other seven AND gates inactive.

The standard graphic symbol for a 8 × 1 multiplexer is shown in Fig. 10.7(a). The qualifying symbol MUX identifies the device as a multiplexer. The symbols inside the block are part of the standard notation, but the symbols marked outside are user-defined symbols. The function table of the 741551 IC can be found in Fig. 9.9. The AND dependency is marked with $G\,\frac{0}{7}$ and is associated with the inputs enclosed in brackets. These inputs have weights of 0, 1, and 2. They are actually what we have called the selection inputs. The eight data inputs are marked with numbers from 0 to 7. The net weight of the active inputs associated with the *G* symbol specifies the number in the data input that is active. For example, if selection inputs $CBA = 110$, then inputs 1 and 2 associated with *G* are active. This gives a numerical value for the AND dependency of $2^2 + 2^1 = 6$, which makes *G* 6 active. Since *G* 6 is ANDed with data input number 6, it makes this input active. Thus, the output will be equal to data input D_6 provided that the enable input is active.

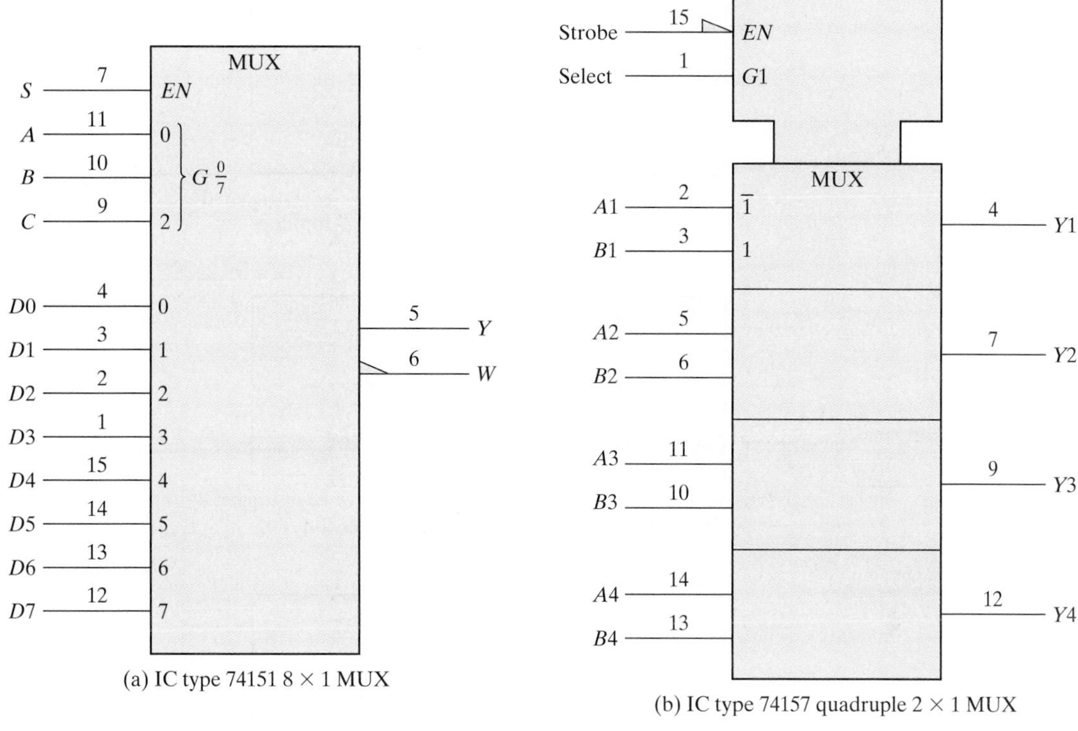

(a) IC type 74151 8 × 1 MUX

(b) IC type 74157 quadruple 2 × 1 MUX

FIGURE 10.7
Graphic symbols for multiplexers

Figure10.7(b) represents the quadruple 2 × 1 multiplexer IC type 74157 whose function table is listed in Fig. 9.17 The enable and selection inputs are common to all four multiplexers. This is indicated in the standard notation by the indented box at the top of the diagram, which represents a *common control block*. The inputs to a common control block control all lower sections of the diagram. The common enable input *EN* is active when in the low-level state. The AND dependency, *G*1, determines which input is active in each multiplexer section. When $G1 = 0$, the *A* inputs marked with $\overline{1}$ are active. When $G1 = 1$, the *B* inputs marked with 1 are active. The active inputs are applied to the corresponding outputs if *EN* is active. Note that the input symbols $\overline{1}$ and 1 are marked in the upper section only instead of repeating them in each section.

10.5 SYMBOLS FOR FLIP-FLOPS

The standard graphic symbols for different types of flip-flops are shown in Fig. 10.8. A flip-flop is represented by a rectangular-shaped block with inputs on the left and outputs on the right. One output designates the normal state of the flip-flop and the

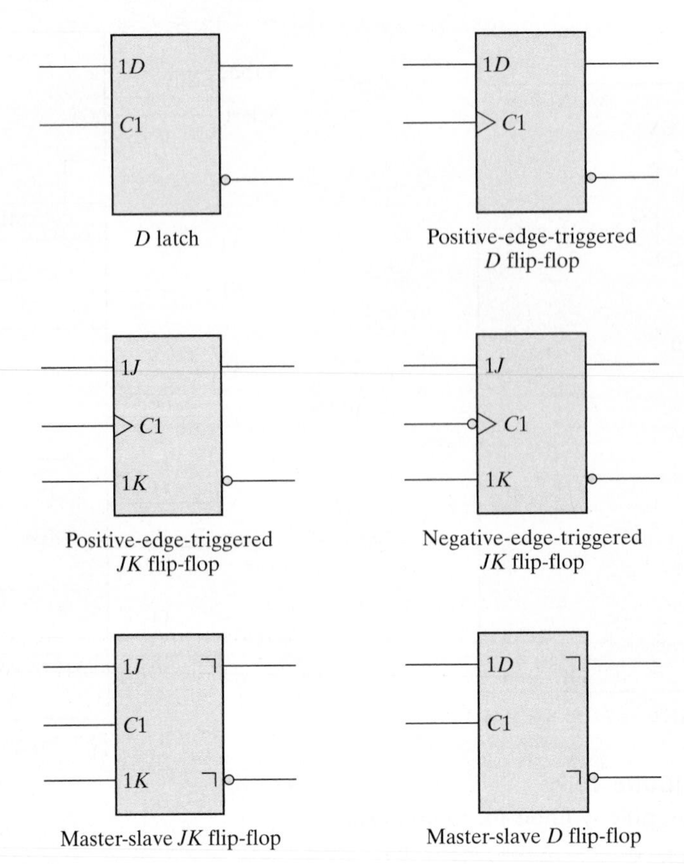

FIGURE 10.8
Standard graphic symbols for flip-flops

other output with a small-circle negation symbol (or polarity indicator) designates the complement output. The graphic symbols distinguish between three types of flip-flops: the *D* latch, whose internal construction is shown in Fig. 6.5; the master–slave flip-flop, shown in Fig. 6.9; and the edge-triggered flip-flop, introduced in Fig. 6.12. The graphic symbol for the *D* latch or *D* flip-flop has inputs *D* and *C* indicated inside the block. The graphic symbol for the *JK* flip-flop has inputs *J, K,* and *C* inside. The notation *C*1, 1*D*, 1*J*, and 1*K* are examples of control dependency. The input in *C*1 controls input 1*D* in a *D* flip-flop and inputs 1*J* and 1*K* in a *JK* flip-flop.

The *D* latch has no other symbols besides the 1*D* and *C*1 inputs. The edge-triggered flip-flop has an arrowhead-shaped symbol in front of the control dependency *C*1 to designate a dynamic input. The dynamic indicator symbol denotes that the flip-flop responds to the positive-edge transition of the input clock pulses. A small circle outside the block along the dynamic indicator designates a negative-edge transition for triggering the flip-flop. The master–slave is considered to be a pulse-triggered flip-flop and is

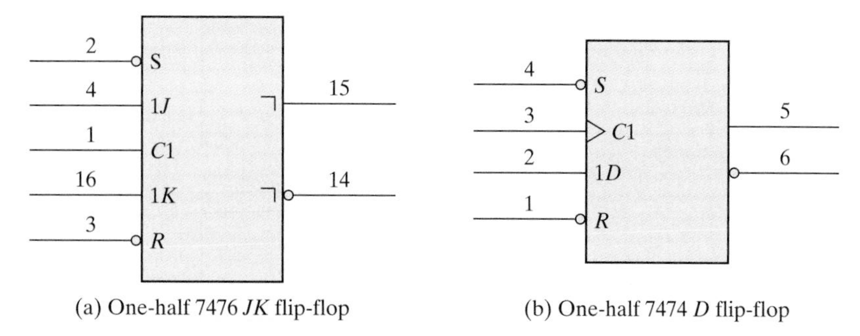

(a) One-half 7476 *JK* flip-flop

(b) One-half 7474 *D* flip-flop

FIGURE 10.9
IC flip-flops with direct set and reset

indicated as such with an upside-down *L* symbol in front of the outputs. This is to show that the output signal changes on the falling edge of the pulse. Note that the master–slave flip-flop is drawn without the dynamic indicator.

Flip-flops available in integrated-circuit packages provide special inputs for setting and resetting the flip-flop asynchronously. These inputs are usually called direct set and direct reset. They affect the output on the negative level of the signal without the need of a clock. The graphic symbol of a master–slave *JK* flip-flop with direct set and reset is shown in Fig. 10.9(a). The notations *C*1, 1*J*, and 1*K* represent control dependency, showing that the clock input at *C*1 controls inputs 1*J* and 1*K*. *S* and *R* have no 1 in front of the letters and, therefore, they are not controlled by the clock at *C*1. The *S* and *R* inputs have a small circle along the input lines to indicate that they are active when in the logic-0 level. The function table for the 7476 flip-flop is shown in Fig. 9.12.

The graphic symbol for a positive-edge-triggered *D* flip-flop with direct set and reset is shown in Fig. 10.9(b). The positive-edge transition of the clock at input *C*1 controls input 1*D*. The *S* and *R* inputs are independent of the clock. This is IC type 7474, whose function table is listed in Fig. 9.13.

10.6 SYMBOLS FOR REGISTERS

The standard graphic symbol for a register is equivalent to the symbol used for a group of flip-flops with a common clock input. Fig. 10.10 shows the standard graphic symbol of IC type 74175, consisting of four *D* flip-flops with common clock and clear inputs. The clock input *C*1 and the clear input *R* appear in the common control block. The inputs to the common control block are connected to each of the elements in the lower sections of the diagram. The notation *C*1 is the control dependency that controls all the 1*D* inputs. Thus, each flip-flop is triggered by the common clock input. The dynamic input symbol associated with *C*1 indicates that the flip-flops are triggered on the positive edge of the input clock. The common *R* input resets all flip-flops when its input is at a low-level state. The 1*D* symbol is placed only once in the upper section instead of

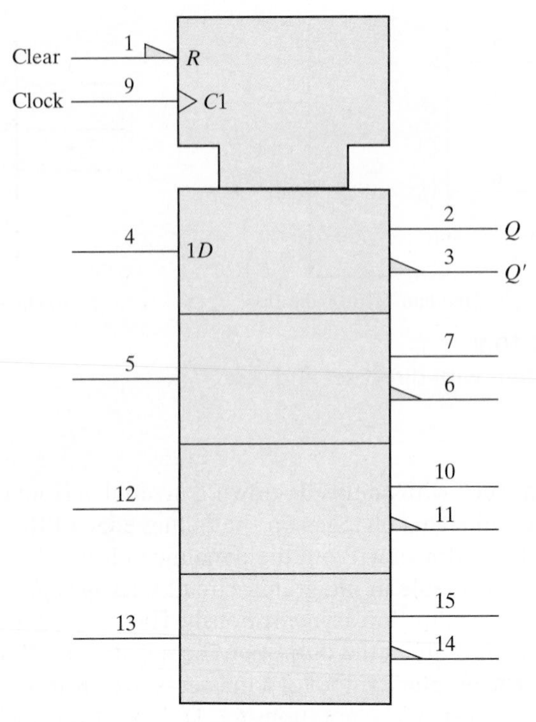

FIGURE 10.10
Graphic symbol for IC type 74175 quad flip-flop

repeating it in each section. The complement outputs of the flip-flops in this diagram are marked with the polarity symbol rather than the negation symbol.

The standard graphic symbol for a shift register with parallel load is shown in Fig. 10.11. This is IC type 74195, whose function table can be found in Fig. 9.16. The qualifying symbol for a shift register is *SRG* followed by a number that designates the number of stages. Thus, *SRG*4 denotes a four-bit shift register. The common control block has two mode dependencies, *M*1 and *M*2, for the shift and load operations, respectively. Note that the IC has a single input labeled *SH*/*LD* (shift/load), which is split into two lines to show the two modes. *M*1 is active when the *SH*/*LD* input is high and *M*2 is active when the *SH*/*LD* input is low. *M*2 is recognized as active low from the polarity indicator along its input line. Note the convention in this symbology: We must recognize that a single input actually exists in pin 9, but it is split into two parts in order to assign to it the two modes, *M*1 and *M*2. The control dependency *C*3 is for the clock input. The dynamic symbol along the *C*3 input indicates that the flip-flops trigger on the positive edge of the clock. The symbol /1 $\rightarrow$ following *C*3 indicates that the register shifts to the right or in the downward direction when mode *M*1 is active.

The four sections below the common control block represent the four flip-flops. Flip-flop *QA* has three inputs: Two are associated with the serial (shift) operation and one

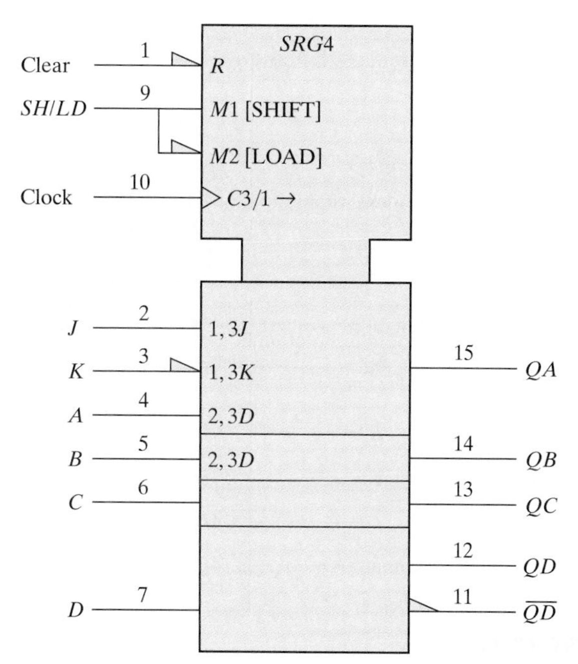

FIGURE 10.11
Graphic symbol for a shift register with parallel load, IC type 74195

with the parallel (load) operation. The serial input label $1, 3J$ indicates that the J input of flip-flop QA is active when $M1$ (shift) is active and $C3$ goes through a positive clock transition. The other serial input with label $1, 3K$ has a polarity symbol in its input line corresponding to the complement of input K in a JK flip-flop. The third input of QA and the inputs of the other flip-flops are for the parallel input data. Each input is denoted by the label $2, 3D$. The 2 is for $M2$ (load), and 3 is for the clock $C3$. If the input in pin number 9 is in the low level, $M1$ is active, and a positive transition of the clock at $C3$ causes a parallel transfer from the four inputs, A through D, into the four flip-flops, QA through QD. Note that the parallel input is labeled only in the first and second sections. It is assumed to be in the other two sections below.

Figure 10.12 shows the graphic symbol for the bidirectional shift register with parallel load, IC type 74194. The function table for this IC is listed in Fig. 9.19. The common control block shows an R input for resetting all flip-flops to 0 asynchronously. The mode select has two inputs and the mode dependency M may take binary values from 0 to 3. This is indicated by the symbol $M\,\frac{0}{3}$, which stands for $M0$, $M1$, $M2$, $M3$, and is similar to the notation for the G dependency in multiplexers. The symbol associated with the clock is

$$C4/1 \rightarrow /2 \leftarrow$$

$C4$ is the control dependency for the clock. The $/1 \rightarrow$ symbol indicates that the register shifts right (down in this case) when the mode is $M1$ ($S_1\ S_0 = 10$). The $/2 \leftarrow$ symbol

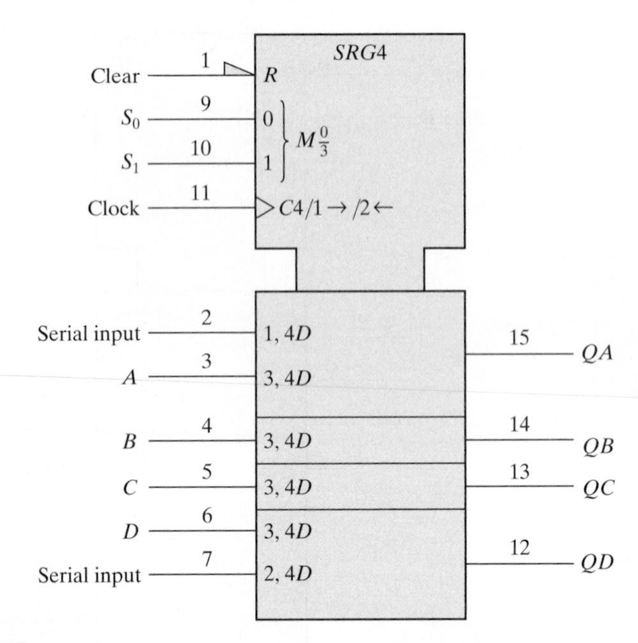

FIGURE 10.12
Graphic symbol for a bidirectional shift register with parallel load, IC type 74194

indicates that the register shifts left (up in this case) when the mode is $M2$ ($S_1\ S_0 = 10$). The right and left directions are obtained when the page is turned 90 degrees counterclockwise.

The sections below the common control block represent the four flip-flops. The first flip-flop has a serial input for shift right, denoted by 1, 4*D* (mode *M*1, clock *C*4, input *D*). The last flip-flop has a serial input for shift left, denoted by 2, 4*D* (mode *M*2, clock *C*4, input *D*). All four flip-flops have a parallel input denoted by the label 3, 4*D* (mode *M*3, clock *C*4, input *D*). Thus, *M*3 ($S_1\ S_0 = 11$) is for parallel load. The remaining mode *M*0 ($S_1\ S_0 = 00$) has no effect on the outputs because it is not included in the input labels.

10.7 SYMBOLS FOR COUNTERS

The standard graphic symbol of a binary ripple counter is shown in Fig. 10.13. The qualifying symbol for a ripple counter is *RCTR*. The designation *DIV*2 stands for the divide-by-2 circuit that is obtained from the single flip-flop *QA*. The *DIV*8 designation is for the divide-by-8 counter obtained from the other three flip-flops. The diagram represents IC type 7493, whose internal circuit diagram is shown in Fig. 9.2. The common control block has an internal AND gate, with inputs *R*1 and *R*2. When both of these inputs are equal to 1, the content of the counter goes to zero. This is indicated by

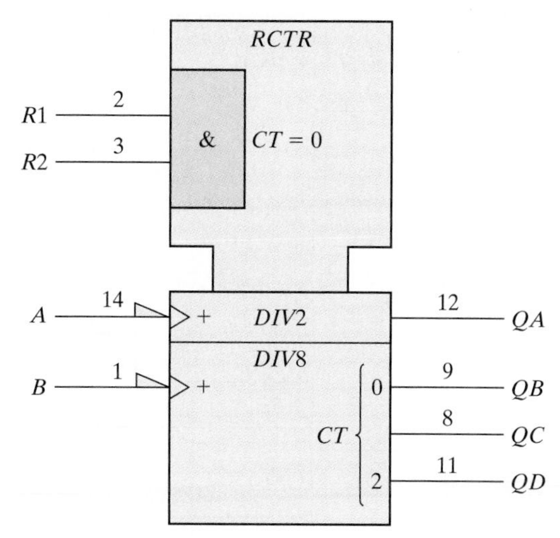

FIGURE 10.13
Graphic symbol for ripple counter, IC type 7493

the symbol $CT = 0$. Since the count input does not go to the clock inputs of all flip-flops, it has no $C1$ label and, instead, the symbol $+$ is used to indicate a count-up operation. The dynamic symbol next to the $+$ together with the polarity symbol along the input line signify that the count is affected with a negative-edge transition of the input signal. The bit grouping from 0 to 2 in the output represents values for the weights to the power of 2. Thus, 0 represents the value of $2^0 = 1$ and 2 represents the value $2^2 = 4$.

The standard graphic symbol for the four-bit counter with parallel load, IC type 74161, is shown in Fig. 10.14. The qualifying symbol for a synchronous counter is CTR followed by the symbol $DIV16$ (divide by 16), which gives the cycle length of the counter. There is a single load input at pin 9 that is split into the two modes, $M1$ and $M2$. $M1$ is active when the load input at pin 9 is low and $M2$ is active when the load input at pin 9 is high. $M1$ is recognized as active low from the polarity indicator along its input line. The count-enable inputs use the G dependencies. $G3$ is associated with the T input and $G4$ with the P input of the count enable. The label associated with the clock is

$$C5/2, 3, 4\ +$$

This means that the circuit counts up (the $+$ symbol) when $M2$, $G3$, and $G4$ are active (load $= 1$, $ENT = 1$, and $ENP = 1$) and the clock in $C5$ goes through a positive transition. This condition is specified in the function table of the 74161 listed in Fig. 9.15. The parallel inputs have the label $1, 5D$, meaning that the D inputs are active when $M1$ is active (load $= 0$) and the clock goes through a positive transition. The output carry is designated by the label

$$3CT = 15$$

This is interpreted to mean that the output carry is active (equal to 1) if $G3$ is active ($ENT = 1$) and the content (CT) of the counter is 15 (binary 1111). Note that the outputs

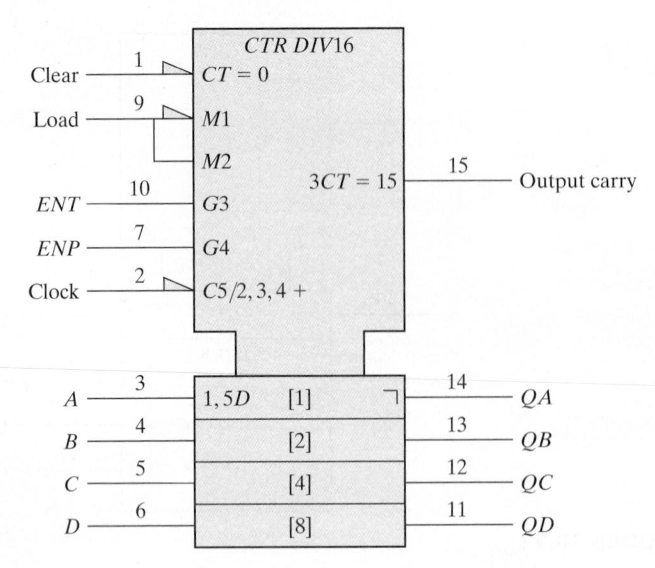

FIGURE 10.14
Graphic Symbol for 4-Bit Binary Counter with Parallel Load, IC Type 74161

have an inverted *L* symbol, indicating that all the flip-flops are of the master–slave type. The polarity symbol in the *C*5 input designates an inverted pulse for the input clock. This means that the master is triggered on the negative transition of the clock pulse and the slave changes state on the positive transition. Thus, the output changes on the positive transition of the clock pulse. It should be noted that IC type 74LS161 (low-power Schottky version) has positive-edge-triggered flip-flops.

10.8 SYMBOL FOR RAM

The standard graphic symbol for the random-access memory (RAM) 74189 is shown in Fig. 10.15. The numbers 16 × 4 that follow the qualifying symbol RAM designate the number of words and the number of bits per word. The common control block is shown with four address lines and two control inputs. Each bit of the word is shown in a separate section with an input and output data line. The address dependency *A* is used to identify the address inputs of the memory. Data inputs and outputs affected by the address are labeled with the letter *A*. The bit grouping from 0 through 3 provides the binary address that ranges from *A*0 through *A*15. The inverted triangle signifies three-state outputs. The polarity symbol specifies the inversion of the outputs.

The operation of the memory is specified by means of the dependency notation. The RAM graphic symbol uses four dependencies: A (address), *G* (AND), *EN* (enable), and *C* (control). Input *G*1 is to be considered ANDed with 1*EN* and 1*C*2 because *G*1 has a 1 after the letter *G* and the other two have a 1 in their label. The *EN* dependency is used

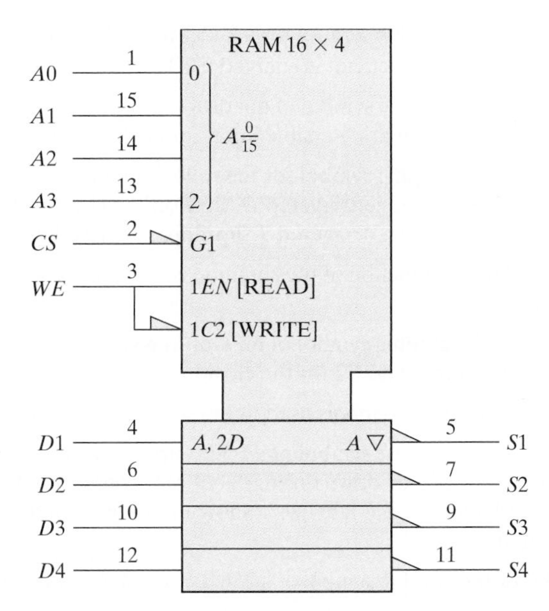

FIGURE 10.15
Graphic symbol for 16×4 RAM, IC type 74189

to identify an enable input that controls the data outputs. The dependency $C2$ controls the inputs as indicated by the $2D$ label. Thus, for a write operation, we have the $G1$ and $1C2$ dependency ($CS = 0$), the $C2$ and $2D$ dependency ($WE = 0$), and the A dependency, which specifies the binary address in the four address inputs. For a read operation, we have the $G1$ and $1EN$ dependencies ($CS = 0$, $WE = 1$) and the A dependency for the outputs. The interpretation of these dependencies results in the operation of the memory as listed in the function table of the 74189 RAM (see Web Search Topics).

PROBLEMS

10.1 Figure 9.1 shows various small-scale integration circuits with pin assignment. Using this information, draw the rectangular-shaped graphic symbols for the 7400, 7404, 7486, 7432, 7408, and 7402 ICs.

10.2 Define the following in your own words:
(a) Positive and negative logic.
(b) Active high and active low.
(c) Polarity indicator.
(d) Dynamic indicator.
(e) Dependency notation.

10.3 Show an example of a graphic symbol that has the three Boolean dependencies—G, V, and N. Draw the equivalent interpretation.

10.4 Draw the graphic symbol of a binary octal-to-decimal decoder. This is similar to a decoder with 3 inputs and 8 outputs.

10.5 Draw the graphic symbol for a BCD-to-decimal decoder with three enable inputs, $E1$, $E2$, and $E3$. The circuit is enabled if $E1 = E2 = 1$, and $E3 = 0$ (assuming positive logic).

10.6 Draw the graphic symbol of quadruple 4-to-1-line multiplexers with common selection inputs and a separate enable input for each multiplexer.

10.7 Draw the graphic symbol for the following flip-flops:
(a) Negative-edge-triggered *D* flip-flop. (b) Master–slave *RS* flip-flop.
(c) Positive-edge-triggered *T* flip-flop. (d) Positive-edge-triggered RS flip-flop.

10.8 Explain the function of the common control block when used with the standard graphic symbols.

10.9 Draw the graphic symbol of an 8-bit register with parallel load using the label $M1$ for the load input and $C2$ for the clock.

10.10 Explain all the symbols used in the standard graphic diagram of Fig. 10.12.

10.11 Draw the graphic symbol of an 5-bit up–down synchronous binary counter with mode input (for up or down) using label M1, count-enable input with *G* dependency and parallel load using label M3. Show the output carries for the up count and the down count.

10.12 Draw the graphic symbol of a 256×4 RAM. Include the symbol for three-state outputs.

REFERENCES

1. *IEEE Standard Graphic Symbols for Logic Functions* (ANSI/IEEE Std. 91-1984). 1984. New York: Institute of Electrical and Electronics Engineers.
2. Kampel, I. 1985. A *Practical Introduction to the New Logic Symbols*. Boston: Butterworth.
3. Mann, F. A. 1984. *Explanation of New Logic Symbols*. Dallas: Texas Instruments.
4. *The TTL Data Book*, Volume 1. 1985. Dallas: Texas Instruments.

WEB SEARCH TOPICS

Bidirectional shift register
Three-state inverter
Three-state buffer
Universal shift register
7483 adder
74151 multiplexer
74155 decoder
74157 multiplexer
7476 flip-flop
7474 flip-flop
74161 flip-flop
74194 shift register
74175 quad flip-flops
74195 shift register
7494 counter
74161 counter
74LS161 flip-flop
74189 RAM
BCD-to-decimal decoder
Random access memory

Appendix

Semiconductors and CMOS Integrated Circuits

Semiconductors are formed by doping a thin slice of a pure silicon crystal with a small amount of a dopant that fits relatively easily into the crystalline structure of the silicon. Dopants are differentiated on the basis of whether they have either three valence electrons or five valence electrons. A silicon crystalline structure is such that each silicon atom shares its four valence electrons with its four nearest neighbors, thereby completing its valence structure. The atoms of a dopant with five valence electrons, referred to as a *n*-type dopant, fit in the physical structure of the crystal, but their fifth electrons are held only loosely by their parent atoms in the bonded structure. Consequently, an applied electric field can cause such electrons to flow as a current. On the other hand, a dopant atom with only three valence electrons, a *p*-type dopant, has a vacant valence site. Under the influence of an applied electric field, an electron from a neighboring silicon atom in the bonded structure can jump from its host and fill a vacant dopant site, leaving behind a vacancy at its host. This migration, visualized as a leapfrogging of electrons from hole to hole, establishes a current.

Current is due to the movement of electrons, which are negative charge carriers. Current is measured, however, in the opposite direction of flow, by convention—since the days of Benjamin Franklin. (Think of current as being the motion of an equivalent positive charge moving in the opposite direction of an electron, whose charge is negative). Holes move in the direction of current, although the underlying physical movement of electrons is in the opposite direction. Thermal agitation causes both types of charge carriers to be present in a semiconductor. If the majority carrier is a hole, the device is said to be a *p*-type device; if the majority carrier is an electron, the device is said to be an *n*-type device. Bipolar transistors rely on both types of carriers. Metal-oxide silicon semiconductors rely on a majority carrier, either an electron or a hole, but not both. The type and relative amount of dopant determine the type of a semiconductor material.

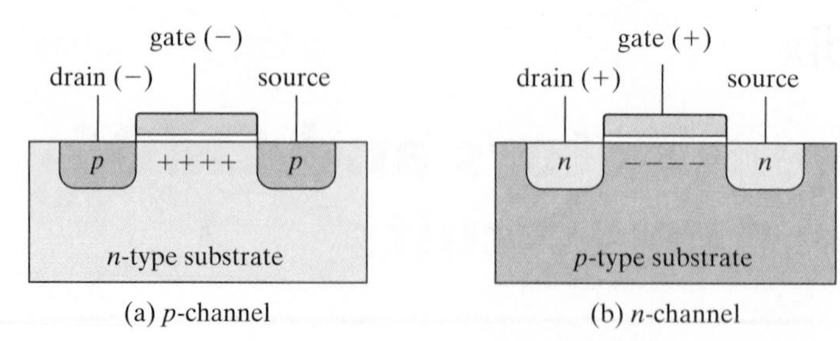

FIGURE A.1
Basic structure of MOS transistor

The basic structure of a metal-oxide semiconductor (MOS) transistor is shown in Fig. A.1. The *p*-channel MOS transistor consists of a lightly doped substrate of *n*-type silicon material. Two regions are heavily doped with *p*-type impurities by a diffusion process to form the *source* and *drain*. The source terminal supplies charge carriers to an external circuit; the drain terminal removes charge carriers from the circuit. The region between the two *p*-type sections serves as the *channel*. In its simplest form, the gate is a metal plate separated from the channel by an insulted dielectric of silicon dioxide. A negative voltage (with respect to the substrate) at the gate terminal causes an induced electric field in the channel that attracts *p*-type carriers (holes) from the substrate. As the magnitude of the negative voltage increases, the region below the gate accumulates more positive carriers, the conductivity increases, and current can flow from source to drain, provided that a voltage difference is maintained between these two terminals.

There are four basic types of MOS structures. The channel can be *p* or *n* type, depending on whether the majority carriers are holes or electrons. The mode of operation can be enhancement or depletion, depending on the state of the channel region at zero gate voltage. If the channel is initially doped lightly with *p*-type impurity (in which case it is called a *diffused channel*), a conducting channel exists at zero gate voltage and the device is said to operate in the *depletion* mode. In this mode, current flows unless the channel is depleted by an applied gate field. If the region beneath the gate is left initially uncharged, a channel must be induced by the gate field before current can flow. Thus, the channel current is enhanced by the gate voltage, and such a device is said to operate in the *enhancement* mode.

The source is the terminal through which the majority carriers enter the device. If the majority carrier is a hole (*p*-type channel), the source terminal supplies current to the circuit; if the majority carrier is an electron (*n*-type channel), the source removes current from the circuit. The drain is the terminal through which the majority carriers leave the device. In a *p*-channel MOS, the source terminal is connected to the substrate and a negative voltage is applied to the drain terminal. When the gate voltage is above a threshold voltage V_T (about -2 V), no current flows in the channel and the drain-to-source path is like an open circuit. When the gate voltage is sufficiently negative below V_T, a channel is formed and *p*-type carriers flow from source to drain. *p*-type carriers are positive and correspond to a positive current flow from source to drain.

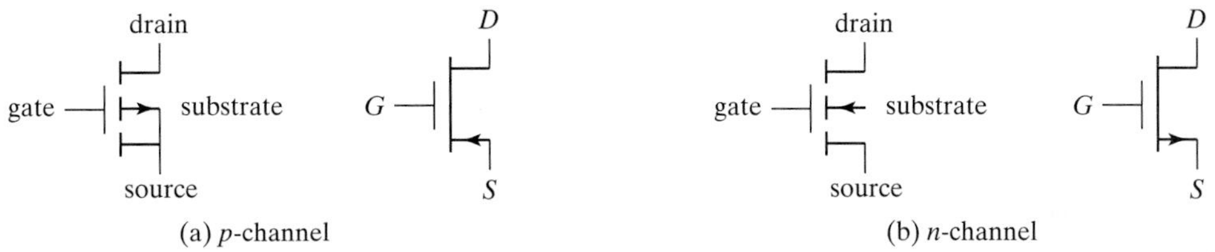

(a) p-channel

(b) n-channel

FIGURE A.2
Symbols for MOS transistors

In the n-channel MOS, the source terminal is connected to the substrate and a positive voltage is applied to the drain terminal. When the gate voltage is below the threshold voltage V_T (about 2 V), no current flows in the channel. When the gate voltage is sufficiently positive above V_T to form the channel, n-type carriers flow from source to drain. n-type carriers are negative and correspond to a positive current flow from drain to source. The threshold voltage may vary from 1 to 4 V, depending on the particular process used.

The graphic symbols for the MOS transistors are shown in Fig. A.2. The symbol for the enhancement type is the one with the broken-line connection between source and drain. In this symbol, the substrate can be identified and is shown connected to the source. An alternative symbol omits the substrate, and instead an arrow is placed in the source terminal to show the direction of *positive* current flow (from source to drain in the p-channel MOS and from drain to source in the n-channel MOS).

Because of the symmetrical construction of source and drain, the MOS transistor can be operated as a bilateral device. Although normally operated so that carriers flow from source to drain, there are circumstances when it is convenient to allow carriers to flow from drain to source.

One advantage of the MOS device is that it can be used not only as a transistor, but as a resistor as well. A resistor is obtained from the MOS by permanently biasing the gate terminal for conduction. The ratio of the source–drain voltage to the channel current then determines the value of the resistance. Different resistor values may be constructed during manufacturing by fixing the channel length and width of the MOS device.

Three logic circuits using MOS devices are shown in Fig. A.3. For an n-channel MOS, the supply voltage V_{DD} is positive (about 5 V), to allow positive current flow from drain to source. The two voltage levels are a function of the threshold voltage V_T. The low level is anywhere from zero to V_T, and the high level ranges from V_T to V_{DD}. The n-channel gates usually employ positive logic. The p-channel MOS circuits use a negative voltage for V_{DD}, to allow positive current flow from source to drain. The two voltage levels are both negative above and below the negative threshold voltage V_T. p-channel gates usually employ negative logic.

The inverter circuit shown in Fig. A.3(a) uses two MOS devices. *Q1* acts as the load resistor and *Q2* as the active device. The load-resistor MOS has its gate connected to V_{DD},

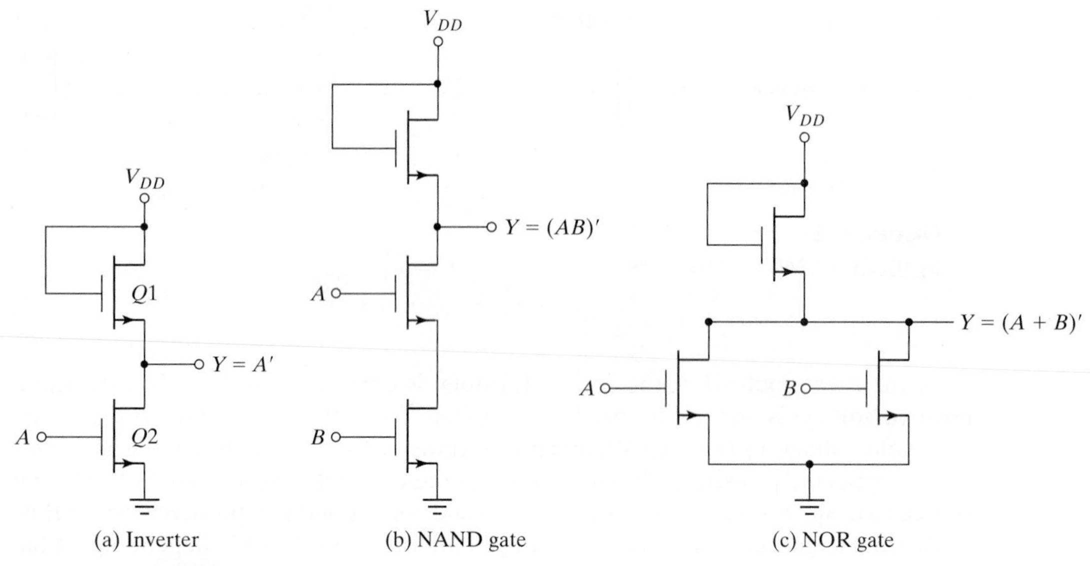

FIGURE A.3
n-channel MOS logic circuits

thus maintaining it in the conduction state. When the input voltage is low (below V_T), *Q2* turns off. Since *Q1* is always on, the output voltage is about V_{DD}. When the input voltage is high (above V_T), *Q2* turns on. Current flows from V_{DD} through the load resistor *Q1* and into *Q2*. The geometry of the two MOS devices must be such that the resistance of *Q2*, when conducting, is much less than the resistance of *Q1* to maintain the output *Y* at a voltage below V_T.

The NAND gate shown in Fig. A.3(b) uses transistors in series. Inputs *A* and *B* must both be high for all transistors to conduct and cause the output to go low. If either input is low, the corresponding transistor is turned off and the output is high. Again, the series resistance formed by the two active MOS devices must be much less than the resistance of the load-resistor MOS. The NOR gate shown in Fig. A.3(c) uses transistors in parallel. If either input is high, the corresponding transistor conducts and the output is low. If all inputs are low, all active transistors are off and the output is high.

A.1 COMPLEMENTARY MOS

Complementary MOS (CMOS) circuits take advantage of the fact that both *n*-channel and *p*-channel devices can be fabricated on the same substrate. CMOS circuits consist of both types of MOS devices, interconnected to form logic functions. The basic circuit is the inverter, which consists of one *p*-channel transistor and one *n*-channel transistor, as shown in Fig. A.4(a). The source terminal of the *p*-channel device is at V_{DD}, and the source terminal of the *n*-channel device is at ground. The value of V_{DD}

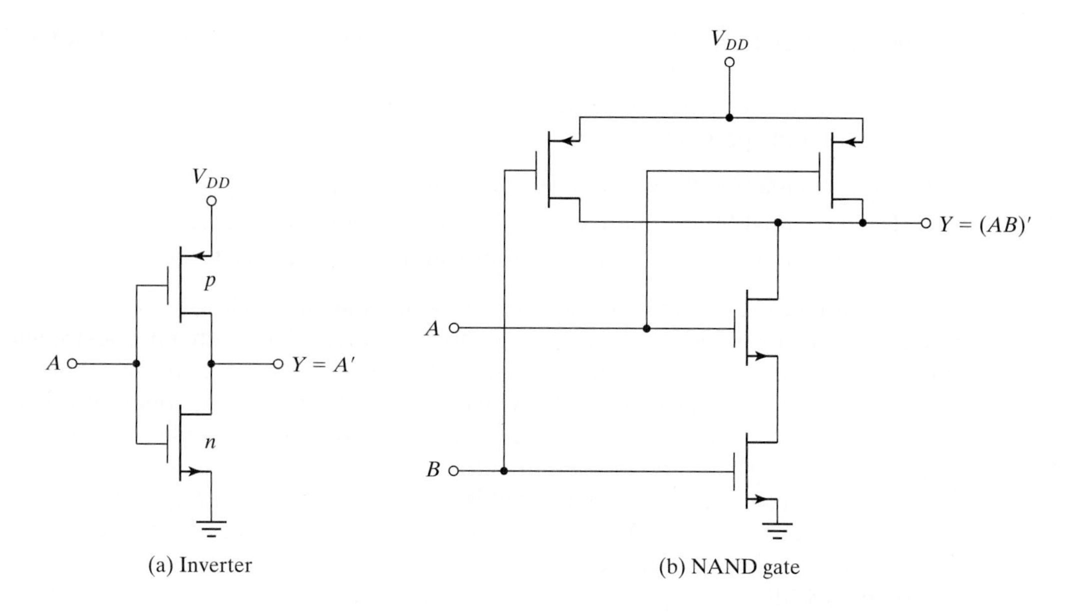

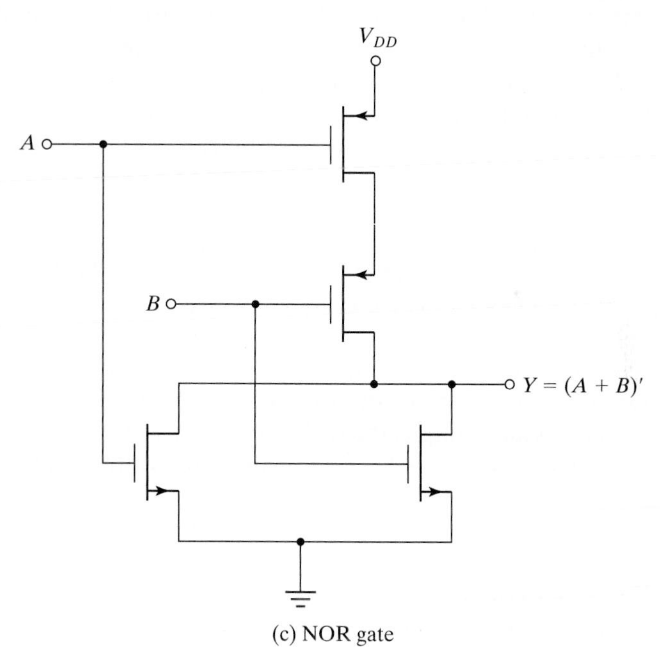

FIGURE A.4
CMOS logic circuits

may be anywhere from +3 to +18 V. The two voltage levels are 0 V for the low level and V_{DD} for the high level (typically, 5 V).

To understand the operation of the inverter, we must review the behavior of the MOS transistor from the previous section:

1. The *n*-channel MOS conducts when its gate-to-source voltage is positive.
2. The *p*-channel MOS conducts when its gate-to-source voltage is negative.
3. Either type of device is turned off if its gate-to-source voltage is zero.

Now consider the operation of the inverter. When the input is low, both gates are at zero potential. The input is at $-V_{DD}$ relative to the source of the *p*-channel device and at 0 V relative to the source of the *n*-channel device. The result is that the *p*-channel device is turned on and the *n*-channel device is turned off. Under these conditions, there is a low-impedance path from V_{DD} to the output and a very high impedance path from output to ground. Therefore, the output voltage approaches the high level V_{DD} under normal loading conditions. When the input is high, both gates are at V_{DD} and the situation is reversed: The *p*-channel device is off and the *n*-channel device is on. The result is that the output approaches the low level of 0 V.

Two other CMOS basic gates are shown in Fig. A.4. A two-input NAND gate consists of two *p*-type units in parallel and two *n*-type units in series, as shown in Fig. A.4(b). If all inputs are high, both *p*-channel transistors turn off and both *n*-channel transistors turn on. The output has a low impedance to ground and produces a low state. If any input is low, the associated *n*-channel transistor is turned off and the associated *p*-channel transistor is turned on. The output is coupled to V_{DD} and goes to the high state. Multiple-input NAND gates may be formed by placing equal numbers of *p*-type and *n*-type transistors in parallel and series, respectively, in an arrangement similar to that shown in Fig. A.4(b).

A two-input NOR gate consists of two *n*-type units in parallel and two *p*-type units in series, as shown in Fig. A.4(c). When all inputs are low, both *p*-channel units are on and both *n*-channel units are off. The output is coupled to V_{DD} and goes to the high state. If any input is high, the associated *p*-channel transistor is turned off and the associated *n*-channel transistor turns on, connecting the output to ground and causing a low-level output.

MOS transistors can be considered to be electronic switches that either conduct or are open. As an example, the CMOS inverter can be visualized as consisting of two switches as shown in Fig. A.5(a). Applying a low voltage to the input causes the upper switch (*p*) to close, supplying a high voltage to the output. Applying a high voltage to the input causes the lower switch (*n*) to close, connecting the output to ground. Thus, the output V_{out} is the complement of the input V_{in}. Commercial applications often use other graphic symbols for MOS transistors to emphasize the logical behavior of the switches. The arrows showing the direction of current flow are omitted. Instead, the gate input of the *p*-channel transistor is drawn with an inversion bubble on the gate terminal to show that it is enabled with a low voltage. The inverter circuit is redrawn with these symbols in Fig. A.5(b). A logic 0 in the input causes the upper transistor to conduct, making the output logic 1. A logic 1 in the input enables the lower transistor, making the output logic 0.

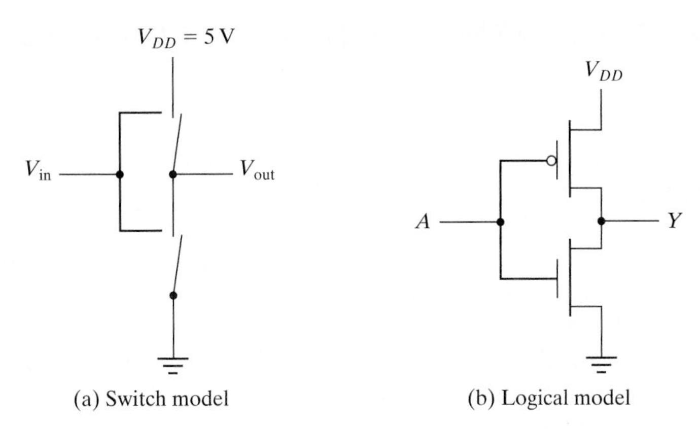

FIGURE A.5
CMOS inverter

CMOS Characteristics

When a CMOS logic circuit is in a static state, its power dissipation is very low. This is because at least one transistor is always off in the path between the power supply and ground when the state of the circuit is not changing. As a result, a typical CMOS gate has static power dissipation on the order of 0.01 mW. However, when the circuit is changing state at the rate of 1 MHz, the power dissipation increases to about 1 mW, and at 10 MHz it is about 5 mW.

CMOS logic is usually specified for a single power-supply operation over a voltage range from 3 to 18 V with a typical V_{DD} value of 5 V. Operating CMOS at a larger power-supply voltage reduces the propagation delay time and improves the noise margin, but the power dissipation is increased. The propagation delay time with $V_{DD} = 5$ V ranges from 5 to 20 ns, depending on the type of CMOS used. The noise margin is usually about 40% of the power supply voltage. The fan-out of CMOS gates is about 30 when they are operated at a frequency of 1 MHz. The fan-out decreases with an increase in the frequency of operation of the gates.

There are several series of the CMOS digital logic family. The 74C series are pin and function compatible with TTL devices having the same number. For example, CMOS IC type 74C04 has six inverters with the same pin configuration as TTL type 7404. The high-speed CMOS 74HC series is an improvement over the 74C series, with a tenfold increase in switching speed. The 74HCT series is electrically compatible with TTL ICs. This means that circuits in this series can be connected to inputs and outputs of TTL ICs without the need of additional interfacing circuits. Newer versions of CMOS are the high-speed series 74VHC and its TTL-compatible version 74VHCT.

The CMOS fabrication process is simpler than that of TTL and provides a greater packing density. Thus, more circuits can be placed on a given area of silicon at a reduced cost per function. This property, together with the low power dissipation of CMOS circuits, good noise immunity, and reasonable propagation delay, makes CMOS the most popular standard as a digital logic family.

A.2 CMOS TRANSMISSION GATE CIRCUITS

A special CMOS circuit that is not available in the other digital logic families is the *transmission gate*. The transmission gate is essentially an electronic switch that is controlled by an input logic level. It is used to simplify the construction of various digital components when fabricated with CMOS technology.

Figure A.6(a) shows the basic circuit of the transmission gate. Whereas a CMOS inverter consists of a *p*-channel transistor connected in series with an *n*-channel transistor, a transmission gate is formed by one *n*-channel and one *p*-channel MOS transistor connected in parallel.

The *n*-channel substrate is connected to ground and the *p*-channel substrate is connected to V_{DD}. When the N gate is at V_{DD} and the P gate is at ground, both transistors conduct and there is a closed path between input X and output Y. When the N gate is at ground and the P gate is at V_{DD}, both transistors are off and there is an open circuit between X and Y. Figure A.4(b) shows the block diagram of the transmission gate. Note that the terminal of the *p*-channel gate is marked with the negation symbol. Figure A.4(c) demonstrates the behavior of the switch in terms of positive-logic assignment with V_{DD} equivalent to logic 1 and ground equivalent to logic 0.

The transmission gate is usually connected to an inverter, as shown in Fig. A.7. This type of arrangement is referred to as a *bilateral switch*. The control input C is connected directly to the *n*-channel gate and its inverse to the *p*-channel gate. When $C = 1$, the

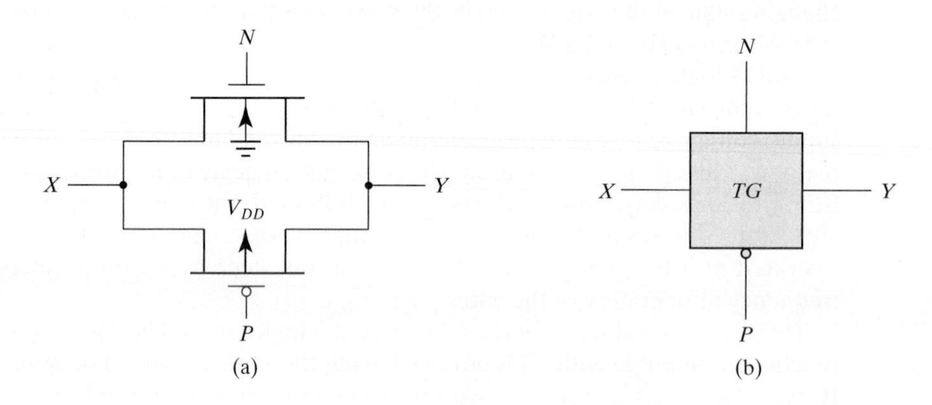

FIGURE A.6
Transmission gate (TG)

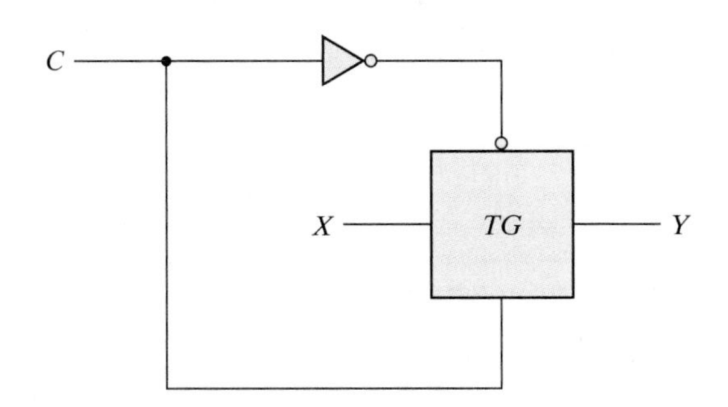

FIGURE A.7
Bilateral switch

switch is closed, producing a path between X and Y. When $C = 0$, the switch is open, disconnecting the path between X and Y.

Various circuits can be constructed that use the transmission gate. To demonstrate its usefulness as a component in the CMOS family, we will show three examples.

The exclusive-OR gate can be constructed with two transmission gates and two inverters, as shown in Fig. A.8. Input A controls the paths in the transmission gates and input B is connected to output Y through the gates. When input A is equal to 0, transmission gate *TG1* is closed and output Y is equal to input B. When input A is equal to 1, *TG2* is closed and output Y is equal to the complement of input B. This results in the exclusive-OR truth table, as indicated in Fig. A.8.

Another circuit that can be constructed with transmission gates is the multiplexer. A four-to-one-line multiplexer implemented with transmission gates is shown in Fig. A.9. The *TG* circuit provides a transmission path between its horizontal input and

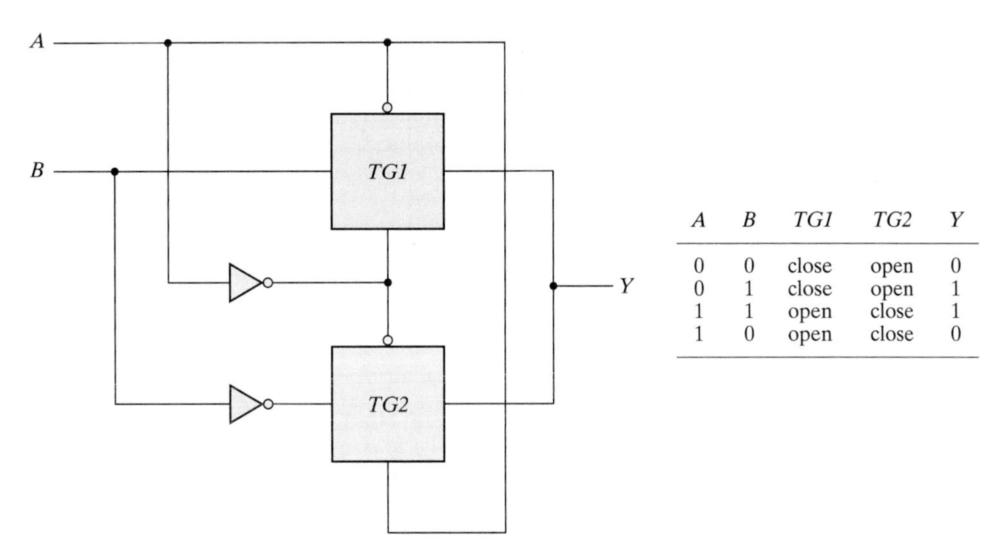

A	*B*	*TG1*	*TG2*	*Y*
0	0	close	open	0
0	1	close	open	1
1	1	open	close	1
1	0	open	close	0

FIGURE A.8
Exclusive-OR constructed with transmission gates

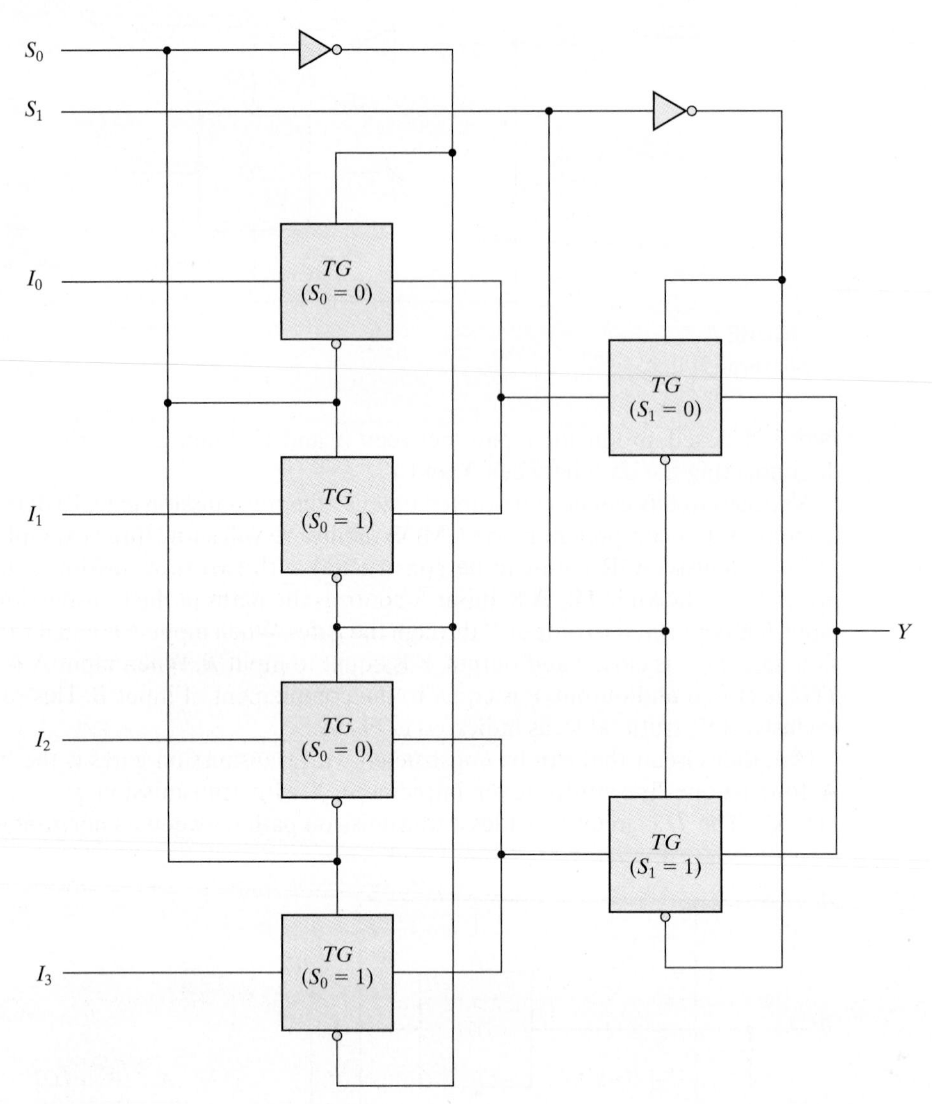

FIGURE A.9
Multiplexer with transmission gates

output lines when the two vertical control inputs have the value of 1 in the uncircled terminal and 0 in the circled terminal. With an opposite polarity in the control inputs, the path disconnects and the circuit behaves like an open switch. The two selection inputs, S_1 and S_0, control the transmission path in the *TG* circuits. Inside each box is marked the condition for the transmission gate switch to be closed. Thus, if $S_0 = 0$ and $S_1 = 0$, there is a closed path from input I_0 to output Y through the two *TG*s marked with $S_0 = 0$ and $S_1 = 0$. The other three inputs are disconnected from the output by one of the other *TG* circuits.

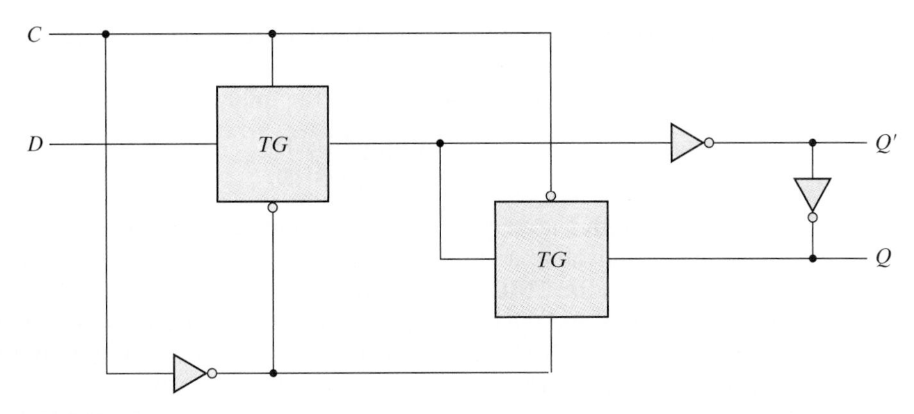

FIGURE A.10
Gated *D* latch with transmission gates

The level-sensitive *D* flip-flop commonly referred to as the gated *D* latch can be constructed with transmission gates, as shown in Fig. A.10. The *C* input controls two transmission gates *TG*. When $C = 1$, the *TG* connected to input *D* has a closed path and the one connected to output *Q* has an open path. This configuration produces an equivalent circuit from input *D* through two inverters to output *Q*. Thus, the output follows the data input as long as *C* remains active. When *C* switches to 0, the first *TG* disconnects input *D* from the circuit and the second *TG* produces a closed path between the two inverters at the output. Thus, the value that was present at input *D* at the time that *C* went from 1 to 0 is retained at the *Q* output.

A master–slave *D* flip-flop can be constructed with two circuits of the type shown in Fig. A.10. The first circuit is the master and the second is the slave. Thus, a master–slave *D* flip-flop can be constructed with four transmission gates and six inverters.

A.3 SWITCH-LEVEL MODELING WITH HDL

CMOS is the dominant digital logic family used with integrated circuits. By definition, CMOS is a complementary connection of an NMOS and a PMOS transistor. MOS transistors can be considered to be electronic switches that either conduct or are open. By specifying the connections among MOS switches, the designer can describe a digital circuit constructed with CMOS. This type of description is called *switch-level modeling* in Verilog HDL.

The two types of MOS switches are specified in Verilog HDL with the keywords **nmos** and **pmos**. They are instantiated by specifying the three terminals of the transistor, as shown in Fig. A.2:

```
nmos (drain, source, gate);
pmos (drain, source, gate);
```

Switches are considered to be primitives, so the use of an instance name is optional.

The connections to a power source (V_{DD}) and to ground must be specified when MOS circuits are designed. Power and ground are defined with the keywords **supply1** and **supply0**. They are specified, for example, with the following statements:

```
supply1 PWR;
supply0 GRD;
```

Sources of type **supply1** are equivalent to V_{DD} and have a value of logic 1. Sources of type **supply0** are equivalent to ground connection and have a value of logic 0.

The description of the CMOS inverter of Fig. A.4(a) is shown in HDL Example A.1. The input, the output, and the two supply sources are declared first. The module instantiates a PMOS and an NMOS transistor. The output *Y* is common to both transistors at their drain terminals. The input is also common to both transistors at their gate terminals. The source terminal of the PMOS transistor is connected to PWR and the source terminal of the NMOS transistor is connected to GRD.

HDL Example A.1

```
// CMOS inverter of Fig. A.4(a)
module inverter (Y, A);
 input A;
 output Y;
 supply1  PWR;
 supply0  GRD;
 pmos (Y, PWR, A);              // (Drain, source, gate)
 nmos (Y, GRD, A);              // (Drain, source, gate)
endmodule
```

The second module, set forth in HDL Example A.2, describes the two-input CMOS NAND circuit of Fig. A.4(b). There are two PMOS transistors connected in parallel, with their source terminals connected to PWR. There are also two NMOS transistors connected in series and with a common terminal *W1*. The drain of the first NMOS is connected to the output, and the source of the second NMOS is connected to GRD.

HDL Example A.2

```
// CMOS two-input NAND of Fig. A.4(b)
module NAND2 (Y, A, B);
 input A, B;
 output Y;
 supply1  PWR;
 supply0  GRD;
 wire W1;                       // terminal between two nmos
 pmos (Y, PWR, A);              // source connected to Vdd
 pmos (Y, PWR, B);              // parallel connection
```

```
 nmos (Y, W1, A);             // serial connection
 nmos (W1, GRD, B);           // source connected to ground
endmodule
```

Transmission Gate

The transmission gate is instantiated in Verilog HDL with the keyword **cmos**. It has an output, an input, and two control signals, as shown in Fig. A.6. It is referred to as a **cmos** switch. The relevant code is as follows:

```
cmos (output, input, ncontrol, pcontrol); // general description
cmos (Y, X, N, P);  // transmission gate of Fig. A.6(b)
```

Normally, ncontrol and pcontrol are the complement of each other. The **cmos** switch does not need power sources, since V_{DD} and ground are connected to the substrates of the MOS transistors. Transmission gates are useful for building multiplexers and flip-flops with CMOS circuits.

HDL Example A.3 describes a circuit with **cmos** switches. The exclusive-OR circuit of Fig. A.8 has two transmission gates and two inverters. The two inverters are instantiated within the module describing a CMOS inverter. The two **cmos** switches are instantiated without an instance name, since they are primitives in the language. A test module is included to test the circuit's operation. Applying all possible combinations of the two inputs, the result of the simulator verifies the operation of the exclusive-OR circuit. The output of the simulation is as follows:

$$A = 0 \quad B = 0 \quad Y = 0$$
$$A = 0 \quad B = 1 \quad Y = 1$$
$$A = 1 \quad B = 0 \quad Y = 1$$
$$A = 1 \quad B = 1 \quad Y = 0$$

HDL Example A.3

```
//CMOS_XOR with CMOS switches, Fig. A.8

module CMOS_XOR (A, B, Y);
 input A, B;
 output Y;
 wire A_b, B_b;
 // instantiate inverter
 inverter v1 (A_b, A);
 inverter v2 (B_b, B);
 // instantiate cmos switch
 cmos (Y, B, A_b, A);               //(output, input, ncontrol, pcontrol)
 cmos (Y, B_b, A, A_b);
endmodule
```

```
// CMOS inverter Fig. A.4(a)
module inverter (Y, A);
 input A;
 output Y;
 supply1  PWR;
 supply0  GND;
 pmos (Y, PWR, A);                    //(Drain, source, gate)
 nmos (Y, GND, A);                    //(Drain, source, gate)
endmodule
// Stimulus to test CMOS_XOR
module test_CMOS_XOR;
 reg A,B;
 wire Y;
 //Instantiate CMOS_XOR
 CMOS_XOR X1 (A, B, Y);
 // Apply truth table
 initial
  begin
  A = 1'b0; B = 1'b0;
 #5 A = 1'b0; B = 1'b1;
 #5 A = 1'b1; B = 1'b0;
 #5 A = 1'b1; B = 1'b1;
 end
// Display results
initial
 $monitor ("A =%b  B= %b  Y =%b", A, B, Y);
endmodule
```

WEB SEARCH TOPICS

Conductor
Semiconductor
Insulator
Electrical properties of materials
Valence electron
Diode
Transistor
CMOS process
CMOS logic gate
CMOS inverter

Answers to Selected Problems

CHAPTER 1

1.2 **(a)** 16,384 **(b)** 33,554,432 **(c)** 3,435,973,837

1.3 **(a)** $(117)_{10}$ **(b)** $(1556)_{10}$

1.5 **(a)** 4 **(b)** 7 **(c)** 11

1.6 6

1.7 $(125715)_8$

1.9 **(a)** $(26.3125)_{10}$ **(b)** $(166.3125)_{10}$ **(c)** $(190.3125)_{10}$

1.12 **(a)** $(10010)_2$ and $(1001000)_2$ **(b)** $(C7)_{16}$ and $(12B4)_{16}$

1.19 **(a)** 010035 **(b)** 008127 **(c)** −8127 **(d)** −10035

1.24 **(a)**

Octal Digit	6311	6421
0	0000	0000
1	0001/0010	0001
2	0011	0010
3	0100	0011
4	0110/0101	0100
5	0111	0101
6	1000	0110/1000
7	1001/1010	1001/0111

1.29 Digital Systems

1.31 $62 + 32 = 94$ printing characters

1.32 bit 6 from the right

1.33 **(a)** 564 **(b)** 231 **(c)** 324 **(d)** $(1380)_{10}$

CHAPTER 2

2.2 **(a)** x' **(b)** x' **(c)** z **(d)** 0

2.3 **(a)** C **(b)** $x'y' + y'z$ **(c)** $x'y'$ **(d)** $x'z'$ **(e)** 0

2.4 **(a)** (A XNOR C) **(b)** $x + y + z$ **(c)** B **(d)** $A'(B + C'A)$

2.9 **(a)** $xy' + x'y$

2.11 $F(x, y, z) = \Sigma(0, 1, 3, 4, 5, 7)$

2.12 **(a)** 00000000 **(c)** 10111111 **(d)** 11110001

2.14 **(b)** $(x + y)' + (x + z')' + (x' + y')'$

2.15 $T_1 = A \oplus B$

$T_2 = (A \oplus B)'$

2.17 **(a)** $\Sigma(1, 4, 5, 6, 7, 14) = \Pi(0, 2, 3, 8, 9, 10, 11, 12, 13, 15)$

2.18 **(c)** $F = y + z'$

2.19 $\Sigma(1, 3, 5, 7, 9, 11, 13, 15) = \Pi(0, 2, 4, 6, 8, 10, 12, 14)$

2.22 **(a)** $AB + BC = (A + C)B$ **(b)** $(x' + y)(x' + z)$

CHAPTER 3

3.1 **(a)** z' **(b)** $x + z'$ **(c)** $x' + z'$ **(d)** $xy' + xz + y'z$

3.2 **(a)** $x'y' + y'z' + xyz$ **(b)** $y + x'z$

3.3 **(a)** y **(b)** $y + xz'$ **(c)** $xz + yz$

3.4 **(a)** y' **(b)** $BC + A'C'$ **(c)** $C'D + ABC' + ABD$

(d) $x'y + x'z'$

3.5 **(a)** $w'xz' + x'y'z + wxy$ **(d)** D'

3.6 **(a)** $B'D' + ABC' + A'BD$ **(b)** $xy' + x'z + wx'y$

3.7 **(a)** $x'y + z$ **(c)** $B'C' + AB' + A'BD + ACD'$

3.8 **(a)** $F(x, y, z) = \Sigma(3, 5, 7, 8, 10, 11, 12, 13, 14, 15)$

(b) $F(A, B, C, D) = \Sigma(0, 2, 4, 5, 6, 9, 10, 11, 13, 14)$

3.9 **(a)** Essential prime implicant: 2′

(b) $F = B'D' + AC + A'BD +$ (CD **or** $B'C$)

3.10 **(c)** $F = C$

3.11 **(a)** $F = (x' + z)(w' + y + z)(w + y' + z)(w' + x + y' + z')$

3.12 **(b)** $F = D'$

3.13 **(a)** (1) $F = z' + xy'$ (2) $F = (x + z')(y' + z')$

3.15 **(b)** $F = D'$

3.17 $F' = B'A' + B'C + AD$

3.19 **(a)** $F = (x' + z)' + (y' + z)' + (x' + y')' + (w' + y')'$

3.30 $F = (A \oplus B)(C \oplus D)$

3.35 The HDL description is available on the Companion Website.

Line 1: Dash not allowed, use underscore: Exmpl_3.
Terminate line with semicolon (;).
Line 2: **inputs** should be **input** (no s at the end).
Change last comma (,) to semicolon (;). Output is declared but does not appear in the port list, and should be followed by a comma if it is intended to be in the list of inputs. If *Output* is a mispelling of **output** and is to declare output ports, C should be followed by a semicolon (;) and *F* should be followed by a semicolon (;).
Line 3: *B* cannot be declared as input (Line 2) and output (Line 3). Terminate the line with a semicolon (;).
Line 4: *A* cannot be an output of the primitive if it is an input to the module
Line 5: Too many entries for the not gate (only two allowed).
Line 6: OR must be in lowercase: change to "or".
Line 7: **endmodule** is mispelled. Remove semicolon (no semicolon after endmodule).

CHAPTER 4

4.1 **(a)** $F_1 = A + B'C + BD' + B'D$
$F_2 = A'B + D$

4.2 $F = A'D + AB' + AC'$
$G = A + D'$

4.3 **(b)** 1024 rows and 14 columns

4.4 **(a)** $F = A \oplus (BC)$

4.6 $F = xy + xz + yz$

4.7 **(a)** $w = A \quad x = A \oplus B \quad y = x \oplus C \quad z = y \oplus D$

4.8 $w = AB + AC'D'$

4.10 Inputs: A, B, C, D; Outputs: w, x, y, z

$z = D$

$y = C \oplus D$

$x = B \oplus C \oplus D$

$w = A \oplus B \oplus C \oplus D$

4.12 **(b)** $Diff = x \oplus y \oplus B_{in}$

$B_{out} = x'y + y'B_{in} + xB_{in}$

4.13

	Sum	C	V
(a)	1101	0	1
(b)	0001	1	1
(c)	0100	1	0
(d)	1011	0	1
(e)	1111	0	0

4.14 30 ns

4.18 $w = A'B'C'$

$x = B \oplus C$

$y = C$

$z = D'$

4.22 $w = AB + ACD$

$x = B'C' + B'D' + BCD$

$y = C'D + CD'$

$z = D'$

4.28 **(a)** $F_1 = \Sigma(2, 4, 6, 7)$

$F_2 = \Sigma(3, 5, 6, 7)$

$F_3 = \Sigma(0, 1, 5, 6, 7)$

4.29 $x = D_7 + D_7'(D_6 + D_6'(D_5 + D_5'(D_4)))$

$y = D_7 + D_7'(D_6 + D_4'D_5'D_6'(D_3 + D_3'(D_2)))$

$z = D_7 + D_7'D_6'(D_5 + D_5'D_4'(D_3 + D_3'D_2'(D_1)))$

4.34 $F(A, B, C, D) = \Sigma\,(0, 1, 2, 4, 7, 9, 10, 14, 15)$

4.35 **(a)** When $AB = 00, F = D$

When $AB = 01, F = (C + D)'$

When $AB = 10, F = CD$

When $AB = 11, F = 1$

4.39 The HDL description is available on the Companion Website.

4.42 **(c)** The HDL description is available on the Companion Website.

4.50 The HDL description is available on the Companion Website.

4.56
```
assign match = (A == B);        // Assumes reg [3: 0] A, B;
```

4.57 The HDL description is available on the Companion Website.

CHAPTER 5

5.4 **(b)** $PN' + PQ(t)' + N'Q(t)$

5.7 $Diff = Q(t)'(x \oplus y) + Q(t)(x \oplus y)$
$= Q(t) \oplus x \oplus y$

5.8 A counter with a repeated sequence of 00, 01, 10

5.9 **(a)** $A(t + 1) = A'x + AB'$
$B(t + 1) = B'x' + AB$

5.10 **(c)** $A(t + 1) = xB + x'A + yA + y'A'B'$
$B(t + 1) = xA'B' + x'A'B + yA'B'$

5.11

State:	*a*	*b*	*d*	*a*	*b*	*d*	*c*	*a*	*d*
Input:	1	1	0	1	1	1	0	0	1
Output:	0	0	1	0	0	0	1	0	0

5.12

Present state	Next state 0	Next state 1	Output 0	Output 1
a	*f*	*b*	0	0
b	*d*	*a*	0	0
d	*g*	*a*	1	0
f	*f*	*b*	1	1
g	*g*	*d*	0	1

5.13 **(a)** State: *a f a f a f g b g b a*
Input: 0 1 0 1 0 0 1 0 1 1 1
Output: 0 0 0 0 0 1 0 1 0 0 0

(b) State: *a b a b a b g b g b a*
Input: 0 1 0 1 0 0 1 0 1 1 1
Output: 0 0 0 0 0 1 0 1 0 0 0

5.15 $Q(t + 1) = T \oplus Q(t)$

5.16 $D_A = Ax' + Bx$

$D_B = A'x + Bx'$

5.18 $J_A = K_A = (BF + B'F')E$

$J_B = K_B = E$

5.19 **(a)** $D_A = A'B'x\_in$

$D_B = A + C'x\_in' + BCx\_in$

$D_C = Cx\_in' + Ax\_in + A'B'x\_in'$

$y\_out = A'x\_in$

5.23 **(a)** $RegA = 125, \quad RegB = 125$

(b) $RegA = 125, \quad RegB = 30$

5.26 **(a)**

$Q(t + 1) = JQ' + K'Q$

When $Q = 0$, $Q(t + 1) = J$

When $Q = 1$, $Q(t + 1) = K'$

```
module JK_Behavior (output reg Q, input J, K, CLK);
 always @ (posedge CLK)
        if (Q == 0)        Q <= J;
        else               Q <= ~K;
endmodule
```

5.31 The HDL description is available on the Companion Website.
Note: The statements must be written in an order that produces the effect of concurrent assignments.

CHAPTER 6

6.4 0110; 1011; 0101; 0010; 1001; 1100

6.8 $A = 0010, 0001, 1000, 1100$. Carry $= 1, 1, 1, 0$

6.9 **(b)** $J_Q = x'y; K_Q = (x' + y)'$

6.14 **(a)** 4

6.15 40 ns; 25 MHz

6.16 $1010 \rightarrow 1011 \rightarrow 0100$

$1100 \rightarrow 1101 \rightarrow 0100$

$1110 \rightarrow 1111 \rightarrow 0000$

6.17 $D_A = Q_A \oplus Q_B \oplus Q_C$

$D_B = Q_B \oplus Q_C$

$D_C = Q_C'$

6.19 **(b)** $D_{Q1} = Q_1'$

$D_{Q2} = Q_2Q_1' + Q_8'Q_2'Q_1$

$D_{Q4} = Q_4Q_1' + Q_4Q_2' + Q_4'Q_2'Q_1$

$D_{Q8} = Q_8Q_1' + Q_4Q_2Q_1$

6.21 $J_{A0} = LI_0 + L'C$

$K_{A0} = LI_0' + L'C$

6.24 $T_A = A \oplus B$

$T_B = B \oplus C$

$T_C = AC + A'C'$ (not self-starting)

$= AC + A'B'C$ (self-starting)

6.26 Use a 3-bit counter to count eight pulses.

6.28 $D_A = A \oplus B$

$D_B = AB' + C$

$D_C = A'B'C'$

6.34 The HDL description is available on the Companion Website. Simulations results for Problem 6.34 follow:

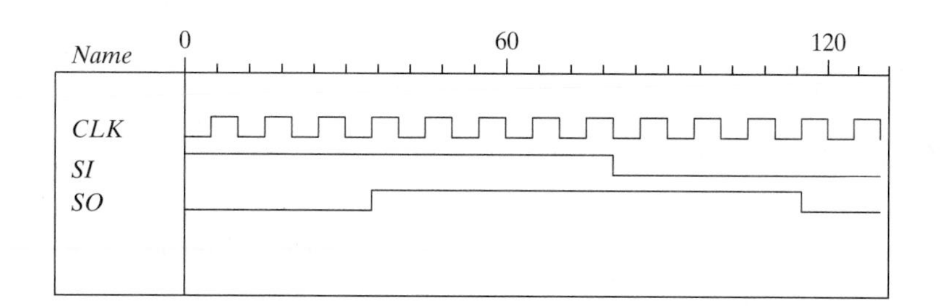

6.35 **(b)** The HDL description is available on the Companion Website.

6.37 The HDL description is available on the Companion Website.

6.38 **(a)** The HDL description is available on the Companion Website.

6.42 Because A is a register variable, it retains whatever value has been assigned to it until a new value is assigned. Therefore, the statement $A <= A$ has the same effect as if the statement was omitted.

6.45 The HDL description is available on the Companion Website. Simulations results for Problem 6.45 follow:

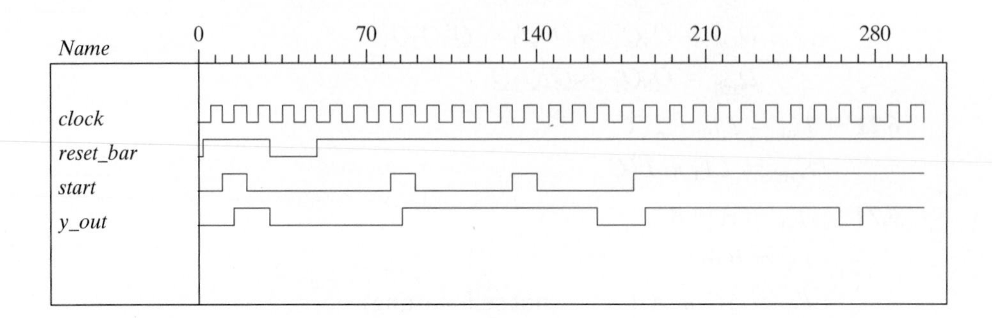

6.50 **(b)** The HDL description is available on the Companion Website. Simulations results for Problem 6.50 follow:

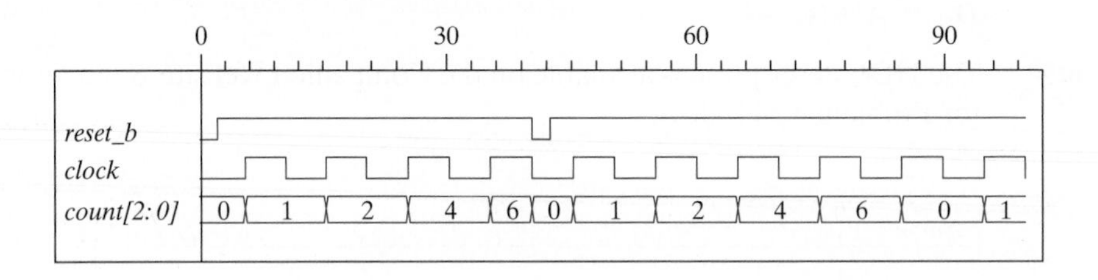

CHAPTER 7

7.2 **(a)** 2^{13} **(b)** 2^{31} **(c)** 2^{26} **(d)** 2^{21}

7.3 Address: 11011011011

Data 16-bit: 1011 0110 0011 1110

7.7 **(a)** Decoder requires 512 AND gates each with 8 inputs. **(b)** $x = 144; y = 88$

7.8 **(a)** 8 chips **(b)** 19; 16 **(c)** 3×8 decoder

7.10 00 0011 0010 1011

7.11 1111101110110

7.12 **(a)** 100110100; **(b)** 101000111; **(c)** 110011010; **(d)** 110110110; **(e)** 010011101

7.13 **(a)** 6 **(b)** 7 **(c)** 8

7.14 **(a)** 010011011

7.16 26 pins

7.20 Product terms: $yz', xz', x'y'z, xy', x'y, z$

7.25 $A = xy' + y'z + yz'$

$B = A + xy$

$C = y'z' + x'y + xz$

$D = z + xy'$

CHAPTER 8

8.1 **(a)** The transfer and increment occur concurrently, i.e., at the same clock edge. After the transfer, *R2* holds the contents that were in *R1* before the clock edge, and *R2* holds its previous value incremented by 1.
(b) Decrement the content of *R3* by one.
(c) If ($S_1 = 1$), transfer content of *R1* to *R0*. If ($S_1 = 0$ and $S_2 = 1$), transfer content of *R2* to *R0*.

8.7 RTL notation:

S0: Initial state: if (start = 1) then ($RA \leftarrow$ data_*A*, $RB \leftarrow$ data_*B*, go to *S*1).
S1: {Carry, *RA*} $\leftarrow RA +$ (2's complement of *RB*), go to *S*2.
S2: If (borrow = 0) go to *S0*. If (borrow = 1) then $RA \leftarrow$ (2's complement of *RA*), go to *S*0.

Block diagram and ASMD chart:

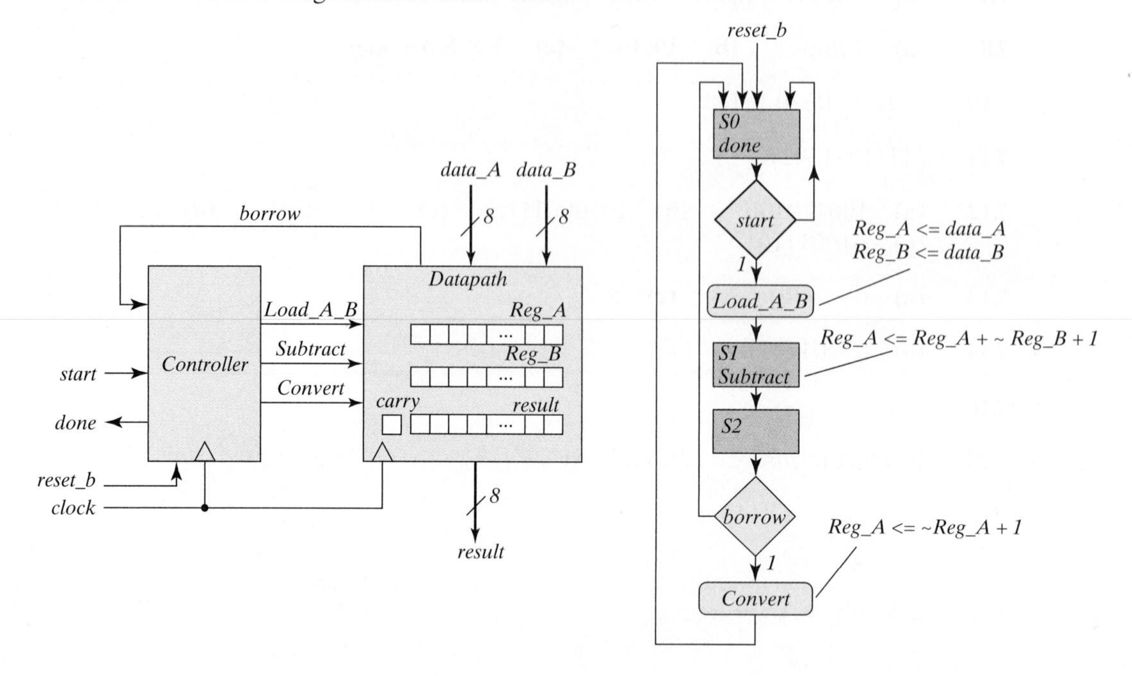

The HDL description is available on the Companion Website. Simulations results for Problems 8.7 follow:

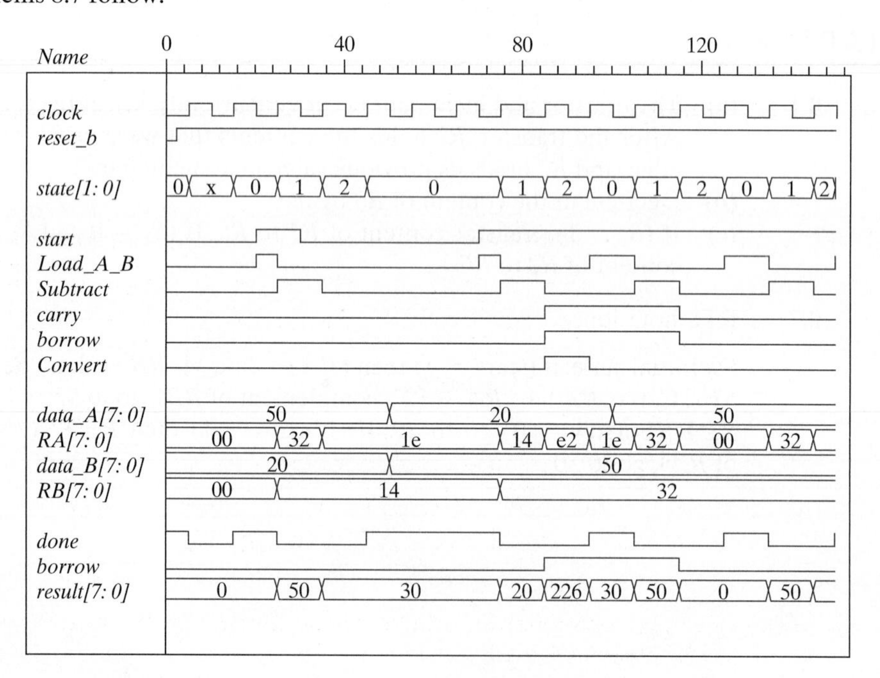

8.8 RTL notation:

S0: if (start = 1) $AR \leftarrow$ input data, $BR \leftarrow$ input data, go to *S1*.
S1: if (AR [15]) = 1(sign bit negative) then $CR \leftarrow AR$(shifted right, sign extension).
else if (positive non-zero) then (Overflow $\leftarrow BR([15] \oplus [14])$, $CR \leftarrow BR$(shifted left)
else if ($AR = 0$) then ($CR \leftarrow 0$).

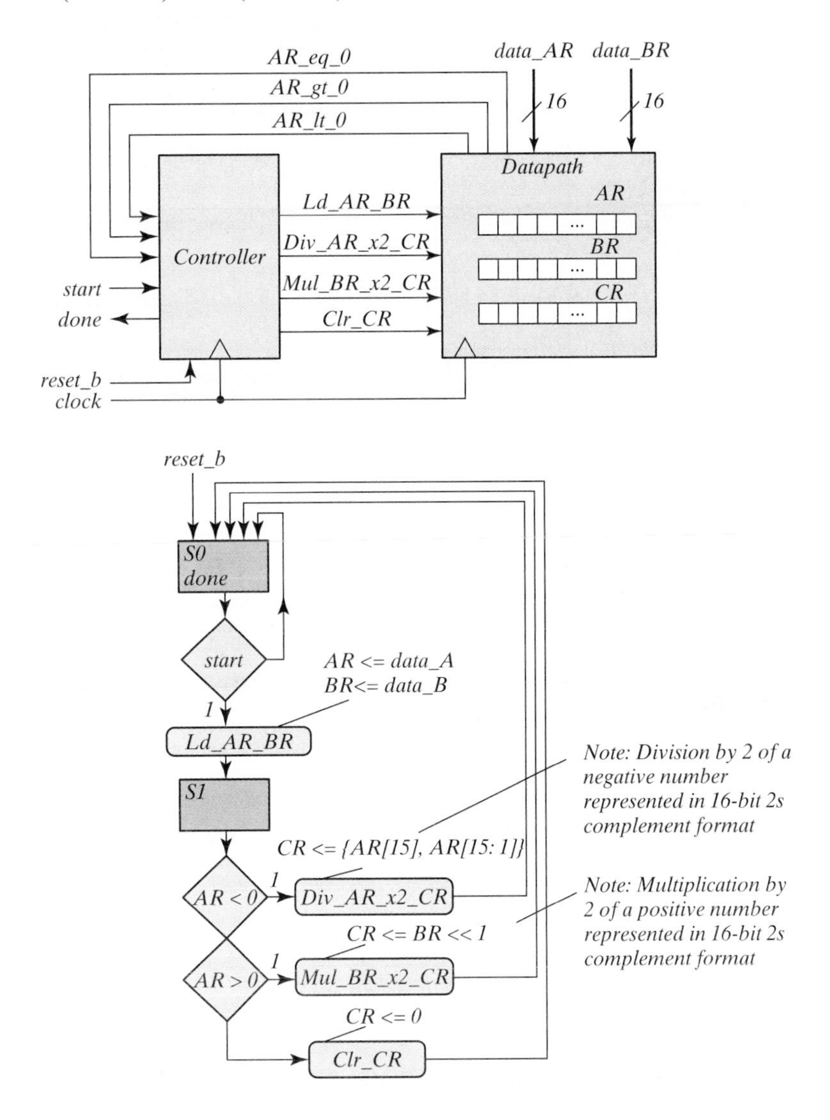

The HDL description is available on the Companion Website. Simulations results for Problem 8.8 follow:

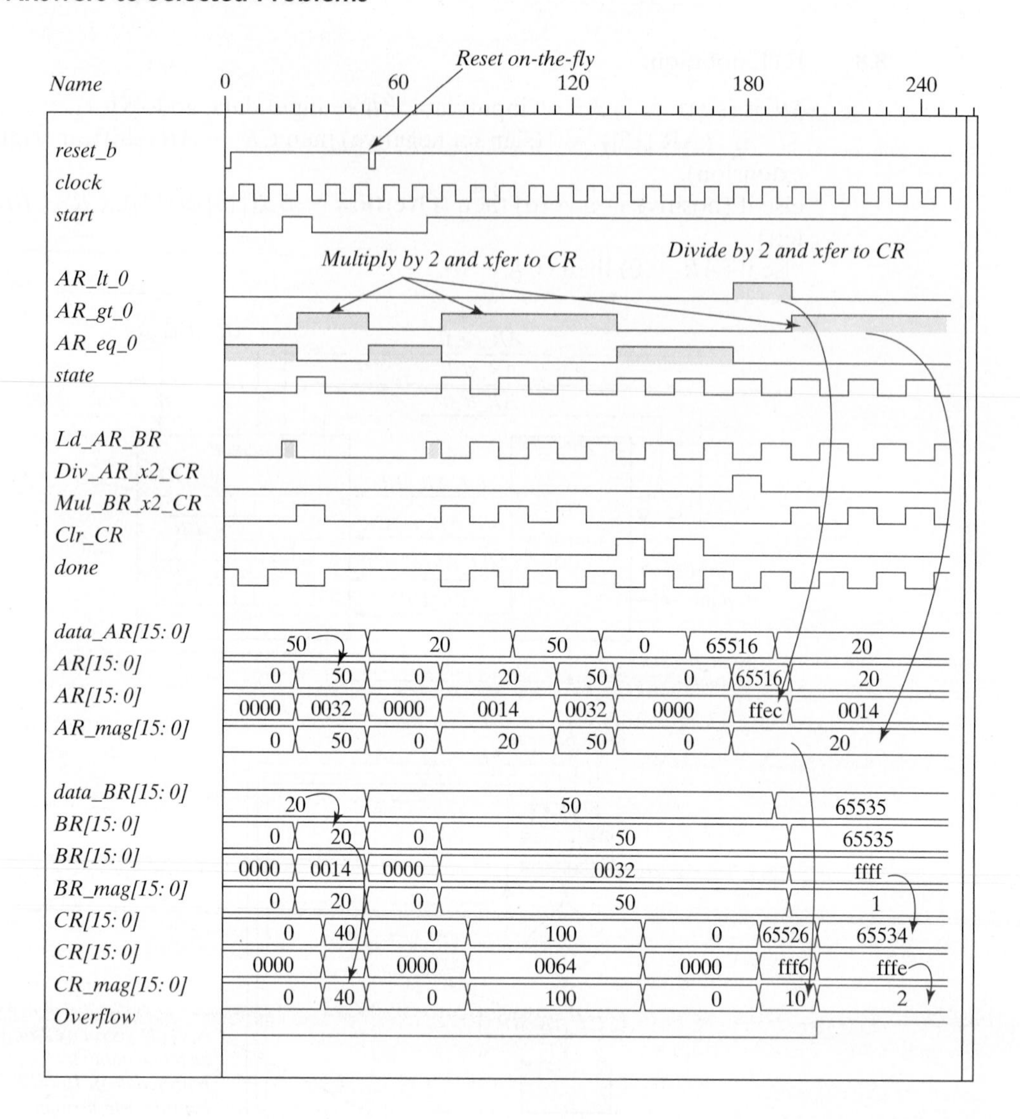

8.9 Design equations:

$D_{S\_idle} = S\_2 + S\_idle\ Start'$

$D_{S\_1} = S\_idle\ Start + S\_1(A2\ A3)'$

$D_{S\_2} = A2\ A3\ S\_1$

The HDL description is available on the Companion Website. Simulations results for Problem 8.9 follow:

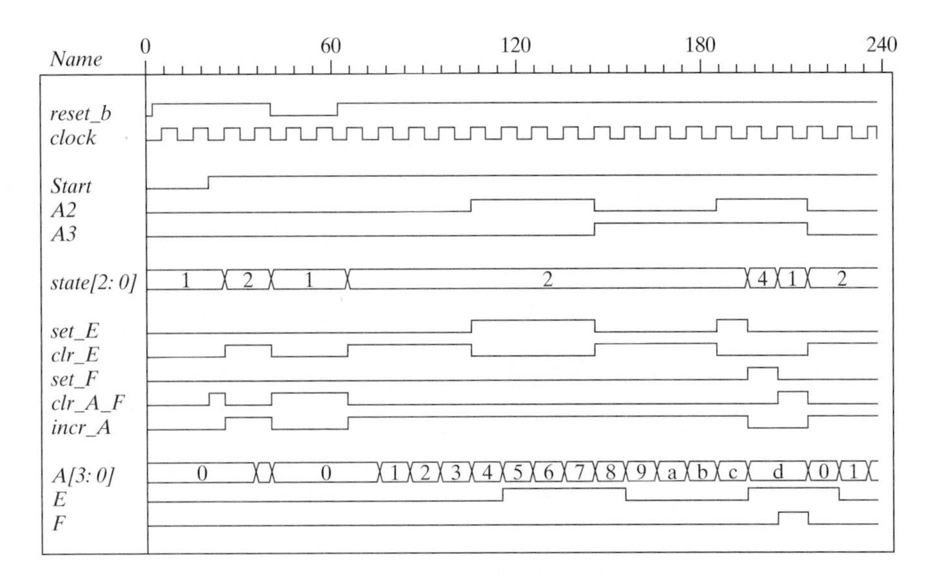

8.11 $D_A = A'B + Ax$

$D_B = A'B'x + A'By + xy$

8.16 RTL notation:

s0: (initial state) If *start* = 0 go back to state *s0*, If(*start* = 1) then $BR \leftarrow$ *multiplicand*, $AR \leftarrow$ *multiplier*, $PR \leftarrow 0$, go to *s*1.

s1: (check *AR* for Zero) *Zero* = 1 if *AR* = 0, if (*Zero* = 1) then go back to *s0* (*done*) If (*Zero* = 0) then go to *s1*, $PR \leftarrow PR + BR$, $AR \leftarrow AR - 1$.

The internal architecture of the datapath consists of a double-width register to hold the product (PR), a register to hold the multiplier (AR), a register to hold the multiplicand (BR), a double-width parallel adder, and single-width parallel adder. The single-width adder is used to implement the operation of decrementing the multiplier unit. Adding a word consisting entirely of 1s to the multiplier accomplishes the 2's complelment subtraction of 1 from the multiplier. Figure 8.16 (a)below shows the ASMD chart, block diagram, and controller of othe circuit. Figure 8.16 (b) shows the internal architecture of the datapath. Figure 8.16 (c) shows the results of simulating the circuit.

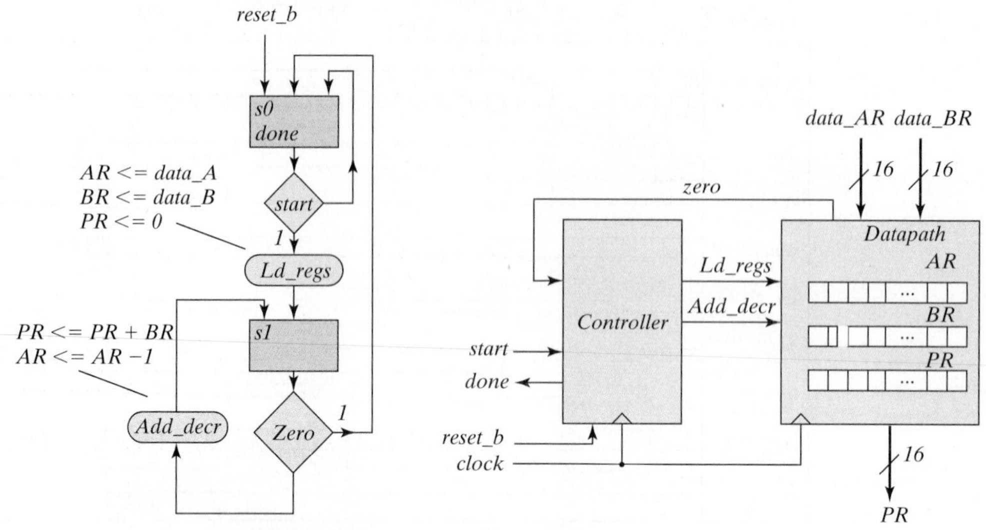

Note: Form Zero as the output of an OR gate whose inputs are the bits of the register AR.

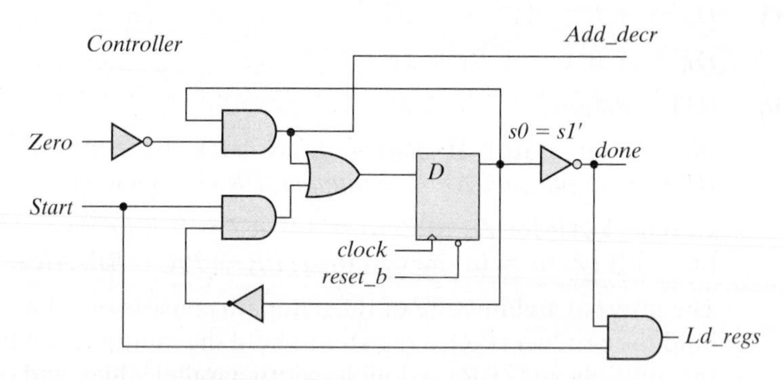

(a) ASMD chart, block diagram, and controller

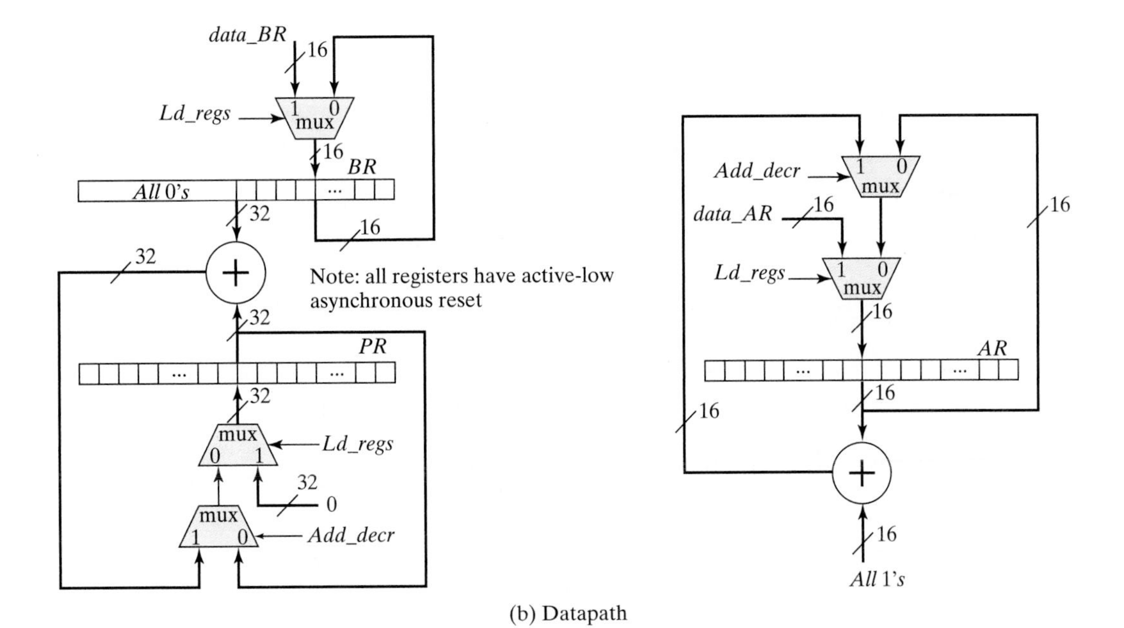

(b) Datapath

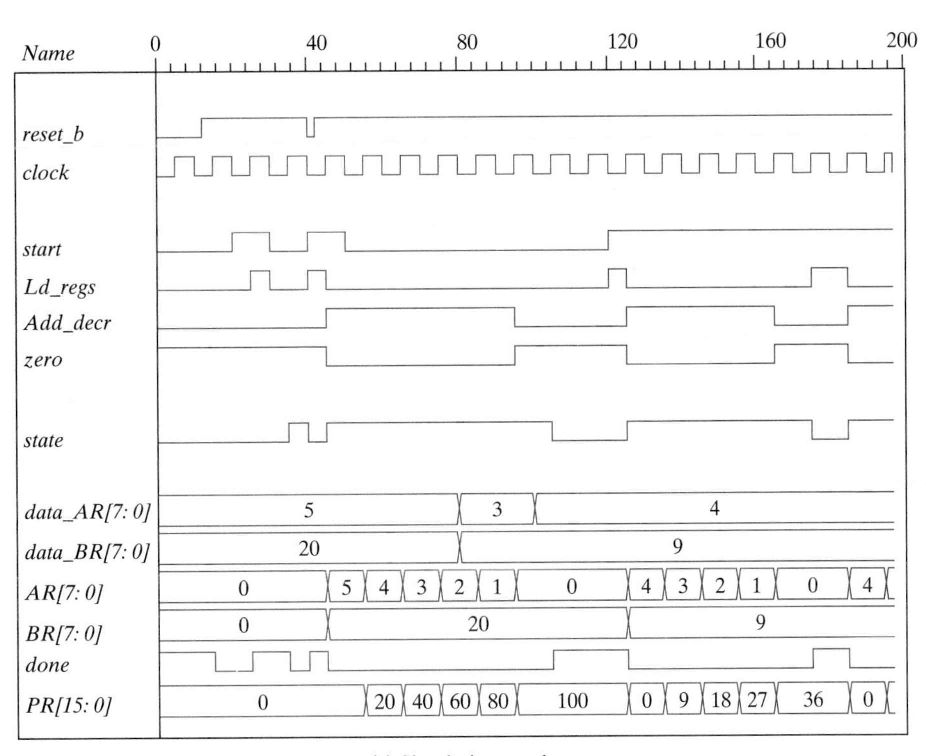

(c) Simulation results

8.17 $(2^n - 1)(2^n - 1) < (2^{2n} - 1)$ for $n \geq 1$

8.18 **(a)** The maximum product size is 32 bits available in registers A and Q.
(b) P counter must have 5 bits to load 16 (binary 10000) initially.
(c) Z (zero) detection is generated with a 5-input NOR gate.

8.20 $2(n + 1)t$

8.21

State codes:	G1	G0
S_idle	0	0
S_add	0	1
S_shift	1	0
unused	0	0

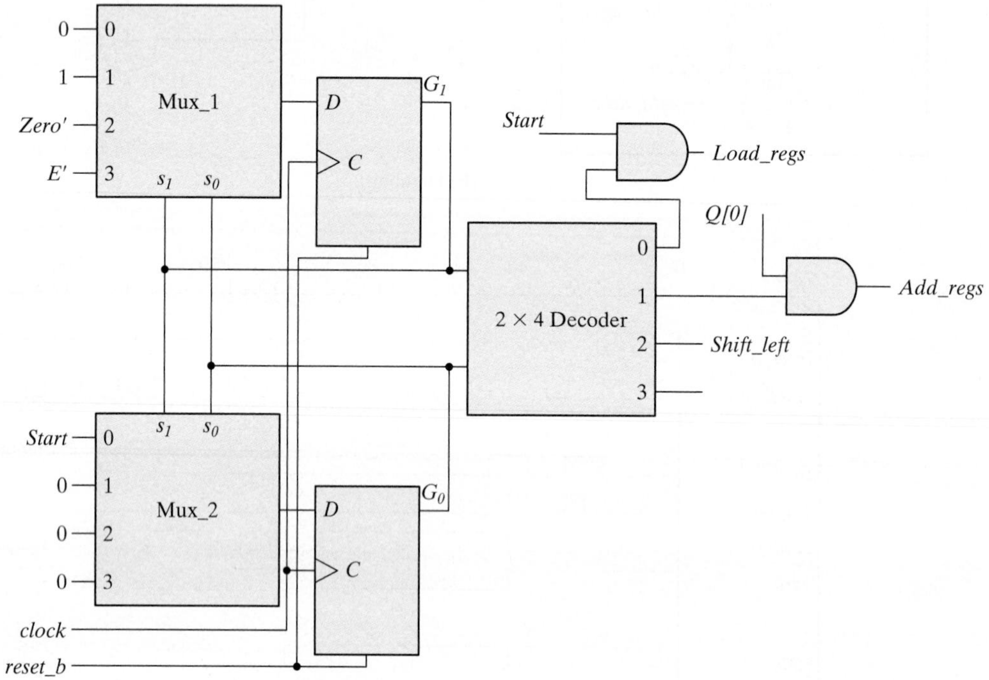

8.30 **(a)** $E = 1$ **(b)** $E = 0$

8.31 A = 0110, B = 0010, C = 0000.
A*B = 1100 A | B = 0110 A && C = 0
A + B = 1000 A^B = 0100 | A = 1
A − B = 0100 &A = 0 A < B = 0
~C = 1111 ~|C = 1 A > B = 1
A & B = 0010 A || B = 1 A != B = 1

8.39

Block diagram and ASMD chart:

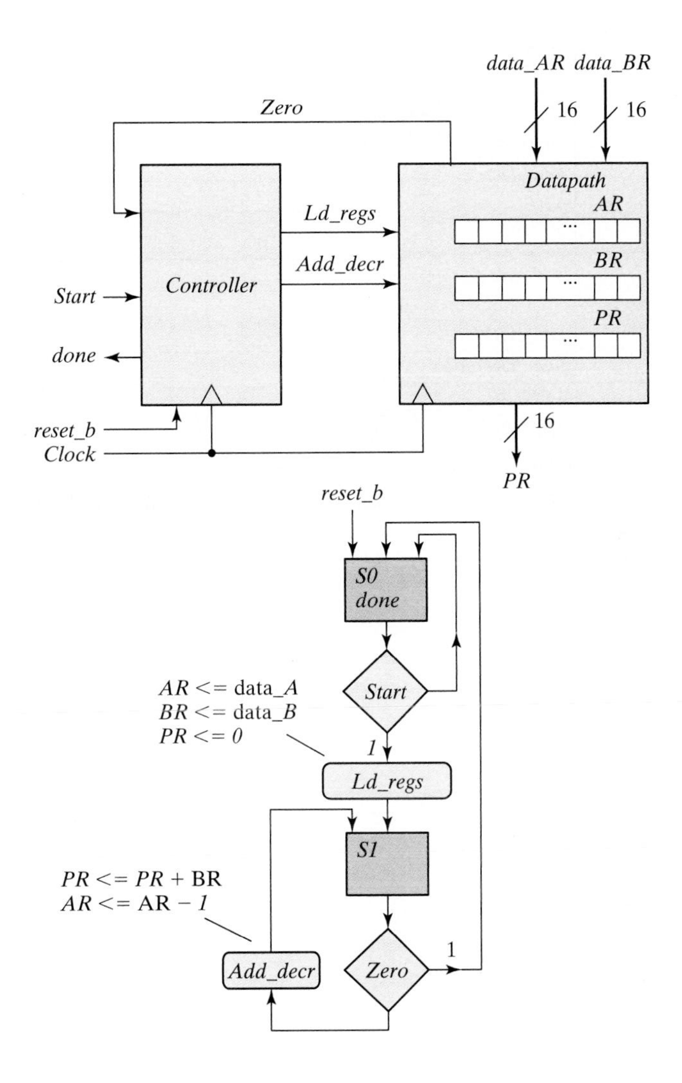

The HDL description is available on the Companion Website. Simulation results for Problem 8.39 follow:

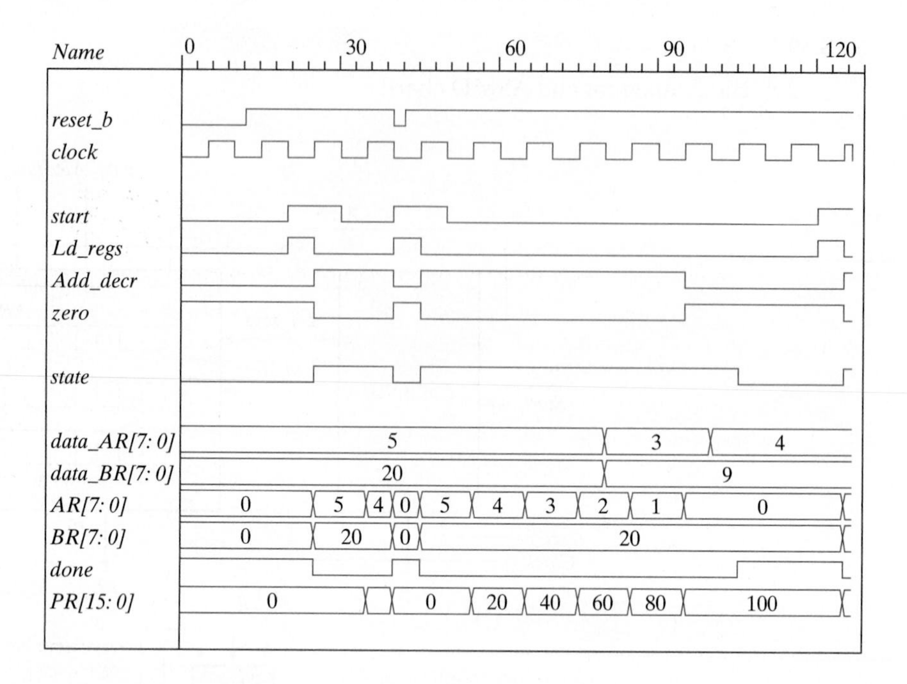
Name
0
30
60
90
120
reset_b
clock
start
Ld_regs
Add_decr
zero
state
data_AR[7: 0]
5
3
4
data_BR[7: 0]
20
9
AR[7: 0]
0
5
4
0
5
4
3
2
1
0
BR[7: 0]
0
20
0
20
done
PR[15: 0]
0
0
20
40
60
80
100

Index